P9-AFH-192

Progress in Molecular and Subcellular Biology · Volume 2

Progress in Molecular and Subcellular Biology

2

Proceedings of the Research Symposium on

Complexes of Biologically Active Substances with Nucleic Acids and Their Modes of Action

Held at the Walter Reed Army Institute of Research
Washington, 16—19 March 1970

Managing Editor

F. E. Hahn

With 158 Figures

Springer-Verlag New York · Heidelberg · Berlin 1971

ISBN 0-387-05321-2 Springer-Verlag New York Heidelberg Berlin
ISBN 3-540-05321-2 Springer-Verlag Berlin Heidelberg New York

Printed in Germany. Typesetting and printing: Carl Ritter & Co., Wiesbaden.
Binding: Karl Hanke, Düsseldorf

Contents

List of Contributors

MICHAEL K. BACH, Department of Hypersensitivity Diseases Research, The Upjohn Company, Kalamazoo, Michigan 49001

GEORGE E. BASS, Department of Molecular and Quantum Biology, College of Pharmacy, University of Tennessee Medical Units, Memphis, Tennessee 38103

WILLIAM R. BAUER, Department of Chemistry, University of Colorado, Boulder, Colorado 80302

KENNETH BEARDSLEY, Departments of Chemistry and Biological Sciences, Columbia University, New York, New York 10027

LERENA W. BLODGETT, Laboratory of Molecular Biology, University of Alabama School of Medicine, Birmingham, Alabama 35233

CHARLES R. CANTOR, Departments of Chemistry and Biological Sciences, Columbia University, New York, New York 10027

CHESTER J. CAVALLITO, Departments of Biochemistry and Medicinal Chemistry, University of North Carolina, Chapel Hill, North Carolina 27514

KENNETH W. CHIN, Departments of Chemistry and Biological Sciences, Columbia University, New York, New York 10027

DONALD M. CROTHERS, Department of Chemistry, Molecular Biophysics and Biochemistry, Yale University, New Haven, Connecticut 06520

SAMUEL J. DE COURCY, JR., Veterans Administration Hospital, Philadelphia, Pennsylvania 19104

MICHAEL DELDIN, Departments of Medicine and Biochemistry, University of Miami School of Medicine, Miami, Florida 33152

YU, V. DUDNIK, Institute of New Antibiotics, Academy of Medical Sciences, Moscow, USSR

LOUIS M. FINK, Institute of Cancer Research and Department of Medicine, Columbia University College of Physicians and Surgeons, New York, New York 10032

DAVID GAUDIN, Laboratory of Molecular Biology, University of Alabama School of Medicine, Birmingham, Alabama 35233

GEORGII F. GAUSE, Institute of New Antibiotics, Academy of Medical Sciences, Moscow, USSR

DEZIDER GRUNBERGER, Institute of Cancer Research and Department of Medicine, Columbia University, College of Physicians and Surgeons, New York, New York 10032

Elizabeth A. Hansen, U.S. Food and Drug Administration, Washington, D.C. 20204

Fred E. Hahn, Department of Molecular Biology, Walter Reed Army Institute of Research, Washington, D.C. 20012

Edward J. Herbst, Department of Biochemistry, University of New Hampshire, Durham, New Hampshire 03824

Erich Hirschberg, Department of Biochemistry, New Jersey College of Medicine and Dentistry, Newark, New Jersey 07103

Hans D. Hoffmann, Department of Radiological Sciences, The Johns Hopkins University, Baltimore, Maryland 21205

David J. Holbrook, Jr., Center for Research in Pharmacology and Toxicology, School of Medicine, University of North Carolina, Chapel Hill, North Carolina 27514

Susan B. Horwitz, Department of Pharmacology, Albert Einstein College of Medicine, New York, New York 10461

Donna R. Hudson, Department of Molecular and Quantum Biology, College of Pharmacy, University of Tennessee Medical Units, Memphis, Tennessee 38103

Herbert G. Johnson, Department of Hypersensitivity Diseases Research, The Upjohn Company, Kalamazoo, Michigan 49001

Helga Kersten, Physiologisch-Chemisches Institut der Universität Erlangen-Nürnberg, 8520 Erlangen, Germany

Walter Kersten, Physiologisch-Chemisches Institut der Universität Erlangen-Nürnberg, 8520 Erlangen, Germany

Anne K. Krey, Department of Molecular Biology, Walter Reed Army Institute of Research, Washington, D.C. 20012

Leonard S. Lerman, Department of Molecular Biology, Vanderbilt University, Nashville, Tennessee 37203

Stephen A. Lesko, Jr., Department of Radiological Sciences, The Johns Hopkins University, Baltimore, Maryland 21205

Sr. Veronica M. Maher, Marygrove College, Detroit, Michigan 48221

Henry R. Mahler, Chemical Laboratories, Indiana University, Bloomington, Indiana 47401

B. D. Mehrotra, Department of Chemistry, Tougaloo College, Tougaloo, Mississippi 39174

Carl R. Morris, Center for Research in Pharmacology and Toxicology, School of Medicine, University of North Carolina, Chapel Hill, North Carolina 27514

Karl H. Muench, Departments of Medicine and Biochemistry, University of Miami School of Medicine, Miami, Florida 33152

James H. Nelson, Departments of Chemistry and Biological Sciences, Columbia University, New York, New York 10027

Jane E. Parker, Department of Molecular and Quantum Biology, College of Pharmacy, University of Tennessee Medical Units, Memphis, Tennessee 38103

P. S. PERLMAN, Chemical Laboratories, Indiana University, Bloomington, Indiana 47401

JULIO C. PITA, JR., Departments of Medicine and Biochemistry, University of Miami School of Medicine, Miami, Florida 33152

WILLIAM P. PURCELL, Department of Molecular and Quantum Biology, College of Pharmacy, University of Tennessee Medical Units, Memphis, Tennessee 38103

HELENE STERNGLANZ, Laboratory of Molecular Biology, University of Alabama School of Medicine, Birmingham, Alabama 35233

ROBERT B. TANGUAY, Department of Biochemistry, University of New Hamsphire, Durham, New Hampshire 03824

TERENCE TAO, Department of Chemistry and Biological Sciences, Columbia University, New York, New York 10027

PAUL O. P. TS'O, Department of Radiological Sciences, The Johns Hopkins University, Baltimore, Maryland 21205

THOMAS A. VICTOR, Department of Molecular Biology, Walter Reed Army Institute of Research, Washington, D.C. 20012

JEROME VINOGRAD, Divisions of Chemistry and Biology, California Institute of Technology, Pasadena, California 91109

THOMAS E. WAGNER, Department of Chemistry, College of Arts and Sciences, Ohio University, Athens, Ohio 45701

MICHAEL WARING, Department of Pharmacology, Medical School, Hills Road, Cambridge CB2 2QD, England

I. BERNARD WEINSTEIN, Institute of Cancer Research and Department of Medicine, Columbia University College of Physicians and Surgeons, New York, New York 10032

ROBERT D. WELLS, Department of Biochemistry, University of Wisconsin, Madison, Wisconsin 53706

LEONA P. WHICHARD, Center for Research in Pharmacology and Toxicology, School of Medicine, University of North Carolina, Chapel Hill, North Carolina 27514

HELEN L. WHITE, Departments of Biochemistry and Medicinal Chemistry, University of North Carolina, Chapel Hill, North Carolina 27514

JAMES R. WHITE, Departments of Biochemistry and Medicinal Chemistry, University of North Carolina, Chapel Hill, North Carolina 27514

LIDIA A. WHITE, Center for Research in Pharmacology and Toxicology, School of Medicine, University of North Carolina, Chapel Hill, North Carolina 27514

ALAN DAVID WOLFE, Department of Molecular Biology, Walter Reed Army Institute of Research, Washington, D.C. 20012

K. LEMONE YIELDING, Laboratory of Molecular Biology, University of Alabama School of Medicine, Birmingham, Alabama 35233

Complexes of Biologically Active Substances with Nucleic Acids — Yesterday, Today, Tomorrow

Fred E. Hahn

I. Introduction

The Symposium whose Proceedings are contained in this volume has been convened after a period of intensive research and increasing knowledge concerning the formation and structures of complexes of biologically active small molecules with nucleic acids and the biological, biochemical and pharmacological effects caused by such complex formation.

Firstly, the decade has seen many detailed investigations of the modes and mechanisms of action of clinical or experimental drugs which form complexes with nucleic acids, especially with DNA, and produce their antiprotozoal, antibacterial, antiviral and antineoplastic effects by interfering with processes in which nucleic acids participate. During this era the term "molecular pharmacology" has been adopted for the field of learning which concerns itself with drug action at the molecular level; knowledge of nucleic acid complexing drugs constitutes perhaps the most advanced area in molecular pharmacology. One result of this advancement is the progressive rationalization of previously empirical structure-activity relationships which provides information for drug design or molecular modification of existing prototype drugs with a view to improving chemotherapy. For example, our (Hahn, this volume) recognition that complex formation with DNA represents the basis of the antimalarial action of quinine has not only explained structure-activity rules which had been discovered some 30 years ago, but also explains the strong antiplasmodial effects of synthetic quinoline methanols which have been designed after quinine and has opened the door to a systematic exploration of this class of potentially useful drugs.

Secondly, the mutagenic effects of aminoacridines or of ethidium bromide led to the discovery of frame-shift mutations in chromosomal genes or of mitochondrial mutations, and have been explained through studies of the binding of these substances to DNA. But the same chemicals, foremost quinacrine, also act as antimutagens (De Courcy; Bach, this volume) which decrease the frequency with which bacteria or plasmodia mutate to resist the action of pharmacopeial chemotherapeutic drugs. We are probably at the beginning of an era in practical chemotherapy in which antimutagens will be administered along with chemotherapeutic drugs to prevent the emergence of microbial drug resistance in patients under therapy. Furthermore, the elimination of episomal resistance factors from enteric bacteria by acridines, known as the "curing" effect, opens the prospect of eliminating R-factor-mediated drug resistance after it has been established through R-factor transfer.

Thirdly, many carcinogenic substances form complexes with DNA. The study of this phenomenon is pursued in several laboratories (for example, LESKO, this volume) in the hope that mechanisms of chemical carcinogenesis will be elucidated and that the nature of genetic lesions, leading to neoplasia, will become understood. This understanding may be one prerequisite to the development of effective antineoplastic drugs.

Finally, it is apparent that substances which complex with DNA and prevent RNA transcription are simple models of genetic repressors. Not only is the flexibility of microorganisms in their response to changing nutritional environments based upon an interplay of repression and derepression on the regulatory segments of coordinated gene clusters, called operons, but the understanding of the developmental biology of higher organisms at the molecular level of organization and causation also depends upon the knowledge of the suspension or actuation of genetic potentials. It is at this point of conceptualization that progress in the study of DNA-complexing substances becomes progress in molecular biology.

The Walter Reed Army Institute of Research owes its original establishment in 1893 and many of its accomplishments to the concern with problems of communicable diseases. The Institute has sponsored this Symposium, whose Proceedings are published here, with the expectation that the knowledge and perspectives presented will ultimately be translated into tangible advances in chemotherapy. The time is approaching when scientifically premeditated drug design will supplant empirical search procedures.

II. Prehistory: Before the Double Helix

Research on the binding of low-molecular ligands to nucleic acids began in the 1940ies and was concerned with cytological staining characteristics of basic dyes. In a classical study, MICHAELIS (1947) recorded the effects produced by nucleic acids in the absorption spectra of basic dyes which "stain" DNA or RNA. His distinction between α-, β- and γ-bands of absorption established physical criteria for the binding of individual molecules, dimers or aggregates of dyes to nucleic acids. Since many DNA-complexing drugs have visible or ultraviolet absorption spectra, the same criteria also are useful in the study of nucleic acid complexes with drugs. MICHAELIS (1947) also anticipated the intercalation hypothesis of dye or drug binding to DNA by speculating that in dye-DNA complexes "each dye cation, combined with one phosphate group, must lie in the space between the planes of the pyrimidine or purine rings".

Interest in cytological staining specificity also led to studies by KURNICK and MIRSKY (1950) of the stoichiometry of the reaction of methyl green with DNA. The stable DNA-methyl green complex was introduced as an experimental substrate for the determination of the activities of deoxyribonucleases (KURNICK, 1950). We have begun to measure the rates of displacement of methyl green from DNA by DNA-complexing drugs and to consider different rates as functions of different affinities of the displacing compounds for DNA.

Unrelated, at the time, to the topic of nucleic acid complexes were extensive investigations by ALBERT and his associates (reviewed by ALBERT, 1968) on the relationships between the structures of aminoacridines and other N-heterocyclic amines and their antibacterial activities. Aminoacridines had been introduced in 1913 by BROWNING as antibacterials for wounds. ALBERT, RUBBO and BURVILL (1949) recognized as essential structural requirements of N-heterocyclic amines to exert antibacterial activity, that such compounds must possess planar areas of 28 $Å^2$ or larger and substituted amino groups which are ionized to at least 50% at physiological pHs. These empirically derived structure-activity rules are now retrospectively recognized as the structural requirements for intercalation binding of antimicrobial substances to double-helical DNA.

III. History: After the Double Helix

WATSON and CRICK proposed the double-helical structure for DNA (1953) based on X-ray diffraction studies in the laboratory of WILKINS, model-building experiments and the logical requirement that DNA's structure must provide a determinant for its correct replication, i.e. for genetic continuity. While it is evident that knowledge of the macromolecular architecture of DNA provides the basis of determining structures of DNA complexes with low-molecular ligands, studies on such complexes were not promptly undertaken after the DNA model had been proposed. Work on the interaction of aminoacridines with DNA by PEACOCKE and SKERRETT (1956) was stimulated by the antibacterial properties of these compounds and measured the extent of binding of proflavine to DNA by spectrophotometric titration and equilibrium dialysis. The study emphasized the need for purines in DNA to bind proflavine and, in reiterating the structure-activity rules of ALBERT et al. (1949), considered them for the first time to be requirements for interaction with DNA; it also distinguished between one strong binding process by which one proflavine molecule is bound per approximately 5 nucleotides and a weaker process which involves the attachment of aggregates of the aminoacridine to DNA.

In comparing the influence of cationic polymers on the tendency of acridine orange to "stack" along such linear macromolecules, BRADLEY and WOLF (1959) concluded that the "stacking coefficient" of DNA was small by comparison to that of single-stranded polynucleotides, of heparin or of polyphosphate. STONE and BRADLEY (1961), in elaborating on these results, concluded that the aggregation of acridine orange, owing to dye-dye interactions, was a function of the conformation of the polymer to which the dye was bound and that the "stacking coefficient" for double-stranded DNA was smaller than for denatured DNA. This work, nevertheless, proposed the first model of the structure of a DNA-ligand complex which was based upon a consideration of the macromolecular architecture of the double helix.

Stimulated by the mutagenic action of aminoacridines and the carcinogenic action of certain benzacridines, LERMAN (as reviewed in 1964) undertook a series of studies on the structure of DNA-acridine complexes which have explained the strong (type 1) binding of aminoacridines to DNA (PEACOCKE and SKERRETT, 1956) and the frameshift mutagenesis by aminoacridines (CRICK, BARNETT, BRENNER and WATTS-TOBIN, 1961) by postulating the intercalation model. Applying a set of hydrodynamic, optical and organic-chemical criteria, it was shown that substituted acridines become inserted between the levels of base pairs into double-helical DNA; the spaces for these insertions are created by local untwisting of the double helix by an estimated 12° of rotation which causes a separation between previously adjacent base pairs of approximately 3.5 Å without a disturbance in the pattern of hydrogen bonds. The resulting lengthening of linear DNA has been measured radioautographically in electron micrographs (CAIRNS, 1962) and has suggested that only every second space between base pairs is available for intercalation. This is in accord with the stoichiometry of strong binding processes for proflavine (PEACOCKE and SKERRETT, 1956) or chloroquine (STOLLAR and LEVINE, 1963) of one intercalant molecule per 4 to 5 component bases of DNA.

The original intercalation model of LERMAN or its modification by PRITCHARD, BLAKE and PEACOCKE (1966) is also based upon the knowledge of the macromolecular architecture of DNA and, indeed, could not have been completely developed before the postulation of the DNA model by WATSON and CRICK (1953).

IV. Antibiotics

In 1960 KIRK, as well as RAUEN, KERSTEN and KERSTEN, reported studies on the mode of action actinomycin D, the prototype of a series of antibiotics discovered by WAKSMAN and WOODRUFF (1940). Actinomycin was found to form a complex with DNA; it cosediments with DNA and its absorption spectrum is altered by DNA. The antibiotic acts as a template poison and inhibits, preferentially, the transscription of RNA from DNA (KIRK, 1960; HURWITZ, FURTH, MALAMY and ALEXANDER, 1962). Extensive studies on the DNA-actinomycin complex (reviewed by REICH, CERAMI and WARD, 1967), and especially X-ray studies of the structure of the complex (HAMILTON, FULLER and REICH, 1963) led to the proposal of a structural model in which actinomycin is lodged in the minor groove of the double helix and requires the amino group in position 2 of guanine or 2-aminopurine for binding to DNA. MÜLLER and CROTHERS (1968) have extensively reinvestigated the properties of DNA complexes with a series of actinomycins and concluded that the hetero-tricyclic chromophore of the antibiotic is intercalated into DNA. The idea of intercalation of actinomycin has been fortified (WARING, this volume) by demonstrating typical conformational changes in supercoiled DNA upon reacting with the antibiotic; the absolute guanine requirement for binding of actinomycin to DNA is placed into doubt by studies of WELLS (this volume).

Many other antibiotics form complexes with DNA, such as daunomycin, cinerubin, nogalamycin, chromomycin, mithramycin and olivomycin (as reviewed in GOTTLIEB and SHAW, 1967), echinomycin (WARD, REICH and GOLDBERG, 1965), quinoxaline antibiotics (SATO, SHIRATORI and KATAGIRI, 1967), hedamycin and rubiflavin (WHITE and WHITE, 1969), kanchanomycin (FRIEDMAN, JOEL and GOLDBERG, 1969), anthramycin (HORWITZ, this volume), sibiromycin (GAUSE, this volume) and distamycin (KREY and HAHN, 1970).

An interesting feature of the binding of antibiotics to DNA is the role of Mg^{++} in the binding process. For some substances, Mg^{++} causes the dissociation of their complexes with DNA, while for others it is an essential requirement for complex formation.

A few antibiotics react with DNA through the formation of covalent bonds. Mitomycin C and some of its congeners (reviewed by SZYBALSKI and IYER, 1967) are reduced *in vivo* and also can be reduced experimentally *in vitro* to active metabolites with very short half-lifes; these metabolites condense with DNA and form covalent cross-links in the double helix. This cross-linked DNA is incapable of serving as a template for its own replication since the component strands can not undergo separation. For this reason, the mitomycins are specific inhibitors of DNA biosynthesis. The structural changes produced in DNA by cross-linking with reduced mitomycin are manifested by spontaneous renaturation of DNA whose linked component strands realign "in register". The detailed structure of the mitomycin-DNA complex is not yet known.

Anthramycin (Kohn and Spears, 1970) also forms a covalent bond with DNA. The nature of the chemical reaction and of the bonds which are formed are also unknown, but the principal biochemical effect of this complex formation is an inhibition of the DNA-dependent RNA and DNA polymerase reactions *in vitro* and of nucleic acid biosynthesis in anthramycin-exposed bacteria (Horwitz, this volume).

V. Synthetic Drugs

While most antibiotics which form complexes with DNA are primarily of investigative interest, certain synthetic drugs of clinical importance exert their chemotherapeutic action also by binding to DNA. This has been studied for quinacrine (Kurnick and Radcliffe, 1962), a compound which Lerman selected for some of his key experiments (1963) to test the intercalation hypothesis of the DNA-acridine complex. Quinacrine acts as a DNA template poison and inhibits the DNA-dependent DNA and RNA polymerase reactions (Hahn et al., 1966); the drug is either bacteriostatic or bactericidal depending upon its concentration, and inhibits, preferentially, DNA biosynthesis *in vivo* (Ciak and Hahn, 1967). A structurally related antimicrobial nitroacridine acts in a similar manner (Wolfe, this volume).

Chloroquine also binds to DNA (Hahn et al., 1966; Yielding, this volume) by intercalation (O'Brien, Allison and Hahn, 1966; Waring, this volume), inhibits the DNA-dependent DNA and RNA polymerase reactions *in vitro* and, prominently, DNA biosynthesis in plasmodia (Polet and Barr, 1968). While antimalarial 8-aminoquinolines also bind to DNA (Holbrook, this volume), they evidently do not form intercalation complexes, and their modes of action have remained unknown. Since it has been suggested that these drugs are converted *in vivo* into chemotherapeutically active metabolites, the meaning of *in vitro* observations of their binding to DNA is not clear.

The antischistosomal drug, miracil D, which also has antibacterial and antitumor activity binds to DNA, probably by intercalation, and inhibits specifically the transcription of RNA from DNA, i.e. RNA biosynthesis (Weinstein, this volume).

Synthetic drugs of lesser medical importance such as ethidium bromide (Waring; Wagner; Mahler, this volume) and quinoline methanols (Hahn, this volume) also form complexes with DNA.

VI. Alkaloids

The first studies of binding of alkaloids to DNA were carried out, beginning in 1964, by Mahler and his associates (cited by Hahn, this volume) on a group of steroidal diamines. This work was undertaken because these compounds possess, at neutral pH, two positive charges with a fixed separation corresponding to the interval between two DNA phosphates across the minor groove of the double helix. Like alipathic diamines of similar spacing of charges, steroidal diamines stabilize DNA to heat. Additionally, these compounds have offered the unique opportunity of studying conformational changes which they produce in DNA by optical methods since their absorption spectra do not occlude the absorption maximum of DNA at 259 nm. One of these substances, irehdiamine, has a warped and non-planar ring structure which appears to eliminate intercalation binding to DNA from consideration; yet, this alkaloid has been found (WARING, this volume) to produce conformational transitions in superhelical DNA which are typical for intercalation.

A second set of studies on DNA-alkaloid complexes was suggested by the anti-malarial action of quinine and by the presumed antimalarial properties of berberine and colchicine. Quinine and berberine were found to form complexes with DNA (HAHN, this volume). This explains the effects of quinine on plasmodial DNA biosynthesis and, hence, on the development of schizonts (POLET and BARR, 1968) as well as the curative action of berberine in cutaneous leishmaniasis and its effect as a mitochondrial mutagen. On the other hand, standard experimental tests for complex formation with DNA were consistently negative for colchicine, and the mutual effects of colchicine and DNA upon each other's specific rotation (ILAN and QUASTEL, 1966) remain unexplained.

Quite recently the binding of the hallucinogenic ergot alkaloid, lysergic acid diethylamide, to DNA (WAGNER; YIELDING, this volume) has been investigated in the hope that this will explain the induction of chromosomal damage in lymphocytes by LSD.

Since alkaloids, in general, are organic amines, it can be expected that additional members of this class of natural compounds will be found to bind to DNA, if only by ionic attraction.

VII. Superhelical DNA

The past four years have seen the emergence of knowledge concerning the wide distribution in nature of a form of circular DNA which is twisted into supercoils because it has a built-in deficiency in the number of helical turns. At the time of the Symposium, 45 different superhelical DNAs were known. These occur in animal viruses, in bacterial viruses, in mitochondria, in bacterial episomes, and in the cytoplasm of animal cells. Neither the mechanism of biosynthesis of superhelical DNAs nor their biological role are understood at this time.

Intercalation binding of synthetic drug or antibiotics (WARING; BAUER, this volume) produces characteristic conformational transitions in superhelical DNA. Progressive intercalation causes progressive unwinding of superhelices by gradual compensation for the natural deficiency in helical turns; at a defined equivalence point, superhelical DNA will have been converted into ordinary circular DNA. Further intercalation, producing further increments in helical turns, twists this circular DNA into unnatural supercoils which owe their existence not to a *deficiency* but to an *excess* of helical turns. If one were to speculate that the supercoiled condition represents a DNA storage form with suspended template function, he might assume that, for example, the antitprypanosomal action of ethidium bromide which leads to the formation of akinetoplastic trypanosomes, results from a suspension of the function of kinetoplastic DNA. It is not impossible that the selective toxicity of intercalative drugs in eliminating bacterial episomes, is not the result of an indiscriminate and massive occupancy of all DNA, but, in contrast, the selective effect of such drugs of changing the conformation of circular episomal DNA and of tying it up into artificial and non-functional supercoils (HAHN and CIAK, 1971).

VIII. The Future

Two structural models of DNA complexes with low-molecular substances have been considered. One involves lateral attachment of such compounds to DNA and is exemplified by the "stacking model" of BRADLEY and WOLF (1959) or by the binding of spermine and other polyamines to the double helix (HERBST, this volume). The other is the intercalation model of LERMAN (1964). With the exception of a few X-ray diffraction studies (HAMILTON et al., 1963; NEVILLE and DAVIES, 1966; LIQUORI, COSTANTINO, CRESCNEZI, ELIA, GIGLIO, PULITI, DE SANTIS SAVINO and VITAGLIANO, 1967; SUWALSKY, TRAUB, SHMUELI and SUBIRANA, 1969), knowledge of the structures of DNA complexes has been produced by statistical rather than deterministic experiments and interpretations.

Investigating a series of low-molecular complexing agents with variations in a given prototype structure (LERMAN, 1964; MÜLLER and CROTHERS, 1968) exemplifies the extent of determinism which has been attained.

Two new approaches, however, promise to identify the specific binding sites for drugs on DNA as well as the structures of the DNA-drug complexes formed. One of these approaches (WELLS, this volume) uses duplex deoxyribopolynucleotides with monotonously repeated base sequences, i.e. it systematically varies the covalent structure of the binding polymer instead of the structure of a prototype ligand which is bound. This approach is capable of considerable extension and refinement, including the use of duplex oligomers with linear sequences of seven or eight nucleotides, i.e. of the critical length for one complete helical turn in DNA.

The other approach is the study of binding of drugs to DNA by nuclear magnetic resonance spectroscopy as exemplified by the work of DANYLUK and VICTOR (1970) on the interaction of actinomycin and DNA. The NMR spectrum of DNA shows an unresolved continuum of signals, but the signals from a binding drug molecule undergo specific changes from that of the free drug depending upon which reactive groups of the drug are involved in interaction with DNA. Preliminary work on the NMR spectrum of chloroquine (VICTOR, this volume) provides the basis for such DNA binding studies.

The unambiguous determination of structures of DNA-drug complexes might not only explain the chemotherapeutic and genetic effects of biologically active substances which form complexes with DNA, but should logically furnish essential information for the premeditated design of substances whose biological target is DNA, and whose biological actions should be predictable. It may well be that the first breakthrough in chemotherapy research to the premeditated design of effective drugs will occur in this area of molecular pharmacology.

References

Albert, A.: Selective toxicity, 4. ed. London: Methuen 1968.

— Rubbo, S. D., Burvill, M. I.: The influence of chemical constitution on antibacterial activity. Part IV: A survey of heterocyclic bases with special reference to benzquinolines, phenanthridines, benzacridines, quinolines and pyridines Brit. J. exp. Path. **30**, 159 (1949).

Bradley, D. F., Wolf, M. K.: Aggregation of dyes bound to polyanions Proc. nat. Acad. Sci. (Wash.) **45**, 944 (1959).

Cairns, J.: The application of radioautography to the study of DNA viruses Cold Spr. Harb. Symp. quant. Biol. **27**, 311 (1962).

Ciak, J., Hahn, F. E.: Quinacrine (Atebrin): Mode of action. Science **156**, 655 (1967).

Crick, F. H. C., Barnett, L., Brenner, S., Watts-Tobin, R. J.: General nature of the genetic code for proteins Nature (Lond.) **192**, 1227 (1961).

Danyluk, S. S., Victor, T. A.: The structure and molecular interactions of actinomycin D, quantum aspects of heterocyclic compounds in chemistry and biochemistry, II. The Israeli Academy of Sciences Humanities, p 394. Jerusalem 1970.

Friedman, P. A., Joel, P. B , Goldberg, I. H.: Interaction of kanchanomycin with nucleic acids. I. Physical properties of the complex. Biochemistry **8**, 1535 (1969).

Gottlieb, D., Shaw, P. D., Eds.: Antibiotics I. Mechanism of action. Berlin-Heidelberg-New York: Springer 1967.

Hahn, F. E., Ciak, J.: Elimination of bacterial episomes by DNA-complexing compounds. Ann. N.Y. Acad. Sci. (1971) (in press).

— O'Brien, R. L., Ciak, J., Allison, J. L., Olenick, J. G.: Studies on modes of action of chloroquine, quinacrine and quinine and on chloroquine resistance. Milit. Med. **131**, 1071 (1966).

Hamliton, L., Fuller, W., Reich, E.: X-ray diffraction and molecular model building studies of the interaction of actinomycin with nucleic acids. Nature (Lond.) **198**, 538 (1963).

Hurwitz, J., Furth, J. J., Malamy, M., Alexander, M.: The role of deoxyribonucleic acid in ribonucleic acid synthesis, III. The inhibition of the enzymatic synthesis of ribonucleic acid and deoxyribonucleic acid by actinomycin D and proflavin. Proc. nat. Acad. Sci. (Wash.) **48**, 1222 (1962).

Ilan, J., Quastel, J. H.: Effects of colchicine on nucleic acid metabolism during metamorphosis of *Tenebrio molitor* and in some mammalian tissues. Biochem. J. **100**, 448 (1966).

Kirk, J. M.: The mode of action of actinomycin D. Biochim. biophys. Acta (Amst.) **42**, 167 (1960).

Kohn, K. W., Spears, C. L.: Reaction of anthramycin with deoxyribonucleic acid. J. molec. Biol. **51**, 551 (1970).

Krey, A. K., Hahn, F. E.: Studies on the complex of distamycin A with calf thymus DNA. FEBS Letters **10**, 175 (1970).

Kurnick, N. B.: Determination of desoxyribonuclease activity by methyl green; application to serum. Arch. Biochem. **29**, 41 (1950).

— Mirsky, A. E.: Methyl green-pyronine. II. Stoichiometry of reaction with nucleic acids. J. gen. Physiol. **33**, 265 (1950).

— Radcliffe, I. E.: Reaction between DNA and quinacrine and other antimalarials. J. Lab. clin. Med. **60**, 669 (1962).

Lerman, L. S.: The structure of the DNA-acridine complex. Proc. nat. Acad. Sci. (Wash.) **49**, 94 (1963).

— Acridine mutagens and DNA structure. J. cell. comp. Physiol. **64**, Suppl. 1, 1 (1964).

Liquori, A. M., Costantino, L., Crescenzi, V., Elia, V., Giglio, E., Puliti, R., de Santis Savino, M., Vitagliano, V.: Complexes between DNA and polyamines: A molecular model. J. molec. Biol. **24**, 113 (1967).

Michaelis, L.: The nature of the interaction of nucleic acids and nuclei with basic dyestuffs. Cold Spr. Harb. Symp. quant. Biol. **12**, 131 (1947).

Müller, W., Crothers, D. M.: Studies on the binding of actinomycin and related compounds to DNA. J. molec. Biol. **35**, 251 (1968).

NEVILLE, D. M., DAVIES, D. R.: The interaction of acridine dyes with DNA: An x-ray diffraction and optical investigation. J. molec. Biol. **17**, 57 (1966).

O'BRIEN, R. L., ALLISON, J. L., HAHN, F. E.: Evidence for intercalation of chloroquine into DNA. Biochim. biophys. Acta (Amst.) **129**, 622 (1966).

PEACOCKE, A. R., SKERRETT, J. N. H.: The interaction of aminoacridines with nucleic acids. Trans. Faraday Soc. **52**, 261 (1956).

POLET, H., BARR, C. F.: Chloroquine and dihydroquinine. In vitro studies of their antimalarial effect upon *Plasmodium knowlesi*. J. Pharmacol. exp. Ther. **164**, 380 (1968).

PRITCHARD, N. J., BLAKE, A., PEACOCKE, A. R.: Modified intercalation model for the interaction of aminoacridines and DNA. Nature (Lond.) **212**, 1360 (1966).

RAUEN, H. M., KERSTEN, H., KERSTEN, W.: Zur Witkungsweise von Actinomycinen. Z. physiol. Chem. **321**, 139 (1960).

REICH, E., CERAMI, A., WARD, D. C.: Actinomycin. In: Antibiotics I. Mechanism of action. Berlin-Heidelberg-New York: Springer 1967.

SATO, K., SHIRATORI, O., KATAGIRI, K.: The mode of action of quinoxaline antibiotics. Interaction of quinomycin A with deoxyribonucleic acid. J. Antibiot. (Tokyo) Ser. A. **20**, 270 (1967).

STOLLAR, D., LEVINE, L.: Antibodies to denatured deoxyribonucleic acid in lupus erythematosus serum. V. Mechanism of DNA-anti DNA Inhibition by chloroquine. Arch. Biochem. **101**, 335 (1963).

STONE, A. L., BRADLEY, D. F.: Aggregation of acridine orange bound to polyanions: The stacking tendency of deoxyribonucleic acids. J. Amer. chem. Soc. **83**, 3627 (1961).

SUWALSKI, M., TRAUB, W., SHMUELI, U., SUBIRANA, J. A.: An X-ray study of the interaction of DNA with spermine. J. molec. Biol. **42**, 363 (1969).

SZYBALSKI, W., IYER, V. N.: The mitomycins and porfiromycins. In: Antibiotics I. Mechanism of action. Berlin-Heidelberg-New York: Springer 1967.

WAKSMAN, S. A., WOODRUFF, H. B.: Bacteriostatic and bactericidal substances produced by a soil actinomyces. Proc. Soc. expt. Biol. (N.Y.) **45**, 609 (1940).

WARD, D. C., REICH, E., GOLDBERG, I. H.: Base specificity in the interaction of polynucleotides with antibiotic drugs. Science **149**, 1259 (1965).

WATSON, J. D., CRICK, F. H. C.: The structure of DNA. Cold Spr. Harb. Symp. quant. Biol. **18**, 123 (1953).

WHITE, H. L., WHITE, J. R.: Hedamycin and rubiflavin complexes with deoxyribonucleic acid and other polynucleotides. Biochemistry **8**, 1030 (1969).

Kinetics of Binding of Drugs to DNA

D. M. CROTHERS

I. Introduction

When a drug molecule binds to DNA, a number of physical properties change. For example, in the case of both proflavine and actinomycin, there is a red shift of the visible absorption band, and in this respect the two compounds seem similar. However, the time axis provides an additional coordinate for distinguishing reactive processes. Using again the example of proflavine and actinomycin, the two compounds differ by orders of magnitude in their rates of reaction with DNA, and the reaction mechanism for the latter is much more complex than for the former. In both cases, kinetic measurements are able to show that at equilibrium there is a mixture of complex forms, and that static measurements determine only the average complex properties. This, then, is the function of kinetic studies of complex formation: to clarify the mechanism or steps in the binding process, and to separate effects due to different complex forms by utilizing resolution along the time axis. In addition, one may find in some cases that biological activity is correlated strongly with kinetic rather than equilibrium properties of the complex. An example is the actinomycin DNA complex, for which a slow dissociation rate seems essential for biological activity (MÜLLER and CROTHERS, 1968).

II. Equilibrium Models

Interpretation of kinetic data requires use of a model for the binding equilibrium. In general, for the binding of substances like actinomycin and phenanthridine or acridine dyes, saturation of the class of strong binding sites occurs at considerably less than one drug molecule per base pair. Two simple models have been used to account for this. On one hand, one can assume that the binding sites are somehow different, perhaps because of the base sequence, and that these sites are completely independent, in the sense that binding at one does not affect the affinity of neighboring sites. Under this assumption, the equation for chemical equilibrium

$$K_{ap} = \frac{C_B}{C_F \cdot C_S} \tag{1}$$

where C_B is the concentration of complexed sites, C_F the free drug concentration, C_S the concentration of free sites, and K_{ap} the association equilibrium constant, is readily transformed to the equation of SCATCHARD (1949).

$$r/C_F = K_{ap} (B_{ap} - r) . \tag{2}$$

In this equation, r is the ratio of bound drug molecules to total monomer units, and B_{ap} is the number of binding sites per monomer unit. The expression (2) predicts a linear variation of r/C_F with r, a relation that is at least approximately obeyed for binding of most drugs when r is small. However, as r approaches the saturation limit, curvature usually appears, which is interpreted in this model as due to the presence of other sites with lesser affinity.

The main trouble with this "independent site" model is that it is difficult to see why, in a highly regular molecule like DNA, one site should be so different from another. For example, in the case of acridines and phenanthridines, one finds little base selectivity for the intercalation reaction. It should therefore be expected that intercalation would occur adjacent to each base pair, producing saturation at one drug per nucleotide pair. The value of B_{ap} from SCATCHARD plots, however, is more like one drug per three base pairs.

An alternative model is that of "neighbor exclusion", in which it is assumed that intercalation at one site prevents binding at neighboring sites (CROTHERS, 1968). The space between every base pair is a potential binding site, but the total saturation level is limited by the neighbor exclusion. The methods of statistical mechanics are required to calculate binding isotherms for this model, but it is found that simple assumptions lead to good agreement between calculated and measured isotherms. Examples are shown in Fig. 1, in the first instance for the binding of ethidium by DNA, taken from the work of BAUER and VINOGRAD (1970). The calculated line is based on the assumption that intercalation at one site blocks intercalation in the potential sites immediately adjacent on each side. Saturation should therefore occur at one drug per four nucleotides (one for each two base pairs), which agrees closely with experiment. Furthermore, the curvature in the binding isotherm is a consequence of the model, and does not require the assumption of additional sites. Fig. 1b shows a comparison between calculated and measured binding isotherms for actinomycin, in this instance with the assumption that each $G \cdot C$ pair can serve as a potential binding site, but that binding at one site prevents binding at $G \cdot C$ pairs that are closer than 6 base pairs away. Again, agreement is close, except at large r, for which the explanation given (MÜLLER and CROTHERS, 1968) was that neighbor exclusion is not absolute. High concentrations of the drug can force binding to occur at more closely spaced intervals.

The one unexplained feature of the neighbor exclusion model is the physical basis of the exclusion interaction. Steric models of the intercalated complexes of proflavine or actinomycin do not reveal any obvious physical overlap. It is possible that the restrictions on the backbone bonds and their preferred orientation in the complex make intercalation at adjacent sites greatly disfavored.

Simple equations apply to the neighbor exclusion model only at high dilution of the bound drug, where neighbor interactions are negligible. Letting B_0 be the number of *potential* binding sites per base pair, the concentration of unoccupied potential sites is $B_0 - r$ times the base pair concentration C_N^0. We can define a "binding constant" $K(r)$, which is a function of r, by analogy with Eq. (1)

$$K(r) = \frac{C_B}{C_N{}^0 (B_0 - r) C_F} = \frac{r}{(B_0 - r) \cdot C_F} . \tag{3}$$

In the limit as r approaches 0, $K(r)$ is just the binding constant for an isolated site, and

$$\lim_{r \to 0} \left\{ \frac{r}{C_F} \right\} = K(0)\, B_0 \,. \tag{4}$$

In an intercalation model for acridines, etc., B_0 is just 1 per base pair, and the intercept of the binding isotherm on the vertical axis gives the intrinsic binding constant $K(0)$. According to this model, the intercept on the horizontal axis (used to determine B_{ap} in

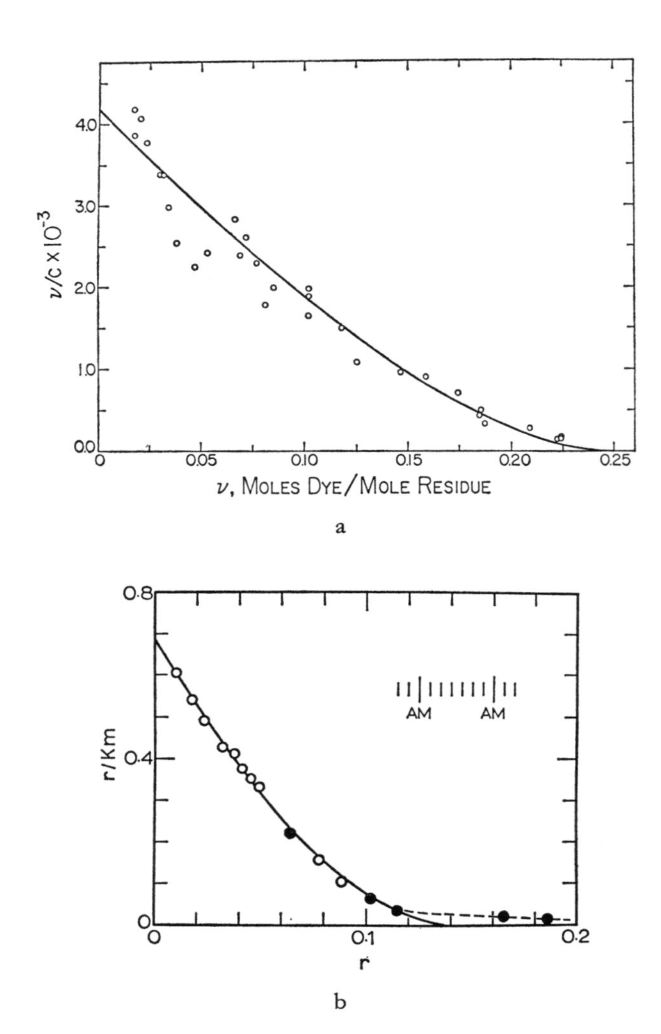

Fig. 1a and b. Comparison of experimental and theoretical binding isotherms using the neighbor exclusion model. a (Taken from BAUER and VINOGRAD, 1970). Ethidium-DNA complex; ν is the ratio of bound dye to nucleotides and C is the same as C_F. The solid line is calculated for exclusion of the next neighbor binding. b (Taken from MÜLLER and CROTHERS, 1968). Actinomycin-DNA complex; r is the ratio of bound actinomycin to base pairs. The inset shows the closest permitted approach of two actinomycins in the theoretical model a Scatchard plot) has no real physical meaning, nor does the slope of the isotherm, usually used to determine K_{ap}.

The neighbor exclusion model is probably the more plausible of these two for description of the binding equilibrium between drugs and DNA. The kinetic data are therefore analyzed in these terms. In particular, experiments are carried out with values of r as small as possible in order to approach the limit of no neighbor interactions.

III. Kinetic Equations

The distinction between independent site and neighbor exclusion models is significant only for the second order step in complex formation. Thus for a reaction between nucleic acid N and drug D,

$$N + D \xrightarrow{k(r)} N - D \tag{5}$$

the concentration of unoccupied potential sites is $C_N{}^\circ (B_o - r)$, and the rate is given as

$$\text{forward rate} = k(r)\, C_N^\circ\, (B_o - r)\, C_F \tag{6}$$

where $k(r)$ is a "rate constant" that depends on r. The limit of $k(r)$ as r approaches 0, $k(0)$, is the rate constant for reaction of an isolated potential binding site. In many cases it is observed that the dissociation rate constant is effectively independent of r, and in this case one can write immediately

$$\frac{K(r)}{K(0)} = \frac{k(r)}{k(0)} \,. \tag{7}$$

Eq. (7) permits extrapolation of measured $k(r)$ to $k(0)$ via the much easier extrapolation of the measured equilibrium constant $K(r)$ to $r = 0$.

IV. Kinetic Methods

The two major kinetic techniques that have been applied to DNA-drug reactions are rapid mixing and relaxation methods. One approach is to mix DNA and drug together, obtaining the second order combination rate constant from the initial slope of the variation of absorbance with time. Examples are the reaction of actinomycin with DNA (MÜLLER and CROTHERS, 1968), the reaction of chromomycin with DNA (BEHR, HONIKEL and HARTMANN, 1969), and the reaction of ethidium with *t*-RNA (BITTMAN, 1969). A method for measuring dissociation rates by mixing the complex with detergent was described by MÜLLER and CROTHERS (1968) and applied to chromomycin by BEHR et al. (1969). The action of the detergent (duponol) seems to be to sequester the free drug, not changing its rate of dissociation, only preventing the reassociation reaction.

The relaxation approach, in the form of temperature jump measurements, has been used to study the interaction of proflavine with DNA (LI and CROTHERS, 1969) and with poly $A \cdot$ poly U (SCHMECHEL and CROTHERS, 1971), and to examine the association of ethidium with *t*-RNA (BITTMAN, 1969). This method involves rapid temperature perturbation of the binding equilibrium, followed by observation of the change in absorbance as a function of time. According to the theory of relaxation techniques (EIGEN and DEMAEYER, 1963), the relaxation response should be described by a set of exponential decay times, called the relaxation times, whose values have a definite relation to rate constants and equilibrium concentrations in the system. In general, the number of relaxation times observed depends on the number of components present. If, for example, proflavine would react with DNA to form only one complex form, a single relaxation time would be observed. Since more relaxation effects are actually seen, the reaction must be more complicated.

V. A Relatively Simple Reaction: Proflavine Binding by Nucleic Acids

Temperature jump measurements on the complex of proflavine with DNA or poly A · poly U show two measureable relaxation times, as illsutrated in Fig. 2 for

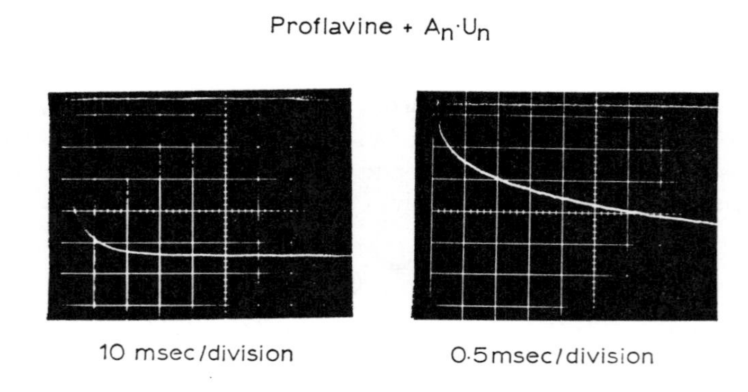

Fig. 2. Two resolvable relaxation times in the response of the poly A · poly U-proflavine complex to a temperature perturbation. (Taken from SCHMECHEL and CROTHERS, 1970)

the poly A · poly U complex (SCHMECHEL and CROTHERS, 1970). The figure shows the variation of light intensity transmitted by the solution following the perturbation. The upper trace in each picture indicates the intensity before the temperature jump. On a time scale of 10 msec per division (Fig. 2a), the slower exponential decay time is observable, but part of the response is much faster than this. The faster relaxation time can be resolved on a time scale of 0.5 msec per division, as seen in Fig. 2b. In addition, there is an extremely rapid optical change (faster than a few microseconds) which LI and CROTHERS (1969) ascribed to a temperature-dependent extinction coefficient of one of the complex forms.

These data have been interpreted in terms of a two step binding mechanism, involving first an external attachment of the dye to the polymer, followed by the insertion or intercalation reaction (LI and CROTHERS, 1969):

$$N + P \underset{k_{21}}{\overset{k_{12}(r)}{\rightleftharpoons}} (N - P)_{\text{out}} \underset{k_{32}}{\overset{k_{23}}{\rightleftharpoons}} (N - P)_{\text{in}} \tag{8}$$

where N is the nucleic acid and P is proflavine. When the first step is much faster than the second, one predicts that the faster relaxation time τ_1 will be given by

$$1/\tau_1 = k_{12}(r)\, X_F + k_{21} \tag{9}$$

where

$$X_F = C_N^o - C_B + C_F \tag{10}$$

and the slower relaxation τ_2 is

$$1/\tau_2 = k_{32} + \frac{k_{23} X_F}{k_{21}/k_{12}(r) + X_F}\,. \tag{11}$$

Fig. 3 shows the variation of the two relaxation times with X_F, for the poly $A \cdot$poly U system. The smooth curves were computed from Eqs. (9) and (11), adjusting the rate constants for best fit.

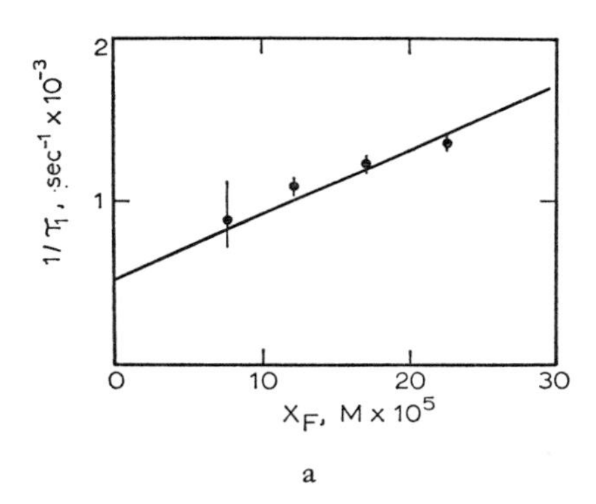

a

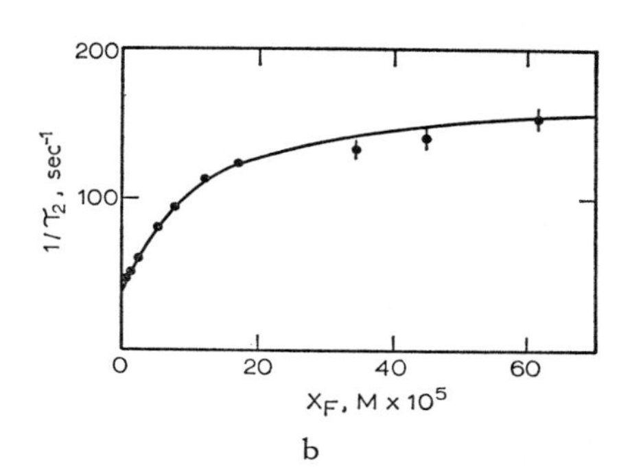

b

Fig. 3. Variation of the fast (Fig. 3a) and slow (Fig. 3b) relaxation times τ_1 with concentration X_F for the poly $A \cdot$ poly U-proflavine complex. The lines are calculated from Eqs. 9 and 11, with rate constants adjusted for best fit.

Data for the rate constants determined in this manner for calf thymus and $T2$ DNA and poly $A \cdot$ poly U are summarized in Tables 1 and 2. There is considerable variation in the mangitude from one nucleic acid to another, but, in general, the second order step has a rate constant of 10^6 to 10^7 M^{-1} sec^{-1} (concentration measured in base pairs), the insertion rate varies from 100 to 1000 sec^{-1}, while the removal of the intercalated dye has a rate constant between 10 and 100 sec^{-1}. It will be seen that these rates are orders of magnitude faster than the reaction of actinomycin with DNA, excepting only the initial second order attachment rate.

One important conclusion from these measurements is that, even when r is small, a measurable fraction of the dye is bound to the outside of the double helix, and not intercalated (see the value of the equilibrium constant K_{23} for the insertion reaction in Table 1). At low salt concentrations the fraction bound in this manner may be greater than 30%. The nature of this external complex is still uncertain. Difference spectra taken following a rapid perturbation reveal that the spectral shift of the outside complex resembles that of intercalation (Li and Crothers, 1969), implying perhaps some interaction with a base pair. Our more recent observations (LI and CROTHERS, unpublished) indicate that ethidium differs strongly from proflavine in the extent of this outside binding. We were unable to detect the outside complex in the case of ehtidium.

Another important question is the mechanism of the insertion reaction. One can imagine two simple limiting cases: either a base pair opens, the drug is inserted and

Table 1. *Comparison of kinetic constants at 10 °C for binding of proflavine to calf thymus DNA, T2-DNA, and $A_n \cdot U_n$*

Type of nucleic acid	$k_{12}(r)$ ($M^{-1}sec^{-1}$)	k_{21} (sec^{-1})	k_{23} (sec^{-1})	k_{32} (sec^{-1})	$K_{12}(r)$ (M^{-1})	K_{23}	$K(r)$ (M^{-1})
Calf thymus DNA, [Na^+] 0.2 M, pH 6.9, $r = 0.092$	2.1×10^6	3500	1500	110	5.9×10^3	13.6	8.8×10^4
T2-DNA, [Na^+] is 0.2 M, pH 6.9, $r = 0.092$	—	—	250	40	1.3×10^4	6.2	9.9×10^4
$A_n \cdot U_n$, [Na^+] 0.2 M, pH 6.9, $r = 0.1466$	4.1×10^7	480	64.8	8.3	8.3×10^3	7.8	7.4×10^4

Table 2. *Energies of the interaction of proflavine with calf thymus DNA, T2-DNA, and $A_n \cdot U_n$ (kcal/M)*

Type of nucleic acid	$E_{12}^\dagger$	$E_{21}^\dagger$	$E_{23}^\dagger$	$E_{32}^\dagger$	$\Delta H_{12}^o(r)$	ΔH_{23}^o	$T\Delta S_{12}^o(r)$ at T 10 °C	$T\Delta S_{23}^o$ at T 10 °C
Calf thymus DNA	4	14	16	14	−9.8	2.0	−5.4	3.4
T2-DNA	—	—	13	17	−3.4	−3.9	1.6	−2.8
$A_n \cdot U_n$	−0.3	0.7	15	23	−1.0	−7.6	3.6	−6.6

$E_{ij}^\dagger$ is the activation energy for transformation from state i to state j; ΔH_{ij}^o and ΔS_{ij}^o are the thermodynamic enthalpy and entropy changes. Measured at [Na^+] = 0.2 M, pH 6.9, in the range 10 °C to 30 °C.

the base pair closes down again, or insertion can occur without opening a base pair, pictured as a simple wedge action of the proflavine forcing its way between two base pairs. The evidence on this is not yet conclusive, but LI and CROTHERS (1969) found that low pH has little effect on the insertion rate. One would expect this condition to increase the average number of base pairs opened and therefore speed up insertion. It was argued on this basis that the more likely mechanism is one in which it is not necessary for a base pair to open. In this connection it is noteworthy that recent preliminary experiments (LI and CROTHERS, unpublished) indicate a substantial speeding up of the insertion rate of ethidium at low pH. It may well be that ethidium is an example of the other limiting mechanism, in which opening of the helix structure is required. It should not be surprising that the preferred insertion mechanism is different in the two cases, since proflavine insertion occurs from an outside bound complex which cannot be detected for ethidium.

VI. Kinetic Properties of the Actinomycin-DNA Complex

Whereas the slowest time constant observed for the interaction of proflavine with DNA is of the order of 0.1 sec, time constants for formation and dissociation of the

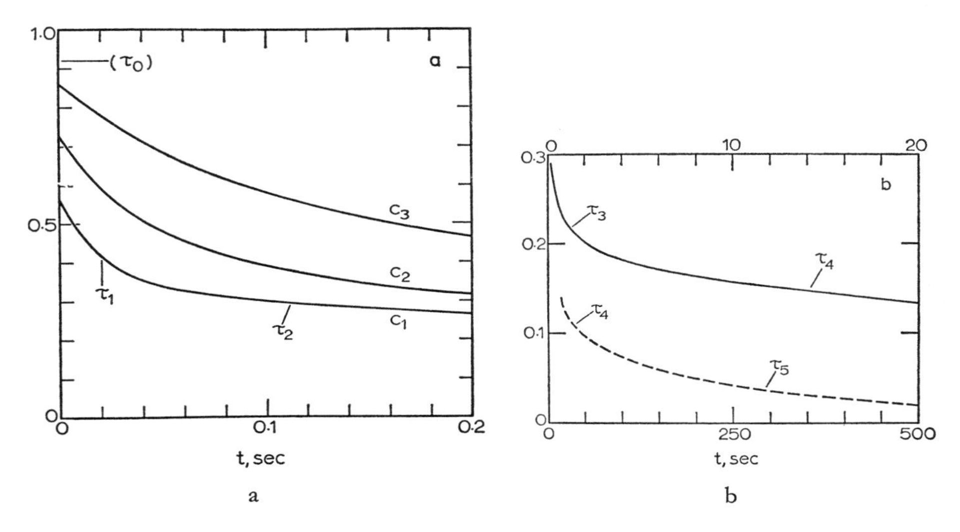

a b

Fig. 4a and b. Kinetic curves for the association of actinomycin with calf thymus DNA. The quantity f, the fraction of the total expected absorbance change, is shown as a function of time after mixing. a Concentration-dependent portions of the reaction at high DNA concentrations; $C_1 = 4 \times 10^{-3}$ M (base pairs), $C_2 = 2 \times 10^{-3}$ M and $C_3 = 10^{-3}$ M. The effect τ_0 is faster than the time resolution of the stopped-flow instrument. The effects labeled τ_1 and τ_2 can be separated reliably only for C_1. b Slower portions of the reaction, which are independent of concentration at these high DNA concentrations. The solid line refers to the upper time scale. Actinomycin concentration: 3 to 5 $\times 10^{-5}$ M. ($\tau_1 = 18.5$ msec at 4×10^{-3} M, $\tau_2 = 0.2$ sec, $\tau_3 = 1.8$ sec, $\tau_4 = 20$ sec, $\tau_5 = 285$ sec, $\tau_o < 5$ msec)

actinomycin-DNA complex range up to hundreds of seconds. If actinomycin and DNA are mixed at low concentration, one observes roughly second order kinetics, with some deviation toward slower rates at long times. However, if the two reactants are mixed at high concentration, a highly complicated reaction is revealed. Fig. 4 shows the fraction of the total expected spectral effect as a function of time after mixing. Only the rate of the initial part of the reaction is concentration-dependent. Thus, there is a second order step, which is rate limiting at low concentrations, but which can be accelerated by raising the concentration of DNA. Under this circumstance, the subsequent first order steps become rate limiting in establishing the final equilibrium complex. Müller and Crothers (1968) found that at least five forms of the complex were needed to explain the kinetic complexity.

The kinetics of the dissociation reaction can be studied by mixing with duponol. Typical results are shown in Fig. 5. Again, more than one time constant is needed to describe the mathematical form of the curves; three dissociation rate constants were

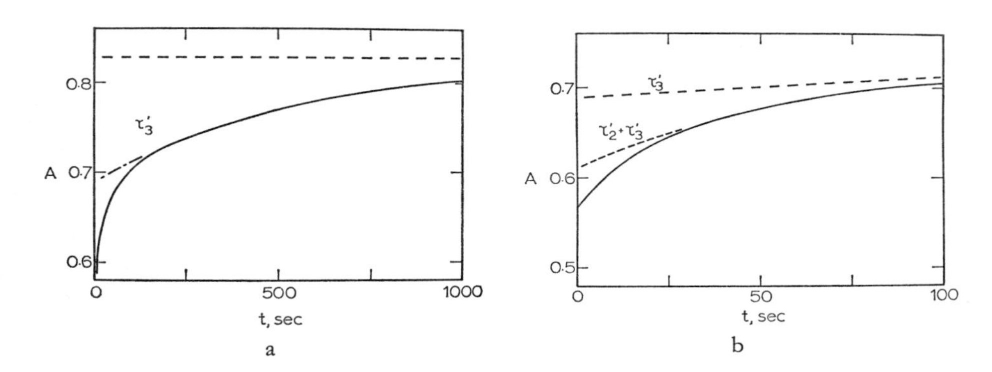

Fig. 5a and b. Kinetic curves for the first-order dissociation of the actinomycin C_3-DNA complex upon mixing with 5% Duponal. a Slow phase of the reaction; the horizontal dashed line shows the equilibrium absorbance A. The dashed curve lebeled τ_3 is an extrapolation of the slowest exponential decay time. b Resolution of the fast portion of the reaction into two separate exponential curves. The dashed curve labeled $\tau_2 + \tau_3$ is an extrapolation of the sum of the slowest two decay curves; the solid line is the experimentally observed curve. ($\tau_1' = 12$ sec, $\tau_2' = 44$ sec, $\tau_3' = 570$ sec)

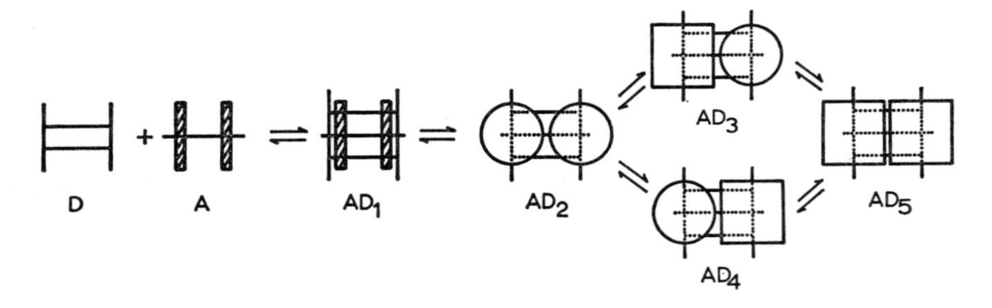

Fig. 6. Schematic drawing of the proposed mechanism of the reaction of actinomycin with DNA. The reaction begins at the left with DNA (D) and actinomycin (A); the DNA base pairs are represented by horizontal lines, as is the actinomycin chromophore. The actinomycin peptide rings are viewed from the edge; it is supposed that their faces interact with each other in solution. In complex AD_1, the chromophore has been inserted between the base pairs, and in complex AD_2 the peptide rings reorient themselves to interact more favorably but non-specifically with the DNA backbones; the rings are now viewed from their faces. In complex forms AD_3 and AD_4, one or the other of the peptide rings undergoes a conformational change which adapts it to interact specifically with the DNA, and in form AD_5 both rings have been so rearranged. At equilibrium, only AD_3, AD_4 and AD_5 are present in significant amounts, with about half the complex present as AD_5

derived from these data. This is clear evidence for the existence of at least three forms of the complex in the equilibrium mixture. Fig. 6 shows the mechanism proposed by MÜLLER and CROTHERS (1968) to account for the kinetic results. The fast parts of the reaction involve attachment of the antibiotic and insertion of the chromophore, followed by slow conformational changes of the peptide rings, giving rise to the several complex forms and multiple time constants. The exact arrangement of the intermediates is hypothetical, since many mechanisms involving the same number of forms can fit the experimental results.

The important qualitative feature of the kinetic model is that the slow time constants arise from interactions involving the peptide rings. This is supported by experiments on analogues of actinomycin that lack these groups or have simpler substituents in their place. An example is the compound actinomine (F. SEELA, to be published),

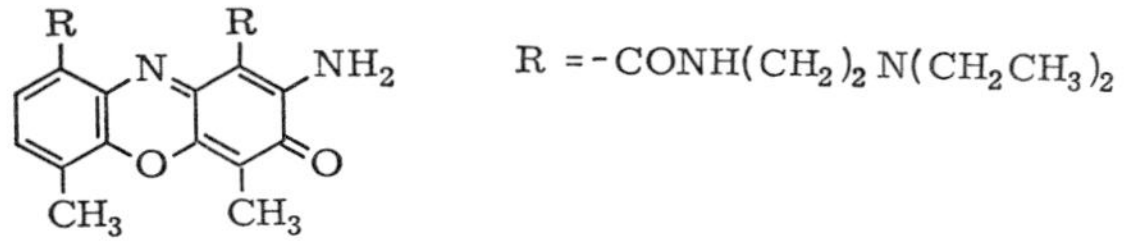

$R = -CONH(CH_2)_2N(CH_2CH_3)_2$

actinomine

which reacts with DNA on a time scale only slightly slower than the reaction of proflavine. The dissociation rate constant is roughly 10 sec^{-1}, compared with about. .002 sec^{-1} for the slowest dissociation rate constant of the actinomycin complex. Thus, even though actinomine has an equilibrium binding affinity roughly equal to that of actinomycin, the dynamic stability of the complex is less by about four orders of magnitude. The source of this difference is the peptide rings, perhaps involving a conformational change of those sterically restricted structures.

VII. Kinetic Evidence for Intercalation

One of the main problems concerning actinomycin binding has been to determine whether the chromophore is hydrogen bonded to the outside of the double helix, as proposed by HAMILTON, FULLER and REICH (1963), or intercalated between the bases as argued by MÜLLER and CROTHERS (1968). Evidence relating to this question can be adduced from kinetic data. Specifically, substitution of the bulky trimethylacetamino group at the 7-position of the actinomycin chromophore,

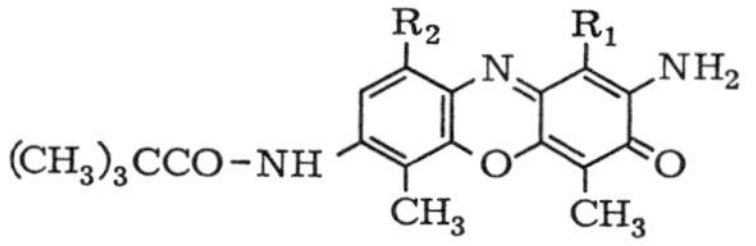

7-trimethylacetamino actinomycin

leads to a reduction by about two orders of magnitude in the rate constants for association and dissociation of the complex. Since the 7-position is not involved in any obvious way in the outside binding model, one would certainly not have predicted the dramatic kinetic effect on the basis of that model. However, intercalation requires getting the chromophore through between the base pairs, and strong steric hindrances to this reaction would be expected. Thus the kinetic findings can be taken in support of an intercalated complex.

VIII. Kinetics and Biological Activity

Substances that to bind to DNA with very similar equilibrium characteristics often have quite different biological activities. For example, of the two intercalating substances proflavine and ethidium, the former is much more active in inducing frameshift mutations. In this connection it is worthwhile noting that kinetic studies reveal strong contrasts in the mechanism of the two binding reactions, although it is still not possible to explain in physical terms the difference in biological activity.

A good example of the importance of kinetic parameters is the difference in kinetic stability of the actinomycin-DNA and actinomine-DNA complexes. Actinomine is inactive in blocking the RNA polymerase. Furthermore, Behr et al. (1969) found a good correlation between the activity of chromomycin derivatives in blocking RNA polymerase activity and the dissociation rate of the complex, those molecules that make slowly dissociating complexes being most active. The proposed reason is that the rate limiting step for progression of the polymerase along the DNA template is the dissociation of the antibiotic complex. In this case the dynamic characteristics of the complex are actually more important than the equilibrium binding affinity.

References

Bauer, W., Vinograd, J.: J. molec. Biol. (1970) (in press).
Behr, W., Honikel, K., Hartmann, G.: Europ. J. Biochem. **9**, 82 (1969).
Bittman, R.: J. molec. Biol. **46**, 251 (1969).
Crothers, D. M.: Biopolymers **6**, 575 (1968).
Eigen, M., de Maeyer, L.: In: Technique of organic chemistry, Vol. 8, part 2 (Freiss, S. L., Lewis, E. S., Weissberger, A., Eds.). New York: Interscience 1963.
Hamilton, L., Fuller, W., Reich, E.: Nature (Lond.) **198**, 538 (1963).
Li, H. J., Crothers, D. M.: J. molec. Biol. **39**, 461 (1969).
Müller, W., Crothers, D. M.: J. molec. Biol. **35**, 251 (1968).
Scatchard, G.: Ann. N.Y. Acad. Sci. **51**, 660 (1949).
Schmechel, D. E. V., Crothers, D. M.: Biopolymers, in press (1971).

The Binding of Actinomycin D to DNA

ROBERT D. WELLS

I. Introduction

The capacity of a nucleic acid to bind actinomycin (AM) is a sensitive indicator of polynucleotide configuration. Double stranded DNA binds the antibiotic [for reviews see REICH and GOLDBERG, 1964, and WARING, 1968 (1)] whereas double-stranded RNA binds only poorly, if at all (HASELKORN, 1964). That double-stranded RNA has a slightly different configuration than double-stranded DNA has been shown by *X*-ray diffraction (ARNOTT, WILKINS, FULLER and LANGRIDGE, 1967). In addition, a hybrid polymer containing one strand DNA and the complementary strand RNA binds little, or no, AM. Again *X*-ray analysis has shown that this molecule possesses a slightly different configuration than bihelical DNA (MILMAN, LANGRIDGE and CHAMBERLIN, 1967). Denatured DNA, or single-stranded DNA, binds only poorly. Thus, the studies herein summarized (WELLS, 1969; WELLS and LARSON, 1970) were undertaken to determine if the AM-binding ability of synthetic polydeoxyribonucleotides could be used as a probe for secondary structure as previously suggested. Other studies (WELLS and BLAIR, 1967; LANGRIDGE, 1969; WELLS, 1970) have already indicated that the DNAs do not all possess identical structures.

A series of DNA model compounds containing defined repeating nucleotide sequences has recently been synthesized by a combination of chemical and enzymatic techniques (BYRD, OHTSUKA, MOON and KHORANA, 1965; WELLS, OHTSUKA and KHORANA, 1965; WELLS, JACOB, NARANG and KHORANA, 1967; WELLS, BÜCHI, KÖSSEL, OHTSUKA and KHORANA, 1967; MORGAN and WELLS, 1968). In addition, a variety of simple DNA polymers are made *de novo* by DNA polymerase (SCHACHMAN, ADLER, RADDING, LEHMAN and KORNBERG, 1960; RADDING, JOSSE and KORNBERG, 1962; INMAN and BALDWIN, 1964; GRANT, HARWOOD and WELLS, 1968). The availability of these DNA models has made possible a variety of chemical, enzymatic and physical studies. In this communication, the AM-binding ability of a number of different DNAs is reported.

Actinomycin is possibly the best known and most thoroughly studied antibiotic which blocks DNA function. Using a variety of analogs of actinomycin, it has been possible to establish the groupings on the molecule critical for binding (REICH, CERAMI and WARD, 1967; MÜLLER and CROTHERS, 1968). On the basis of kinetic studies with AM analogs, as well as other considerations, MÜLLER and CROTHERS (1968) made predictions concerning the structure of the complex and the mechanism of its formation. We have, in a sense, done the complementary experiments by studying a variety of analogs of DNA. Our results demonstrate that the presence of deoxyguanylic acid is *neither necessary nor sufficient* for the binding reaction. These results are contrary to the predictions from previous studies (REICH and GOLDBERG, 1964). Also a sequence preference for the binding process is established. Predictions on the structure of the AM-DNA complex as well as on the apparent role of deoxyguanylic acid can be made from these studies.

II. Results

1. Equilibrium Dialysis

A typical binding isotherm, from equilibrium dialysis studies, demonstrating an actinomycin D-DNA interaction is shown in Fig. 1. Expression of experimental data in this manner (SCATCHARD, 1949) readily provides the number of binding sites per

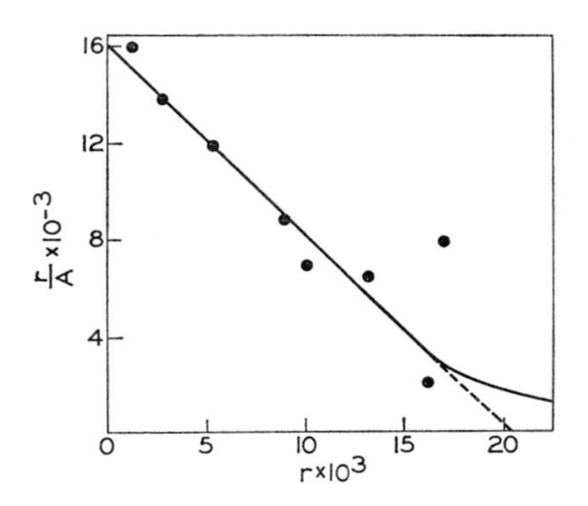

Fig. 1. Scatchard plot of the binding of actinomycin D to poly *d(T-A-C)* · poly *d(G-T-A)*. *r*, moles bound AM/total moles DNA nucleotide; *A*, moles free AM. The intercept on the abscissa is the number of binding sites per nucleotide and the intercept on the ordinate is the apparent binding constant *times* the number of binding sites. Data was obtained from equilibrium dialysis studies using ^{3}H-actinomycin D.

nucleotide from the extrapolated abscissa intercept and the equilibrium constant from the ordinate intercept. The observed equilibrium constants and number of binding sites for ten different DNAs are listed in Table 1. The equilibrium constant observed for the *Micrococcus luteus* DNA — actinomycin *D* interaction is 2.5×10^6 M^{-1} and the number of nucleotides per binding site is 9; these results are in excellent agreement with previously reported values (GELLERT, SMITH, NEVILLE and FELSENFELD, 1965). Likewise the results observed for salmon sperm DNA (Table 1) are comparable to previous data on similar DNAs (GELLERT et al., 1965; MÜLLER and CROTHERS, 1968). The equilibrium constants observed for the interactions with poly *d(G-C)* · poly *d(G-C)** and poly *dG* · poly *dC* are 3.2×10^6 and 2.0×10^6 M^{-1} respectively, however almost eight times as much actinomycin is bound to the former DNA as to the poly *dG* · poly *dC*. Our value of 91 nucleotides per binding site for poly *dG* · poly *dC* is considerably greater than previously reported values (KAHAN, KAHAN and HUR-

* The abbreviations for the polynucleotides used in this study have been described previously [WELLS et al., 1967 (1)] but have been modified to conform to IUPAC conventions.

WITZ, 1963; GELLERT et al., 1965). The reason for this discrepancy may be that somewhat different experimental techniques were used to derive the data or may be due to subtle differences in the preparations of this DNA. The preparations used by both KAHAN et al. and GELLERT et al. are reported to contain 65% *d*GMP and 35% *d*CMP; our poly *dG* · poly *dC* sample was reconstituted from the previously separated single-strands (INMAN and BALDWIN, 1964; WELLS and BLAIR, 1967). That different authentic preparations of this DNA possess slightly different properties has been recognized (WELLS 1970; GRAY and TINOCO, personal communication).

The two repeating dinucleotide DNAs which contain all four common bases poly *d(T-G)* · poly *d(C-A)* and poly *d(T-C)* · poly *d(G-A)*, bind actinomycin (Table 1) somewhat less tightly than naturally occurring DNAs or the DNAs containing 100%

Table 1. *Equilibrium constants and number of binding sites for DNA-actinomycin interactions*

DNA	% $G+C$	K_{app}	Nucleotides per site
Micrococcus luteus DNA	72	2.5×10^6	9
Salmon sperm DNA	42	2.0×10^6	20
poly *d(G-C)* · poly *d(G-C)*	100	3.2×10^6	12
poly *dG* · poly *dC*	100	2.0×10^6	91
poly *d(T-G)* · poly *d(C-A)*	50	1.3×10^6	25
poly *d(T-C)* · poly *d(G-A)*	50	0.6×10^6	37
poly *d(T-T-G)* · poly *d(C-A-A)*	33	1.2×10^6	45
poly *d(T-T-C)* · poly *d(G-A-A)*	33	0.7×10^6	67
poly *d(T-A-C)* · poly *d(G-T-A)*	33	0.8×10^6	50
poly *dI*	—	1.3×10^6	111

Binding parameters were obtained from equilibrium dialysis studies. The experimental error for values of K_{app} and number of binding sites was ± 15%.

G and *C*; also they bind appreciably less AM. The unpublished data of HYMAN and DAVIDSON on poly *d(T-G)* · poly *d(C-A)* of $K = 2 \pm 1 \times 10^6$ M^{-1} and 1 site per 18 ± 8 nucleotides are in excellent agreement with our results (Table 1). The same general trend is observed from the data for the two sequence isomeric repeating trinucleotide DNAs, poly *d(T-T-G)* · poly *d(C-A-A)* and poly *d(T-T-C)* · poly *d(G-A-A)*.

Comparing the data for the three pairs of sequence isomeric DNAs, that is poly *d(G-C)* · poly *d(G-C)* versus poly *dG* · poly *dC*, poly *d(T-G)* · poly *d(C-A)* versus poly *d(T-C)* · poly *d(G-A)* and poly *d(T-T-G)* · poly *d(C-A-A)* versus poly *d(T-T-C)* · poly *d(G-A-A)*, the following is observed: (1) the poly $d(pur_n\text{-}pyr_m)$ · poly $d(pur_m\text{-}pyr_n)$ DNA binds more AM than does the poly *dpur* · poly *dpyr* sequence isomer and (2) the former isomer binds AM somewhat more tightly than does the latter DNA.

Table 1 shows that poly *d(T-A-C)* · poly *d(G-T-A)* binds 1 AM for each 50 DNA nucleotides with an observed equilibrium constant of 0.8×10^6 M^{-1}. The sequence isomer of this DNA, poly *d(A-T-C)* · poly *d(G-A-T)* shows no detectable binding of AM under identical conditions. That is, at levels of actinomycin (2 to 3×10^{-6} M) that are saturating for poly *d(T-A-C)* · poly *d(G-T-A)* (as well as the other DNAs in Table 1) approximately 60,000 cpm/reaction were bound to this DNA; under identical conditions with poly *d(A-T-C)* · poly *d(G-A-T)* as the DNA, less than 500 cpm ^{3}H-actinomycin *D* per reaction were bound. Our assay was designed to detect as little as 1 AM per 100 nucleotides and was successfully used (Table 1) for poly *dG* · poly *dC* and poly *dI* at this sensitivity. The limit of the sensitivity of the assay is roughly double this value, i.e. 1 AM per 200 nucleotides.

Five other DNAs were tested for their ability to bind actinomycin D. None of the following showed detectable binding: poly *d(A-T)* · poly *d(A-T)*, poly *dA* · poly *dT*, poly *d(I-C)* · poly *d(I-C)*, poly *dI* · poly *dC* and *Cancer productus* crab *d(A-T)*. Poly *d(A-T)* · poly *d(A-T)* and poly *dI* · poly *dC* have been previously reported (Reich and Goldberg, 1964) not to bind AM. In addition, poly *d(I-C)* · poly *d(I-C)* induces no change in the visible spectrum of AM under the conditions used by Reich and Goldberg (1964). Hyman and Davidson (1967) have reported the binding of AM to *Cancer antenarrius* crab *d(A-T)* (containing 3% *G*+ *C*) at a level of 1 AM per 112 DNA nucleotides. If *Cancer productus* crab *d(A-T)* contains as much as 3% *G*+ *C* we might have expected detection of a small amount of binding. A possible reason for this discrepancy is that different experimental techniques were used as well as somewhat different DNAs; Hyman and Davidson (1967) used a centrifugation technique for determining binding whereas equilibrium dialysis was used herein. A slight amount of binding to *Cancer productus d(A-T)* is observed using different techniques (Wells and Larson, 1970).

2. Poly *dI*-Actinomycin D Complex

The most unexpected result of this study is that poly *dI* binds actinomycin D. Equilibrium dialysis studies (Table 1) show that the binding reaction has an apparent equilibrium constant of 1.3×10^6 M^{-1} and one AM is bound per 111 nucleotides. *In vitro* transcription studies also demonstrate that actinomycin D binds to poly *dI*. Fig. 2 shows that the transcription of poly *dI* in the presence of 4.5 μM AM proceeds at only 45% the rate observed in the absence of the antibiotic. At this level of AM, the rate of transcription of poly *d(I-C)* · poly *d(I-C)*, a DNA that exhibits no detectable AM binding by equilibrium dialysis, spectral or buoyant density studies, is virtually identical to that observed in the absence of AM. At higher levels of AM, the transcription of poly *dI* is inhibited by as much as 94%. Indeed, this extent of inhibition is at least as great as observed for DNAs which contain deoxyguanosine. In addition poly *dI* causes a hypochromic shift in the visible spectrum of actinomycin D (Wells and Larson, 1970) which is characteristic of AM- DNA complexes under the conditions employed.

To determine if this unexpected binding of AM to poly *dI* is due to a previously unrecognized property of the base inosine, the possible binding of AM by poly *rI* was also studied by equilibrium dinlysis. Poly *rI* binds no detectable AM under identical conditions used for poly *dI*.

The observation that poly *dI* binds AM is clearly not in accord with the proposal (Hamilton, Fuller and Reich, 1963; Cerami, Reich, Ward and Goldberg, 1967) that actinomycin binds to DNA by hydrogen-bonding in the narrow groove of the double-helix to deoxyguanosine.

3. Other Measurements of the Binding Reaction

The *in vitro* inhibition of RNA synthesis by actinomycin is well documented (GOLDBERG and RABINOWITZ, 1962; HURWITZ, FURTH, MALAMY and ALEXANDER, 1962; GOLDBERG, RABINOWITZ and REICH, 1962; KAHAN et al., 1963); this effect is the basis for the bacteriostatic and antitumor activities of the molecule (REICH and GOLDBERG, 1964).

In order better to understand the interaction of AM with DNA, we have studied (WELLS and LARSON, 1970) the ability of AM to inhibit the transcription of all the above DNA's. A 260 fold concentration range of AM was employed in these studies.

Fig. 2 has already shown that the transcription of poly *dI*, a DNA that binds the drug, is markedly inhibited whereas the transcription of poly *d(I-C)* · poly *d(I-C)*,

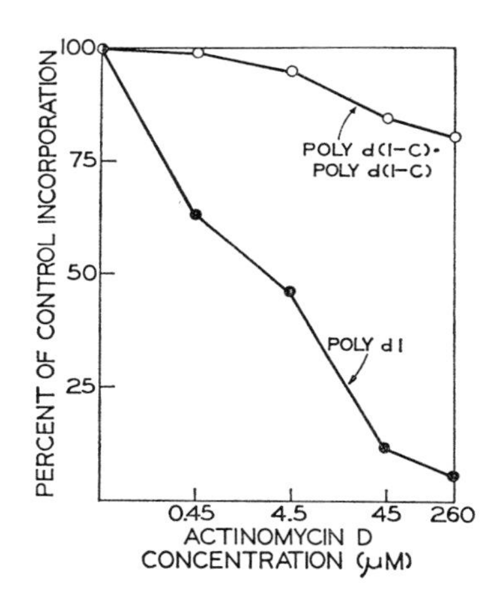

Fig. 2. The effect of AM on the transcription of poly *d(I-C)* · poly *d(I-C)* and poly *dI* by *E. coli* RNA polymerase. GTP and CTP were present in all reactions; CTP was the ^{14}C-labeled substrate. In the control experiments (no AM added), the following extents of RNA synthesis were observed in the 30 min time period: poly *d(I-C)* · poly *d(I-C)*, 2.9 fold; poly *dI*, 3.0 fold

a DNA that does not bind the drug, is not appreciably inhibited except at high AM concentrations. The conclusions from the studies with 16 different DNA's are: (1) the transcription of DNA's which bind AM (Table 1) is inhibited; (2) the transcription of DNA's which do not bind AM, as judged by equilibrium dialysis, is not appreciably inhibited except at very high concentrations of AM; (3) the extent of inhibition of DNA polymers is less than that observed for naturally-occurring DNA's containing the same $G+C$ content; (4) in comparing the extent of inhibition of binding DNA's with the same composition but different nucleotide sequences, invariably the transcription of the poly *d pur* · poly *d pyr* DNA is less inhibited than the transcription of the poly $d(pur_n\text{-}pyr_m)$ · poly $d(pur_m\text{-}pyr_n)$ DNA. This is in agreement with the equilibrium dialysis studies (Table 1); (5) AM inhibits equally the transcription of both strands of the double-stranded DNA's.

DNA-antibiotic interactions can be qualitatively monitored by equilibrium density gradient centrifugation studies (Kersten, Kersten and Szybalski, 1966). This additional technique was used (Wells and Larson, 1970) to study the binding of AM to the seventeen different DNA's. The results are similar to those obtained from other measurements (described above) and have been described in detail (Wells and Larson, 1970).

III. Discussion

1. Effect of DNA Composition and Sequence on Binding of AM

The three major conclusions that may be drawn from this work are the following: (1) the presence of deoxyguanosine is not necessary for the binding of AM to DNA, (2) the presence of deoxyguanosine is not sufficient for binding and (3) a marked nucleotide sequence preference exists for the binding reaction. Each of these conclusions are discussed separately.

a) The Non-necessity of Guanine for Binding

Poly *dI* binds actinomycin almost as tightly as does naturally occurring DNA as measured by equilibrium dialysis (Table 1); also *in vitro* transcription studies (Fig. 2) in the presence of AM and spectral analysis both confirm the binding reaction. Thus, the presence of the 2-amino group on deoxyguanosine, which has been inferred to be necessary for the binding reaction (Hamilton et al., 1963; Cerami et al., 1967), is not essential in all cases.

The configuration of poly *dI* which is responsible for the binding is unknown; however, from the studies presented above, it seems certain to be an ordered structure as opposed to a random coil configuration. Inman (1964) has reported that poly *dI* possesses an ordered structure above room temperature at NaCl concentrations greater than approximately 0.2 M. A two stranded poly *dI* structure was suggested by this author. However, it is probable that the ribo analog of this polynucleotide, poly *rI*, exists as a three-stranded ordered structure (Rich, 1958). Appropriate studies have not yet been performed to determine the exact molecular configuration of the ordered form of poly *dI*. That an ordered structure of poly *dI* is responsible for the antibiotic binding is suggested by the spectral studies (Wells and Larson, 1970); in the absence of divalent metal ions no actinomycin spectral change was induced by the polymer whereas a pronounced spectral change was found in the presence of metal ions. The RNA polymerase inhibition studies (Fig. 2) are also consistent with this suggestion. The transcription of poly *dI* was inhibited by AM to a greater extent than any other repeating-sequence polymer tested. These studies must be performed in the presence of divalent metal ions to observe enzymatic activity, thus it was not possible to assay for the effect of $MgCl_2$-$MnCl_2$ by this technique. Conversely, all equilibrium dialysis studies were performed in 0.01 M NaCl to 0.001 M sodium phosphate (pH 7.4) and divalent metal ions were rigorously excluded. By this assay, poly *dI* binds the antibiotic quite well (Table 1). However for these studies, AM and the polymer were equilibrated for 5 to 9 days and it is conceivable that the presence of the antibiotic slowly facilitates the formation of a suitable configuration of poly *dI* for binding. That this is a slow process is suggested by the spectral analysis in the same ionic environment. Whatever the nature of the structure of poly *dI* responsible for the binding reaction, it is clear that a similar structure cannot be formed when the polymer is complexed with poly *dC* since poly *dI* · poly *dC* binds no AM as judged by equlibrium dialysis, *in vitro* transcription or buoyant density studies.

b) The Insufficiency of Guanine for Binding

Poly *d(A-T-C)* · poly *d(G-A-T)*, which contains 33% *G-C*, binds virtually no AM as measured by equilibrium dialysis, spectral, absorbance-temperature, *in vitro* transcription and buoyant density studies (WELLS, 1969; WELLS and LARSON, 1970). Thus, contrary to predictions from previous studies (CERAMI et al., 1967; REICH and GOLDBERG, 1964), the presence of guanine in a DNA is not a sufficient requisite for binding. The role of *G-C* pairs in influencing the binding with other DNAs is, in no way, contradicted by this finding. In fact the bulk of the studies in this paper (Table 1) buttress the generally accepted observation that *G-C* pairs play some role in the binding reaction. The reason that poly *d(A-T-C)* · poly *d(G-A-T)* does not bind is unknown at present. However physical and enzymatic studies [WELLS et al., 1967 (1), WELLS and BLAIR, 1967; WELLS, 1970] show that this DNA has somewhat different properties from its sequence isomer, poly *d(T-A-C)* · poly *d(G-T-A)*, which does bind AM (Table 1). Hence, poly *d(A-T-C)* · poly *d(G-A-T)* must possess a molecular configuration which does not permit AM binding.

Thus, what is the role of deoxyguanine in the binding reaction? It was suggested (WELLS, 1969) that the presence of deoxyguanosine in a DNA may induce a suitable configuration in a small region of the DNA chain to permit binding. Thus, it was possibly fortuitous that, in all previous studies (CERAMI et al., 1967; REICH and GOLDBERG, 1964), polymers which contain a purine 2-amino group do bind AM whereas polymers without a purine 2-amino group do not bind AM. That is, for all DNAs examined to date except one [poly *d(A-T-C)* · poly *d(G-A-T)*], the presence of a purine 2-amino group was sufficient to generate an appropriate configuration for the binding of AM. We are suggesting that, in most cases, the presence of *G* in a DNA *is* sufficient to generate an appropriate configuration, however the presence of *G* is not necessarily sufficient or obligatory. For example, in certain cases (i.e. poly *dI*) a suitable configuration may be formed without the presence of *G*.

The details of the molecular nature of the appropriate configuration for binding are obscure at present. Such information may only be obtained by *X*-ray diffraction analyses of a variety of DNAs or of a suitable AM-polynucleotide complex.

c) The Nucleotide Sequence Preference for Binding

The majority of the studies reported herein were performed with DNA polymers with defined repeating nucleotide sequences. By complete nearest neighbor frequency analyses it was possible to assign with certainty the primary nucleotide sequence in each case [RADDING et al., 1962; BYRD et al., 1965; WELLS et al., 1965; WELLS et al., 1967 (1); GRANT et al., 1968]. Thus it was possible to study the influence of DNA nucleotide sequence on the binding reaction. Equilibrium dialysis studies (Table 1), as well as all other studies reported herein, indicate the following: in comparing sequence isomeric DNAs, the isomer which contains both purines and pyrimidines on both complementary strands binds more AM and binds AM more tightly than does the isomer which contains all purines on one strand and all pyrimidines on the complementary strand. Thus poly *d(G-C)* · poly *d(G-C)*, poly *d(T-G)* · poly *d(C-A)* and poly *d(T-T-G)* · poly *d(C-A-A)* bind more AM and bind AM more tightly than do poly *dG* · poly *dC*, poly *d(T-C)* · poly *d(G-A)* and poly *d(T-T-C)* · poly *d(G-A-A)*, respectively. Such a comparison is not possible with poly *dA* · poly *dT*, poly *d(A-T)* · poly *d(A-T)*, poly *dI* · poly *dC* and poly *d(I-C)* · poly *d(I-C)* since none of these polymers binds AM. A variety of physical and enzymatic data indicates that the poly *d pur* · poly *d pyr* isomer is somewhat anomalous when compared to the poly $d(pur_n\text{-}pyr_m)$ · poly $d(pur_m\text{-}pyr_n)$ isomer and to naturally occurring DNA. Thus, if the apparent DNA specificity demonstrated herein is a function of primary nucleotide sequence or is a matter of DNA configuration cannot be ascertained at present. Indeed it may be that these two variables are inseparable, hence such a discussion is meaningless, since it appears that the DNA nucleotide sequence determines the three-dimensional configuration (Wells, 1970; Mitsui, Langridge and Wells, 1969).

Unquestionably the most marked effect of nucleotide sequence was observed in comparing poly *d(T-A-C)* · poly *d(G-T-A)*, which binds AM, with poly *d(A-T-C)* · poly *d(G-A-T)*, which does not bind AM. A longer abbreviation for these DNAs is: poly *d(T-A-C)* · poly *d(G-T-A)*

3'-end -----*A-T-G-A-T-G-A-T*----- 5'-end
5'-end -----*T-A-C-T-A-C-T-A*----- 3'-end

poly *d(A-T-C-)* · poly *d(G-A-T)*

3'-end -----*T-A-G-T-A-G-T-A-G*----- 5'-end
5'-end -----*A-T-C-A-T-C-A-T-C*----- 3'-end

Thus, focusing on a *G-C* pair in both DNAs, the difference is simply the relative orientation of the *A-T* pairs. The binding properties of these two DNAs are discussed above.

Previous studies on the binding of AM to DNA have indicated the importance of nucleotide sequence. Gellert et al. (1965) found that the stoichiometry of the binding reaction was relatively constant for a variety of naturally-occurring DNAs, ranging in composition from 35 to 73% *G-C*. Also, the studies of Hyman and Davidson (1967) on the binding of AM to the *Cancer antenarrius* crab *d(A-T)* DNA suggest a sequence specificity. Indeed, such a specificity has been established for the bindingof nogalamycin (a tetracycline-like molecule) to DNA (Bhuyan and Smith, 1965).

3. Molecular Nature of DNA-AM Complex

The molecular nature of the complex formed between AM and DNA has been the subject of considerable study (Reich and Goldberg, 1964; Gellert et al., 1965; Cerami et al., 1967; Müller and Crothers, 1968). Two models have been proposed: (1) Hamilton et al. (1963) proposed that the AM chromophore is hydrogen-bonded to the outside of the DNA helix; stabilization of the complex is provided by a hydrogen bond between the actinomycin quinone oxygen and the 2-amino group of guanine as well as hydrogen bonds from the AM amino group to the N_3 of guanine and to the deoxyribose ring oxygen. The peptide lactones were considered to provide additional hydrogen-bonds with phosphodiester oxygens. The major justifications for the proposal were: (a) the apparent specificity for deoxyguanosine residues, (b) the importance of an unsubstituted chromophore amino group, (c) the inability to detect intercalation by *X*-ray diffraction studies (REICH et al., 1967). (2) MÜLLER and CROTHERS (1968) proposed that the AM chromophore is intercalated into the DNA chain with the peptide lactones projecting into the DNA minor groove. The basis for this proposal was: (a) the marked decrease in the rate of complex formation when AM analogs contained a bulky substituent on the 7 position of the chromophore (the side of the chromophore distal to the helix in the above model) and (b) a careful reexamination of the hydrodynamic properties of DNA of various molecular weights in the presence of AM.

Recent studies by WARING [1968 (2)] are consistent with the intercalation model; actinomycin acts virtually identically to ethidium (an acknowledged intercalating agent) in its effect on the supercoiled structure of the replicative form of $\phi \times 174$ DNA. Also, the recent NMR studies on AM-nucleotide complexes of DANYLUK and VICTOR (1970) are consistent with this notion. A complete review of the justification for each of these models cannot be presented and the reader is referred to the literature cited. That multiple modes of binding exist has been established by virtually all laboratories concerned with this problem (REICH and GOLDBERG, 1964; GELLERT et al., 1965; MÜLLER and CROTHERS, 1968; CAVALIERI and NEMCHIN, 1968). Hence, when discussing the molecular nature of the complex, it should be clear that only the very tight binding of AM to DNA is considered.

As mentioned above, one reason HAMILTON et al. (1963) proposed the hydrogen-bonded "outside-binding" model was the apparent necessity of deoxyguanine for binding. Apurinic acid, poly *d(A-T)* · poly *d(A-T)*, poly *dA* · poly *dT* and poly *dI* · poly *dC* do not form complexes with AM (REICH et al., 1967). Conversely, DNAs which contain deoxyguanine, or a purine bearing a 2-amino group, do bind AM and include naturally occurring DNA, poly *dG* · poly *dC*, poly *d(DAP-T)* · poly *d(DAP-T)* (CERAMI et al., 1967), a *d(A-T)*-like polymer containing a proportion of 2-aminopurine residues (REICH et al., 1967) and *Cancer antenarrius* crab *d(A-T)* (HYMAN and DAVIDSON, 1967). Hence it was suggested that the presence of a purine 2-amino group is sufficient, and perhaps the sole requirement, for AM binding (CERAMI et al., 1967). Our results are clearly not in agreement with this suggestion since poly *dI* binds AM (Table 1) and inosine bears no 2-amino group. Furthermore our results do not indicate that the presence of deoxyguanine (or a purine 2-amino group) is a sufficient requirement for binding since poly *d(A-T-C)* · poly *d(G-A-T)*, which contains 33% *G-C*, binds no AM (WELLS, 1969; WELLS and LARSON, 1970).

We propose that the apparent specificity of guanine (or a purine 2-amino group) previously observed, and corroborated in some cases herein, is a secondary effect. That is, we propose that the presence of a purine 2-amino group in a polymer is sufficient, in most cases, to confer on a DNA a suitable steric and electronic environment to permit the binding of AM. Thus, the presence of a suitable DNA configuration may be the primary determinant for AM binding rather than the presence of a purine 2-amino group. That poly *d(A-T-C)* · poly *d(G-A-T)* does not bind AM although it contains *G-C* pairs may be an effect of other structural features which do not allow the polymer to assume a suitable configuration for binding (WELLS, 1970). We suggest that other DNAs will be found that contain *G* but do not bind AM and also DNAs that do not contain *G* but do bind AM. The molecular nature of the critical binding environment is, at present hypothetical. One feasible structure, which is in agreement with our results, has been proposed (MÜLLER and CROTHERS, 1968) on the basis of model building.

Summary

The ability of seventeen different DNAs to bind actinomycin D(AM) was studied by a variety of techniques including equilibrium dialysis, *in vitro* transcription and analytical buoyant density centrifugation. The major conclusions are: (1) The presence of deoxyguanylic acid in a DNA is not necessary for complex formation. Poly *dI* binds approximately 1/4 as much AM as DNAs which contain 50% G-*C*; the equilibrium constant for the poly *dI*-AM complex is as large as that observed for 50% *G-C* DNAs. (2) The presence of deoxyguanylic acid in a DNA is not sufficient for complex formation. Poly *d(A-T-C)* · poly *d(G-A-T)*, which contains 33% *G-C*, binds little or no AM as judged by five different techniques. (3) A marked nucleotide sequence preference exists for the binding reaction. When comparing sequence isomeric DNAs, the poly $d(pur_n\text{-}pyr_m)$ · poly $d(pur_m\text{-}pyr_n)$ isomer binds more AM and binds AM more tightly than does the poly *d pur* · poly *d pyr* isomer. (4) The guanine-containing DNAs tested, with the exception of poly *d(A-T-C)* · poly *d(G-A-T)*, bind AM and are: *M. luteus* DNA, salmon sperm DNA, poly *d(G-C)* · poly *d(G-C)*, poly *dG* · poly *dC*, poly *d(T-G)* · poly *d(C-A)*, poly *d(T-C)* · poly *d(G-A)*, poly *d(T-T-G)* poly *d(C-A-A)*, poly *d(T-T-C)* · poly *d(G-A-A)* and poly *d(T-A-C)* · poly *d(G-T-A)*. (5) The DNAs which are devoid of deoxyguanine, with the exception of poly *dI*, do not bind AM and are: poly *d(A-T)* · poly *d(A-T)*, poly *dA* · poly *dT*, poly *dI* · poly *dC* and poly *d(I-C)* · poly *d(I-C)*. The results are discussed in relation to two models for the AM-DNA complex, the hydrogen-bonded, "outside-binding" model of HAMILTON, FULLER and REICH (1963) and the intercalation model of MÜLLER and CROTHERS (1968). The data are not consistent with the hydrogen bonded model.

Acknowledgement

This work has been supported by grants from the National Science Foundation (GB-6629 and GB-8786) and the Life Insurance Medical Research Foundation.

References

ARNOTT, S., WILKINS, M. H. F., FULLER, W., LANGRIDGE, R.: Molecular and crystal structures of double-helical RNA. J. molec. Biol. **27**, 535 (1967).

BHUYAN, B. K., SMITH, C. G.: Differential interaction of nogalamycin with DNA of varying base composition. Proc. nat. Acad. Sci. (Wash.) **54**, 566 (1965).

BYRD, C., OHTSUKA, E., MOON, M. W., KHORANA, H. G.: Synthetic deoxyribooligonucleotides as templates for the DNA polymerase of *E. coli*: New DNA-like polymers containing repeating nucleotide sequences. Proc. nat. Acad. Sci. (Wash.) **53**, 79 (1965).

CAVALIERI, L. F., NEMCHIN, R. G.: The binding of actinomycin D and F to bacterial DNA. Biochim. biophys. Acta (Amst.) **166**, 722 (1968).

CERAMI, A., REICH, E., WARD, D. C., GOLDBERG, I. H.: The interaction of actinomycin with DNA: Requirement for the 2-amino group of purines. Proc. nat. Acad. Sci. (Wash.) **57**, 1036 (1967).

DANYLUK, S. S., VICTOR, T. A.: The structure and molecular interactions of actinomycin D. In: Quantum aspects of heterocyclic compounds in chemistry and biochemistry (BERGMAN, E. D., PULLMAN, B., Eds.) (1970) (in press).
GELLERT, M., SMITH, C. E., NEVILLE, D., FELSENFELD, G.: Actinomycin binding to DNA: Mechanism and specificity. J. molec. Biol. **11**, 445 (1965).
GOLDBERG, I. H., RABINOWITZ, M.: Actinomycin D inhibition of deoxyribonucleic acid-dependent synthesis of ribonucleic acid. Science **136**, 315 (1962).
— REICH, E.: Basis of actinomycin action, I. DNA binding and inhibition of RNA-polymerase synthetic reactions by actinomycin. Proc. nat. Acad. Sci. (Wash.) **48**, 2094 (1962).
GRANT, R. C., HARWOOD, S. J., WELLS, R. D.: The synthesis and characterization of poly *d(I-C)* · poly *d(I-C)*. J. Amer. chem. Soc. **90**, 4474 (1968).
HAMILTON, L. D., FULLER, W., REICH, E.: X-ray diffraction and molecular model building studies of the interaction of actinomycin with nucleic acids. Nature (Lond.) **198**, 538 (1963).
HASELKORN, R.: Actinomycin D as a probe for nucleic acid secondary structure. Science **143**, 682 (1964).
HURWITZ, J., FURTH, J. J., MALAMY, M., ALEXANDER, M.: The role of deoxyribonucleic acid in ribonucleic acid synthesis, III. The inhibition of the enzymatic synthesis of ribonucleic acid and deoxyribonucleic acid by actinomycin D and proflavin. Proc. nat. Acad. Sci. (Wash.) **48**, 1222 (1962).
HYMAN, R. W., DAVIDSON, N.: The binding of actinomycin to crab dAT; the nature of the DNA binding site. Biochem. biophys. Res. Commun. **26**, 116 (1967).
INMAN, R. B.: Transitions of DNA homopolymers. J. molec. Biol. **9**, 624 (1964).
— BALDWIN, R. L.: Helix-random coil transitions in DNA homopolymer pairs. J. molec. Biol. **8**, 452 (1964).
KAHAN, E., KAHAN, F. M., HURWITZ, J.: The role of deoxyribonucleic acid in ribonucleic acid synthesis. VI. Specificity of action of actinomycin D. J. biol. Chem. **238**, 2491 (1963).
KERSTEN, W., KERSTEN, H., SZYBALSKI, W.: Physicochemical properties of complexes between deoxyribonucleic acid and antibiotics which affect ribonucleic acid synthesis. Biochemistry **5**, 236 (1966).
LANGRIDGE, R.: Nucleic acids and polynucleotides. J. Cell Physiol. **74**, 1 (1969).
MILMAN, G., LANGRIDGE, R., CHAMBERLIN, M. J.: The structure of a DNA-RNA hybrid. Proc. nat. Acad. Sci. (Wash.) **57**, 1804 (1967).
MITSUI, Y., LANGRIDGE, R., WELLS, R. D.: Poly *d(I-C)*: An unusual double helical DNA. Acta Cryst. **25**, 177 (1969).
MORGAN, A. R., WELLS, R. D.: Specificity of the three-stranded complex formation between double-stranded DNA and single-stranded RNA containing repeating nucleotide sequences. J. molec. Biol. **37**, 63 (1968).
MÜLLER, W., CROTHERS, D. M.: Studies of the binding of actinomycin and related compounds to DNA. J. molec. Biol. **35**, 251 (1968).
RADDING, C. M., JOSSE, J., KORNBERG, A.: Enzymatic synthesis of deoxyribonucleic acid. XII. A polymer of deoxyguanylate and deoxycytidylate. J. biol. Chem. **237**, 2869 (1962).
REICH, E., GOLDBERG, I. H.: Actinomycin and nucleic acid function. In: Progress in nucleic acid research and molecular biology **3**, p. 183. New York: Academic Press, Inc. 1964.
— CERAMI, A., WARD, D. C.: Actinomycin (GOTTLIEB, D., SHAW, P. D., Eds.). Antibiotics **1**, 714 (1967).
RICH, A.: The molecular structure of polyinosinic acid. Biochim. biophys. Acta (Amst.) **29**, 502 (1958).
SCATCHARD, G.: The attractions of proteins for small molecules and ions. Ann. N.Y. Acad. Sci. **51**, 660 (1949).
SCHACHMAN, H. K., ADLER, J., RADDING, C. M., LEHMAN, I. R., KORNBERG, A.: Enzymatic synthesis of deoxyribonucleic acid. VII. Synthesis of a polymer of deoxyadenylate and deoxythymidylate. J. biol. Chem. **235**, 3242 (1960).
WARING, M. J.: (1) Drugs which affect the structure and function of DNA. Nature (Lond.) **219**, 1320 (1968).

— (2) Uncoiling of bacteriophage $\phi \times 174$ replicative form deoxyribonucleic acid by ethidium, daunomycin and actinomycin. Biochem. J. **109**, 28 (1968).
Wells, R. D.: Actinomycin binding to DNA: Inability of a DNA containing guanine to bind actinomycin D. Science **165**, 75 (1969).
— Studies on the effect of nucleotide sequence on DNA physicochemical properties. 8th International Congress of Biochemistry, Switzerland, September 1970.
— Blair, J. E.: Studies on polynucleotides. LXXI. Sedimentation and buoyant density studies of some DNA-like polymers with repeating nucleotide sequences. J. molec. Biol. **27**, 273 (1967).
— Larson, J. E.: Studies on the binding of actinomycin D to DNA and DNA model polymers. J. molec. Biol. **49**, 319 (1970).
— Ohtsuka, E., Khorana, H. G.: Studies on polynucleotides. L. Synthetic deoxyribopolynucleotides as templates for the DNA polymerase of *Escherichia coli'* A new double-stranded DNA-like polymer containing repeating dinucleotide sequences. J. molec. Biol. **14**, 221 (1965).
— Jacob, T. M., Narang, S. A., Khorana, H. G.: (1) Studies on polynucleotides. LXIX. Synthetic deoxyribopolynucleotides as templates for the DNA polymerase of *Escherichia coli:* DNA-like polymers containing repeating trinucleotide sequences. J. molec. Biol. **27**, 237 (1967).
— Büchi, H., Kössel, H., Ohtsuka, E., Khorana, H. G.: (2) Studies on polynucleotides. LXX. Synthetic deoxyribopolynucleotides as templates for the DNA polymerase of *Escherichia coli:* DNA-like polymers containing repeating tetranucleotide sequences. J. molec. Biol. **27**, 265 (1967).

Interaction of Antitumor Antibiotics with DNA: Studies on Sibiromycin

GEORGII F. GAUSE and YU. V. DUDNIK

Sibiromycin is a new antitumor antibiotic produced by an actinomycete, *Streptosporangium sibiricum* (GAUSE et al., 1969). This antibiotic was first detected by its capacity to inhibit the multiplication of suspended ascites tumor cells. For example, the culture liquid of *Streptosporangium sibiricum*, diluted 100 times, inhibited the multiplication of cells of reticulo-endothelial sarcoma of mice (strain RAB-1), as well as multiplication of cells of mouse lymphoadenoma (strain NK/Ly). In the culture liquid of the producing strain sibiromycin is present in low concentrations and inhibits only the growth of tumor cells but not that of bacteria. However, when sibiromycin was later concentrated it was observed that at higher concentrations the substance inhibited also the growth of various bacteria, notably that of *Bacillus mycoides*. This microorganism was then used for the biological assay of sibiromycin at an early stage of research, before the chemical method of determining the concentration of sibiromycin was established (KONSTANTINOVA et al., 1970).

Sibiromycin was obtained in pure form; it possesses amphoteric properties and is soluble in dilute acids and alkalies (BRAZHNIKOVA et al., 1970). The empirical formula of sibiromycin is $C_{24}H_{34}N_3O_6$; the antibiotic contains one amino group, three C-methyl groups and two groups that can be acetylated. A crystalline derivative of sibiromycin was prepared which contained sulphur and retained all biological activity of the original compound.

Sibiromycin possesses considerable antitumor activity against a number of transplantable tumors of animals (SHORIN et al., 1970). Its activity, however, is most pronounced against squamous cell carcinoma. In the treatment of mice with inoculated squamous praegastric cancer cells (strain OG-5) it was observed that after two injections of sibiromycin in maximal tolerated doses the tumors disappeared completely. Preliminary clinical studies indicate that intravenous injections of sibiromycin are effective in the treatment of some forms of squamous cell carcinoma in men.

Sibiromycin possesses a characteristic capacity for interaction with DNA and exerts a selective action on the synthesis of DNA *in vivo*. In a first series of experiments the effect of sibiromycin upon the synthesis of macromolecules was measured by analysis for DNA, RNA and protein the cultures of *Staphylococcus aureus* 209 with methods described in detail (LAIKO, 1964). DNA was determined by the reaction with diphenylamine, RNA by the reaction with orcinol, and protein by the method of LOWRY. Fig. 1 shows that sibiromycin selectively inhibited the synthesis of DNA in *S. aureus* 209. Complete inhibition of DNA synthesis was caused by a concentration of sibiromycin of 1 μg/ml. At this concentration the synthesis of RNA as well as the synthesis of protein were inhibited by 70%. At lower concentrations of sibiromycin the selective inhibition of synthesis of DNA was also observed, and the synthesis of RNA was inhibited more strongly than the synthesis of protein. At 0.25 μg/ml sibiromycin, the synthesis of DNA was inhibited by 83%, the synthesis of RNA by 57%, and the synthesis of protein only by 20%.

An investigation was also made of the inhibitory action of sibiromycin upon the synthesis of macromolecules in *Bacillus subtilis* 168 T^-, met^-. In these experiments was measured the incorporation of C^{14}-thymidine into DNA, the incorporation of C^{14}-uracil into RNA, and the incorporation of C^{14}-lysine into protein of bacteria, suspended in nutrient medium containing graded concentrations of sibiromycin. It should be pointed out that at the concentrations used in these experiments, sibiromycin did not at first inhibit the growth of the organisms. For example, at 1 μg/ml of

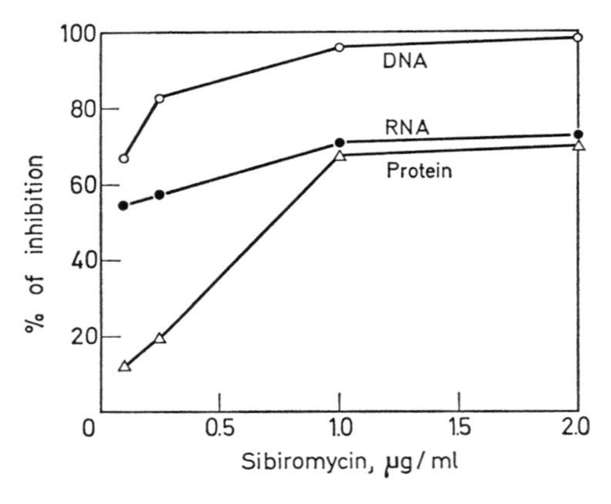

Fig. 1. Effect of sibiromycin on the synthesis of DNA, RNA and protein in *Staphylococcus aureus* 209

sibiromycin the rate of growth of the culture, during the first twenty minutes, did not differ from that of the control. Fig. 2 indicates that sibiromycin selectively inhibited the synthesis of DNA. With sibiromycin at 0.3 μg/ml the synthesis of DNA was completely inhibited in 15 min, whereas the synthesis of RNA was inhibited but slightly and that of protein not at all.

In common with other antitumor antibiotics, selectively inhibiting the synthesis of DNA, sibiromycin was found to be an inducer of phage production in lysogenic bacteria. The capacity for induction was observed in the lysogenic strain of *Micrococcus lysodeikticus* 53-40 (N-5 phage production), as well as with the lysogenic strain of *Escherichia coli* (lambda phage production). The methods used in these experiments have been described previously (DUDNIK, 1965). The results of the study are shown in Fig. 3. While sibiromycin had significant inducing capacity, the response to this antibiotic is quantitatively not so strong as that observed with some other antineoplastic antibiotics, *viz.*, mitomycin C, bruneomycin (streptonigrin), and rubomycin (daunomycin). It should be mentioned that certain antitumor antibiotics such as actinomycin or olivomycin do not induce phage production in lysogenic bacteria.

Turning now to interaction of sibiromycin with nucleic acids *in vitro*, it is of considerable interest that this antibiotic forms a complex with DNA, but not with RNA. Table 1 indicates that upon adding DNA to solutions of sibiromycin the

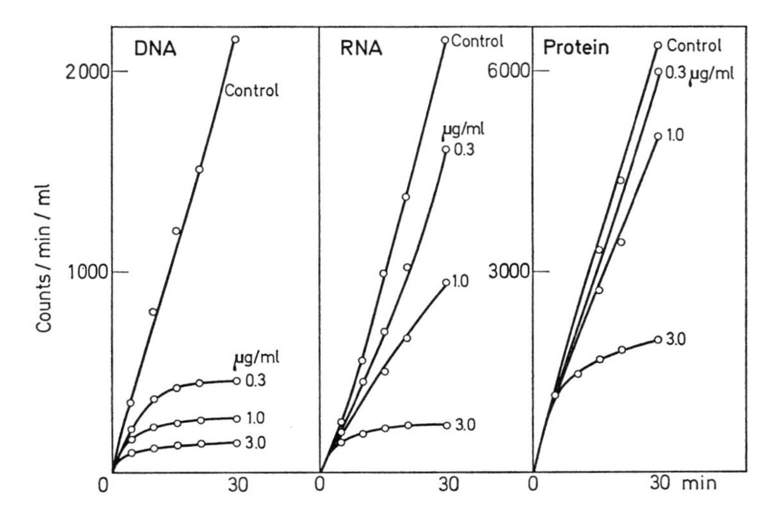

Fig. 2. Effect of sibiromycin on the incorporation of C^{14}-thymidine into DNA, C^{14}-uracil into RNA, and C^{14}-lysine into protein of *Bacillus subtilis* 168 (strain T^-, met^-). Concentrations of sibiromycin are 0.3, 1.0, 3.0 μg/ml

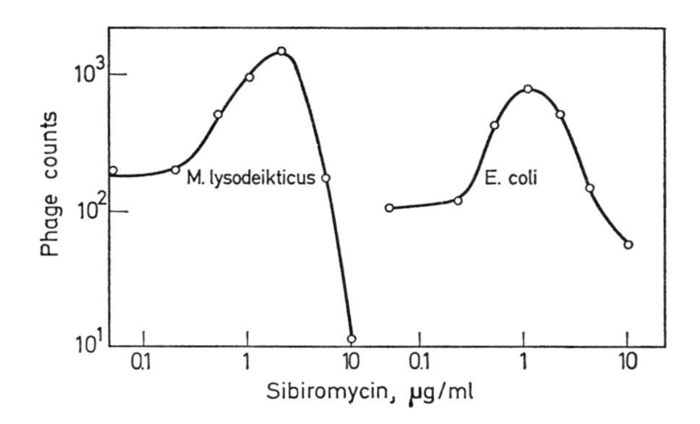

Fig. 3. Phage-inducing action of sibiromycin in lysogenic strains of *Micrococcus lysodeikticus* and *Escherichia coli*

antibacterial action of the drug disappears as a result of binding of sibiromycin to DNA. However, upon adding yeast RNA to solutions of sibiromycin under similar conditions, the antibacterial effect of the antibiotic is not decreased. On the basis of these observations it was deduced that sibiromycin binds to DNA but not to RNA.

Studies of the effects of DNA and RNA upon the spectral properties of sibiromycin confirmed this idea.

Interaction between sibiromycin and nucleic acids was studied in 0.01 M Tris, pH 7.2, containing 0.01 M NaCl. Sibiromycin possesses a characteristic absorption

Table 1. *Effect of DNA and RNA upon the inhibitory action of sibiromycin on growth of Staphylococcus aureus 209*

	Optical density[a]		Optical density[a]
1. Nutrient broth (NB)	0.398	6. Same + RNA	0.186
2. NB + DNA (300 μg/ml)	0.463	7. NB + sibiromycin (0.5 μg/ml)	0.065
3. NB +RNA (300 μg/ml)	0.342		
4. NB + sibiromycin (0.25 μg/ml)	0.100	8. Same + DNA	0.320
		Same + RNA	0.086
5. Same + DNA	0.375		

[a] Optical density of bacterial suspension after 2.5 h of growth at 37°. Initial optical density attained 0.081.

maximum at 310 nm (Fig. 4). When native DNA isolated from *Escherichia coli* was added to solutions of sibiromycin, the spectrum of the antibiotic was changed, its extinction maximum was shifted to longer wavelengths (320 nm), and absorption

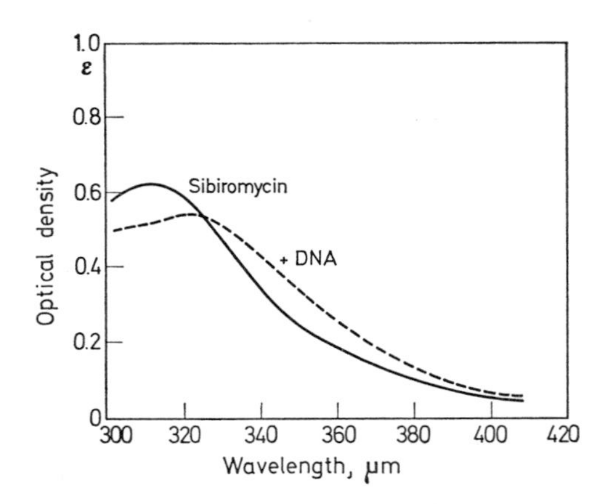

Fig. 4. Absorption spectrum of sibiromycin (15 μg/ml) and its change by added DNA (56 μg/ml)

at 310 nm was reduced as shown in Fig. 4. This shift in the absorption spectrum of sibiromycin in the presence of DNA indicated the formation of a complex between the antibiotic and DNA.

The formation of the complex between sibiromycin and DNA proceeds rather slowly. The rate of reaction of sibiromycin with native DNA from *E. coli* was measured, to study the kinetics of this process. The decrease of absorption of sibiromycin at 310 nm with time after addition of DNA is shown in Fig. 5. The decrease of absorption is most rapid during the first 2 min and is completed in 6 to 10 min. The rate of this process is comparable to that observed in the reaction of chromomycin with DNA (BEHR et al., 1969) or of actinomycin with DNA. By contrast, the antibiotic anthramycin interacts with DNA much more slowly, and this process continues for about 90 min (KOHN et al., 1968).

It should be pointed out that sibiromycin does not interact with RNA or with serum albumin. When high concentrations of RNA or of albumin were added to solutions of sibiromycin, the spectrum of the antibiotic was not changed. These studies employed yeast RNA as well as RNA isolated from ribosomes of *E. coli.*

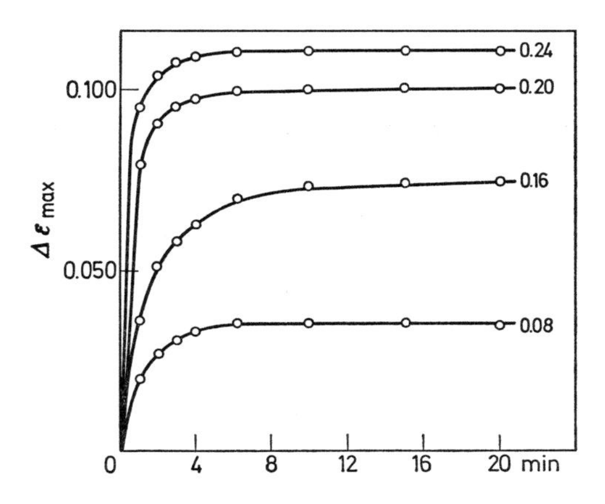

Fig. 5. Reactions rates of sibiromycin (15 μg/ml) with DNA at 20°. Concentrations of DNA (0.08, 0.16, 0.20, 0.24) are expressed in μmoles of DNA phosphorus

No alterations in the spectrum of sibiromycin were observed with the additions of bases of nucleic acid (adenine, thymine, uracil, guanine), ribonucleosides (adenosine, guanosine, cytidine, uridine), ribonucleotides (adenylic, uridylic, cytidylyc, guanylic acids), and deoxyribonucleotides (deoxyadenylic, deoxycytidylyc, thymidylic, deoxyguanylic acids).

Further studies were concerned with the effect of various agents upon the complex formation between sibiromycin and DNA, as well as with the stability of this complex. First was investigated the effect of bivalent cations, such as Mg^{++}, upon the binding of sibiromycin to DNA. It is known that antibiotics of the olivomycin-chromomycin group interact with DNA only in the presence of bivalent cations, notably Mg^{++}. In contrast, sibiromycin binds to DNA in the absence of bivalent cations in buffer solutions of various composition, including buffers which contain EDTA.

Magnesium ions do not interact with the molecule of sibiromycin. Even upon addition of Mg^{++} in molar concentration, the spectrum of sibiromycin is not changed. This is helpful in the study of the effect of Mg^{++} on the complexing of sibiromycin with DNA and on the stability of the complex formed.

Mg^{++} inhibits the formation of the complex of sibiromycin and native DNA isolated from *E. coli:* At 2×10^{-3}M, magnesium ions inhibit the decrease in the absorption of sibiromycin at 310 nm by added DNA as well as the shift of the absorption maximum to longer wavelengths. Magnesium ions at 10^{-1}M completely prevent any alteration in the spectrum of antibiotic by added DNA. However, the addition of Mg^{++} to the complex between sibiromycin and DNA already formed, does not lead to any dissociation of the latter.

The complexing of sibiromycin with DNA produces an increase in the melting temperature of DNA, as it is shown in Fig. 6. Sibiromycin has a stabilizing action upon the secondary structure of DNA to the effect of heat.

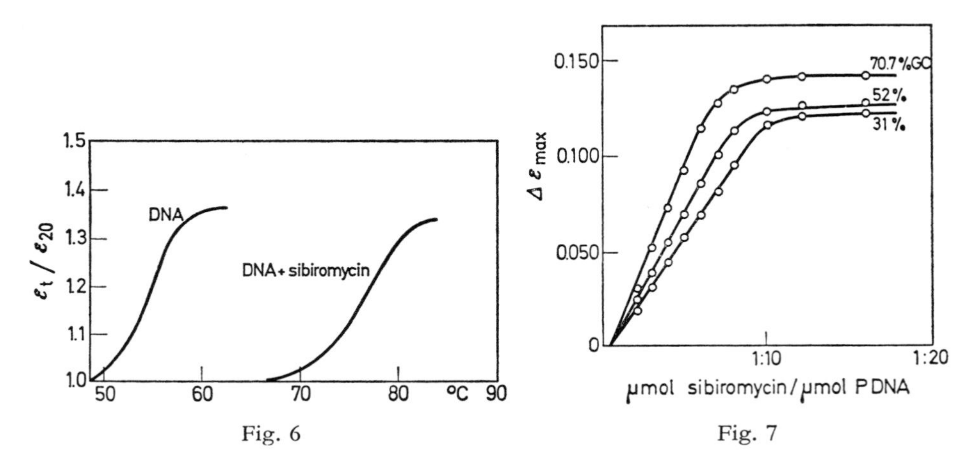

Fig. 6 Fig. 7

Fig. 6. Curve of melting of native DNA from *E. coli* (20 μg/ml) and increase of the melting temperature after complexing with sibiromycin (5 μg/ml). 0.001 M Phosphate buffer, pH 7.0, containing 0.0001 M EDTA. ε at 260 nm

Fig. 7. Effect of nucleotide composition of DNA upon its complexing with sibiromycin

Of considerable interest is the effect of the nucleotide composition of DNA upon complexing with sibiromycin. The following types of DNA were used in this study:

1. Native DNA from *Flavobacterium esteroaromaticum*, containing 70.7% of *GC* pairs. When the molecular weight of sibiromycin is calculated to be 460, one can estimate, according to the data given in Fig. 7, that one molecule of sibiromycin is bound on the average to 7.2 nucleotides of DNA.

2. Native DNA from *Escherichia coli* which contains less *GC* pairs (52%) complexes with sibiromycin less frequently. In this case one molecule of sibiromycin is bound to 8.5 nucleotides of DNA.

3. Native DNA from *Flavobacterium breve* with even lower content of *GC* pairs (31%) shows still less frequent complexing with sibiromycin. One molecule of sibiromycin binds here on the average to 9.9 nucleotides of DNA.

Sibiromycin complexes not only with native DNA, but also with DNA denatured by heat (15 min at 100°). However, in the latter case the complexing is much less frequent. For example, in the native DNA from *E. coli* about 8.5 nucleotides bind one molecule of sibiromycin, whereas in the denatured DNA from the same source 24 nucleotides are required to bind one molecule of sibiromycin.

Summary

Sibiromycin interacts specifically with DNA. In the cultures of *Staphylococcus aureus* and *Bacillus subtilis* the antibiotic selectively inhibits the synthesis of DNA. Sibiromycin induces phage production in lysogenic bacteria. The antibiotic complexes with DNA, but not with RNA, ribonucleotides and deoxyribonucleotides. The absorption spectrum of the antibiotic is changed by adding DNA, and these spectral changes were used for kinetic measurements. Magnesium ions prevent the binding of sibiromycin to DNA. The complexing of sibiromycin with DNA increases the melting temperature of the latter. It was observed that DNA with high *GC* content binds more antibiotic than DNA with low *GC* content.

References

Behr, W., Honikel, K., Hartmann, G.: Interaction of the RNA polymerase inhibitor chromomycin with DNA. Europ. J. Biochem. **9**, 82 (1969).

Brazhnikova, M. G., Kovsharova, I. N., Konstantinova, N. V., Mesentsev, A. S., Proshljakova, V. V., Tolstych, I. V.: Chemical study of antitumor antibiotic sibiromycin. Antibiotiki **15**, 297 (1970).

Dudnik, Ju. V.: Induction of lysogenic *Micrococcus lysodeikticus* by antibiotics with the ability to affect DNA synthesis. Antibiotiki **10**, 112 (1965).

Gause, G. F., Preobrazhenskaya, T. P., Ivanitskaya, L. P., Sveshnikova, M. A.: Production of antibiotic sibiromycin by *Streptosporangium sibiricum*. Antibiotiki **14**, 963 (1969).

Kohn, K., Bono, V., Kann, H.: Anthramycin, a new type of DNA-inhibiting antibiotic. Biochim. biophys. Acta (Amst.) **155**, 121 (1968).

Konstantinova, N. V., Tolstych, I. V., Brazhnikova, M. G.: Quantitative chemical assay of antitumor antibiotic sibiromycin in preparations and culture liquids. Antibiotiki **15**, 304 (1970).

Laiko, A. V.: Selective action of some antitumor antibiotics on nucleic acids of staphylococci and their mutants. Antibiotiki **9**, 711 (1964).

Shorin, V. A., Rossolimo, O. K.: Experimental study of antitumor activity of sibiromycin. Antibiotiki **15**, 300 (1970).

Anthramycin

SUSAN B. HORWITZ*

In 1963, it was reported that the fermentation broth of *Streptomyces refuineus* var. *thermotolerans* contained a component referred to as 'refuin' which had activity against experimental tumors (TENDLER and KORMAN, 1963). The active component of this broth was identified as anthramycin [LEIMGRUBER, STEFANOVIC, SCHENKER, KARR and BERGER, 1965 (2)] an antibiotic subsequently found to possess antitumor, antimicrobial, amebicidal and chemosterilant properties. The following paper reviews the biochemical pharmacology of anthramycin with particular reference to its interaction with DNA.

I. Chemistry

The structure and stereochemistry (Fig. 1) of anthramycin and related compounds have been elucidated [LEIMGRUBER, BATCHO and SCHENKER, 1965 (1)], and a total synthesis of anthramycin has been accomplished (LEIMGRUBER, BATCHO and CZAJKOWSKI, 1968). Anthramycin crystallizes from hot methanol-water as the 11-methyl ether, and this derivative is more stable than free anthramycin [LEIMGRUBER et al., 1965 (1)]. Anthramycin methyl ether rapidly hydrolyzes in aqueous solution via anhydroanthramycin (III)** to yield an equilibrium mixture of anthramycin (I) and epianthramycin (II) (W. LEIMGRUBER, private communication)***.

Anthramycin has an unusual structure which differs from those of the many antibiotics and alkaloids known to interact with DNA. It does not contain the large side chains of actinomycin and chromomycin, and its interaction with DNA should, therefore, be less complex than that of other molecules of natural origin. Since anthramycin contains a seven-membered ring, it cannot exist as a completely planar molecule.

II. Effects on Intact Cells

The antibacterial, chemotherapeutic and pharmacologic properties of anthramycin have been reported (TENDLER and KORMAN, 1963; GRUNBERG, PRINCE, TITSWORTH, BESKID and TENDLER, 1966; ADAMSON, HART, DEVITA and OLIVERIO, 1968). Anthramycin inhibits the growth of Ehrlich carcinoma, Sarcoma 180, Walker carcinosarcoma

* This work was supported by the American Cancer Society, T-418 B.

** Roman numerals refer to the structural formulas in Fig. 1.

*** In all of the experiments done in our laboratory, anthramycin, in the form of its methyl ether, was dissolved in dimethyl sulfoxide just prior to use at a concentration of 20 mg/ml. Dimethyl sulfoxide was included in all control reactions, although it had no apparent inhibitory effects at the concentrations used.

256, human epidermoid carcinoma No. 3, and human adenoma No. 1 in rodents. The antibiotic exerts a chemotherapeutic effect in experimental infections with *Trichomonas vaginalis*, *Entamoeba histolytica* and *Syphacia obvelata* (GRUNBERG et al., 1966).

Anthramycin penetrates the cell wall of certain bacteria and inhibits the growth of *Sarcina lutea*, *Bacillus subtilis* and *E. coli* B at concentrations of 10^{-5}M (cf. Fig. 2). Anthramycin inhibits the biosynthesis of deoxyribonucleic acid (DNA) and ribonucleic acid (RNA) in HeLa cells by 90% at concentrations which have no effect on protein synthesis (Table 1). Inhibition of nucleic acid synthesis was complete within 10 min after anthramycin was added to the cultures (HORWITZ and GROLLMAN, 1969). Similar effects on nucleic acid synthesis in L1210 cells have been reported by KOHN, BONO and KANN (1968) and in Ehrlich ascites cells by BATES, KUENZIG and WATSON (1969).

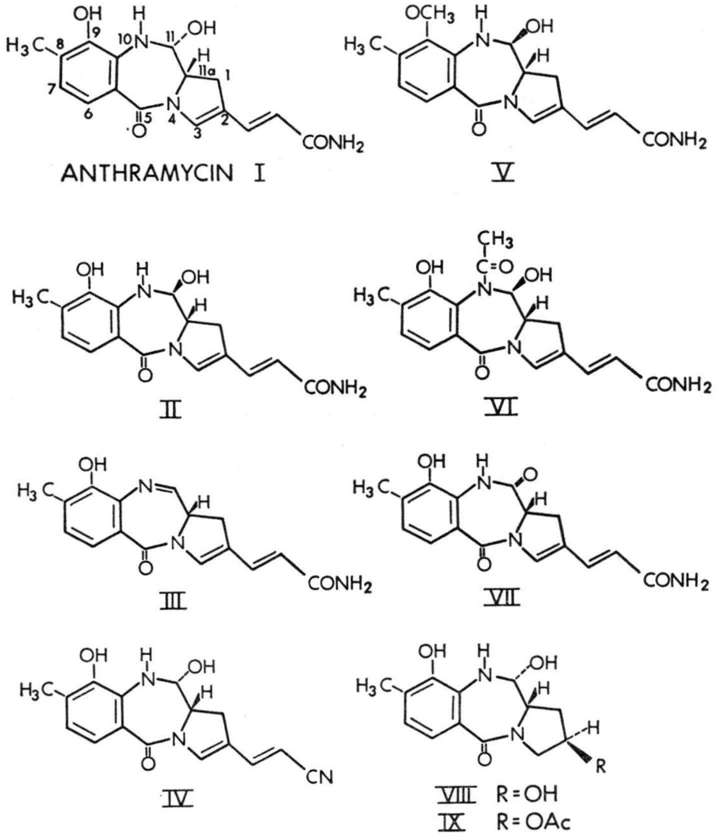

Fig. 1. Structural formula of anthramycin and related derivatives

Anthramycin is a potent chemosterilant in *Drosophila melanogaster* (BARNES, FELLIG and MITROVIC, 1969). At low levels of ingestion, anthramycin methyl ether induces complete and permanent sterilization of the female *D. melanogaster*. Other antitumor antibiotics, such as actinomycin D, novobiocin and echinomycin, have no chemosterilant activity when tested under similar conditions (BARNES et al., 1969).

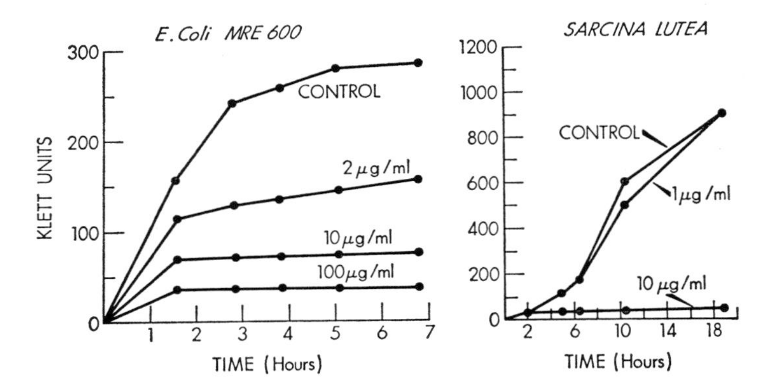

Fig. 2. Effect of anthramycin on the growth of *E. coli* and *S. lutea* (from HORWITZ and GROLLMAN, 1969. Reproduced by permission of the American Society for Microbiology)

Table 1. *Effect of anthramycin on nucleic acid synthesis in HeLa cells. Methods employed have been described elsewhere* (cf. HORWITZ and GROLLMAN, 1969)

Anthramycin concentration M	Inhibition of RNA synthesis %	Inhibition of DNA synthesis %
1×10^{-6}	50	—
3×10^{-6}	90	50
2×10^{-5}	>90	90

III. Mode of Action

All of the biological activities of anthramycin can be attributed to its effects on nucleic acid synthesis. Anthramycin inhibits the DNA-dependent RNA polymerase reaction from *E. coli* (HORWITZ and GROLLMAN, 1969). The DNA primer must be incubated with anthramycin for 5 min at 37° before adding the enzyme to achieve the maximal inhibitory effect. The *E. coli* DNA polymerase reaction is also inhibited, but to a lesser extent than the RNA polymerase reaction. Anthramycin appears to act by virtue of its ability to bind to DNA and to interfere with the function of DNA.

IV. Structure-Activity Relationships

Derivatives and analogs of anthramycin have been tested for their ability to inhibit the activity of RNA polymerase (Table 2). The same degree of inhibition was observed with either epimer of anthramycin (I, II) or with the anhydro-compound (III). This result is not unexpected since the three forms equilibrate easily [LEIMGRUBER et al., 1965 (2)]. The derivative in which a nitrile group on the side chain is substituted for the amide (IV) retains most of its inhibitory activity.

Changes in the molecule at positions 2, 9, 10 and 11 decrease the inhibitory effects of anthramycin. Introduction of an hydroxy or an O-acetyl group at position 2 (VIII, IX) in place of the side chain, O-methylation at position 9 (V), N-acetylation at position 10 (VI) or introduction of a keto group at position 11 (VII), results in loss of inhibitory activity.

Table 2. *Effect of anthamycin and its derivatives on RNA polymerase activity and on melting temperature* (T_m) *of calf thymus DNA. The concentration of inhibitors was* 2×10^{-4} *M for the enzyme assay and* 2×10^{-5} *M for the determination of* T_m
(cf. HORWITZ and GROLLMAN, 1969, 1970)

Derivatives[a]	RNA polymerase inhibition %	ΔT_m °C
Active		
Anthramycin (I)	>75	+7.0
II	>75	+7.0
III	>75	+7.0
IV	>50	+4.0
Inactive		
V	<10	<1.0
VI	<10	<1.0
VII	<10	+1.5
VIII	<10	<1.0
IX	<10	<1.0

[a] Roman numerals refer to the structural formulas in Fig. 1.

The capacities of anthramycin and its derivatives to affect the melting temperature (T_m) of calf thymus DNA were also compared. The T_m of calf thymus DNA increased in the presence of anthramycin and its nitrile derivative (Table 2). Other compounds tested had no effect. Results of these thermal denaturation studies are in good agreement with those on inhibitions of the RNA polymerase reaction.

V. Interaction of Anthramycin with DNA

The interaction of anthramycin and DNA has been studied by several groups of investigators (HORWITZ and GROLLMAN, 1968; KOHN et al., 1968; STEFANOVIC, 1968; BATES et al., 1969; and KOHN and SPEARS, 1970). When calf thymus DNA is added to a solution of anthramycin, there is a shift of the antibiotics absorption maximum from 333 mμ to 343 mμ, and the extinction coefficient decreases by approximately 15%. The extent of the change in absorption maximum and extinction is dependent on the relative concentrations of anthramycin and DNA. These spectroscopic changes were reported as being complete within several seconds (STEFANOVIC, 1968), but other studies have shown that 45 min are required for the reaction to come to completion

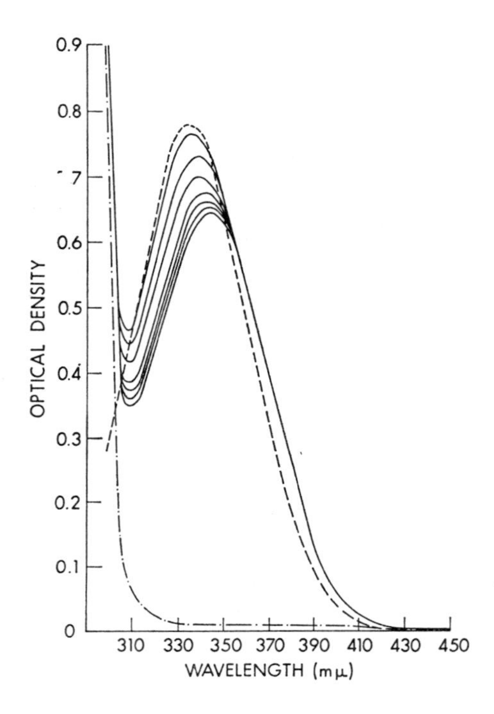

Fig. 3. Change in absorption spectrum of anthramycin at various times after the addition of DNA. (- - - - -) 2.2×10^{-5} M anthramycin in 0.0015 M NaCl—0.00015 M sodium citrate pH 7.0. (—.—.) 2×10^{-3} M calf thymus DNA in 0.0015 M NaCl—0.00015 M sodium citrate pH 7.0. (———) A mixture of 2×10^{-3} M calf thymus DNA and 2.2×10^{-5} M anthramycin in 0.0015 M NaCl—0.00015 M sodium citrate pH 7.0. Spectra shown were recorded at 1, 5, 10, 20, 30, 40 and 60 min. There was no further change in the spectrum after 60 min

at room temperature (KOHN et al., 1968). The slow rate of spectral change that occurs during the interaction of DNA and anthramycin has been repeatedly observed in our laboratory (Fig. 3).

In contrast to the effect of sodium lauryl sulfate on the complex formed between DNA and actinomycin (MÜLLER and CROTHERS, 1968) or chromomycin (BEHR, HONIKEL and HARTMANN, 1969) the spectral change of anthramycin cannot be reversed by adding the detergent to the reaction (HORWITZ and GROLLMAN, 1970).

There is no change in the spectrum of the antibiotic when free bases or the mononucleotides which constitute DNA are added to the solution.

The structural requirements for the interaction of DNA with anthramycin were established by determining the effect of altering the primer of the RNA polymerase reaction (Table 3). Synthetic double-stranded primers, such as poly *d(AT)* or poly *(dGdC)* were used as a primer for RNA polymerase in place of calf thymus DNA. Anthramycin failed to inhibit the reaction at concentrations 100 times greater than that required to inhibit the enzyme when native calf thymus DNA was used as a primer. Anthramycin did not inhibit the activity of RNA polymerase when the primer, poly *(dAdT)*, was used. Nogalamycin, which is an excellent inhibitor or RNA polymerase primed with poly *d(AT)* (Bhuyan and Smith, 1965), and actinomycin, which inhibits this enzyme when primed with poly *(dGdC)* (Goldberg, Rabinowitz and Reich, 1962) served as controls for these experiments.

Table 3. *Effect of primer and inhibitor on RNA polymerase activity*
(cf. Horwitz and Grollman, 1969, 1970)

Primer	Anthramycin concentration M	RNA polymerase inhibition %
Calf thymus DNA	2×10^{-6}	50
Heat denatured calf thymus DNA	2×10^{-6}	<10
dGdC	2×10^{-4}	<10
dAT	2×10^{-4}	<10
dAdT	2×10^{-4}	<10

Inhibition of RNA polymerase by anthramycin requires the presence of double-stranded DNA as primer for the enzyme. Inhibition is abolished when heat denatured calf thymus DNA is substituted for native DNA.

VI. Properties of the DNA-Anthramycin Complex

Equilibrium dialysis studies, using a wide range of anthramycin concentrations and calf thymus DNA, have demonstrated that one molecule of anthramycin is bound for approximately each eight base pairs of calf thymus DNA (Horwitz and Grollman, 1970). Anthramycin binds tightly to DNA, and an anthramycin-DNA complex can be isolated by precipitation with alcohol or by gel filtration. When a mixture of anthramycin and DNA, previously incubated for 1 h, is applied to a Sephadex-G-75 column, a portion of the anthramycin is eluted with DNA. This anthramycin has an absorption maximum at 343 mμ. Unreacted anthramycin with an absorption maximum at 333 mμ is eluted from the column in later fractions (Horwitz and Grollman, 1970).

The complex of *Escherichia coli* DNA and anthramycin is characterized by a reduced buoyant density in CsCl (Kohn et al., 1968). The thermal denaturation temperature of calf thymus DNA is increased in the presence of anthramycin (Kohn et al., 1968; Bates et al., 1969; Horwitz and Grollman, 1970).

Anthramycin protects DNA from the action of nucleases, and the enzymatic hydrolysis of DNA-anthramycin complexes is slower than that of free DNA. Complexes containing 1 mole of anthramycin for each twelve base pairs of DNA or 1 mole of anthramycin for each thirty-seven base pairs of DNA were prepared and hydrolyzed with pancreatic deoxyribonuclease I and snake venom phosphodiesterase. After 48 h, the complex in which the ratio of DNA base pairs to anthramycin was originally 37:1 was hydrolyzed to 85% compared to uncomplexed DNA, and the complex in which

Table 4. *Rate of degradation of anthramycin-DNA complexes by DNA nucleases. Two anthramycin-DNA complexes were prepared with a DNA base pair/anthramycin ratio of 37: 1 and 12:1, respectively. Each complex was digested with DNA nucleases within a dialysis bag at 25°. A 10 ml solution containing 2.1 mM calf thymus DNA complexed with anthramycin, 50 mM Tris-Cl pH 8.5, 1 mM $MgCl_2$, 5 μg of pancreatic DNase I, and 500 μg of snake venom phosphodiesterase was dialyzed against 50 ml of 50 mM Tris-Cl pH 8.5 containing 1 mM $MgCl_2$. After 24 h of hydrolysis, DNA nucleases were again added to the solution inside of the dialysis bag. The absorbance of the complex at 260 mμ and 343 mμ inside the bag was followed and the extent of hydrolysis of the complexes was compared to that of free DNA*

Time (h)	DNA/Anthramycin[a]	% Hydrolysis
Expt. 1		
0	37	0
24	19	49
48	15	85
Expt. 2		
0	12	0
24	11.5	16
48	11	36

[a] Expressed as ratio of base pairs per mole of anthramycin.

the ratio of DNA base pairs to anthramycin was 12:1 had been hydrolyzed 36% (Table 4).

VII. Summary

Anthramycin is an antibiotic with potent antitumor, antimicrobial and chemosterilant activities. Its primary mode of action is the inhibition of syntheses of nucleic acids. By virtue of its ability to form complexes with DNA, anthramycin inhibits the activity of DNA-dependent RNA polymerase and certain DNA nucleases. One molecule of anthramycin binds to eight base pairs of calf thymus DNA, and the involvement of at least three groups on the alkaloid in binding to DNA has been established. Anthramycin requires native helical DNA, and, possibly, the presence of all four bases to inhibit the activity of DNA-dependent RNA polymerase.

Acknowledgements

I am indebted to Dr. W. LEIMGRUBER and associates of Hoffman-La Roche, Inc. for samples of anthramycin methyl ether and related compounds and also for most helpful discussions. I am grateful to Dr. ARTHUR P. GROLLMAN for many useful suggestions during the progress of this work.

References

ADAMSON, R. H., HART, L. G., DE VITA, V. T., OLIVERIO, V. T.: Cancer Res. **28**, 343 (1968).
BARNES, J. R., FELLIG, J., MITROVIC, M.: J. Econ. Entomol. **62**, 902 (1969).
BATES, H. M., KUENZIG, W., WATSON, W. B.: Cancer Res. **29**, 2195 (1969).
BEHR, W., HONIKEL, K., HARTMANN, G.. Europ. J. Biochem. **9**, 82 (1969).
BHUYAN, B. K., SMITH, C. G.: Proc. nat. Acad. Sci. (Wash.) **54**, 566 (1965).
GOLDBERG, I. H., RABINOWITZ, M., REICH, E.: Proc. nat. Acad. Sci. (Wash.) **48**, 2094 (1962).
GRUNBERG, E., PRINCE, H. N., TITSWORTH, E., BESKID, G., TENDLER, M. D.: Chemotherapia (Basel) **11**, 249 (1966).
HORWITZ, S. B., GROLLMAN, A. P.: Abstracts of 156th Meeting of the American Chemical Society, Atlantic City, Biol. p. 244, 1968.
— — Antimicrob. Agents. Chemother. **1968**, 21.
— — Manuscript in preparation (1970).
KOHN, K. W., BONO, V. H., Jr., KANN, H. E., Jr.: Biochim biophys. Acta (Amst.) **155**, 121 (1968).
— SPEARS, C. L.: J. Molec. Biol. **51**, 551 (1970).
LEIMGRUBER, W., BATCHO, A. D., SCHENKER, F.: (1) J. Amer. chem. Soc. **87**, 5793 (1965).
— STEFANOVIC, V., SCHENKER, F., KARR, A., BERGER, J.: (2) J. Amer. chem. Soc. **87**, 5791 (1965).
— BATCHO, A. D., CZAJKOWSKI, R. C.: J. Amer. chem. Soc. **90**, 5641 (1968).
MÜLLER, W., CROTHERS, D. M.: J. molec. Biol. **35**, 251 (1968).
STEFANOVIC, V.: Biochem. Pharmacol. **17**, 315 (1968).
TENDLER, M. D., KORMAN, S.: Nature (Lond.) **199**, 501 (1963).

Inhibition of RNA Synthesis by Quinone Antibiotics

WALTER KERSTEN

I. Introduction

During the last decade, a considerable number of drugs, many of them antibiotics, have been shown to interfere with the replication of DNA or the information transfer from gene to protein at the transcriptional level.

Several of the antibiotics shown in Fig. 1 contain quinone ring systems. The anthracyclines are tetrahydrotetracenquinones. They interact with DNA *in vitro* and

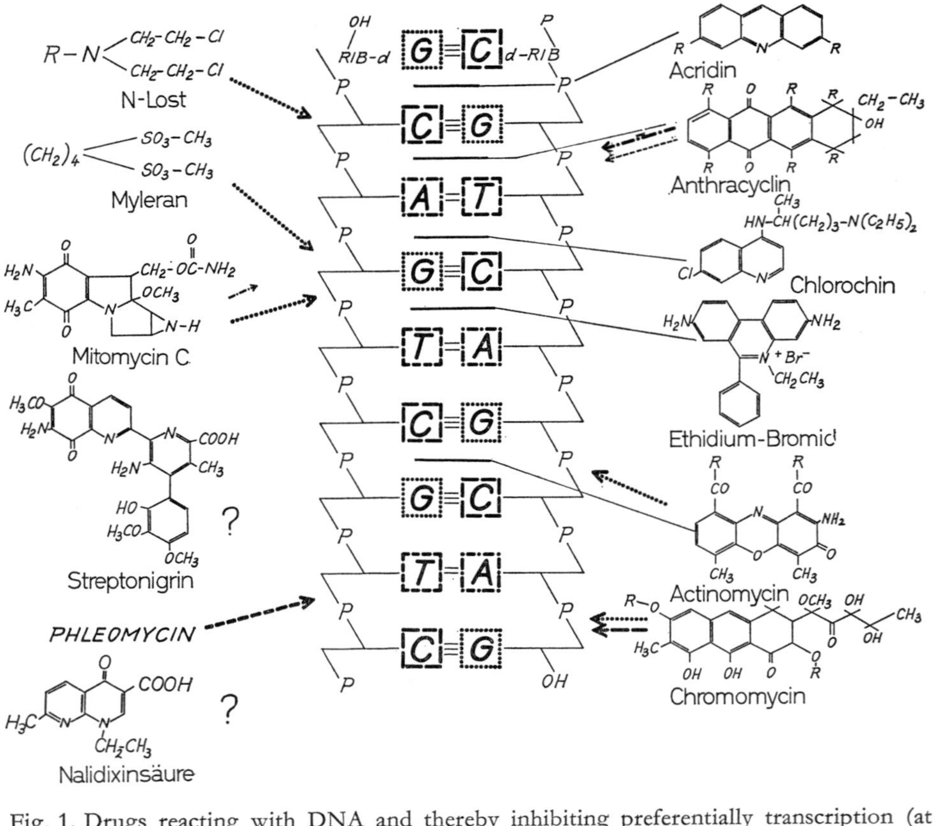

Fig. 1. Drugs reacting with DNA and thereby inhibiting preferentially transcription (at right) or replication (at left). Bars indicate intercalation; arrows indicate preferential binding to certain DNA bases.

in vivo and inhibit preferentially the DNA-dependent synthesis of RNA (Kersten, Kersten, 1965; Hartmann, Goller, Koschel, Kersten, Kersten, 1964; Bhuyan, Smith, 1965). Physico-chemical measurements indicate that the anthracyclines intercalate between base pairs of DNA. The structure of the anthracycline-DNA complex is still unknown. Whereas actinomycins, mithramycin and chromomycin preferentially bind to native double-stranded DNA of high *GC* content, it is still uncertain whether the anthracyclines show any base preference in binding to native DNA (Kersten, Kersten, Szybalski, 1965; Bhuyan, Smith, 1965). As has been shown by Wells in this symposium, the preferential binding of actinomycin to DNA of high *GC* content depends probably on a certain conformation rather than on a specificity towards guanine or the *GC* pair, since actinomycin does associate with some synthetic deoxypolynucleotides of defined base sequence, not containing *GC* pairs.

In the first part of this report (II) evidence will be presented that DNA induces conformational changes in the chromophor of the anthracyclines. These conformational changes, measured by circular dichroism, do not show gross dependence on the *GC* content of the DNA (Fey and Kersten, 1970).

Besides the tetrahydroanthracenquinones, other quinonecontaining antibiotics are known to interfere with the synthesis of nucleic acid. The antibiotic streptonigrin, a quinolinquinone, interacts with DNA and inhibits preferentially the DNA-dependent synthesis of DNA (Levine and Bothwick, 1963). The mitomycins are indoloquinones containing, in addition, alkylating groups which are formed upon reduction of the mitomycins within the cell. Therefore, mitomycins behave like difunctional alkylating agents; they establish crosslinks between the two strands of a DNA molecule and in this manner prevent DNA replication (Iyer and Szybalski, 1963; White and White, 1964). Derivatives of the mitomycin antibiotics, devoid of the alkylating aziridine ring, still interfere with nucleic acid synthesis, however preferentially inhibit the RNA transcription step. From these observations we suggested that the quinone ring of the mitomycins might be involved in the effect on RNA synthesis (Kersten and Kersten, 1969). In an approach to this question we compared the effect of quinone-containing antibiotics with that of synthetic quinones on the synthesis of RNA. The results of these experiments will be presented in section III.

II. Complexes between DNA, Tetrahydroanthracene-quinones and Related Compounds

Neville and Bradley (1961) first showed induced Cotton effects in the absorption bands of amino acridines upon complex formation with DNA. The induced optical activity was thought to be the result of an association of the acridine molecules rather than of an interaction between individual acridines and DNA. Induced Cotton effects in the absorption bands of proflavin and other amino acridines upon addition of native or denatured DNA or RNA were studied by Blake and Peacocke (1966, 1967), and revealed in agreement with the statistical theory that in the case of proflavin three or four molecules of the dye can associate with each other at neighbouring binding sites of native DNA. When denatured DNA or RNA were used only two molecules associate at neighbouring binding sites. Gardener and Mason (1967) observed three induced bands in the *CD* spectra of acridine orange. Two bands are attributed to acridine dimers bound to DNA. The third band originates from molecules associated with DNA as monomers. Anomalies in the optical rotatory dispersion spectra of antimalarials induced by DNA were reported by HAHN and KREY (1968). The association of luteoskyrin with DNA in the presence of Mg^{2+} was studied using optical rotatory dispersion by OHBA and FROMAGEOT (1968). The ORD spectra indicate that luteoskyrin molecules associate in aqueous solution. The associative forces decrease upon complex formation with single-stranded purinepolydeoxynucleotides and increase upon complexing with single-stranded purine-free polynucleotides or double-stranded native DNA.

PERMOGOROW and LAZURKIN (1965) concluded from similarities in the ORD spectra of the DNA-actinomycin-complex with that of the acridine-DNA-complex at high ratios of DNA-P to dye that actinomycins intercalate. In accordance with the findings of MÜLLER and EMME (1965), COURTOIS, GUSCHLBAUER and FROMAGEOT (1968) concluded from ORD measurements a dimerisation of actinomycin in solution. The chromophore itself seems not to be involved in this association. The molecular amplitude of the Cotton effect observed upon complex formation of actinomycin with DNA increases markedly with increasing ratio of DNA-P to actinomycin indicating that the change in optical activity is not a consequence of an interaction among drug molecules but is caused by the specific interaction between actinomycin and the nucleotides of DNA. ZIFFER, YAMAOKA and MAUGER (1968) came to the same conclusion by their measurements whereas HOMER (1969) favors the hypothesis that actinomycin binds to DNA as a dimer, and probably crosslinks two DNA molecules.

The tetrahydrotetracene-quinones daunomycin and nogalamycin differ in their mode of interaction with DNA from the related tetrahydroanthracene antibiotics chromomycin and mithramycin. Mithramycin and chromomycin preferentially bind to native DNAs with high content of *GC* base pairs, and do not change the physicochemical properties of poly-*dAT*, indicating that there is no association to poly-*dAT*. Nogalamycin and daunomycin interact with poly-*dAT*. Studies with DNA of different *GC* content revealed that nogalamycin and daunomycin showed slight preference for DNA with high content of *GC* (KERSTEN, KERSTEN, SZYBALSKI, 1965). BHUYAN and SMITH (1969) however postulated that nogalamycin shows preference in binding for *AT* pairs in DNA.

Mithramycin and chromomycin have both characteristic and intensive absorption bands at 405 nm and 280 nm. In the *CD* spectrum two nearly symmetrical doublets correspond to these absorption bands. On both sides of the doublets there are additional maxima at 445 nm and at 380 nm respectively (Fig. 2).

Mithramycin and chromomycin bind to DNA only in the presence of Mg^{2+}. In accordance with these findings the *CD* spectra of the antibiotics do not change upon addition of DNA in the absence of Mg^{2+}. The *CD* spectra of the antibiotics are altered quantitatively but not qualitatively upon addition of Mg^{2+}. The amplitudes decrease but the bands are not shifted in wavelength.

The Mg^{2+} complexes of these two antibiotics interact with DNA; this is accompanied by drastic changes in the *CD* spectra depending on the ratio of DNA-P to antibiotic. The maximum at 445 nm is pronounced, the doublet at about 400 nm decreases, and at 333 nm a new absorption band of constant height is formed and at 320 nm a new maximum arises. The negative amplitude of the doublet at 280 nm disappears (see Fig. 2).

The two doublets in the *CD* spectra indicate that in buffer solution at pH 7 and at a concentration of 1×10^{-4}M, chromomycin and mithramycin are dimers. The association occurs at the chromophores. At high ratios of DNA-P to drug (24:1) they are bound as monomers. This is concluded from the observation that the doublets disappear and nearly all the mithramycin and chromomycin molecules are bound to DNA. The *CD* spectra of the complexes of chromomycin with DNA of different *GC* content revealed that the amount of drug which associates with DNA increased with increasing *GC* content of the DNA. We have tested *Cytophaga johnsonii* DNA with

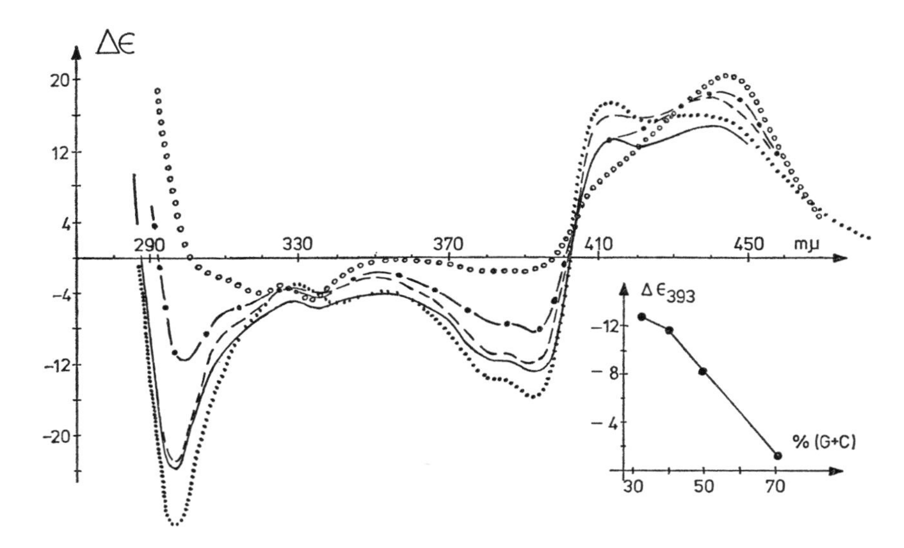

Fig. 2. Effect of DNAs of increasing GC content on the CD spectrum of mithramycin: ...... Mithramycin + Mg^{2+} 1:2, ——— Mithramycin + Mg^{2+} + DNA *Cytophaga johnsonii* 34% GC, — — — Mithramycin + Mg^{2+} + DNA *B. subtilis* 43% GC, —·—·— Mithramycin + Mg^{2+} + DNA *E. coli* 50% GC, oooooo Mithramycin + Mg^{2+} + DNA *B. luteus* 72% GC. Molar ratios: Mithramycin: Mg^{2+}:DNA-P 1:2:6

33% *GC*, *B. subtilis* DNA with 43% *GC*, *E. coli* DNA with 50% *GC* and *Sarcina lutea* with 72% *GC* (see Fig. 2) (Fey and Kersten, 1970).

The complex formation of DNA with nogalamycin or daunomycin is also accompanied by changes in the *CD* spectra of these dyes. Upon complex formation of DNA with nogalamycin the *CD* bands at 343 nm and at 455 nm are shifted towards longer wavelengths by about 10 nm. The amplitudes increase by a factor of 1.2 and 1.9 respectively. A new *CD* band occurs at 315 nm and the negative maximum at 305 nm decreases. The changes in the *CD* spectra of nogalamycin upon complex formation with DNA are not so pronounced as in the case of the chromomycins. However, contrary to chromomycin and mithramycin, there is no evidence that nogalamycin forms dimers in solution. *CD* spectra of the complexes of nogalamycin with DNA of varying *AT* content (28% to 67%) were measured. If we assume that the degree of increase in the amplitude at 460 nm and the degree of decrease of the negative maximum at 305 nm are indicative of complex formation, then we find rather a slight preference of nogalamycin binding to DNA with high *GC* content, if the ratio of DNA-P per drug is in the order of 3.5. These and earlier findings (KERSTEN et al., 1965) indicate that the tetrahydroanthracene-quinones do not show preference or specificity for certain bases or base sequences when interacting with native DNA.

III. Inhibition of RNA Synthesis by Quinone Antibiotics and Synthetic Quinones

Originally it was thought that the antibiotic granaticin was related in structure to the anthracyclines. This assumption was made at a time when the conformation of granaticin was not yet known. In the concentration range between 0.1 and 0.8 mμg/ml the effect of granaticin is bacteriostatic. Higher concentrations are bactericidal. At bactericidal concentrations morphological changes occur which could be demonstrated by electron microscopy. The cell wall and the cytoplasmic membrane are broken up, ribosomes are degraded and lost, and eventually the cell lyses.

The effect of this antibiotic on the incorporation of radioactive precursors into RNA, DNA and protein was measured in *B. subtilis* as test organism. Granaticin preferentially inhibits the synthesis of RNA (KERSTEN et al., 1969; KERSTEN et al., 1970). This effect could not be reversed by the addition of DNA to the culture medium. Granaticin does not sediment with DNA during ultracentrifugation. The spectrum of granaticin is not changed upon addition of DNA. Granaticin has no influence on the buoyant density of DNA in CsCl. Treatment of *B. subtilis* with different doses of granaticin has no effect on the transforming activity of the DNA. From these results we concluded that granaticin does not inhibit the synthesis of RNA by the same mechanism as the anthracyclines.

Granaticin contains a quinone ring system which might be involved in the inhibitory action on RNA synthesis. This suggestion is based on our observation that the derivatives of mitomycins which retain the quinone ring but lack the alkylating aziridine ring do not interact with DNA upon reduction. These mitomycin derivatives are still potent inhibitors of cell growth and affect preferentially the synthesis of RNA (KERSTEN and KERSTEN, 1969). To answer the question whether quinones can influence the transcription process *in vivo*, synthetic quinones like quinolin-quinones or naphtoquinones were included in these studies. In all cases concentrations of the quinones were used which inhibit the growth of *B. subtilis* by not more than 20%. The cultures were pre-incubated with quinones for 5 min, 20 min and 60 min. ^{3}H-uridine, ^{3}H-thymidine or ^{3}H-leucine were then added and the incorporation of the precursors into the macromolecules was measured 10 min later. The percentage of inhibition of RNA, DNA and protein synthesis was calculated and plotted as a function of the time of pre-incubation (see Fig. 3). Granaticin, the mitomycin derivative and the synthetic quinoline- or naphtoquinone behave quite similar with respect to their effect on RNA, DNA and protein synthesis. During the first 15 min protein synthesis proceeds nearly unaffected, the synthesis of RNA is inhibited 90 to 100%, that of DNA 40 to 60%. The inhibitory effect of the quinones on RNA synthesis decreased on further incubation, and is thus reversible. Also the inhibition of DNA synthesis was reversed. The striking phenomenon, observed in all cases, is the initial uncoupling effect of the quinones on RNA and protein synthesis.

Granaticin inhibits the synthesis of *t*RNA and *r*RNA equally well (KERSTEN et al., 1969). Further investigations on the effect of granaticin and other quinones on the

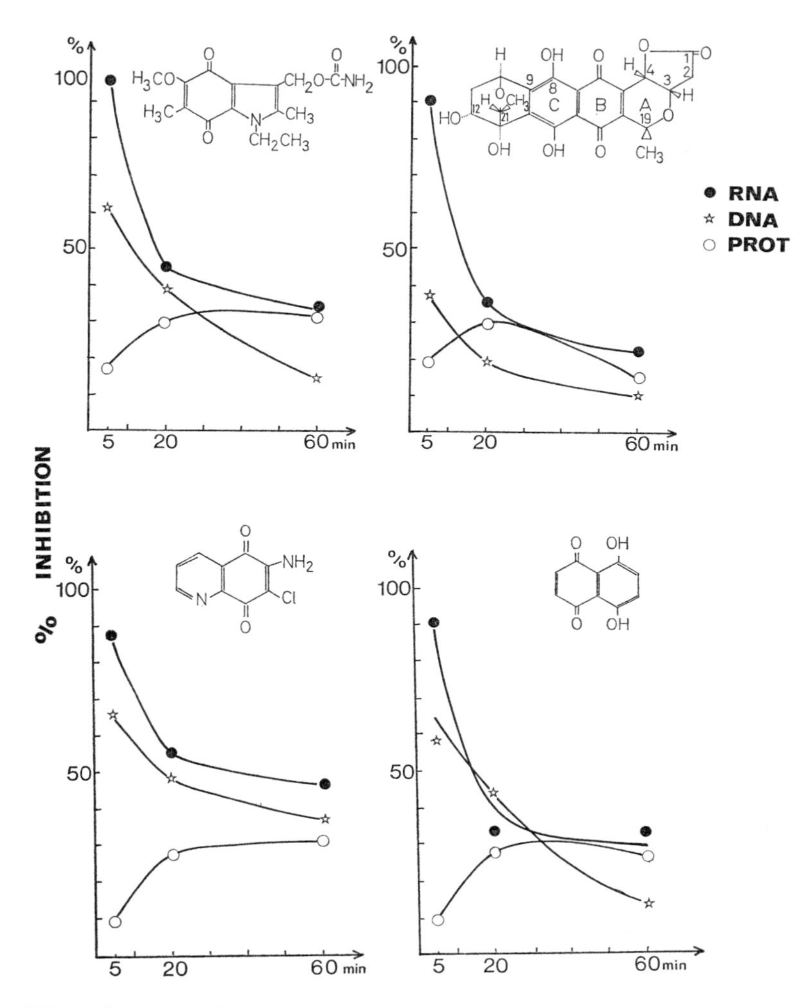

Fig. 3. Effect of quinone derivatives on the incorporation of ^{14}C uracil, ^{3}H thymidine and ^{14}C leucine in *B. subtilis*. The cultures were incubated for 5, 20, 60 min with the drugs. The radioactive precursors were then added and incorporation measured for 10 min

incorporation of ^{14}C-uracil into polysomes showed that the formation of *m*RNA was also blocked when the cells were pre-treated for 5 min with granaticin or other quinones at concentrations which inhibit protein synthesis by not more than 10%. After a short pulse with ^{14}C-uracil the polysomal and ribosomal fractions are not labelled (see Fig. 4 A). From this result we conclude that the synthesis of *m*RNA was totally inhibited and that no ribosomal precursors were made. However, when the cells were first labelled for 1.5 min with ^{14}C-uracil and then treated with the quinone for 15 min, considerable amounts of radioactivity were found within the ribosomal and polysomal fractions (see Fig. 4 B). Since under the influence of the quinones the

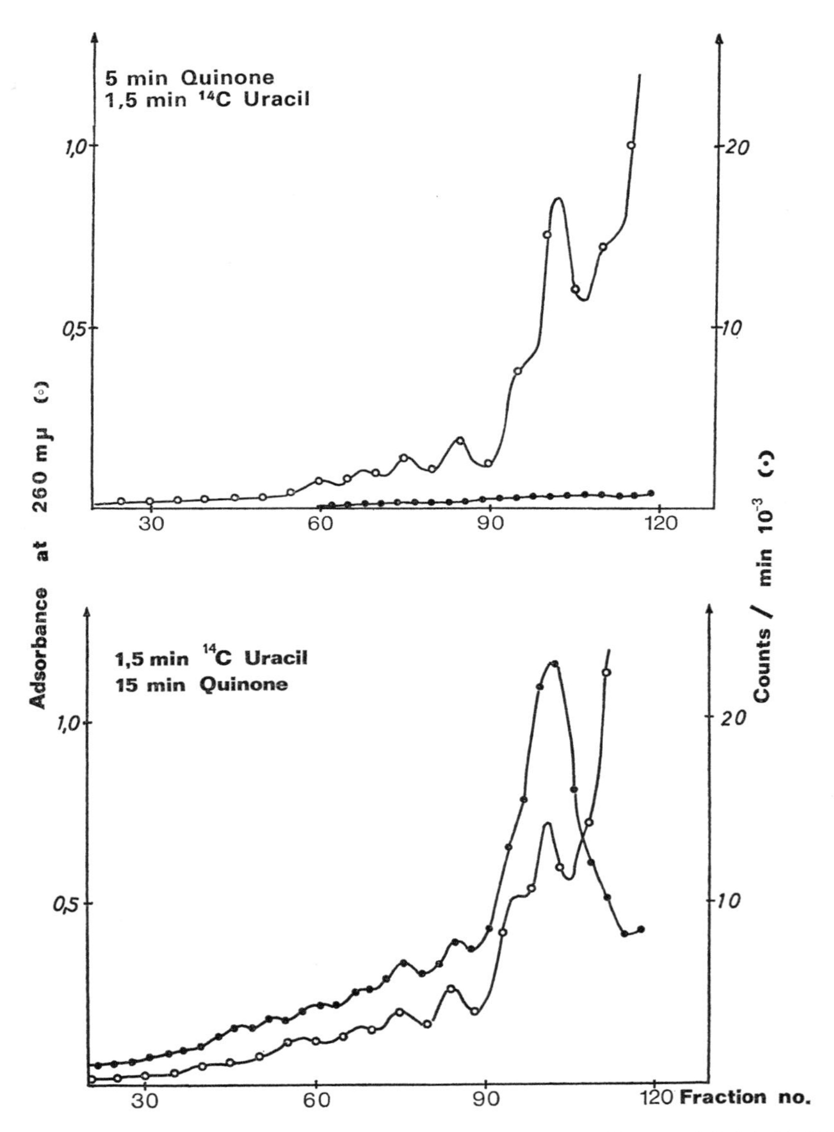

Fig. 4. Effect of quinolin quinone on the incorporation of ^{14}C uracil into rapidly labelled RNA. Above pulse was given for 1.5 min, 5 min after treatment with the quinone. Below: after a pulse of 1.5 min the cells were treated for 15 min with the quinone. The cells were rapidly cooled, ground and the polysome fraction separated in a sucrose gradient by centrifugation

synthesis of all types of RNA was blocked, the radioactivity must have been incorporated into *m*RNA before the addition of the drug. In the presence of quinones the *m*RNA is stable for at least 15 min, in contrast no stable *m*RNA was found in *B. subtilis* treated with DNA-complexing inhibitors of RNA synthesis, e.g. actinomycin.

IV. Discussion

According to the target molecules of the cell, inhibitors of RNA synthesis can be divided into different classes: To the first group belong all those inhibitors which form stable complexes with DNA *in vitro* and *in vivo*, thereby preventing the RNA polymerase to function at the DNA template. Knowledge of the structure and conformation of complexes between DNA and small molecules may give us some information to the as yet unanswered question how larger molecules, such as repressors, interact with DNA.

From the *CD* spectra of the tetrahydrotetracene-quinones (e.g. nogalamycin) and the related tetrahydroanthracene antibiotics (chromomycin and mithramycin) and the changes in these spectra which are induced upon association with DNA, the following conclusions can be drawn:

Nogalamycin and daunomycin exist in aqueous solutions as monomers, chromomycin and mithramycin form dimers. The complex formation of nogalamycin and daunomycin with DNA is accompanied by a conformational change of their chromophores. This change in conformation is only slightly dependent on the *GC* content of the added DNA. Chromomycin and mithramycin are bound to DNA in the presence of Mg^{2+} as monomers. The conformational change in the chromophore increases markedly with raising *GC* content of the DNA.

Inhibitors of RNA synthesis in the second group affect RNA polymerase and specifically bind to the enzyme: the rifamycines, for example, bind to bacterial polymerase and prevent chain initiation (HARTMANN, HONIKEL, KNÜSEL and NUESCH, 1967; UMEZAWA, MIZUNO, YAMAZAKI and NITTA, 1968; WEHRLI, KNÜSEL, SCHMID and STAEHELIN, 1969); streptovaricin acts similar to rifampicin (MIZUNO, YAMAZAKI, NITTA and UMEZAWA, 1968); streptolydigin also reacts with the bacterial enzyme and inhibits chain elongation (SCHLEIF, 1969); α-amanitin inhibits chain elongation but only interferes with the RNA polymerase of eucaryotes (SEIFART and SEKERIS, 1969; NOVELLO, FIUME and STIRPE, 1970; KEDINGER, GNIAZDOWSKI, MANDEL, JR., GISSINGER and CHAMBON, 1970).

The third group which has been shown to inhibit preferentially the synthesis of RNA in microorganisms includes the quinone-antibiotic granaticin, mitomycin-derivatives and synthetic quinolin- or naphtoquinones. These substances do not form stable complexes with DNA *in vitro*. Granaticin does not inhibit the DNA-dependent synthesis of RNA with the RNA polymerase *in vitro* (HARTMANN, personal communication).

We therefore assume that the target molecule within the cell which is affected by these is neither the DNA- nor the RNA-polymerase. The mechanism by which the quinones influence the synthesis of RNA has not been clarified.

The quinones might interfere with some natural occurring quinones which are important for the phosphorylation of the RNA precursors. This interpretation is in accord with the observation that a naphtoquinone-containing antibiotic "naphtomycin" inhibits the synthesis of RNA and protein and that this effect can be reversed by vitamin K. The effect of granaticin and of other quinones, however, could not be antagonized by vitamin K. We have not tested whether other naturally occurring quinones can reverse the inhibitory effect of quinone-containing antibiotics on RNA synthesis.

The antibiotic naphtomycin interacts with cystein (Balerna, Keller-Schierlein, Martius, Wolf and Zähner, 1969). The effect of this antibiotic on growth and nucleic acid synthesis can be reversed by cystein. Likewise, the effect of granaticin on RNA synthesis can be antagonized by the addition of cystein.

Our current hypothesis of the mechanism of quinone action is that quinones of different structures interfere with different SH-containing proteins and thus interfere with certain steps of cell metabolism. Quinones which cause the stabilization of polysomes might interfere with an SH-containing protein involved in the regulation of RNA and protein synthesis.

References

Balerna, M., Keller-Schierlein, W., Martius, C., Wolf, H., Zähner, H.: Naphtomycin, ein Antimetabolit von Vitamin K. Arch. Mikrobiol. **65**, 303 (1969).

Bhuyan, B. K., Smith, C. G.: Differential interaction of nogalamycin with DNA of varying base composition. Proc. nat. Acad. Sci. (Wash.) **54**, 566 (1965).

Blake, A., Peacocke, A. P.: Extrinsic Cotton effects of aminoacridines bound to DNA. Biopolymers **4**, 1091 (1967).

— — Extrinsic Cotton effects of proflavin bound to polynucleotides. Biopolymers **5**, 383 (1967).

— — Induced optical activity of various aminoacridines bound to DNA. Biopolymers **5**, 871 (1967).

Courtois, I., Guschlbauer, W., Fromageot, P.: Interaction entre pigments et acides nucléiques. 6. Etude de l'interaction de l'actinomycine avec la DNA par dispersion optique rotatoire. Europ. J. Biochem. **6**, 106 (1968).

Fey, G., Kersten, H.: Circulardichrographie an Komplexen von Deoxyribonucleinsäuren mit Antibiotica. H 5. Z. Physiol. Chem. **351**, 111 (1970).

Gardener, B. J., Mason, S. F.: Structure and optical activity of the DNA-aminoacridine-complex. Biopolymers **5**, 79 (1967).

Hahn, F. E., Krey, A.: DNA induced anomalous ORD of antimalarial drugs and dyes. Antimicrobial Agents and Chemotherapy, p. 15 (1968).

Hartman, G., Goller, K., Koschel, K., Kersten, W., Kersten, H.: Hemmung der DNA-abhängigen RNA- und DNA-Synthese durch Antibiotica. Biochem. Z. **341**, 126 (1964).

— Honikel, K. O., Knüsel, F., Nuesch, J.: The specific inhibition of the DNA-directed RNA synthesis by rifamycin. Biochim. biophys. Acta (Amst.) **145**, 843 (1967).

Homer, R. B.: The circular dichroism of actinomycin D and its complexes with DNA and d-GMP 5'. Arch. Biochem. **129**, 405 (1969).

Iyer, V. N., Szybalski, W.: A molecular mechanism of mitomycin action: Linking of complementary DNA strands. Proc. nat. Acad. Sci. (Wash.) **50**, 355 (1963).

Kedinger, C., Gniazdowski, M., Mandel, J. L., Jr., Gissinger, E., Chambon, P.: α amanitin: A specific inhibitor of one of two DNA dependent RNA polymerase activities from calf thymus. Biochem. biophys. Res. Commun. **38**, 167 (1970).

Kersten, H., Kersten, W.: Inhibitors acting on DNA and their use to study DNA replication and repair. Inhibitors, tools in cell research (Bücher, Th., Sies, H., Eds.). Berlin-Heidelberg-New York: Springer 1969.

Kersten, W., Kersten, H.: Die Bindung von Daunomycin, Cinerubin und Chromomycin A an Nucleinsäuren. Biochem. Z. **341**, 174 (1965).

— — Szybalski, W.: Physico-chemical properties of complexes between DNA and antibiotics which affect RNA synthesis. Biochemistry **5**, 236 (1965).

— Ogilvie, A., Wanke, H.: Influence of granaticin and quinones on RNA and protein metabolism. VI. FEBS Meeting Madrid 1969, Abstr. 676.
— Kersten, H., Wanke, H., Ogilvie, A.: Einfluß von Antibiotica auf Nucleinsäuren und den Nucleinsäurestoffwechsel. Zbl. Bakt. I. **212**, 259 (1970).
Levine, M., Bothwick, M.: Action of streptonigrin on genetic recombinations between bacteriophages. Proc. XI. Int. Congr. Genet., The Hague 1963.
Mizuno, S., Yamazaki, H., Nitta, K., Umezawa, H.: Inhibition of DNA-dependent RNA polymerase reaction of *Escherichia coli* by an antimicrobial antibiotic, streptovaricin. Biochim. biophys. Acta (Amst.) **157**, 322 (1968).
Müller, W., Emme, I.: Zum Verhalten der Actinomycine in wäßrigen Lösungen. Z. Naturforsch. **20 b**, 835 (1965).
Neville, D. M., Bradley, D. F.: Anomalous rotatory dispersion of acridine-orange-native DNA-complexes. Biochim. biophys. Acta (Amst.) **50**, 397 (1961).
Novello, F., Fiume, L., Stirpe, F.: Inhibition by α amanitin of RNA polymerase solubilized from rat liver nuclei. Biochem. J. **116**, 177 (1970).
Ohba, Y., Fromageot, P.: Interactions entre pigments et acides nucléiques. 5. Stoechiométrie du complex I formé entre la lutéoskyrine les iones magnésium et les acides nucléiques. Europ. J. Biochem. **6**, 98 (1968).
Permogorow, V. J., Lazurkin, Y. S.: Mechanism of binding of actinomycin with DNA. Biofizika **10**, 17 (1965).
Schleif, R.: Isolation and characterization of a streptolydigin resistant RNA polymerase. Nature (Lond.) **223**, 1068 (1969).
Seifart, K. H., Sekeris, C. E.: α Amanitin, a specific inhibitor of transcription by mammalian RNA polymerase. Z. Naturforsch. **24 b**, 1538 (1969).
Umezawa, H., Mizuno, S., Yamazaki, H., Nitta, K.: Inhibition of DNA-dependent RNA synthesis by rifamycins. J. Antibiot. (Tokyo) **21**, 2345 (1968).
Wehrli, W., Knüsel, F., Schmid, K., Staehelin, M.: Interaction of rifamycin with bacterial RNA polymerase. Proc. nat. Acad. Sci. (Wash.) **61**, 667 (1969).
White, I. R., White, H. C.: Effect of intracellular redox environment on bacteriocidal action of mitomycin C and streptonigrin. Antimicrobial Agents and Chemotherapy, p. 495 (1964).
Ziffer, A., Yamaoka, K., Mauger, A. B.: Optical properties of actinomycin D. I. Influence of the lactone rings on its optical activity. Biochemistry **7**, 996 (1968).

Changes in the Properties of Ribosomes and Transfer RNA Induced by Inhibitors of Protein Synthesis

HELGA KERSTEN

I. Introduction

Since the discovery that chloramphenicol binds to ribosomes (VAZQUEZ, D., 1964; WOLFE and HAHN, 1965) other inhibitors of protein synthesis also have been found to associate with ribosomes *in vitro*. These antibiotics can be classified according to their specificity towards the ribosomes (VAZQUEZ, STAEHELIN, CELMA, BATTANER, FERNANDEZ-MUNOZ and MONRO, 1969). Two main classes exist, one affecting only the 70*S* ribosomes, the other interacting with both, 70*S* and 80*S* ribosomes. The substances which bind to the 70*S* ribosomes can be subdivided further into one group which specifically interacts with the 30*S* subunit and another group which binds to the 50*S* subunit. Likewise inhibitors affecting the 80*S* ribosomes can be classified depending on whether they associate with the 40*S* or with the 60*S* subunit.

In vitro studies on binding of antibiotics to ribosomes and their inhibitory effect on cell free synthesis of protein suggest that these substances affect protein synthesis *in vivo* by the same molecular mechanisms as *in vitro*. However, we cannot exclude the possibility that in the living cell their actions are more complicated. We therefore focused our attention on the effect of inhibitors of protein synthesis within the cell.

The antibiotics to be discussed are pactamycin and tetracyclines (e.g. oxytetracycline). The structure of pactamycin has been elucidated by WILEY, JAHNKE, MACKELLAR, KELLY and ARGOUDELIS (1970). This antibiotic is a cyclopentane derivative which contains the following substituents: in position 1, hydroxyethyl- and dimethylurea residues, in position 2, an amino group, in position 3, an *m*-acetoanilide-residue, in position 4, a hydroxyl and a methyl group, and in position 5, a methyl ester of 6-methylsalicylic acid. The structures of the tetracyclines are well known.

Both, pactamycin and tetracyclines specifically bind to the 30*S* subunit (COHEN, GOLDBERG and HERNER, 1969; VAZQUEZ et al., 1969). COHEN et al. [1969 (2)] showed that pactamycin interferes with the formation and stability of the initiating complex consisting of poly*U*,N-acetyl-1-phenylalanyl-*t*RNA and ribosomes. Tetracyclines inhibit the binding of aminoacyl-*t*RNA to the specific binding site at the 70*S* ribosome, whereas peptidyl-*t*RNA does bind to the ribosomes in the presence of tetracyclines (SUZUKA, KAJI and KAJI, 1966; GOTTESMAN, 1967). SARKAR and THACH (1968) concluded that tetracyclines, like pactamycin, inhibit the formation of the initiator complex.

This presentation is subdivided into the following parts:

Section II deals with sedimentation behavior, chemical properties and biological activity of ribosomes and of ribosomal RNA isolated from normal and from pactamycin-treated microorganisms. Section III describes the chromatographic behaviour, methyl-accepting properties and the biological activity of *t*RNA from untreated and from pactamycin- or tetracycline treated microorganisms. In section IV evidence will be presented that in the presence of pactamycin or tetracycline, methyl-deficient RNA is synthesized. The results are discussed in section V with respect to the observations of other groups.

II. Changes in the Properties of Ribosomes and Ribosomal RNA

Inhibitors which bind to ribosomes *in vitro* may also interact with ribosomes *in vivo*, and as a consequence changes in the properties and functional integrity of the ribosomes might occur. To test this hypothesis ribosomes were isolated from normal and from pactamycin-treated *B. subtilis* strain W 23. The sedimentation patterns of ribosomes from untreated and from treated cells were compared. The ribosomal fractions showed considerable differences. Ribosomes isolated from *B. subtilis* (resting phase) at a Mg^{2+} concentration of 1×10^{-2}M contained mainly 100*S* particles and 70*S* ribosomes in about equal amounts and small amounts of 30*S* and 50*S* subunits. The 100*S* dimers are probably formed during the isolation procedure. Upon treatment of the bacteria with pactamycin the ratio of 70*S* monomeric ribosomes to 100*S* dimers increases with increasing doses of the drug and with time of incubation. The 70*S* ribosomes lose their property to form the 100*S* dimer (KERSTEN, H., KERSTEN, W., EMMERICH and CHANDRA, 1967). In no instance of pactamycin treatment have we found chloramphenicol like particles which have been suggested to be the normal precursors of ribosomes (HOSOKAWA and NOMURA, 1965). That the "CM-particles" are normal precursors of ribosomes must be questioned because it has been shown that non-ribosomal proteins can combine with "CM *r*RNA" during the preparation of the cell extracts (YOSHIDA and OSAWA, 1968).

Comparative studies on the poly*U*-directed polymerization of phenylalanine with enzymes from normal and ribosomes from pactamycin-treated bacteria showed that the ribosomal fractions from treated cells were considerably less active, although they consisted mainly of ribosomes with a sedimentation coefficient of 70*S* (KERSTEN et al., 1967). Similar results have been reported by BHUYAN (1967) using ribosomes from livers of pactamycin treated rats in a cell free protein synthesizing system.

Ribosomal RNA isolated from chloramphenicol-treated *E. coli* contains less methyl groups than normal *r*RNA from E. coli (DUBIN, 1964; DUBIN and GÜNALP, 1967; HAYASHI, OSAWA and MIURA, 1966). Pactamycin exhibits the same effect on the methylation of *r*RNA in *B. subtilis* (H. KERSTEN, CHANDRA, TANCK, WIEDEMHÖVER and W. KERSTEN, 1968). RNA was prepared from the ribosomal fractions and tested in a cell free system for its methyl-accepting properties. According to HURWITZ, GOLD and ANDERS (1964) this system contained the *r*RNA as substrate [^{14}C] S-adenosylmethionine as a donor of methyl groups, the homologous methylating enzyme system from *B. subtilis* or the heterologous methylating enzymes from *E. coli* (78.000 × g supernatant).

With both enzyme systems we have found an increase in the amount of methyl groups, incorporated into the *r*RNA from treated cells (Fig. 1) by comparison to *r*RNA from untreated bacteria.

Similar to our results are the findings of CHENG and HARTMAN (1968) on the properties of ribosomes and *r*RNA formed by the relaxed mutant of *E. coli* m^- during growth with ethionine. Ribosomes formed in this strain of *E. coli* when grown with ethionine instead of with methionine have the same properties as ribosomes from pactamycin treated cells: the extent of association of 70*S* ribosomes to 100*S* dimers was markedly decreased and the *r*RNA isolated from ribosomal fractions was submethylated.

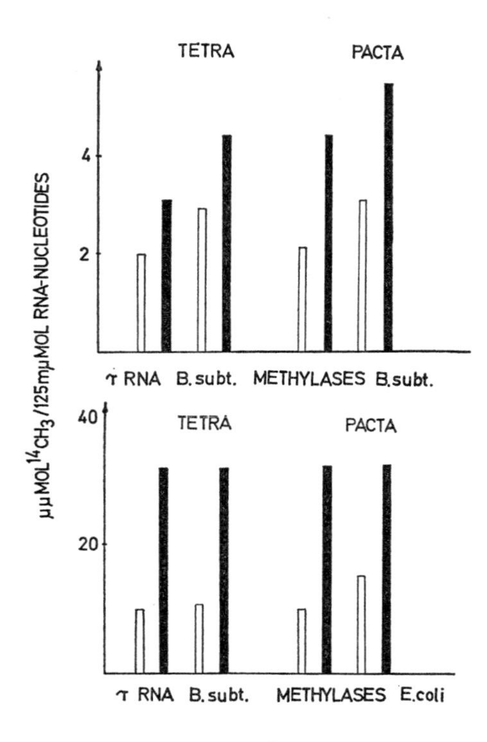

Fig. 1. In vitro methylation according to HURWITZ et al. (1964) of ribosomal RNA from untreated (open columns), from tetracycline or pactamycin-treated (filled columns) B. subtilis. The homologous B. subtilis and the heterologous E. coli methylating enzyme systems were used

The complete methylation of *r*RNA could be important for the last steps of the maturation of the ribosomes or for maintaining the integrity of the ribosomal structure (OSAWA, OTAKA, ITOH, FUKUI, 1969). The question of whether differences between sedimentation behavior and biological activity between ribosomes from normal and pactamycin-treated *B. subtilis* are related to differences in the methylation of ribosomes has to be answered by further experiments. We cannot exclude the possibility that the decrease in biological activity might be caused by an irreversible binding of pactamycin to the ribosomes rather than by an insufficient methylation of *r*RNA.

III. Changes in the Properties of Transfer RNA

Since pactamycin causes incomplete methylation of ribosomal RNA the question arose as to whether this antibiotic and other inhibitors of protein synthesis also interfere with the methylation of *t*RNA and cause alterations in the properties and function of *t*RNA.

Logarithmically growing cultures were treated with pactamycin or oxytetracycline at different concentrations and for different periods of time. Cultures were selected

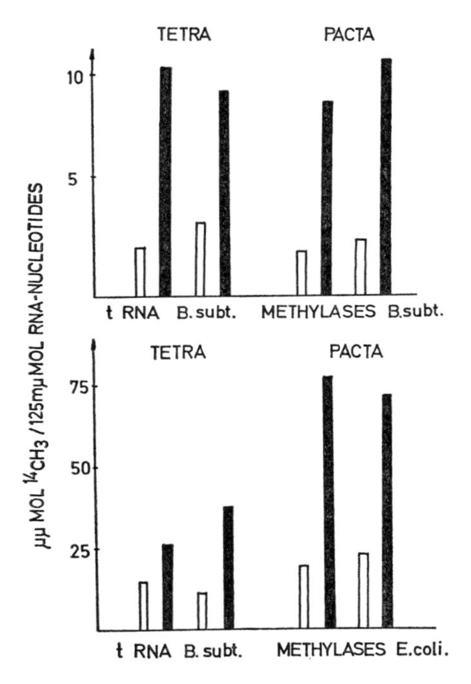

Fig. 2. In vitro methylation according to HURWITZ et al. (1964) of transfer RNA from untreated (open columns), from tetracycline or pactamycin-treated (filled columns) B. subtilis. The homologous B. subtilis and the heterologous E. coli methylating enzyme systems were used

for further experiments in which overall growth was inhibited by not more than 20 to 30%. After 2 to 3 h of treatment the cells were harvested, disrupted and *S*-100 fractions were used as a source of *t*RNA. The *t*RNAs were isolated from these fractions by a modification of Zubay's method (KERSTEN, W., KERSTEN, H., STEINER and EMMERICH, 1967). The *t*RNAs from tetracycline- or pactamycin-treated cells were tested as substrates in the *in vitro* methylating system. The results of these experiments clearly show an increased ability to accept methyl groups with the homologous as well as with the heterologous methylating enzyme system (Fig. 2).

The acceptor avtivities of *t*RNAs from normal or from pactamycin treated *B. subtilis* were measured for phenylalanine, lysine, alanine and valine. The acceptor activity of *t*RNAs from treated cultures was decreased for phenylalanine by 30% but unchanged for lysine, alanine and valine (KERSTEN et al., 1968). In these experiments only the *t*RNAs were isolated from treated cultures whereas all the enzymes originated from untreated *B. subtilis*.

The *t*RNAs were further tested in cell-free incorporation systems with several amino acids in the presence of synthetic polynucleotides. As a source for the enzymes and the cofactors a homologous system was used. The results of these experiments

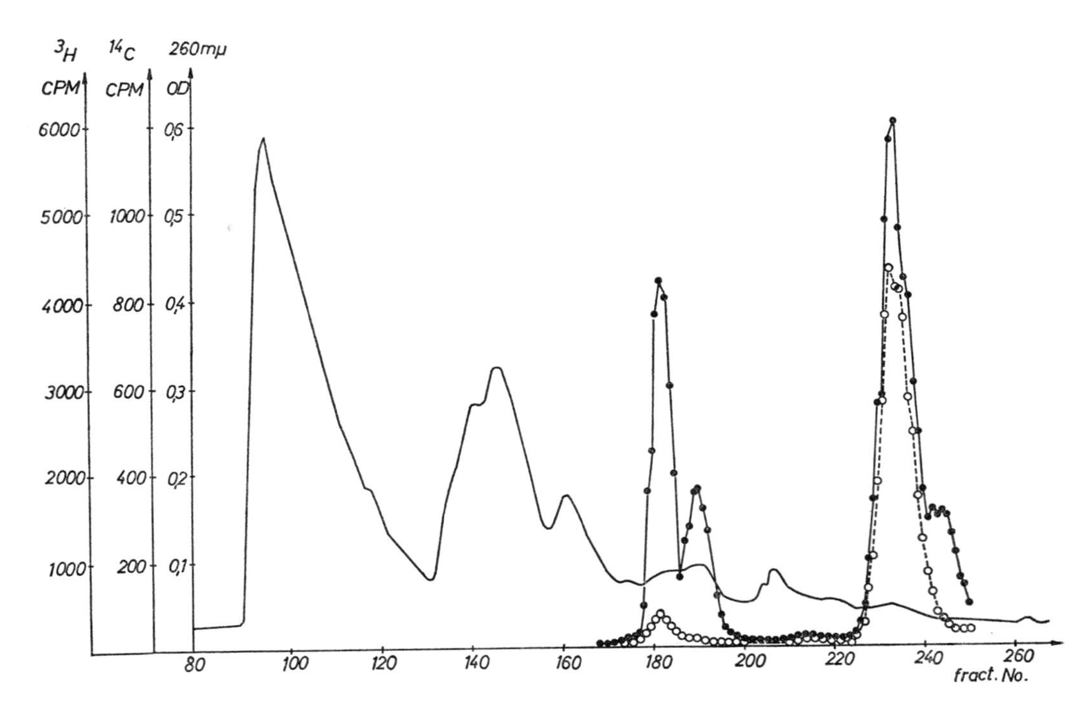

Fig. 3. Elution profile of tRNAs upon reversed phase chromatography according to KELMERS et al. (1965 ——— OD 260 mμ. tRNA from untreated B. subtilis was charged with ^{14}C phenylalanine ○-○-○, tRNA from pactamycin-treated B. subtilis was charged with ^{3}H phenylalanine ●-●-●. The tRNAs were cochromatographed. These experiments were done in cooperation with M. STAEHELIN and W. WEHRLI. Ciba, Basel, Switzerland 1969

revealed no change in the incorporation of valine, lysine, proline and serine by using poly *UG*, poly *A*, poly *C* or poly *UC* as templates. However, the binding of leucine was increased by about 30% with poly *UC* as template, whereas phenylalanine incorporation with *t*RNA from treated cells was inhibited by 30%. Since the *t*RNA in these experiments was not precharged with amino acids, the inhibition of phenylalanine incorporation had perhaps taken place already at the charging step (KERSTEN et al., 1968).

Another aspect of the biological activity of *t*RNA is its ambiguity of codon recognition. That the correct methylation of *t*RNA can influence the ambiguity of codon recognition has been reported for phenylalanyl-*t*RNA by REVEL and LITTAUER (1966) and for leucyl-*t*RNA by CAPRA and PETERKOFSKY (1968). The leucyl-*t*RNA from pactamycin-treated *B. subtilis* differed from normal leucyl-*t*RNA: The stimulation of leucine incorporation in the presence of poly *UC* by *t*RNA from treated cells could not be observed by using as a template poly *UG* instead of poly *UC*. The extent of decrease in transfer activity for phenylalanine was the same with poly *U* and poly *UC* (KERSTEN et al., 1968).

In all experiments described so far the inhibitors were added to logarithmically growing cells. If the *t*RNAs formed upon addition of e.g. pactamycin, differed from the normal *t*RNAs, one might expect alterations in the chromatographic behaviour of certain species of *t*RNA. To test this hypothesis the *t*RNA fractions from normally grown *B. subtilis* charged with ^{14}C phenylalanine and those from pactamycin-treated cells were charged with ^{3}H phenylalanine. The *t*RNAs were cochromatographed according to the method of reversed phase chromatography (KELMERS, NOVELLI and STULBERG, 1965). The results of these experiments (Fig. 3) (KERSTEN, 1969) show that the *t*RNA from treated cells contained beside the normal fraction at least two iso-accepting phenylalanyl-*t*RNAs.

IV. On the Mechanism of Formation of Methyldeficient Ribosomal and Transfer RNA

Neither pactamycin nor tetracycline inhibited the *in vitro* methylation when methyl-deficient *t*RNA from methionine starved *E. coli* meth^{-} RC^{rel} was used as a substrate. Therefore, it seems unlikely that the antibiotics interact directly with either the methylating enzymes or with the substrates. This, however, has to be tested with purified methylating enzymes.

Inhibitors of protein synthesis stimulate the synthesis of RNA. As a consequence RNA can accumulate in treated cultures up to twofold compared to non-treated controls. As a working hypothesis we assume that the increased rate of RNA synthesis is not accompanied by a corresponding increase in the rate of methylation of RNA, i.e. the synthesis of RNA and the subsequent methylation might be uncoupled. This hypothesis was tested as follows: *E. coli* meth^{-} cultures were grown with supplemented methionine. Pactamycin or oxytetracycline were added during the early logarithmic phase of growth together with tritium-labelled uridine and ^{14}C methionine. The specific activities of ^{3}H labelling and ^{14}C labelling of the total RNA were measured after 10, 20 or 30 min of further incubation. Different concentrations of the drugs were used. In untreated controls the incorporation of ^{3}H uridine and of $^{14}CH_3$ into RNA increased concomittantly. At increasing doses of the drugs only the incorporation of ^{3}H uridine was stimulated. The RNA formed under the influence of the antibiotics did not become methylated (Fig. 4). Thus pactamycin or tetracycline at certain concentrations caused an uncoupling of the synthesis of RNA and the subsequent methylation.

V. Discussion

The results presented in this report show that in microorganisms under the influence of pactamycin or tetracycline altered ribosomes and *t*RNAs are produced.

The increase in methyl-accepting properties of *r*RNA and *t*RNA from treated bacteria is probably not caused by addition of an excess of methyl groups because the inhibitors *in vivo* uncoupled the synthesis of RNA from the subsequent methylation. We conclude that the RNA, which accumulates under the influence of these antibiotics, is methyl-deficient. The increased methyl-accepting properties of the RNAs found with hetero-

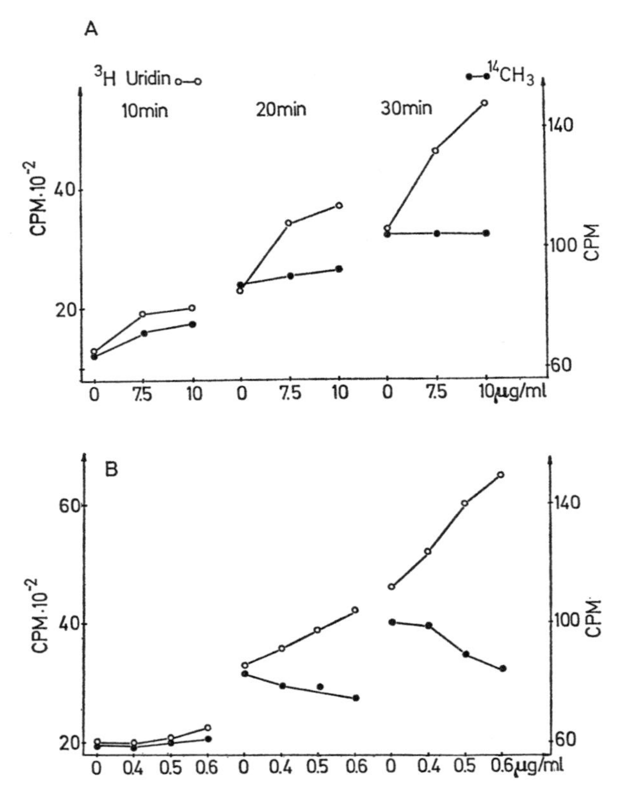

Fig. 4. Effect of pactamycin (A) and oxytetracycline (B) on the synthesis and methylation of RNA in B.subtilis. Incorporation of radioactive ^{3}H uridine O-O-O, and radioactive $^{14}CH_3$ into acid precipitable RNA were measured 10 min, 20 min and 30 min upon treatment with the antibiotics. For pactamycin each 7.5 or 10 μg/ml were added (A). For oxytetracycline each 0.4, 0.5 and 0.6 μg/ml were administered to logarithmically growing cultures. The term 0 μg/ml was used to show incorporation of ^{3}H uridine and $^{14}CH_3$ into RNA of untreated B. subtilis

logous methylating enzymes may depend on structural changes causing availability of positions which are normally not recognized.

The ribosomes and certain species of *t*RNAs of treated bacteria show alteration in the biological activity. The question must be asked as to how the decrease in biological activities of the ribosomes and of the *t*RNAs from treated cells is related to the deficiency in methyl groups.

Methylated bases occur in both 16*S* and 23*S* *r*RNAs. The 16*S* species contains about 20% more methyl groups than the 23*S* species (STARR, 1963; SVENSSON, BOMAN, ERIKSSON, KJELLIN, 1963). In addition to the methylated purines and pyrimidines, RNA contains 2'-O-methylribosides in all of the mature bases (HALL, 1964). When nascent *r*RNA is converted to ribosomes the RNA becomes fully methylated (SYPHERD and FANSLER, 1967). It is not yet clear, whether the methylation of *r*RNA takes place gradually during the process of ribosome formation or occurs at particular steps of development. MACDONALD, TURNOCK and FORCHHAMMER (1967) isolated a mutant strain of E. coli in which a considerable amount of a 43*S* component is accumulated in the cells during growth. The ability to support protein synthesis *in vitro* of 70*S* ribosomes from these mutants was found to be greatly impaired. It is therefore likely that a modification of *r*RNA molecules occurs at the end of ribosome maturation. This modification might be responsible for the proper binding of proteins to the last intermediate to complete the ribosome. It was suggested by MACDONALD et al. (1967) that the decreased activity of the 70*S* ribosome could be due to disordered methylation of *r*RNA at the final step of the formation of the 50*S* subunit.

The ribosomal fractions from pactamycin-treated cells contain submethylated ribosomal RNA and are less active in cell free synthesis of protein. This also favors the proposal that at least one or a few methyl groups are incorporated during the last steps of ribosome-maturation, and that the decrease in biological function is caused by a deficiency in methyl groups in the 70*S* ribosomes.

Transfer ribonucleic acids contain several methylated bases distributed primarily in single-stranded loops. The *t*RNA methylating enzymes are species-specific, they know exactly the position of a base which has to be methylated in a certain *t*RNA (HURWITZ et al., 1964; FLEISSNER and BOREK, 1963). BAGULEY and STAEHELIN (1968) demonstrated that the methylating enzymes can recognize a certain base by virtue of its surrounding nucleotide sequence.

The high specificity of *t*RNA methylating enzymes suggests that the methyl groups in *t*RNA play an important role in maintaining the proper structure and function of the *t*RNA within the cell. The first studies were made by STARR (1963) who measured the amino acid acceptor activity of submethylated *t*RNAs from a methionine-requiring *E. coli* mutant, grown at limiting concentrations of methionine. Since among several amino acids tested only the acceptor activities for phenylalanine and arginine were decreased by about 20%, the author concluded that the methyl groups in *t*RNA do not influence the acceptor activity. REVEL et al. (1966) by using methyl-deficient *t*RNA from the *E. coli* mutant observed after MAK-column chromatography a new peak for submethylated phenylalanyl *t*RNA. The normal and the submethylated *t*RNAs were tested for codon recognition, using poly *U*, poly *UC* and poly *UA* as templates. It was found that the methyl-deficient *t*RNA responded better to poly *UC* and also recognized poly *UA* which is not recognized by normal phenylalanyl *t*RNA. From these experiments it was concluded that the methyl groups may have an influence on the fidelity of codon recognition. Alterations in chromatographic behavior and in coding properties of methyldeficient leucyl-*t*RNA were reported by CAPRA et al. (1968). Submethylated fractions of leucine *t*RNA were obtained by chromatographic separation according to KELMERS et al. (1965). The submethylated fractions showed preference in binding to the ribosomes in the presence of poly *UC* compared to the binding capacity in the presence of poly *UG*. After remethylation of this fraction, recognition was the same for poly *UC* as for poly *UG*. These results also suggest an influence of methyl groups in the case of leucyl *t*RNA on recognition properties.

Shugart, Novelli and Stulberg (1968) used again the submethylated *t*RNA from methionine-starved *E. coli*. Fully methylated *t*RNA when charged with [^{14}C] phenylalanine and separated according to Kelmers et al. (1965) contains one radioactive peak. The *t*RNAs from starved cells comprise in addition two iso-accepting fractions, one containing a mixture of normal and submethylated phenylalanine *t*RNA and another consisting only of submethylated phenylalanine *t*RNA. The submethylated *t*RNA fractions were found to have a reduced acceptor activity. Concerning the chromatographic properties, Biezunski, Gireon and Littauer (1970) obtained the same results. However, on further purification the submethylated iso-accepting phenylalanine *t*RNA showed the same acceptor activity as the normal *t*RNA.

The *t*RNA fractions from pactamycin or tetracycline treated *B. subtilis* differ from the methyl-deficient *t*RNAs from the *E. coli* mutant: they contain still a tenfold amount of methyl groups, thus only very few methyl groups are missing. However, the changes in biological activity of the *t*RNA fractions were found to be similar to those of the submethylated *t*RNA from the *E. coli* mutant: the acceptor activity for phenylalanine was reduced, leucyl *t*RNA binding to ribosomes was increased in the presence of poly *UC*, but not in the presence of poly *UG*. Whether the alteration in biological activity is indeed related to an insufficient methylation of the *t*RNA species has to be investigated.

Waters (1969) reported on iso-accepting *t*RNAs for phenylalanine from chloramphenicol-treated *E. coli*. The unusual phenylalanyl *t*RNA peak was not found to be submethylated. Iso-accepting phenylalanyl *t*RNAs were recently isolated by Wettstein and Stent (1968) from *E. coli* which were grown under conditions of iron starvation. These authors reported that the abnormal phenylalanyl *t*RNA species had the same chromatographic properties as methyl-deficient phenylalanyl *t*RNA. However, a deficiency in methyl groups could not be demonstrated.

A failure to show methyl deficiency when perhaps only very few methyl groups are missing might occur when crude extracts are used as source for the methylating enzymes. Some of the methylating enzymes are very unstable and could be lost during isolation. Therefore, it might be possible that the iso-accepting phenylalanyl *t*RNAs are indeed methyldeficient.

An alteration in the extent of methylation has also been discovered for iso-accepting *t*RNA fractions occurring in *E. coli* infected with phage T_2 (Waters and Novelli, 1967) in the livers of chicks inocculated with the oncogenic viral agent of Marek's disease (Hacker and Mandel, 1969) and in L M tumors developing in C_3H mice upon injection with L M cells (Yang, Hellman, Martin, Hellman and Novelli). That iso-accepting *t*RNAs from tumors differ in the extent of methylation from *t*RNA of normal cells has first been proposed by Tsutui, Srinivasan, and Borek (1966).

The uncoupling effect of inhibitors of protein synthesis on RNA-synthesis and on the subsequent methylation in whole cells was observed under conditions of increased rate of RNA-synthesis and decreased rate of protein-synthesis. So far studied, the antibiotics do not influence the methylation of RNA by interacting either with RNA or with the methylating enzymes. Since some of the methylating enzymes are unstable, they probably have high turnover rates. If an inhibition of protein synthesis caused by pactamycin or tetracycline results in a corresponding decrease in the level of RNA methylating enzymes the synthesis of RNA and the subsequent methylation could become uncoupled. Methylating enzymes may play a role in the regulation of RNA and protein synthesis.

Pactamycin or tetracyclines, and probably other inhibitors of protein synthesis, provide useful tools to affect the methylation of *r*RNA and *t*RNA *in vivo*. Thus, submethylated RNA can be obtained from different microorganisms and probably also from mammalian cells. (Pactamycin acts on microorganisms as well as on mammalian systems.) The biological activities of submethylated RNA can be tested by using homologous enzyme systems. These antibiotics may help to gain insight into the significance of methylating enzymes and into the functional role of methyl groups in RNA.

References

BAGULEY, B. C., STAEHELIN, M.: The specificity of transfer ribonucleic acid methylases from rat liver. Biochemistry **7**, 45 (1968).

BIEZUNSKI, N., GIREON, D., LITTAUER, U. Z.: Purification and properties of *Escherichia coli* methyl-deficient phenylalanine tRNA. Biochim. biophys. Acta (Amst.) **199**, 382 (1970).

BHUYAN, B. K.: Pactamycin, an antibiotic that inhibits protein synthesis. Biochem. Pharmacol. **16**, 1411 (1967).

CAPRA, J. D., PETERKOFSKY, A.: Effect of in vitro methylation on the chromatographic and coding properties of methyl-deficient leucine transfer RNA. J. molec. Biol. **33**, 591 (1968).

CHENG, T.-Y., HARTMAN, K. A., Jr.: Properties of ribosomes and ribosomal RNA formed by a relaxed mutant of *Escherichia coli* during growth with ethionine. J. molec. Biol. **31**, 191 (1968).

COHEN, L. B., HERNER, A. E., GOLDBERG, I. H.: Inhibition by pactamycin of the initiation of protein synthesis. Binding of N-acethylphenylalanyl transfer ribonucleic acid and polyuridylic acid to ribosomes. Biochemistry **8**, 1312 (1969).

— GOLDBERG, I. H., HERNER, A. E.: Inhibition of pactamycin of the initiation of protein synthesis. Effect on the 30S ribosomal subunit. Biochemistry **8**, 1327 (1969).

DUBIN, D. T., ELKORT, A. T.: Some abnormal properties of chloramphenicol RNA. J. molec. Biol. **10**, 508 (1964).

— GÜNALP, A.: Minor nucleotide composition of ribosomal precursor, and ribosomal ribonucleic acid in *Escherichia coli*. Biochim. biophys. Acta (Amst.) **134**, 106 (1967).

FLEISSNER, E., BOREK, E.: Studies on the enzymatic methylation of soluble RNA. I. Methylation of the S-RNA polymer. Biochemistry **2**, 1093 (1963).

GOTTESMAN, M. E.: Reaction of ribosome-bound peptidyl transfer ribonucleic acid with amino acyl transfer ribonucleic acid or puromycin. J. biol. Chem. **242**, 5564 (1967).

HACKER, B., MANDEL, L. R.: Altered transfer RNA methylase patterns induced by an avian oncogenic virus. Biochim. biophys. Acta (Amst.) **190**, 38 (1969).

HALL, R. H.: On the 2′-O-methylribonucleoside content of ribonucleic acids. Biochemistry **3**, 876 (1964).

HAYASHI, Y., OSAWA, S., MIURA, K.: The methyl groups in ribosomal RNA from *Escherichia coli*. Biochim. biophys. Acta (Amst.) **129**, 519 (1966).

HURWITZ, J., GOLD, M., ANDERS, M.: The enzymatic methylation of ribonucleic acid and deoxyribonucleic acid. J. biol. Chem. **239**, 3462 (1964).

HOSOKAWA, K., NOMURA, M.: Incomplete ribosomes produced in chloramphenicol- and puromycin-inhibited *Escherichia coli*. J. molec. Biol. **12**, 225 (1965).

KELMERS, A. D., NOVELLI, G. D., STULBERG, M. P.: Separation of transfer ribonucleic acids by reverse phase chromatography. J. biol. Chem. **240**, 3979 (1965).

KERSTEN, W., KERSTEN, H., STEINER, F. E., EMMERICH, B.: The effect of chromomycin and mithramycin on the synthesis of deoxyribonucleic acid and ribonucleic acids. Hoppe-Seylers Z. physiol. Chem. **348**, 1415 (1967).

KERSTEN, H., KERSTEN, W., EMMERICH, B., CHANDRA, P.: Studies on the mode of action of pactamycin. Hoppe Seylers Z. physiol. Chem. **348**, 1424 (1967).

— CHANDRA, P., TANCK, W., WIEDEMHÖVER, W., KERSTEN, W.: Effect of pactamycin on methylation of RNA and protein synthesis. Hoppe Seylers Z. physiol. Chem. **349**, 659 (1968).

— Formation of methyl-deficient tRNA in microorganisms in the presence of pactamycin and tetracyclines. The 6th international congress of chemotherapy Tokyo, August 1969.

MACDONALD, R. E., TURNOCK, G., FORCHHAMMER, J.: The synthesis and function of ribosomes in a new mutant of *Escherichia coli*. Proc. nat. Acad. Sci. (Wash.) **57**, 141 (1967).

OSAWA, S., OTAKA, E., ITOH, T., FUKUI, T.: Biosynthesis of 50S ribosomal subunit in *Escherichia coli*. J. molec. Biol. **40**, 321 (1969).

REVEL, M., LITTAUER, U. Z.: The coding properties of methyl-deficient phenylalanine transfer RNA from *Escherichia coli*. J. molec. Biol. **15**, 389 (1966).

SARKAR, S., THACH, R. E.: Inhibition of formyl methionyl-transfer RNA binding to ribosomes by tetracycline. Proc. nat. Acad. Sci. (Wash.) **60**, 1479 (1968).

SHUGART, L., NOVELLI, G. D., STULBERG, M. P.: Isolation and properties of undermethylated phenylalanine transfer ribonucleic acids from a relaxed mutant of *Escherichia coli*. Biochim. biophys. Acta (Amst.) **157**, 83 (1968).

STARR, J. L.: The incorporation of amino acids into "methyl-poor" amino acid transfer ribonucleic acid. Biochem. biophys. Res. Commun. **10**, 181 (1963).

SUZUKA, J., KAJI, H., KAJI, A.: Binding of specific sRNA to 30S ribosomal subunits. Effect of 50S ribosomal subunits. Proc. nat. Acad. Sci. (Wash.) **55**, 1483 (1966).

SVENSSON, J., BOMAN, H. G., ERIKSSON, K. G., KJELLIN, K.: Studies on microbial RNA. I. Transfer of methyl groups from methionine to soluble RNA from *Escherichia coli*. J. molec. Biol. **7**, 254 (1963).

SYPHERD, P. S., FANSLER, B. S.: Structural transitions in ribonucleic acid during ribosome development. J. Bact. **93**, 920 (1967).

TSUTSUI, E., SRINIVASAN, P. R., BOREK, E.: tRNA methylases in tumors of animal and human origin. Proc. nat. Acad. Sci. (Wash.) **56**, 1003 (1966).

VAZQUEZ, D.: Uptake and binding of chloramphenicol by sensitive and resistant organisms. Nature (Lond.) **203**, 257 (1964).

— STAEHELIN, T., CELMA, M. L., BATTANER, E., FERNANDEZ-MUNOZ, R., MONRO, R. E.: Inhibitors as tools in elucidating ribosomal function. 20th Colloquium der Gesellschaft für Biologische Chemie, p. 100. Berlin-Heidelberg-New York: Springer 1969.

WATERS, L. C.: Altered chromatographic properties of tRNA from chloramphenicol-treated *Escherichia coli*. Biochem. biophys. Res. Commun. **37**, 296 (1969).

— NOVELLI, G. D.: A new change in leucine transfer RNA observed in *Escherichia coli* infected with bacteriophage T2. Proc. nat. Acad. Sci. (Wash.) **57**, 979 (1967).

WETTSTEIN, F. O., STENT, G. S.: Physiologically induced changes in the property of phenylalanine tRNA in *Escherichia coli*. J. molec. Biol. **38**, 25 (1968).

WILEY, P. F., JAHNKE, H. K., MACKELLAR, F., KELLY, R. B., ARGOUDELIS, A. D.: The structure of pactamycin. Research Laboratories, The Upjohn Company Kalamazoo, Michigan 49001 (to be published).

WOLFE, A. D., HAHN, F. E.: Effects of chloramphenicol upon a ribosomal amino acid polymerization system and its binding to bacterial ribosome. Biochim. biophys. Acta (Amst.) **95**, 146 (1965).

YANG, W.-K., HELLMAN, A., MARTIN, D. H., HELLMAN, K. B., NOVELLI, G. D.: Isoaccepting transfer RNAs of L-M cells in culture and after tumor induction in C_3H mice. Proc. nat. Acad. Sci. (Wash.) **64**, 1411 (1969).

YOSHIDA, K., OSAWA, S.: Origin of the protein component of chloramphenicol particles in *Escherichia coli*. J. molec. Biol. **33**, 559 (1968).

Chloroquine Binding to Nucleic Acids: Characteristics, Biological Consequences, and a Proposed Binding Model for the Interaction*

K. Lemone Yielding, Lerena W. Blodgett, Helene Sternglanz and David Gaudin

I. Introduction

Chloroquine and other antimalarials have served as useful tools for studying drug action both at the molecular and gross biological level. First of all, such drugs are of therapeutic usefulness both in parasitic as well as certain non-parasitic diseases. Thus, the search for new and better antiparasitic agents can best be guided through an understanding of action mechanisms. Secondly, these drugs have proved valuable for probing the nature of specific biological systems. Finally, biochemical studies of these drugs have served as the models for exploring the structure-function relationships and molecular interactions of a variety of other drugs.

The binding of chloroquine to nucleic acids was first examined from 1949 to 1952 by Parker and the Irvins who described a change in the drug's spectrum upon binding. They also observed other changes in physical properties such as solubility of the complex. As a result of these studies they suggested that the antimalarial effects could well depend on this interaction with nucleic acids. Subsequently, aminoacridine binding received considerable attention while chloroquine was not studied in additional detail until the 1960's. Kurnick and Radcliffe (1962) observed that binding of chloroquine or quinacrine caused an increase in the viscosity of DNA. They also discovered that the drug-DNA complex was less susceptible to DNAase hydrolysis than was native DNA. Stollar and Levine (1963) furthermore observed that complex formation inhibited the DNA antibody reaction as perhaps related to its use in treating lupus erythematosus and DNA hypersensitivity (Levin and Pinkus, 1961).

Studies by Cohen and Yielding (1963, 1964, 1965) and by Hahn and associates (1965, 1966) characterized the binding process in some detail. Cohen and Yielding reported that binding involved purine residues as well as the backbone phosphates, was modified by secondary structure and by ions, and provided a pronounced stabilizing effect on helical DNA (Cohen and Yielding, 1963, 1965). The drug interfered with the priming action of DNA for the DNA polymerase, and to a lesser extent the RNA polymerase reaction (Cohen and Yielding, 1964, 1965). These

* Supported by contract No. DA-49-193-MD-3040, from the United States Army Research Program on Malaria. Portions of this work were also supported by a project grant from the American Cancer Society.

findings were extended by HAHN and his associates who emphasized that the preferential effect on DNA over RNA synthesis occurred even under the same conditions of ionic strength, pH, etc. and could be shown for other antimalarials as well (O'BRIEN et al., 1966; HAHN et al., 1966). These results led both groups of workers to propose that the biological effects of the drug might be based on inhibition of DNA synthesis (COHEN and YIELDING, 1965; HAHN et al., 1966). Considerable attention therefore, has been directed in several laboratories toward the specificity of the binding process and its possible biological consequences.

The present discussion summarizes findings and conclusions from our own laboratories over the past few years on the nature, specificity and consequences of chloroquine interactions with DNA, mono-, and polynucleotides. A specific binding model is presented based on these studies and the construction of molecular models. Some biological effects will be presented to illustrate possible directions in which exploitation of the knowledge of drug interactions may be pursued. Of particular interest are the control of virus replication and modification of the effects of certain cancer treatment agents.

II. Studies on Binding of Chloroquine to Nucleic Acids

1. Methods for Studying Drug Binding

Study of nucleic acid induced changes in the chloroquine spectrum, first used by PARKER and IRVIN (1949, 1950) has provided the most convenient means of following the binding process. When the extinction coefficients of the free and bound drug are known, the extent of binding can be calculated from the absorbance of a mixture. Data are usually analyzed according to the methods devised by KLOTZ and URQUHART (1949) and by SCATCHARD et al. (1957) in which linear plots yield both k (intrinsic equilibrium constant) and n (number of potential binding sites). There are limitations to this approach, however, which seem to apply to the majority of DNA binding drugs which have been studied. First, the determination of the absorption of the bound drug must be precise. This is usually carried out experimentally by measuring drug absorption in the presence of a large excess of the polymer or with a series of polymer concentrations followed by extrapolation of the results to yield the absorption when polymer is infinite and all drug molecules are bound. This value, in fact, represents the strongest binding sites and/or the most common sites. Thus, if absorbancy of bound dye varies as a function of the extent of saturation of the polymer, or if multiple types of sites exist, use of the extrapolated value to determine the amount of bound dye under a variety of conditions is not valid and will produce erroneous results. The assumption is often made that the presence of an isosbestic point precludes the presence of more than two spectral species (bound and unbound). A isosbestic point, however, can exist with more than two species, as has been emphasized before, if the binding function is linear over the range studied and if the spectra of all bound species intersect the free ligand spectrum at approximately the same point (COHEN and FISCHER, 1962). For simple linear plots to be obtained by the KLOTZ or SCATCHARD procedures, the additional conditions must be met that binding sites must be identical and must not interact with each other (i.e. through electrostatic repulsion or assistance, or through configurational changes in the polymer). In this

Symposium, Dr. CROTHERS has referred to a special case of negative site interactions as "neighbor exclusion".

The fact is that curved data plots are obtained with chloroquine and with most other agents which interact with DNA. Although most often interpreted in terms of heterogeneous sites, this interpretation is not unambiguous in the absence of additional information. Even with these limitations, spectral studies are useful but must be interpreted with caution, particularly when comparing limited data from different studies. The most reliable data for determining intrinsic equilibrium constants would appear to be those involving sufficiently low bound ligand/total polymer ratios so that negative site interference can be avoided and only the tightest binding is evaluated. However, under these circumstances, determination of the number of potential binding sites (n) involves a long extrapolation. Thus it is often difficult to obtain an accurate estimate of both equilibrium constant and the number of available binding sites by the spectral method.

Equilibrium dialysis is the most direct way to obtain a measure of the amount of bound and free ligand. However, it must be remembered that *all* binding is measured and may not correlate directly with spectral changes. Sedimentation of DNA ligand complex in the ultracentrifuge has provided a simple alternate to equilibrium dialysis, and eliminates the problem of DNA or drug binding (ALLISON et al., 1965; LLOYD et al., 1968; BLODGETT and YIELDING, 1968) to the dialysis membrane.

The evaluation of the role of specific atoms in the molecular binding process poses a more difficult problem. The use of drug and polymer analogs is the simplest approach and has found wide application. Similarly, the effect of complex formation on the chemical reactivity of specific substituents on the interacting drugs has been useful. In the present studies, nuclear magnetic resonance spectroscopy has been applied to learn the fate of specific protons as a result of binding.

2. Binding of Chloroquine to DNA

Our interest in chloroquine binding started in collaboration with Dr. STANLEY COHEN (see Table 1). His studies employing UV spectroscopy produced curved SCATCHARD plots, and were interpreted by means of a curve fitting program, in terms of two types of sites, "strong" (K ca. 1.1×10^4) and "weak" (K ca. 1.2×10^3). There were about one strong and as many as two weak sites per two base pairs. The conclusion that purines were preferred by the drug was based on the observations that apyrimidinic acid but not apurinic acid showed substantial binding. Deoxyribonucleic acids with $G+C$ contents which varied from 38% to 67% showed essentially the same interaction, indicating the same extent of binding to A or G, in contrast to the results reported for actinomycins (REICH, 1964). There were differences both in equilibrium constants and the extent of binding observed, when the chloroquine interaction with heated and rapidly cooled DNA was compared to that with native DNA. Conversely, extensive stabilization of the DNA helix was seen in the presence of the drug. Furthermore, divalent cations antagonized binding. These findings, together with the observation that the drug binding resulted in inhibition of DNA polymerase has prompted additional studies into the nature, specificity, and the consequences of the chloroquine-DNA association.

Owing to the uncertainties in spectral methods and their interpretations, use was made of centrifugation to confirm certain aspects of the binding process. As shown in Table 2, at a low ligand/polymer ratio, the equilibrium constant obtained was in good agreement with that for the "strong" process in earlier studies, and was found to be relatively insensitive to variations in temperature in the range of 15° to 26°.

The effect of 1×10^{-3} M $MgCl_2$ was also re-examined; $MgCl_2$ was found to antagonize drug binding with an apparent reduction of the equilibrium constant from

Table 1. *Summary of studies on chloroquine binding*
(by COHEN and YIELDING, 1963, 1964, 1965, 1965)

1. Characteristic change in drug spectrum resulted from interaction with natural and synthetic nucleic acid polymers and mononucleotides. Molar extinction of bound drug varied with base composition of the polymer.
2. Curved Scatchard plots were consistent with a solution for 2 classes of sites for natural polymers.
   Native DNA $K_1 = 1.1 \times 10^4$, $n = 0.22$,
   $K_2 = 1.2 \times 10^3$, $n = 0.44$;
   "Denatured" DNA $K_1 = 0.98 \times 10^4$, $n = 0.15$,
   $K_2 = 1.1 \times 10^3$, $n = 0.75$.
3. "Strong" binding favored by purines (both A and G).
   Apyrimidinic acid $K = .9 \times 10^4$, $n = .55$.
   Apurinic acid $K = 1 \times 10^3$, $n = 0.49$.
   Binding strength was unchanged by varying $G + C$ content from 38 to 67%.
4. Binding was reversed by $MgCl_2$, and depended on doubly protonated species of the drug.
5. DNA helix was stabilized by binding.
6. *Binding Model Proposed:* Specific ring-ring interactions plus electrostatic binding of the side chain.
7. *Binding* — resulted in inhibition of DNA polymerase and to a lesser extent, RNA polymerase.
8. DNA binding and polymerase inhibition of drug analogs were well correlated with antimalarial activity.
9. Findings, therefore, provided the basis for the biological effects of this class of drugs.

1.8×10^4 to 3.2×10^3. The question of whether chloroquine binds to "denatured" DNA has received considerable attention. Binding does indeed occur extensively to heated and recooled DNA (Table 1 and 2) and seemingly more than could be accounted for simply by limited reassembly of the DNA helix. It was reported, however, that binding does not occur to formaldehyde-treated DNA (helix reassembly prevented) nor at 98° (HAHN et al., 1966). Formaldehyde reacts with amino groups of purines and pyrimidines and changes primary as well as secondary structure. Experiments at 98° are conclusive only if such elevations in temperature do not influence the binding process or solvent structure, either of which is a major assumption. In our experiments binding did not occur at pH 3 where protonation of purines had occurred. Binding to native and heated-recooled DNA was also compared with regard to the temperature sensitivity of the complex, by observing the absorbance of chloroquine at 343 μm. As shown in Fig. 1, chloroquine did not dissociate from native DNA until collapse of the helix had occurred, while dissociation was essentially complete above 50° for the heated-recooled sample. Thus, while binding occurred to the denatured sample it was, in contrast to that to native DNA, more sensitive to temperature. Whether this sensitivity was due to collapse of secondary structure to which the drug binds, or due directly to an effect on the binding process is not clear. If the drug binds only to the ordered structure it would be expected that the drug should stabilize such order and should antagonize the hyperchromicity which can be observed on heating denatured DNA. However, no appreciable effect on "melting" of "denatured" DNA could be detected. It was concluded, therefore that binding to denatured DNA occurs but is different from that to native DNA.

Table 2. *Chloroquine DNA interactions measured by centrifugation of the complex. These experiments were performed as described previously (*BLODGETT *and* YIELDING, *1968) by centrifuging the chloroquine-DNA complex at 40,000 rpm for 4 h from a buffer solution into a dense layer of sucrose in buffer with direct measurement of free ligand in the supernatant*

	Temperature	Association ($\times 10^4$)
Native DNA	15°	1.82 ± 0.12
	20°	1.84 ± 0.15
	25°	1.83 ± 0.13
	30°	1.80 ± 0.15
	30° + 1×10^{-3} M $MgCl_2$	0.32
Heated-recooled DNA	25°	1.63 ± 0.03

Binding of chloroquine to synthetic polynucleotides has also been reported as indicated by spectral changes (COHEN and YIELDING, 1965; ALLISON et al., 1965), inhibition of polymerase priming activity (COHEN and YIELDING, 1965), and stabilization of polynucleotide helical configuration (ALLISON et al., 1965). We have extended such studies to gain additional information on the specificity and mechanism of the interactions. As emphasized before (COHEN and YIELDING, 1965; BLODGETT and YIELDING, 1968), care must be exercised in comparing spectral studies for different polymers, because the change in drug absorption which occurs is a function both of the specific change per bound drug, and the total amount of binding which occurs. Thus, comparisons of 'strength' of association require extensive data. The data in Table 3 clearly indicate that significant binding occurs to both single stranded purine homopolymers and helical copolymers. The only pyrimidine examined in these studies

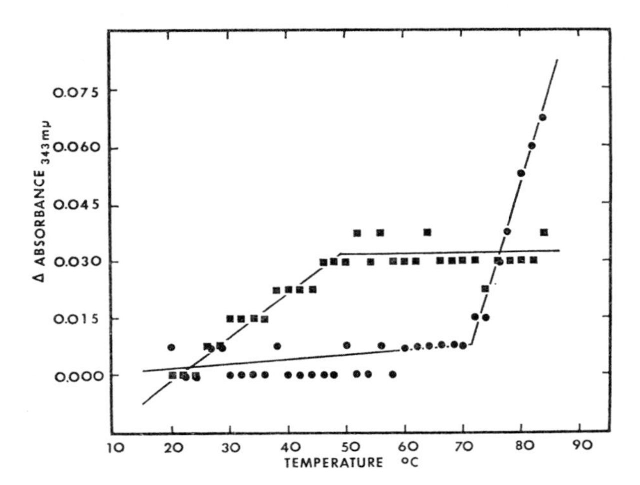

Fig. 1. The effect of temperature on release of chloroquine from chloroquine-DNA complex (● = native DNA, ■ = heated-recooled DNA). The increase in free chloroquine concentration was followed spectrophotometrically by the increase in OD at 343 mμ. The experiment employed 9×10^{-6} M DNA and 3×10^{-5} M chloroquine in 0.02 M phosphate buffer pH 6.0

Table 3. *Binding of chloroquine to polynucleotides. Binding parameters were calculated from least squares fits of spectrophotometric data from a minimum of two experiments to double reciprocal plots of drug bound/mole of DNA versus free drug. Experiments were done at 26°, using a chloroquine concentration of 4×10^{-5} M in 0.02 M phosphate buffer pH 7. For DNA the centrifuged values were obtained from Table 2 and those done by spectroscopy from* COHEN*'s paper* (COHEN and YIELDING, 1965)

Polymer	n drug/base	Ka
poly *A*	1	$2.40 \pm 0.32 \times 10^3$
poly *G*	1	$3.50 \pm 0.40 \times 10^3$
poly *dA*	1	$6.00 \pm 1.3 \times 10^3$
poly *dAT*	1	$3.20 \pm 0.13 \times 10^3$
poly *dG:dC*	2	$3.00 \pm 0.05 \times 10^3$
Native DNA (centrifuged)	—	$1.80 \pm 0.13 \times 10^4$
Heated-recooled DNA (centrifuged)	—	$1.63 \pm 0.03 \times 10^4$
Native DNA[a] (spectrophotometry)	0.22	1.10×10^4
	0.44	1.2×10^3
Heated-recooled DNA[a] (spectrophotometry)	0.15	0.98×10^4
	0.15	1.2×10^3

[a] Reproduced from COHEN and YIELDING (1965).

was poly *U* which showed insignificant binding at 26°, and limited binding at lower temperatures. These findings agree with our previous conclusion that strong binding of chloroquine is related to purine content, both *A* and *G*, of DNA (COHEN and YIELDING, 1965). Thus, this drug differs from the actinomycins both in binding specificity (REICH, 1964) and in its preferential inhibition of DNA polymerase (COHEN and YIELDING, 1965). The comparison of binding curves for poly *A* and poly *G* are particularly instructive (Fig. 2) since they illustrate that even when binding occurs to the same extent on both polymers, the magnitude of the spectral change is quite different. This is reasonable since the spectral change simply reflects the difference in the excited electronic state of the drug and does not provide any detailed information on the nature of the interaction.

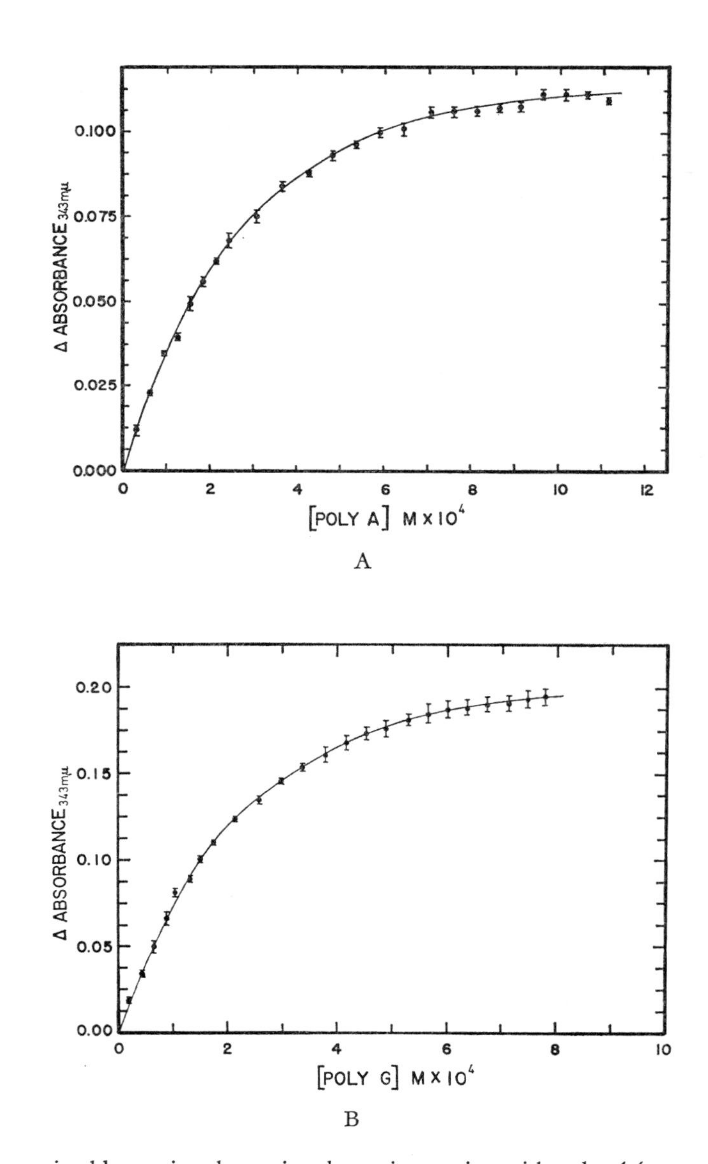

Fig. 2. Change in chloroquine absorption due to interaction with poly *A* (curve A) and poly *G* (curve B). Chloroquine concentration was 4×10^{-5} M in each experiment in 0.02 M phosphate buffer 7.0 at 26°

Although it may be inferred that a helical structure is not required for drug binding, single-stranded homopolymers like "denatured" DNA and synthetic homopolymers are not devoid of secondary structure, since extensive stacking interactions exist between the bases, particularly the purines (EPAND and SCHERAGA, 1967). It is particularly noteworthy that poly *U* showed limited binding at lower temperatures where it is ordered, but essentially none at 26° where there is little or no secondary structure (LIPSETT, 1960). A brief attempt was made to determine the chain length at which a polynucleotide begins to show strong interaction with the drug by employing a series of oligoadenylates. Fig. 3 compares the effects of $(A)_n$ ($n = 2-6$), and poly *A*

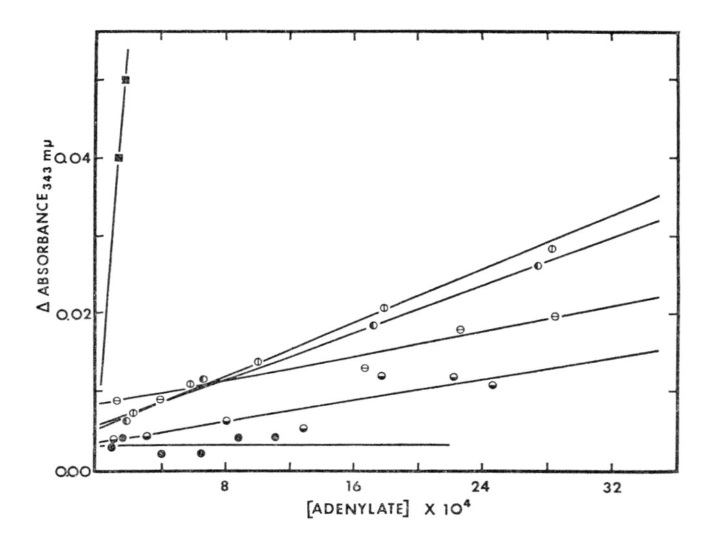

Fig. 3. Binding of various oligoriboadenylates (n = 2 to 6) and poly *A* to chloroquine. Experiments were done at 26° using 4×10^{-5} M chloroquine in 0.02 M phosphate buffer pH 7. Interaction was followed by the change in absorption on addition of the oligonucleotides. Nucleotide concentrations were based on the repeating unit. (● = *ApA*; ○ = $(Ap)_2A$; ⊖ = $(Ap)_3A$; ◐ = $(Ap)_4A$; ⦶ = $(Ap)_5A$; and ■ = poly *A*].

(mol. wt. ca. 100,000) on the drug spectrum. As shown, the spectral change was negligible for $(A)_2$ and increased progressively with chain length. These findings suggest that the extent of intramolecular order is important since substantial stacking should occur at a chain length of about 5 to 6 bases. It will be interesting to examine longer oligonucleotides both with homologous and heterologous base composition, since longer sequences would be required to form extensive regions of a stable base stacked conformation.

3. NMR Studies of Chloroquine-Nucleotide Interactions

The characteristic change in the drug's absorption spectrum resulting from binding to nucleic acids has suggested a specific interaction involving the drug's ring system. Absorption spectra, however, merely reflect the electronic environment of the molecule and do not give specific information about the type of interaction. For example, the change in chloroquine spectrum from binding to nucleic acids is not vastly different from that obtained on ionization, or in alcoholic solution. The base specificity involved in binding may be argued to provide some evidence for direct ring interaction, but such specificity could also result from spatial and steric considerations based on secondary structure. It was, therefore, decided to seek

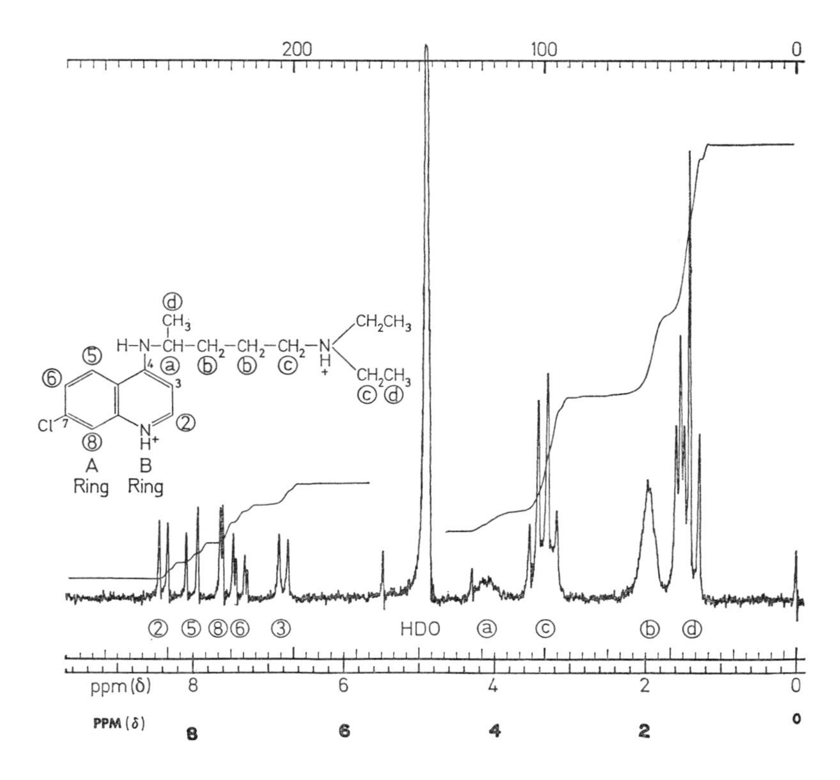

Fig. 4. NMR spectrum for chloroquine. (We are indebted to Dr. T. A. VICTOR for pointing out an error in our earlier assignment of protons 5 and 6. This has been corrected in the present spectrum)

specific evidence for ring-ring interactions between chloroquine and the nucleotide bases. For this purpose nuclear magnetic resonance spectroscopy is ideally suited since it permits observation of specific proton signals of each molecule involved. This technique, although ideal from the standpoint of specificity, is severely limited in sensitivity. Our initial studies have been carried out on mononucleotides (STERNGLANZ et al., 1969), and short chain oligonucleotides owing to the limited solubility of long chain polymers. These, of course, must be considered as models and may differ from interactions with macromolecular DNA.

The chemical shifts of proton signals in the NMR spectra of solutions containing both chloroquine and purine nucleotides are sufficiently different so that they can be clearly identified. Fig. 4 shows the spectrum for chloroquine with the proton assign-

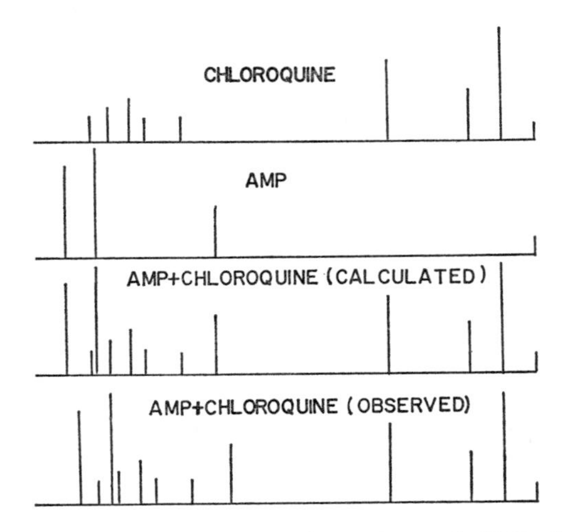

Fig. 5. NMR peak positions for spectra of chloroquine and AMP alone, in a theoretical mixture with no intermolecular interactions, and in an actual mixture

Table 4. *Changes in NMR spectrum of 0.058 M chloroquine from addition of nucleotides. Nucleotide concentrations were 0.071 M except for GMP, which was not completely soluble at this concentration. All spectra were done with a Varian A-60A spectrometer equipped with a variable temperature probe and the V-6040 variable temperature controller using an internal standard of 3-(trimethylsilyl) propanesulfonic acid. The experiments were performed at an apparent pH of 7.8 to 8.0 and a temperature of 41°. Differences for each proton are given as chemical shift (with respect to the standard) in the mixture minus the shift of the proton in chloroquine alone*

Plus	Choroquine							
	ΔCH_3-C ppm	ΔCH_2-C	ΔCH_2-N	ΔH-2	ΔH-3	ΔH-5	ΔH-6	ΔH-8
AMP	0	0	0	−0.08	−0.12	−0.13	−0.15	−0.13
GMP[a]	0	0	0	−0.07	−0.05	−0.05	−0.08	−0.08
UMP	0	0	0	0	0	−0.02	[b]	+0.02
CMP	0	0	0	−0.03	−0.05	+0.02	[b]	0

[a] Lower concentration. [b] Not discernible.

ments used. Fig. 5 is a simplified sketch of approximate peak positions for the drug and AMP singly, as would be expected in a non-interacting mixture, and as observed in the actual mixtures. The results for chloroquine in mixtures containing AMP, GMP, UMP or CMP are shown in Table 4. Both purines produced upfield shifts characteristic of ring stacking while the interaction with the pyrimidine was negligible. Table 5 shows the changes in the nucleotide spectra which resulted from addition of chloroquine. Again, changes in chemical shift were observed characteristic for ring stacking.

In view of the lack of binding of the drug to DNA at elevated temperature, NMR studies were also performed at different temperatures. Table 6 illustrates that the interaction was not measurable as the temperature was raised above 80 °C. Consequently, the ring-stacking component of the drug interaction does not appear at such elevated temperatures. The question was also raised as to what extent stacking

Table 5. *Effects of 0.058 M chloroquine on the NMR spectrum of nucleotides. Data expressed as the change in chemical shift from that of nucleotides alone*
(Reprinted from STERNGLANZ et al., 1969)

	$\Delta\delta$ of nucleotide protons on addition of chloroquine		
	$H_8(H_0)$	$H_2(H_5)$	H_1
AMP	−0.10	−0.15	−0.11
GMP	−0.10		−0.12
UMP	−0.03	−0.06	−0.03
CMP	−0.05	−0.06	−0.03

Table 6. *Effects of temperature on the chloroquine induced changes in the NMR spectrum of AMP. Experiments performed as in Table 4*
(Reprinted from STERNGLANZ et al., 1969)

	AMP + Chloroquine		
	H_8 $\Delta\delta$	H_2	H_1
11 °C	−0.14	−0.23	−0.20
41 °C	−0.10	−0.15	−0.11
93 °C	0	0	0

might occur between drug molecules, and between nucleotide molecules. To answer this question chemical shifts were examined as a function of concentration for chloroquine and for AMP singly. The upfield shift which occurred in each instance was indicative of ring-stacking at higher concentrations. However, the effects seen in solutions of the nucleotide and drug are not explained by such homogeneous stacking alone.

These experiments employing NMR demonstrate that ring-ring interactions can occur between purine nucleotides and chloroquine and also between chloroquine molecules. These interactions may provide the basis for the binding specificity which is seen with polynucleotides. Preliminary studies have also revealed that quinacrine stacks with AMP but not with CMP.

4. Effect of Drug Charge on DNA Binding

The charged nature of chloroquine has been shown to be important in its binding to nucleic acids. Its ionization properties, therefore, are of interest. It was concluded earlier that the drug binds as the doubly protonated species (COHEN and YIELDING, 1965). Since the *pK* for the ring protonation is 8.1, the characteristics of this prontonation are of particular interest at physiological pH. In the present studies it was found that a change in temperature from 25° to 65° resulted in a lowering of this *pK* by 1 pH unit. Thus, binding of the drug near physiological pH is quite sensitive to changes in temperature. At pH 7.1, for example, substantial binding should occur at 25°, but would be virtually abolished at 65°. This effect can be noted in the pH

Table 7. *The effect of pH on the stabilization of helical DNA by chloroquine. Experiments were done at the pH values indicated in 0.02 M phosphate buffer using 9 × 10^{-5} M DNA and 3 × 10^{-5} M chloroquine*

pH	*Tm* DNA alone	DNA-chloroquine
4.5	58.0	73.0
5.5	70.0	81.5
6.0	72.0	83.0
7.0	74.0	79.5
8.0	76.0	79.0

dependence of helix stabilization by the drug (Table 7). As expected, at pH values above 7, the drug is largely dissociated from the DNA before the transition temperature of the nucleic acid is reached and minimal stabilization is apparent.

5. A Proposed Model for Chloroquine Binding to DNA

In considering a physical model for the binding process, a number of features of the interaction must be accounted for. The drug binding: a) Exhibits a preference for purines in an ordered polynucleotide structure (COHEN and YIELDING, 1965). b) Is antagonized by divalent cations (COHEN and YIELDING, 1965). c) Causes a shift in drug spectrum probably due to ring-ring interactions (PARKER and IRVIN, 1952); and d) results in orientation of the drug molecules in the same plane as the bases (HAHN et al., 1966). As a consequence of binding there is stabilization of the DNA helix (COHEN and YIELDING, 1963, 1965; ALLISON et al., 1965), inhibition of the DNA primed polymerase reactions (COHEN and YIELDING, 1965; O'BRIEN et al., 1966), an increase in viscosity of the complexed DNA (HAHN et al., 1966), and a concomitent decrease in sedimentation velocity (HAHN et al., 1966).

Previous proposals have centered around the combined involvement of electrostatic forces and ring-ring interactions particularly involving purines (COHEN and YIELDING, 1965). The precise spatial orientation of the drug was not specified in most instances.

The most popular explanation for binding of heterocyclic drugs to DNA is that of intercalation between adjacent base pairs. Although certainly an attractive proposal, particularly in well studied cases such as acridine binding, the evidence supporting this mode of binding is not completely unambiguous. There is ample reason to consider alternate binding modes to extend our understanding of the processes involved. Intercalation of a drug is deduced mainly from three lines of evidence: (1) Greater chain length inferred from studies showing increased viscosity on binding (LUZZATI et al., 1961); (2) *X*-ray diffraction and scattering data of the spacing and of decrease in mass per unit length of DNA (LUZZATI et al., 1961; NEVILLE and DAVIES, 1966); and (3) unwinding by drug of superhelical closed circular DNA (BAUER and VINOGRAD, 1968). For intercalation to occur, the stacking interactions between adjacent

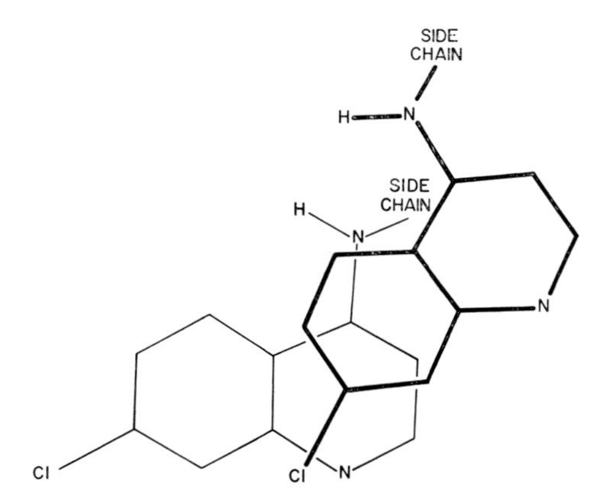

Fig. 6. Chloroquine binding model, relationships between chloroquine rings. Rotation represents that from stacking snugly into major groove of the *B* configuration of DNA

bases must be disrupted and rotations around multiple backbone bonds are required to open the helix for entry of the drug molecule. In the case of chloroquine, the observation that binding causes an increase in viscosity and a decrease in sedimentation velocity has prompted the proposal that the drug intercalates (HAHN et al., 1966). In addition to the question of whether intercalation can and does occur with a particular drug, the frequency of intercalation as compared with other modes of binding must be considered. For example, acridine binding has been divided into "internal" and "external" binding with the "internal" attributed to intercalation (LI and CROTHERS, 1969). Similarly, it has been proposed that actinomycin binds both by intercalation and by some other mechanism (MÜLLER and CROTHERS, 1968). Thus, intercalation has gained wide attention as a possible binding mode for a variety of drugs and has been proposed specifically for chloroquine (HAHN et al., 1966).

We have considered a possible binding model for chloroquine which might explain the characteristics of the binding on a basis other than intercalation. Our proposal, based on model building studies, takes special note of the tendency of heterocyclic systems in general, and of nucleotide and chloroquine specifically to participate in ring stacking (both self stacking and with other molecules). Our model would produce a minimal disruption of the native DNA, in contrast to that produced by extensive intercalation, and depends on drug-drug stacking into the major groove of DNA assisted by specific stacking interaction between the edges of the drug molecules and the edges of the nucleotides that project into the groove. Crystallographic studies of a variety of ring systems have revealed that there is a great tendency for stacking interactions to involve only limited overlap of the most polarizable or polarized groups on the molecules (BUGG et al.). Thus we have arranged the chloro-

Fig. 7. Chloroquine binding model. A close view of a space filling model with placement of drug molecules in the relationship shown in Fig. 6: (1) Ring nitrogen of chloroquine; (2) 7-Cl of chloroquine; and (3) 7-ring nitrogen of adjacent purine ring. The orientation of the DNA chains is such that the left hand chain proceeds in the 3′5′ direction from top to bottom, and the view is of the major groove. In the right lower corner, the side chain amino group may be seen in contact with the negative group on the phosphate backbone. The proton on the amino group has been omitted so that amino group may be seen more easily

quine molecules with the Cl projecting over the positively charged *B* ring as shown in Fig. 6. This permits stacking of a series of drug molecules in the major groove with additional contact for each drug molecule with an adjacent base in the DNA. Fig. 7 shows the precise fit that occurs with such an arrangement. Note that not only are the rings stacked in a highly favorable position both with respect to the drugs and the bases, but the charged amino group on the side chain of chloroquine is aligned to contact the negatively charged phosphate backbone. Close examination of this model also shows that the purine nucleotides provide the best contact with the drug. Purines also provide more favorable stacking interactions in simpler systems as shown experimentally (EPAND and SCHERAGA, 1967). This could mean that regions of DNA rich in purines would provide optimal stacking with both drug and nucleotide base, while those rich in pyrimidines would depend to a larger proportion on the drug-drug stacking. Fig. 8 of a larger region of DNA involved in binding shows

Fig. 8. Chloroquine binding model. Large scale view of space filling model with portion of major groove filled with stacked chloroquine molecules. The arrows identify contact between amino groups of chloroquine side chains and phosphate in the DNA backbone. Stars identify the chlorine on each drug molecule

the structure of a DNA-chloroquine complex which would result. The major groove is completely filled with the aliphatic side chains of chloroquine, and each phosphate group of one of the DNA strands is in close contact with the charged amino group of the side chain. This model, in agreement with previous proposals, stresses the important role both for electrostatic interactions and for ring-ring interactions (PARKER and IRVIN, 1952; LEVIN and PINKUS, 1961; ALLISON et al., 1965). It differs from previous discussions in that the binding process is pictured as a specific orientation of the drug in the major groove to permit optimum drug-drug and drug-nucleotide stacking without disruption of the normal positions of the base pairs. According to this reasoning stacking might also occur, though perhaps less favorably, with single stranded polymers. In fact, weak binding might also be expected to polyanions of the right spacing to permit chroloquine-chloroquine interactions, and such has been observed by COHEN and YIELDING for polydeoxyribosephosphate.

No direct evidence exists for the model proposed here but a brief consideration of the consequences of such binding are in order. Of particular interest, in addition to the specific orientation and space filling nature of the binding, is the change in the character of the major groove. Since H_2O would be displaced by less dense chloroquine molecules, there should be a decrease in density of the complex over that of free DNA as well as a considerable increase in tendency for side to side aggregation. The latter prediction is in keeping with the observation (PARKER and IRVIN, 1952) that highly complexed chloroquine-DNA is insoluble. The effect on the viscosity of DNA is more difficult to predict, but is critical for consideration of the model. Filling of the major groove by drug with displacement of water plus the consistent neutralization of charges on *one* of the backbones might be expected to produce considerable distortion of the shape of the helix in its progression around the screw axis. There would also be a considerable increase in rigidity. Each of these factors could result in an increase in viscosity. Major assumptions of shape, hydration, and porosity must be made for relating molecular length to viscosity (DRUMMOND et al., 1966). Thus, although an increase in viscosity might be predicted due to rigidity, shape, charge and porosity parameters, at this point it is impossible to predict the magnitude of such changes.

The biological consequences of the stacking model proposed here might be quite different from those resulting from intercalation. If intercalation does indeed permit frame shift mutations as has been proposed (BRENNER et al., 1961), there is an obvious limit to the usefulness of intercalating drugs in treating disease. In contrast, if DNA functions are merely blocked through occupation of a major portion of the DNA helix, considerably more leeway exists for the design of clinically useful drugs while avoiding the hazard of extensive mutagenicity.

We wish to propose, therefore, that the stacking model presented here be considered as a possible alternate (or additional) mode of drug-DNA interaction.

III. Biological Consequences of Chloroquine Binding to Nucleic Acids

It was proposed almost 20 years ago that binding of chloroquine to DNA might account for its biological properties (PARKER and IRVIN, 1952). Moreover, antimalarial effects were attributed to inhibition of nucleic acid synthesis in 1961 (SCHELLENBERG and COATNEY, 1961), although no mechanism was envisaged at the time. In 1963, COHEN and YIELDING reported that binding of the drug to the DNA template interfered with its ability to prime the DNA polymerase reaction, thus providing a basis for a specific bioinhibitory mechanism based on DNA binding (COHEN and YIELDING, 1964, 1965). These findings were confirmed for chloroquine and a similar inhibitory mechanism identified for quinacrine and quinine (O'BRIEN et al., 1966). There is correlation between DNA binding, inhibition of the polymerase and antimalarial effects. Other biological consequences of chloroquine which have been explained on the basis of DNA binding include inhibition of the *LE* cell phenomenon (LEVIN and PINKUS, 1961), the anti-DNA antibody reaction specifically (STOLLAR and LEVINE, 1963) and, more generally, the nonspecific inhibition of the immune response (ZELEZNICK et al., 1969). Although inhibition of cell proliferation in bacterial and tissue cultures (CIAK and HAHN, 1966; LIPP, 1964) can be produced, chloroquine is not particularly toxic to cells, and is not a potent inhibitor of cell division at the usual therapeutic dose levels.

Although not in the scope of the present discussion, it should be noted that chloroquine can also serve to inhibit cellular oxidative reactions by inhibiting the electron transport chain (MUSHINSKI et al., 1962; ARDUESSER and HEIM, 1967; SKELTON et al., 1968). This effect is being explored in some detail in our laboratory at the present time.

The question may be raised whether chloroquine and similar agents could be used as inhibitors of limited types of DNA synthesis (such as parasite replication) under circumstances where normal cell division proceeds. In the case of malaria, the differential effect to the parasite can be rationalized from the knowledge that the parasitized red cell is non-dividing, and furthermore that drug accumulates to substantial levels in these cells (HAHN et al., 1966). It may also be that the availability of DNA for drug binding varies over a wide range in the intracellular environment due to various factors such as the presence of RNA, protein, polyamines, and ions. If chloroquine can distinguish in a more subtle way between the DNA synthesis of normal cell division and various limited types of syntheses it may provide a major means of understanding such events, and lead to their selective control.

1. Inhibition of Virus Replication by Chloroquine

We have reported that replication of the DNA virus, $\phi X-174$, is inhibited by chloroquine and certain of its analogs under circumstances where growth of the *E. coli* host proceeds without interruption (YIELDING, 1967).

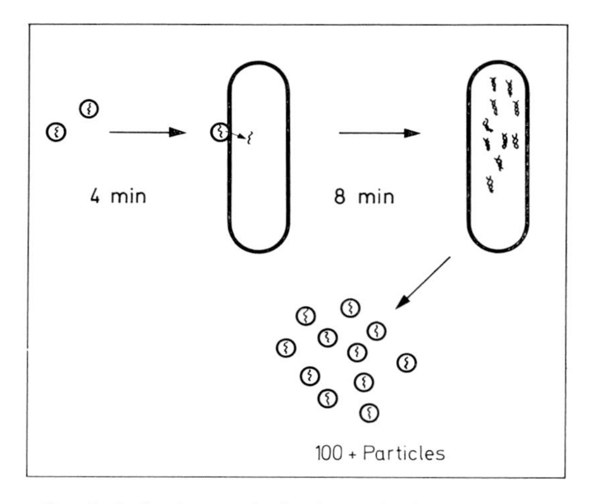

Fig. 9. Infection cycle for bacteriophage ϕ X-174

Thus, this system has been considered as a model for differential control of an intracellular parasite. Fig. 9 is a sketch of the phases in the infection of E. coli by $\phi X-$ 174. When chloroquine was added at varying times after infection (Table 8) an early post-absorptive phase in the cycle was inhibited. Furthermore, artificial lysis of cells following chloroquine inhibition confirmed that the drug was not merely interfering with lyses by the phage, since there was not a substantial increase in the phage content of intact cells. Table 9 shows a comparison of the inhibitory effect of chloroquine with that of several other drugs also found to be effective. These experiments suggest that DNA complexing drugs should be explored in detail as possible antiviral agents. Furthermore, insight into why the differential effect on viral synthesis is observed should assist in understanding the host parasite relationship.

Table 8. *Effect of the addition of chloroquine (5×10^{-4} M) at various times after infection on the production of bacteriophage ϕX to 174*

	Increase in virus titer at 30 min
Control	40 ×
Chloroquine at 0 time	10 ×
Chloroquine 4 min after infection	11 ×
Chloroquine 7 min after infection	24 ×

Table 9. *Effects of various antimalarial drugs as inhibitors of replication bacteriophage ϕX-174. Drugs were added at the time of infection*

	Concentration to achieve 50 to 75% Inhibition
Chloroquine	5×10^{-4}
Propoquine	2×10^{-4} M
Camoquin	5×10^{-4} M
Primaquine	5×10^{-4} M
Quinacrine	1×10^{-3} M

2. Inhibition by Chloroquine of 'Excision' Repair (Mutation Repair) and Possible Biological Consequence

Most living cells appear to have the capacity to recognize, excise, and replace certain types of induced damage (chemical alteration) to their DNA by a "cut and patch" repair mechanism. Agents whose effects are repaired include X-rays, alkylating agents, UV light, and certain antibiotics (HANAWALT and HAYNES, 1965; CRATHORN and ROBERTS, 1966; PAINTER and CLEAVER, 1967; RASMUSSEN and PAINTER, 1964).

The repair apparently consists of four steps: a) chain opening; b) excision of defective region; c) reinsertion of appropriate bases using the complementary strand as a template and d) rejoining of the backbone of the repaired sequence. According to our present concepts, the net result of this process is to convert chemically altered DNA back to a biologically active form which is presumably identical to that existing prior to this damage. Since this process involves at least three and possibly four enzyme steps, including limited resynthesis of a short segment, it was of interest to determine whether chloroquine acts as an inhibitor by virtue of its DNA binding properties, and thus provides a tool for manipulating this biological mechanism. These experi-

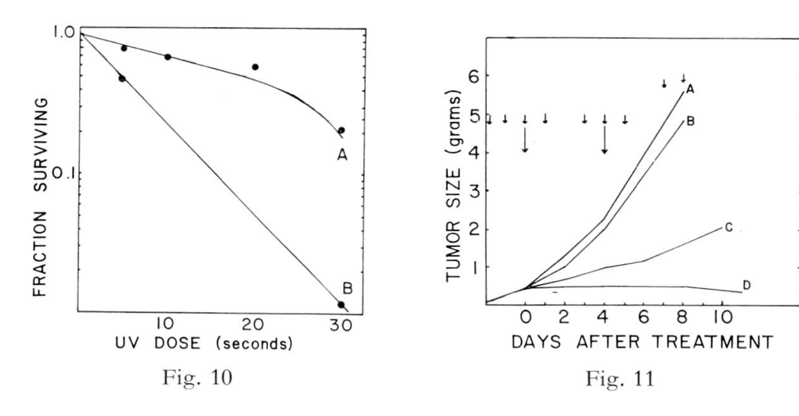

Fig. 10. Effect of 5×10^{-4} M chloroquine on survival of E. coli *B* on minimal medium following UV irradiation (253 mμ) for time intervals shown. Log of surviving fraction is plotted on the *Y* axis. ((Top curve = control; bottom curve = chloroquine)

Fig. 11. Effects of chloroquine on the response by transplanted melanomas in hamsters to treatment with phenylalanine mustard. Each curve represents the growth progress of transplanted melanomas in at least 6 hamsters. Curve A = no treatment; curve B = chloroquine alone; curve C = phenylalanine mustard alone at 3/kg on days indicated by large arrows; and curve D = mustard and chloroquine 30 mg/kg on days indicated by the small arrows

ments have employed both bacterial cells (Yielding et al., 1970) and human lymphocytes. Fig. 10 illustrates the effect of chloroquine on the UV sensitivity of a strain of E. coli B which possesses the "cut and patch" system. The shoulder region seen on the UV killing curve when cells are grown on minimal media may be presumed evidence for an active repair process (Haynes, 1966). Chloroquine abolished this region and converted the killing curve to a simple exponential function. This drug effect was confirmed by the additional finding that chloroquine interfered with the time-dependent increase in survival of UV irradiated cells that occurred on holding the irradiated cells in non-nutrient medium prior to growth (Yielding et al., 1970).

Chloroquine inhibition of the repair mechanism was further studied in human lymphocytes by measuring the uptake of tritiated thymidine following UV irradiation under circumstances where normal DNA synthesis did not occur. A concentration of 1×10^{-4} M chloroquine inhibited this uptake.

Finally we used chloroquine as a tool to explore the repair process in determining the sensitivity or resistance by certain tumor cells to the cytotoxic effects of alkylating agents and *X*-ray. Since the effects of both *X*-ray and alkylating agents are subject to repair, it was speculated that the DNA repair mechanism may play a role in the resistance of tumor cells to such therapy. If this were correct, it should be possible to manipulate cytotoxicity through selective control of the repair enzymes. Thus, an inhibitor of repair such as chloroquine might be expected to enhance the sensitivity of cells to both alkylating agents and *X*-rays. The effects of chloroquine have been explored in three lines of transplantable tumors in hamsters. Our first experiments

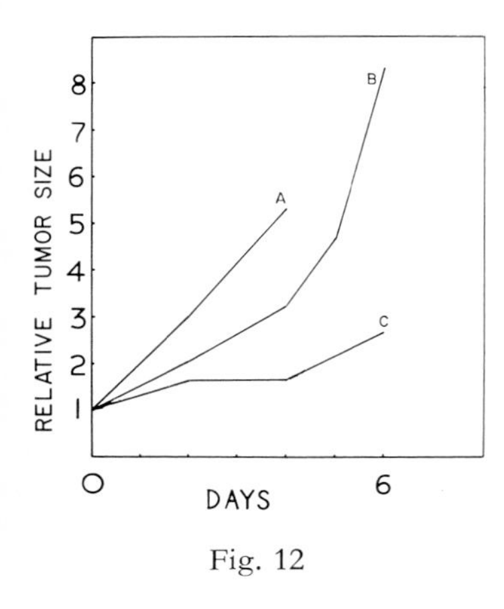

Fig. 12

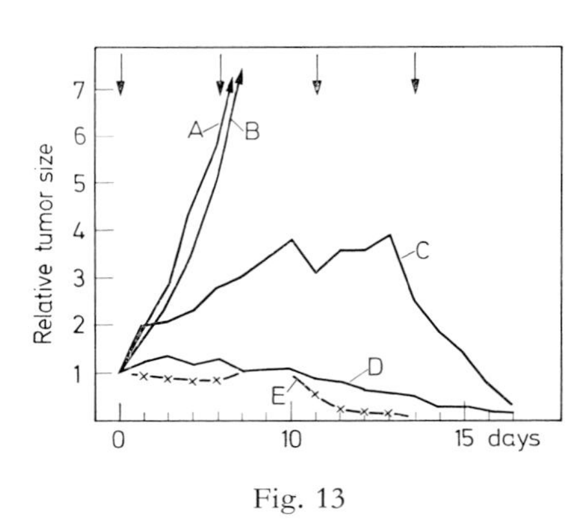

Fig. 13

Fig. 12. Effects of chloroquine, 25 mg/kg/day on the response of transplanted melanomas in hamsters to 800 r X-ray. Each curve represents combined results on at least 6 animals: Curve A = no treatment; curve B = X-ray alone; curve C = X-ray and chloroquine 30 mg/kg

Fig. 13. Effects of chloroquine on response by transplanted plasmacytomas in hamsters to cytoxan. Each curve represents at least 6 animals: Curve A = no treatment; curve B = chloroquine alone; curve C = cytoxan, 6.5 mg/kg on days indicated by arrows; and D = cytoxan plus chloroquine 30 mg/kg

employed a plasmacytoma, resistant to cytoxan at 25 mg/kg or to 800 *r* of *X*-ray. These tumors showed substantial regression in size when chloroquine was added to the treatments at 30 mg/kg (GAUDIN and YIELDING, 1969). Similar experiments are shown in Fig. 11 and 12 for a melanoma line using phenylalanine mustard or *X*-rays as the treatment agent. Again, a substantial effect was observed from addition of chloroquine to the treatment schedule, while chloroquine alone failed to produce a response. Experiments were then performed using a sub-optimal dose of cytoxan in a sensitive line of plasmacytoma. Fig. 13 shows that chloroquine enhanced the effectiveness of cytoxan in this system also. Although it cannot be concluded from these observations that the chloroquine effect resulted from repair inhibition, there is a suggestion that repair was an important parameter in determining tumor response. It was particularly interesting (if somewhat surprising) to find no increase in cytoxan host-toxicity from chloroquine administration as indicated by the peripheral white blood cell count and the cytology of bone marrow in mice.

These experiments, therefore, suggest a new approach to chemotherapy through selective modification of the cell mechanisms which determine resistance to cytotoxic agents.

IV. Summary

1. Chloroquine binds to DNA and other polynucleotides. Binding involves a preferential affinity for and ring interaction with purine bases; it is most extensive for ordered polynucleotide structures, is antagonized by divalent cations, is specific for the protonated form of the drug and stabilizes the double-helical form of DNA.

2. As a consequence of binding chloroquine inhibits the involvement (both substrate and template) of DNA in enzymatic reactions.

3. A cooperative binding model has been proposed in which chloroquine stacks into the major groove of DNA by electrostatic interaction between the side chain amino and backbone phosphate groups, and by ring stacking both between adjacent chloroquine molecules and between chloroquine and the bases as they protrude into the major groove.

4. A number of biological consequences may be attributed to chloroquine-DNA interactions. In addition to the well-known antiplasmodial effect we have demonstrated inhibition of bacterial virus replication and of normal excision (cut and patch) repair of DNA damage in bacterial and mammalian cells. The drug has been used to modify the effectiveness of *X*-ray and drug therapy of certain experimental tumors. The results illustrate rational approaches to drug therapy which can be derived from an understanding of the molecular mechanisms of drug action.

References

Allison, J. L., O'Brien, R. L., Hahn, F. E.: Science **149**, 1111 (1965).
— — — Antimicrobial Agents and Chemotherapy **1965**, 310.
Arduesser, G. A., Heim, H. C.: J. pharm. Sci. **56**, 254 (1967).
Bauer, W., Vinograd, J.: J. molec. Biol. **33**, 141 (1968).
Blodgett, L. W., Yielding, K. L.: Biochim. biophys. Acta (Amst.) **169**, 451 (1968).
Brenner, S. Barnett, L., Crick, F. H. C., Orgel, A.: J. molec. Biol. **3**, 121 (1961).
Bugg, C. E., Thomas, J. M., Sundaralingam, M., Rao, S. T.: Biopolymers (in press).
Ciak, J., Hahn, F. E.: Science **151**, 347 (1966).
Cohen, M. D., Fischer, F.: J. chem. Soc. 3044 (1962).
Cohen, S. N., Yielding, K. L.: Arth. and Rheum. **6**, 767 (1963); **7**, 302 (1964).
— — J. biol. Chem. **240**, 3123 (1965).
— — Proc. nat. Acad. Sci. (Wash.) **54**, 521 (1965).
Crathorn, A. R., Roberts, J. J.: Nature (Lond.) **211**, 150 (1966).
Drummond, D. S., Pritchard, N. J., Simpson-Gildemeister, Peacocke, A. R.: Biopolymers **4**, 971 (1966).
Epand, R. M., Scheraga, H. A.: J. Amer. chem. Soc. **89**, 3888 (1967).
Gaudin, D., Yielding, K. L.: Proc. Soc. exp. Biol. (N.Y.) **131**, 1413 (1969).
Hahn, F. E., O'Brien, R. L., Ciak, J., Allison, J. L., Olenick, J. G.: Milit. Med. **131**, 1071 (1966).
Hanawalt, P., Haynes, R. H.: Biochem. biophys. Res. Commun. **19**, 462 (1965).
Haynes, R. H.: Radiat. Res. Suppl. **6**, 1 (1966).
Irvin, J. L., Irvin, E. M., Parker, F. S.: Science **110**, 426 (1949).

Klotz, I. M., Urquhart, J. M.: J. Amer. chem. Soc. **71**, 847 (1949).
Kurnick, N. B., Radcliffe, I. E.: J. Lab. clin. Med. **60**, 669 (1962).
Levin, M. B., Pinkus, H.: New Engl. J. Med. **264**, 533 (1961).
Li, H. J., Crothers, D. M.: J. molec. Biol. **39**, 461 (1969).
Lipp, R.: Arch. klin. exp. Derm. **218**, 228 (1964).
Lipsett, M. N.: Proc. nat. Acad. Sci. (Wash.) **46**, 445 (1960).
Lloyd, P. H., Prutton, R. N., Peacocke, A. R.: Biochem. J. **107**, 353 (1968).
Luzzati, V., Masson, F., Lerman, L. S.: J. molec. Biol. **3**, 634 (1961).
Muller, W., Crothers, D. M.: J. molec. Biol. **35**, 251 (1968).
Mushinski, J. F., Yielding, K. L., Munday, J. S.: Arth. and Rheum. **5**, 118 (1962).
Neville, D. M., Jr., Davies, D. R.: J. molec. Biol. **17**, 57 (1966).
O'Brien, R. L., Olenick, J. G., Hahn, F. E.: Proc. nat. Acad. Sci. (Wash.) **55**, 1511 (1966).
Painter, R. B., Cleaver, J. E.: Nature (Lond.) **216**, 369 (1967).
Parker, F. S., Irvin, J. L.: J. biol. Chem. **199**, 897 (1952).
Rasmussen, R. E., Painter, R. B.: Nature (Lond.) **203**, 1360 (1964).
Reich, E.: Science **143**, 684 (1964).
Scatchard, G., Coleman, J. S., Shen, A. L.: J. Amer. chem. Soc. **79**, 12 (1957).
Schellenberg, K. A., Coatney, G. R.: Biochem. Pharmacol. **6**, 143 (1961).
Skelton, F. S., Pardini, R. S., Heidker, J. C., Folkers, K.: J. Amer. chem. Soc. **90**, 5334 (1968).
Sternglanz, H., Yielding, K. L., Pruitt, K. M.: Molec. Pharmacol. **5**, 376 (1969).
Stollar, D., Levine, L.: Arch. Biochem. **101**, 335 (1963).
Yielding, K. L.: Proc. Soc. exp. Biol. (N.Y.) **125**, 780 (1967).
— Yielding, L., Gaudin, D.: Proc. Soc. exp. Biol. (N.Y.) **133**, 999 (1970).
Zeleznick, L. D., Crim, J. A., Gray, G. D.: Biochem. Pharmacol. **18**, 1823 (1969).

A Preliminary Nuclear Magnetic Resonance Study of the Effect of pH on the Structure of Chloroquine Diphosphate

THOMAS A. VICTOR, FRED E. HAHN and ELIZABETH A. HANSEN

I. Introduction

Chloroquine diphosphate (CD), a synthetic drug discovered by ANDERSAG (ANDERSAG, BREITNER and JUNG, 1939), has a number of interesting biological properties. One of its most important uses has been as a therapeutic drug in the treatment and

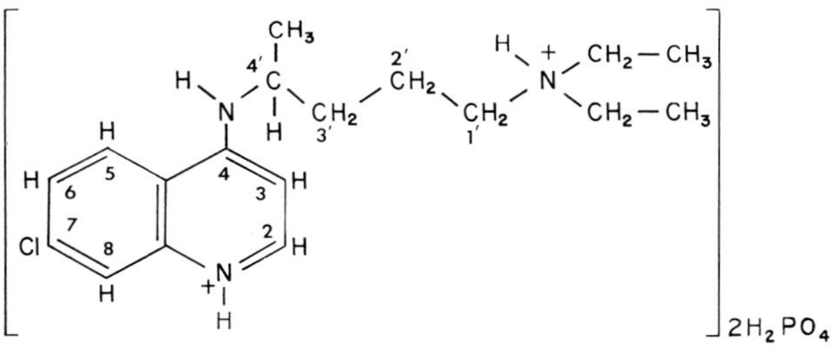

Fig. 1. Structure of CD

prevention of malaria (COATNEY, 1963). However, strains of the malarial parasite, *Plasmodium* falciparum, have recently proven to be almost completely resistant to CD as well as to other antimalarial drugs (WHO, 1964). Solving the problem concerning CD-resistant malaria involves understanding the basis of (1) the chemical and physical nature of the drug's antimalarial activity and (2) *Plasmodium* falciparum's resistance to CD. The physical and chemical properties of CD are now under investigation in our laboratories.

The chemical structure of CD has been established (ANDERSAG et al.), Fig. 1, but the conformation of the molecule in solution is not known. Chloroquine itself consists of a quinoline chromophore with a chlorine atom attached at position 7 of the ring and a diamino aliphatic chain attached at position 4. All of these structural features have been implicated in the drug's antimalarial activity (O'BRIEN and HAHN, 1966), but their precise roles are not yet understood.

The mode of action of chloroquine has been investigated in micro-organisms including *Plasmodia*. The antimicrobial action of the drug has been attributed to its inhibition of both RNA and DNA synthesis (HAHN, O'BRIEN, CIAK, ALLISON and OLENICK, 1966). *In vitro* experiments also indicate that DNA and RNA polymerase reactions are inhibited (O'BRIEN, OLENICK and HAHN, 1966) in the presence of the drug. One explanation which accounts for all of these effects is an interaction of CD with DNA (ALLISON, O'BRIEN and HAHN, 1965; PARKER and IRVIN, 1952; COHN and YIELDING, 1965).

Optical and hydrodynamic measurements provide evidence for the occurrence of such interactions *in vitro* (HAHN et al., ALLISON et al., PARKER et al., COHN et al.). The measurements indicate that CD forms a complex with DNA which is highly

Table 1. *Chemical shifts and coupling constants of CD in CF_3COOH*[a]

	δ	Type	J Coupling constants
Aromatic protons			
C(2)H	8.31	23	7.4
C(3)H	6.84	23	7.4
C(5)H	8.18	56	9.2
		58	0
C(6)H	7.70	56	9.2
		68	1.9
C(8)H	7.89	58	0
		68	1.9
Diethylamino Protons			
CH_2	–(3.28)[b]	H_2CCH_3	3.7
CH-	1.45	H_2CCH_3	3.7
Aliphatic protons			
C(1′)H	–(3.28)[b]	—	—
C(2′)H	–(2.02)[b]	—	—
C(3′)H	–(2.02)[b]	—	—
C(4′)H	4.16	HCNH	3.8
		$HCCH_3$	2.7
$C(4')CH_3$	1.55	$HCCH_3$	2.7
NH	7.22	HCNH	3.8

[a] Chemical shifts (δ) are measured in ppm and are accurate to $\pm$ 0.01 ppm. Coupling constants (J) are measured in Hz and are accurate to $\pm$ 0.1 Hz. All measurements were made on 0.3 M solutions of CD in CF_3COOH at 34 °C.

[b] Since the chemical shifts for these protons cannot be reported, the center of the absorption bands in which the proton signals are located is reported in parentheses.

specific for native, double-stranded DNA containing guanine (HAHN et al.). The structural requirements for CD-DNA complexes have been further defined by studying binding between chloroquine analogues and DNA (O'BRIEN and HAHN, 1966). On the basis of these studies a hypothetical structure for the CD-DNA complex has been proposed (O'BRIEN and HAHN).

In this structure the quinoline ring system is intercalated between the base pairs of double-stranded DNA, and the aliphatic diamine stabilizes the complex by spanning the minor groove and electrostatically binding to the phosphate oxygen atoms. Structural features of CD which correlate with high antimalarial activity and fit well with this type of complex include (1) a planar, heterocyclic ring with an area of 28 Å^2 (consistent with intercalation and ring-stacking interactions) and (2) a diamino aliphatic side-chain with a distance of 7.5 Å separating the nitrogen atoms.

The next step in determining the precise structural requirements for binding of CD to DNA and for antimalarial activity is to establish definitively the conformation of CD and CD-DNA complexes. Proton magnetic resonance spectroscopy (PMR) has recently been used to study the structure of actinomycin D-nucleic acid complexes in solution (Victor, 1970; Danyluk and Victor, 1970). This method can be used in a similar manner to study CD-DNA complexes.

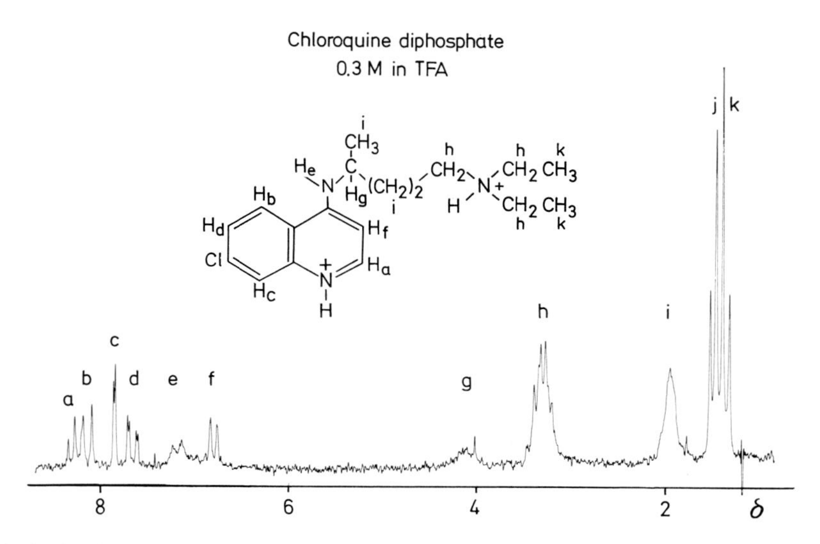

Fig. 2. A single resonance proton spectrum of CD in CF_3COOH at 34 °C. Chemical shifts are in ppm to low-field of internal tetramethyl silane (TMS)

Since the pH dependence of the optical spectrum of CD (Irvin and Irvin, 1947) indicates that the molecule behaves as both a strong base and a weak base, the cationic species of CD interacting with DNA will definitely be established by the pH of the aqueous solution in which the study is performed. Before any attempt to determine what ionic species of CD interacts with DNA, the dissociation properties of the molecule must first be clearly defined.

PMR has been successfully utilized to study the protonation of N-heterocyclic molecules (Blears and Danyluk, 1967). The preliminary results of a similar investigation of CD as a weak base are reported in this paper.

II. The PMR Spectra of Chloroquine Diphosphate

It is necessary to analyze the PMR spectrum of CD before studying its protonation reactions. The PMR spectrum of CD at 60 MHz in D_2O has been previously reported

(STERNGLANZ, YIELDING and PRUITT, 1969); however, pD effects were not investigated. We have re-examined the spectrum of CD in D_2O at varying pD values and in the strongly protonating solvents CF_3COOD and CF_3COOH. Using multiple resonance and deuterium exchange, we were able to interpret the entire spectrum of CD

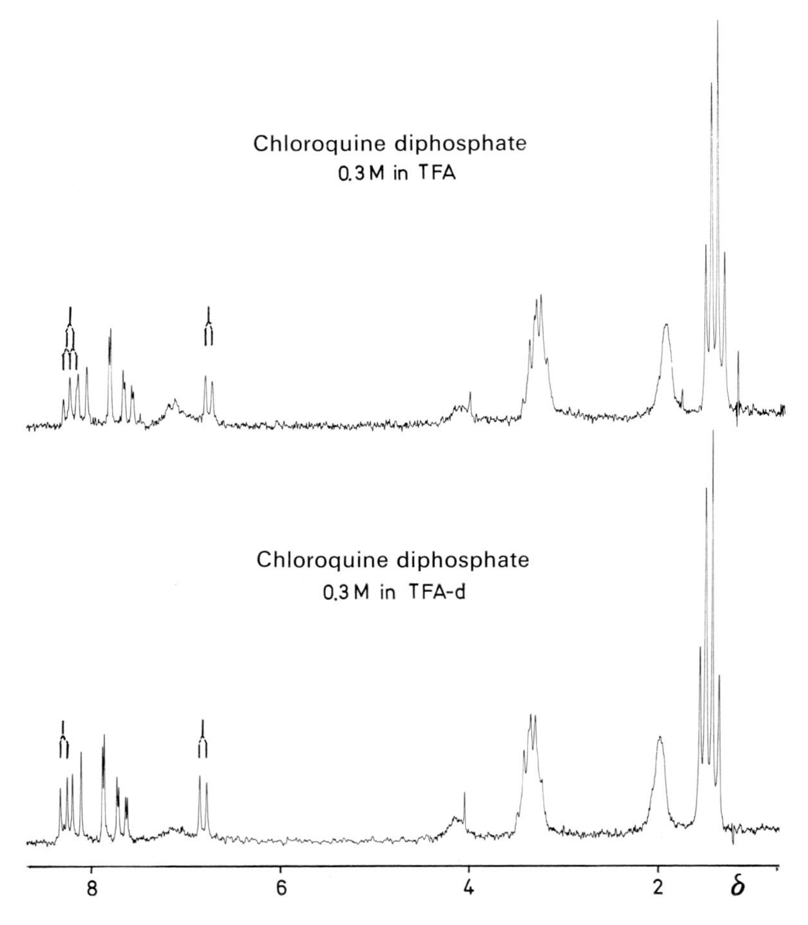

Fig. 3. Single resonance proton spectra of CD in CF_3COOH and in CF_3COOD at 34 °C demonstrating those signals affected by the presence of deuterium. Chemical shifts are in ppm relative to internal TMS

in these solvents with the exception of the details of the bands corresponding to the methylene protons in the aliphatic chain. Table 1 summarizes the PMR parameters for a solution of CD in CF_3COOH; the single-resonance spectrum at 100 MHz is shown in Fig. 2. The spectrum is quite similar to that previously reported for CD in D_2O. The assignment of the carbocyclic protons of the quinoline ring and their treatment as an ABX spin system differ from those previously reported (HAIGH, PALMER and

SEMPLE, 1965). While a complete discussion of the assignments will be published elsewhere, it might be well to discuss, briefly, several points here.

PMR parameters for exchangeable protons could be determined by analyzing a CF_3COOH solution of CD. Thus the 4-amino proton gives rise to an observable doublet; decoupling experiments reveal that this splitting arises from spin-coupling with the $C_{4'}$ proton. In CF_3COOD this proton exchanges completely in about 24 h.

Although optical studies indicate that the diethylamino nitrogen atom and the heterocyclic nitrogen atom should be protonated in strong acid (IRVIN et al.), signals

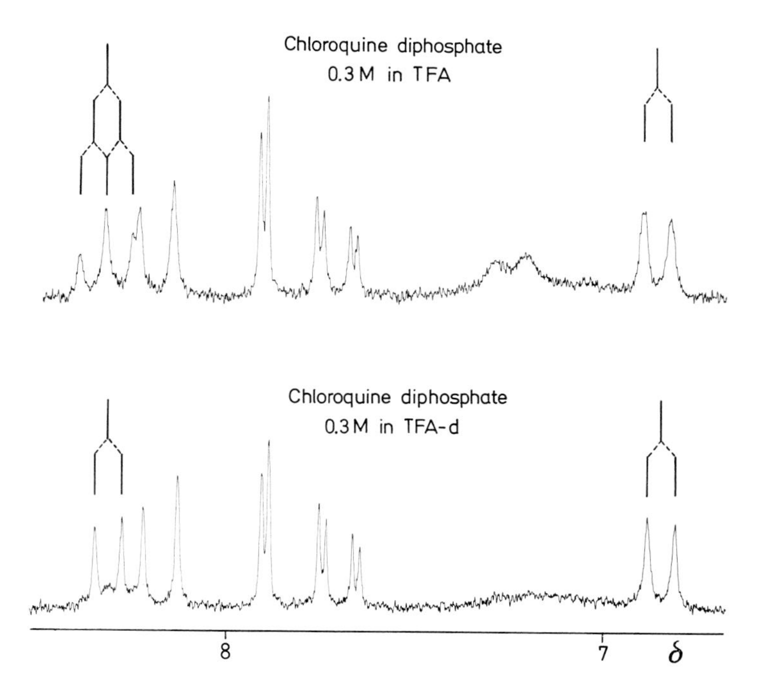

Fig. 4. Low-field regions of the single resonance spectra of CD shown in Fig. 3. Chemical shifts are in ppm relative to internal TMS

for these protons do not appear in the CF_3COOH spectrum. However, the C_2 proton of CD gives rise to three lines in the spectrum of a CF_3COOH solution which collapse to a doublet in the spectrum of a CF_3COOD solution; this is shown in Fig. 3 and 4. Since this observation can only be explained on the basis of deuterium-proton exchange, the additional splitting of the C_2 proton signal in CF_3COOH solution must arise as a result of spin-coupling with a proton on the heterocyclic nitrogen atom which is exchanging slowly with the solvent. Decoupling experiments indicate that the C_2 and C_3 protons behave as a first order spin system; it should then be possible to decouple the three line C_2 proton band by irradiation of the proton signal of CF_3COOH. A double resonance experiment resulted in the collapse of the C_2 proton multiplet to a doublet, in line with the above expectation.

As pointed out previously, a band corresponding to the diethylamino proton is not observed for CD in CF_3COOH solution. Although the residence of a proton on the diethylamino nitrogen might be detected by the proton's coupling to vicinal

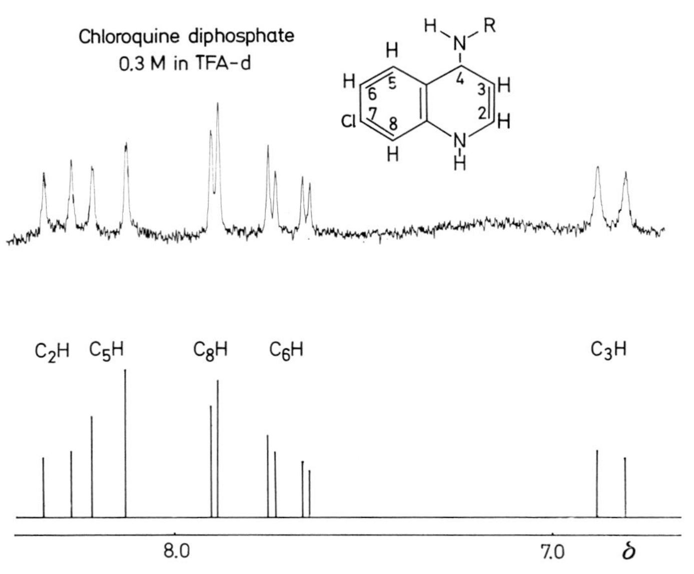

Fig. 5. Upper trace, low-field region of the single resonance spectrum of CD in CF_3COOD at 34 °C; lower trace, calculated spectrum. Chemical shifts are in ppm relative to internal TMS

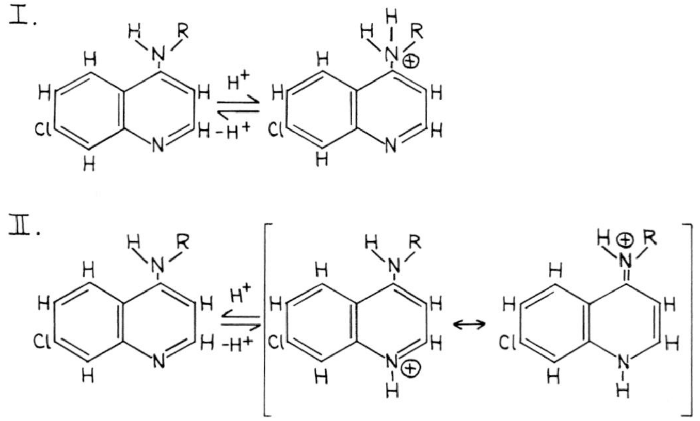

Fig. 6. Possible protonation reactions of CD with a pK of 8.06

methylene protons, there is no apparent difference in signals due to the methylene protons attached to this nitrogen for the spectra of CD in CF_3COOH and CF_3COOD, Fig. 3. That no coupling is resolved is probably due to very rapid proton exchange at this site.

The assignments of the C_5, C_6, and C_8 protons should also be mentioned. These protons are treated as an ABX spin system. Only eight lines appear in the spectrum for these protons; the simple line pattern is due to one J being approximately zero. Then $J_{ortho} = J_{56} = 9.0$ Hz and $J_{meta} = J_{68} = 1.9$ Hz, and the carbocyclic ring protons can be assigned as shown in Fig. 5. Since the spectrum calculated with the above

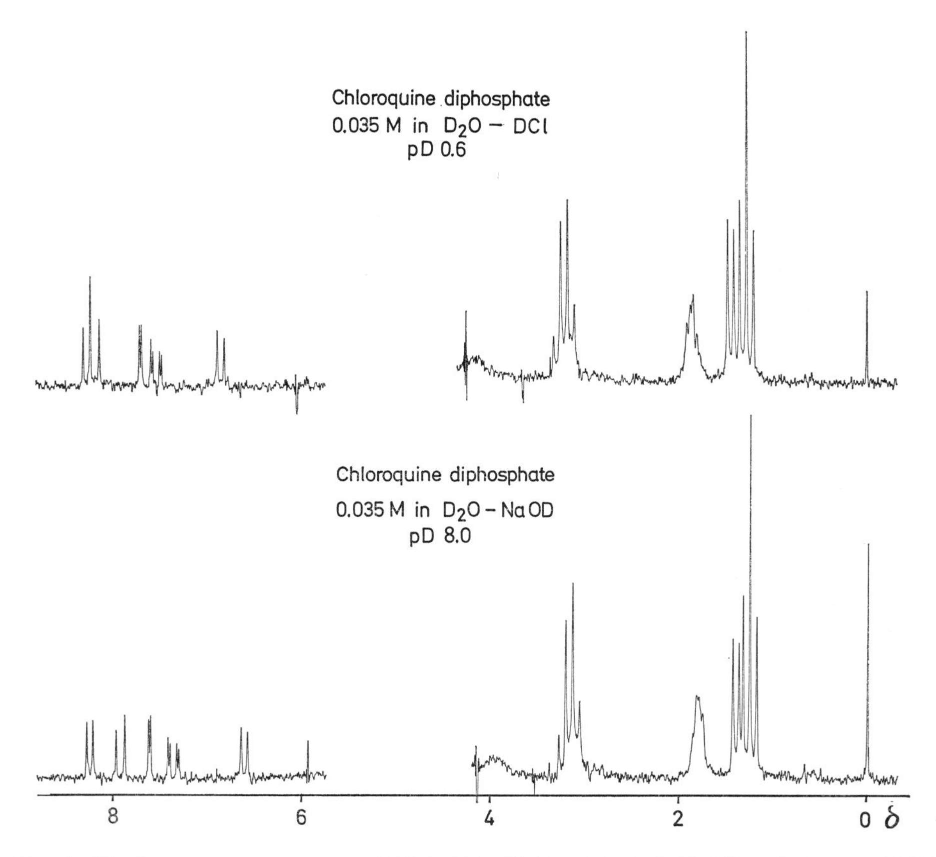

Fig. 7. Single resonance spectra of CD in D_2O-DCl and in D_2O-NaOD solutions at 34 °C demonstrating those signals affected by different pD values. Chemical shifts are in ppm relative to internal TMS

δ and J values is consistent, within experimental error, with the observed spectrum, we consider this interpretation and assignment of the spectrum the most likely.

III. Deductions Regarding the pD Dependence of the CD Spectrum

The pH dependence of the optical spectrum of CD in the range 0 to 14 has been reported previously (Irvin et al.) and has been used to derive its apparent acid dissociation constants. The dissociation with a pK of 10.1 has been attributed to a proton reaction involving the 1-diethylamino nitrogen atom of the aliphatic side-chain. A second proton reaction is also measured in this pH range with an apparent pK of 8.06. Of the remaining two acceptor-centers (4-amino nitrogen and heterocyclic nitrogen atoms) of the aromatic nucleus, it has been hypothesized that protonation

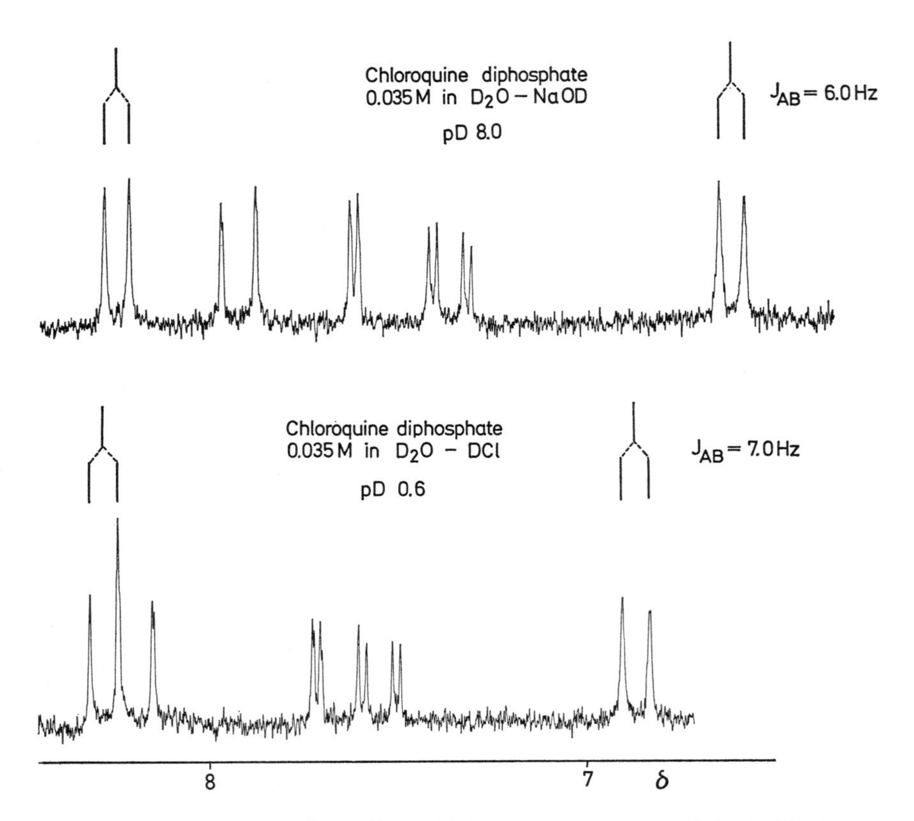

Fig. 8. Low-field regions of the single resonance spectra of CD in Fig. 7

occurs at the ring nitrogen atom and that the resulting cation is stabilized as a resonance hybrid (IRVING et al.). The two possible protonation reactions with a pK of 8.06 are shown in Fig. 6.

Analysis of the PMR spectrum of CD in CF_3COOH discussed in the previous section indicates that protonation occurs at the heterocyclic nitrogen atom as has been predicted from the optical studies (IRVIN et al.). The pD dependence of various PMR parameters provides additional information about the resonance hybrid.

Since the spectrum of CD in D_2O is concentration-dependent, solutions of varying pD's were all made 0.035 M with respect to CD. Fig. 7, 8, 9, and 10 show the pD dependence (the graphs depict a deprotonation) of the chemical shifts for the quinoline ring system protons and the $C_{4'}$ methyl protons of CD. The sigmoid curves characteristic of proton ionization are not observed because shift changes above pD 8.2 (ionization of CD as a weak base has a pK of 8.06) at 0.035 M CD cannot be measured: the solubility of CD decreases sharply in deuterium oxide above pD 8.2. However, the ring and aliphatic proton signals are shifted upfield with increasing pD values. Since the shift changes are greatest in magnitude close to the pK of the ring deprotonation (or protonation), presumably they are associated with this reaction, Fig. 9 and 10. The magnitudes of the downfield shifts for the ring protons as pD decreases (protonation) vary in the order $\Delta\delta C_3H > \Delta\delta C_5H > \Delta\delta C_6H > \Delta\delta C_8H > \Delta\delta C_2H \approx O$.

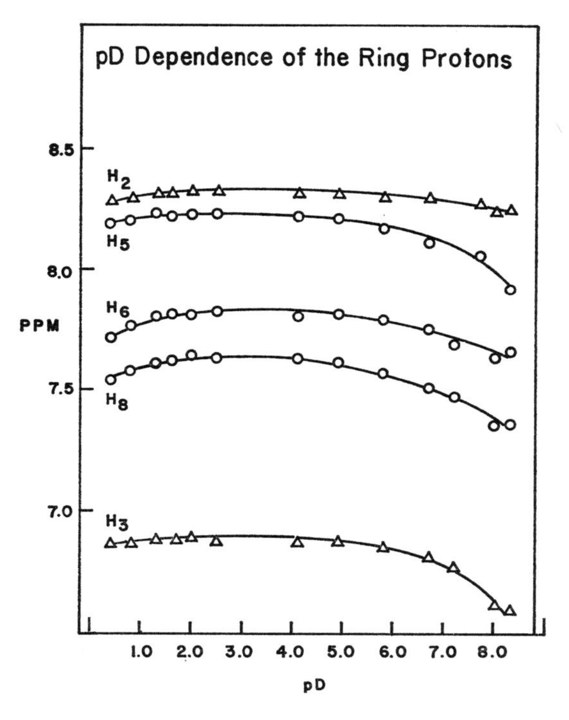

Fig. 9. Proton chemical shifts of the aromatic protons of CD in D_2O-NaOD and in D_2O-DCl solutions at 34 °C vs pD. Chemical shifts are in ppm relative to internal TMS

As discussed above, the downfield shifts of the ring proton signals and the $C_{4'}$ methyl proton signal can be attributed to protonation of the heterocyclic ring nitrogen atom (Blears et al.). The small shift of the C_2 proton signal in contrast to the larger downfield shifts of the other ring proton signals is somewhat surprising; however, this lack of deshielding has been observed in other heterocyclic molecules and with formation of the pyridinium cation (Smith and Schneider, 1961). This result suggests that the protonated heterocyclic ring exists as a pyridinium-like cation rather than a quinolimine, because the C_2 proton chemical shift would be expected to approach a value similar to that of an analogous proton in a vinylamine [e.g. δC_2H for pyrrole as compared to δC_2H for pyridine (Jackman and Sternhell, 1969)], and this change is not observed with protonation. The slow exchange rate of the 4-amino proton of CD in CF_3COOH solution also supports the hypothesis that the heterocyclic ring exists as a pyridinium-like cation rather than a quinolimine cation.

Qualitatively, the downfield shifts observed for protons in the cation indicate a delocalization of the charge deficiency to the carbocyclic and heterocyclic rings as well as to the 4-amino nitrogen atom (IRVIN et al.; DANYLUK and SCHNEIDER, 1962). Certain structural implications arise with this conclusion. In order for the charge deficiency of the aromatic cation to delocalize partially at the 4-amino nitrogen atom,

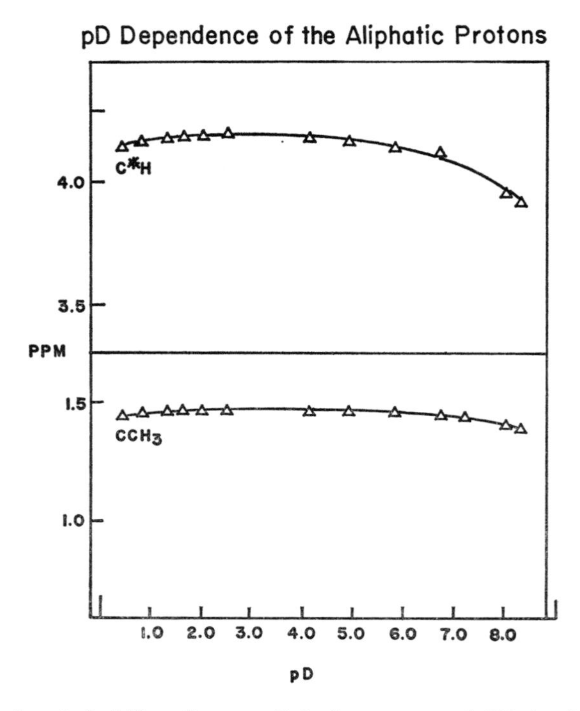

Fig. 10. Proton chemical shifts of some aliphatic protons of CD in the same solutions described for Fig. 9

the amino group must be coplanar with the quinoline ring. Delocalization of the charge deficiency throughout the benzocyclic and heterocyclic rings creates a large positive region, and CD then exists as a rather asymmetric dipolar cation. Because of the repulsive forces between the positively charged centers of the CD di-cation, the molecule would be expected to adopt a molecular conformation with the maximum separation possible between the charge centers.

IV. Conclusions

Several important conclusions about the structure and conformation of CD in aqueous solutions at pD values below its pK of 8.06 can be drawn from the results of this PMR study of its protonation reactions. First, with respect to the structure of CD, the 4-amino quinoline moiety is protonated at the heterocyclic nitrogen, and second, the resulting charge deficiency is delocalized to the entire quinoline ring system and to the 4-amino nitrogen.

Conformational features contingent upon the delocalized charge deficiency include (1) planarity of the 4-amino substituent with the quinoline ring system, and (2) a conformation in which the positively charged diethylamino nitrogen atom is separated as much as possible from the positively charged quinoline ring.

These results point to the fact that at pH 7.4 the species of CD which interacts with DNA is most likely a di-cation with the above-mentioned structure. Furthermore, as a consequence of this structure, the quinoline ring and aliphatic side-chain will most probably exist in a coplanar conformation with a maximum separation between the positive charge centers of the molecule. This conformation satisfies the requirements postulated for an intercalation model of the CD-DNA complex.

At present the conformation of CD in solution and the structural features of the molecule stabilizing this conformation are not completely defined. Further studies are now in progress in our laboratory.

Acknowledgements

The authors are indebted to Dr. Ernest Lustig, Food and Drug Administration, and Dr. Charles L. Bell, University of Illinois at the Medical Center, Chicago, Illinois, for their constructive suggestions and criticism.

References

Allison, J. L., O'Brien, R. L., Hahn, F. E.: DNA: Reaction with chloroquine. Science **149**, 1111 (1965).

Andersag, H., Breitner, S., Jung, H.: Verfahren zur Darstellung von in 4-Stellung basisch substituierte Aminogruppen enthaltenden Chinolinverbindungen. German Patent **683**, 692 (1939).

Blears, D. J., Danyluk, S. S.: A nuclear magnetic resonance study of the protonation of nitrogen heterocyclic molecules. Tetrahedron **23**, 2927 (1967).

Coatney, G. R.: Pitfalls in a discovery: The chronicle of chloroquine. Amer. J. trop. Med. **12**, 121 (1963).

Cohn, S. N., Yielding, K. L.: Spectrophotometric studies of the interaction of chloroquine with deoxyribonucleic acid. J. biol. Chem. **240**, 3123 (1965).

Danyluk, S. S., Schneider, W. G.: The proton resonance spectra and structures of substituted azulenes in trifluoracetic acid. Can. J. Chem. **40**, 1777 (1962).

— Victor, T. A.: The structure and molecular interactions of actinomycin D, quantum aspects of heterocyclic compounds in chemistry and biochemistry. The Jerusalem Symposia on Quantum Chemistry and Biochemistry, II, The Israel Academy of Sciences Humanities, Jerusalem, 1970, pp. 394—411.

Hahn, F. E., O'Brien, R. L., Ciak, J., Allison, J. L., Olenick, J. G.: Studies on modes of action of chloroquine, quinacrine, and quinine and on chloroquine resistance. Milit. Med. **131**, 1071 (1966).

Haigh, C. W., Palmer, M. H., Semple, B.: Preparation and nuclear magnetic resonance spectra of some halogenoquinolines. Nearly degenerate ABX spectra. J. Chem. Soc. **1965**, 6004.

Irvin, J. L., Irvin, E. M.: Spectrophotometric and potentiometric evaluation of apparent acid dissociation exponents of various 4-aminoquinolines. J. Amer. Chem. Soc. **69**, 1091 (1947).

Jackman, L. M., Sternhell, S.: Applications of nuclear magnetic resonance spectroscopy in organic chemistry, p. 209. Oxford: Pergamon Press 1969.

O'Brien, R. L., Hahn, F. E.: Chloroquine structural requirements for binding to deoxyribonucleic acid and antimalarial activity. Antimicrob. Agents and Chemother. **1966**, 315.

— Olenick, J. G., Hahn, F. E.: Reactions of quinine, chloroquine, and quinacrine with DNA and their effects on the DNA and RNA polymerase reactions. Proc. nat. Acad. Sci. (Wash.) **55**, 1511 (1966).

Parker, F. S., Irvin, J. L.: The interaction of chloroquine with nucleic acids and nucleoproteins. J. biol. Chem. **199**, 897 (1952).

Smith, I. C., Schneider, W. G.: The proton magnetic resonance spectrum and the charge distribution of the pyridinium ion. Can. J. Chem. **39**, 1158 (1961).

Sternglanz, H., Yielding, K. L., Pruitt, K. M.: Nuclear magnetic resonance studies of the interaction of chloroquine diphosphate with adenosine-5-phosphate and other nucleotides. Mol. Pharmacol. **5**, 376 (1969).

Victor, T. A.: Thesis, The structure and interactions of actinomycin D. Chicago, Ill.: University of Illinois at the Medical Center 1970.

WHO: Resistance of malaria parasites to drugs. Wld. Hlth. Org. techn. Rep. Ser. 296 (1964).

Effects of Chloroquine and Some Related Compounds on Aminoacylation of Transfer Ribonucleic Acids

KARL H. MUENCH*, MICHAEL DELDIN and JULIO C. PITA, JR.**

I. Introduction

Transfer RNA presents certain theoretical advantages over DNA for studies of selective effects of ligand binding. The number of unique ligand-binding sites in DNA is limited by the rarity of unusual bases and by the regular secondary structure. Moreover, many intercalating drugs, such as chloroquine (O'BRIEN, ALLISON, and HAHN, 1966; SUTHERLAND and SUTHERLAND, 1969) bind rather nonspecifically to both purine and pyrimidine residues in DNA [COHEN and YIELDING, 1965 (1); BLODGETT and YIELDING, 1968]. Finally, the enzymes most used in studies of ligand-treated DNA, for example DNA and RNA polymerases [COHEN and YIELDING, 1965 (2); O'BRIEN, OLENICK and HAHN, 1966), in order to act must move along DNA for hundreds of base pairs with consequently increased chance for collision with ligands and thus inhibition.

In contrast, tRNA's contain a relatively high concentration of minor bases (HALL, 1965) and not all tRNA's have the same minor bases. Moreover, instead of a regular secondary structure tRNA's have regions of unpaired bases and regions of base pairing with an important and partially defined tertiary structure superimposed (LEVITT, 1969). Many drugs, including chloroquine, bind more strongly to nucleic acids with secondary structure than to nucleic acids without secondary structure [PARKER and IRVIN, 1952; ALLISON, O'BRIEN and HAHN, 1965; COHEN and YIELDING, 1965 (1)] and that rule may apply to regions with and without secondary structure in a single tRNA chain. Finally, tRNA's have at least three distinct active sites, and the macromolecules interacting with each of these sites need not move along the tRNA chain for proper function.

For these reasons one might expect chloroquine to alter one function (for example, aminoacylation) more strongly than others in a single tRNA or to alter functions of one specific tRNA more than the same functions of other tRNA's. Therefore, the chloroquine-mediated conversion of $tRNA^{Trp}$ of *Escherichia coli* from an inactive into an active state [MUENCH, 1966, 1969 (1, 2)] received our attention.

---

* Markle Scholar in Academic Medicine. Supported by Grant NIH-AM-09001-06 from the National Institutes of Health, U.S. Public Health Service and Grant PRA-21 from the American Cancer Society.

** Fellow of the Life Insurance Medical Research Fund.

II. Materials and Methods

Details of the preparation of aminoacyl-tRNA synthetases, tRNA, aminoacyl-RNA, L-^{14}C-tryptophan, ^{32}P-PP_i, and measurement of chloroquine, tryptophan, protein, aminoacyl-tRNA synthetases, and acceptor capacities of tRNA's, are as previously described [MUENCH, 1966, 1969 (1, 2); MUENCH and SAFILLE, 1968]. Spermine, spermidine, tryptamine, 5-hydroxytryptamine (serotonin), quinine, quinidine, quinacrine, and primaquine, were purchased from Sigma, cadaverine from Nutritional Biochemicals, and proflavine and putrescine from MANN. The other chloroquine analogs were kindly provided by Dr. ALEXANDER R. SURREY, Sterling-Winthrop Research Institute, with the exception of Camoquin (amodiaquin), generously provided by Dr. EDWARD F. ELSLAGER, Research Laboratories, Parke Davis and Company. Chloroquine dihydrochloride was purchased from Winthrop as a sterile solution, 127 mM.

III. Results and Discussion

1. Active and Inactive $tRNA^{Trp}$

Most unfractionated tRNA prepared from *E. coli* by phenol extraction contains both active and inactive conformations of $tRNA^{Trp}$, designated $tRNA_a^{Trp}$ and $tRNA_i^{Trp}$, respectively. Kinetic evidence indicates $tRNA_i^{Trp}$ is neither a substrate nor an inhibitor of Trp-tRNA synthetase [MUENCH, 1969 (1)], whereas $tRNA_a^{Trp}$ is a substrate and can be measured in the forward reaction catalyzed by that enzyme:

$$\text{Trp} + \text{ATP} + \text{tRNA}_a^{\text{Trp}} \rightleftharpoons \text{Trp-tRNA}_a + \text{AMP} + \text{PP}_i \quad (1)$$

We study the reaction at 37° in 0.50 ml of 100 mM sodium cacodylate buffer, pH 6.9, 10 mM $MgCl_2$, 10 mM KCL, 4 mM reduced glutathione, 1.0 mM ATP, and 0.040 to 0.10 mM L-^{14}C-tryptophan, containing 200 to 400 pmoles of $tRNA^{Trp}$ in 10 to 20 A_{260} units of whole tRNA, and excess Trp-tRNA synthetase. Under these conditions $tRNA_a^{Trp}$ is stable, and $tRNA_i^{Trp}$ is nearly stable but is slowly converted into $tRNA_a^{Trp}$, the conversion being less than 10% complete in 30 min, the usual maximum time for assays. When chloroquine ,which binds reversible to tRNA (MUENCH, 1969 [2)], is added to a final concentration of 2.5 mM in the reaction mixture, the conversion

$$\text{tRNA}_i^{\text{Trp}} \xrightarrow{\text{2.5 mM CQN, 37°}} \text{tRNA}_a^{\text{Trp}} \quad (2)$$

is complete within 10 min at 37°. In the presence of excess enzyme and substrates as described the $tRNA_a^{Trp}$ formed is immediately and completely charged to become Trp-$tRNA_a$ by reaction (1). At lower temperatures or chloroquine concentrations the conversion (2) is slower but still complete when coupled with forward reaction (1).

Whereas Trp-$tRNA_a$ is a substrate for Trp-tRNA synthetase in the reverse reaction (1), Trp-$tRNA_i$ is not. These two forms can be resolved by hydroxylapatite chromatography at pH 5.8 or by methylated-albumin-Kieselguhr chromatography at pH 6.7 [MUENCH, 1969 (2)]. Therefore, the active and inactive conformations of the charged moieties are distinct both by enzymatic and chromatographic criteria. Trp-$tRNA_a$ can be converted into Trp-$tRNA_i$ by treatment at 23° with 1 mM EDTA and phenol (GARTLAND and SUEOKA, 1966), and the reverse change can be effected by 2.5 mM chloroquine at 37° [MUENCH, 1966, 1969 (1, 2)].

Several other tRNA's from both yeast and *E. coli* undergo conformational changes to inactive or denatured states stable enough to be studied (LINDAHL, ADAMS and FRESCO, 1966). The inactivation is accomplished most easily by heating the tRNA to 60° for 5 min in 1 mM EDTA, 10 mM Tris · HCl buffer, at pH 8.0 (LINDAHL et al., 1966). Other means include treatment of tRNA with EDTA and phenol (GARTLAND and SUEOKA, 1966), hexadecyltrimethyl ammonium chloride [MUENCH, 1969 (2)] or various salt and hydrogen ion concentrations (ISHIDA and SUEOKA, 1968). The effect of phenol on tRNA probably accounts for the presence of substantial quantities of inactive tRNA in preparations extracted from bacteria and yeast by phenol (SUEOKA and HARDY, 1968).

The inactivation of tRNA has been characterized by hydrodynamic and optical properties (ADAMS, LINDAHL and FRESCO, 1967; ISHIDA and SUEOKA, 1967). Thus, denatured tRNA has increased viscosity, decreased sedimentation coefficient, and an increased Stokes radius. Ultraviolet absorbance, optical rotatory dispersion, and circular dichroism indicate a loss of several base pairs as well as changes in tertiary structure upon conversion of yeast native $tRNA^{Leu}$ into the denatured state. Ultraviolet absorbance indicated formation of base pairs during conversion of coli $tRNA_i^{Trp}$ into $tRNA_a^{Trp}$ (ISHIDA and SUEOKA, 1967). Functionally, denatured tRNA has been characterized by decreases in acceptor activity, in codon response by the ribosomal binding assay, in activity in protein-synthesizing systems, and in ability to act as a substrate for tRNA adenylyl transferase (LINDAHL, ADAMS, GEROCH and FRESCO, 1967; GARTLAND, ISHIDA, SUEOKA and NIRENBERG, 1969).

In general, denatured tRNA's are converted into the active state by heating for 5 min at 60° in 20 mM $MgCl_2$, 10 mM Tris · HCl buffer, pH 8.0 (LINDAHL et al., 1966). In this respect coli $tRNA_i^{Trp}$ is not unusual. Such treatment results in an activation of $tRNA_i^{Trp}$ indistinguishable by the acceptor assay from that caused by presence of 2.5 mM chloroquine in the assay medium [MUENCH, 1969 (2)].

Although chloroquine activates inactive conformations of *E. coli* $tRNA^{Glu}$, $tRNA^{His}$ and $tRNA^{Leu}$, the effect is difficult to observe, because these inactive forms are unstable under our conditions of assay.

2. Specificity

To approach possible, specific base requirements for chloroquine-$tRNA^{Trp}$ binding and/or the conformational change we studied submethylated tRNA and tRNA with 84% of uracil residues replaced by fluorouracil [MUENCH, 1969 (2)]. In both instances chloroquine converted $tRNA_i^{Trp}$ into $tRNA_a^{Trp}$. However, these experiments failed to demonstrate that the $tRNA^{Trp}$ had shared in the submethylation or fluorouracil substitution characteristic of the whole tRNA used.

Another approach to the structural requirements for effects of chloroquine on $tRNA^{Trp}$ is the study of chloroquine analogs. However, caution must be exercised in interpretation of such results, because analog concentration, analog pK relative to the pH, and competing effects of other divalent and even monovalent cations must be considered. Conceivably, some analogs known to bind to nucleic acids would not convert $tRNA_i^{Trp}$ into $tRNA_a^{Trp}$. A wide variety of 4-aminoquinoline derivatives at concentrations of 1 to 3 mM and at 37° activated $tRNA_i^{Trp}$: the iodo- and bromo-analogs of chloroquine, amodiaquin, the "folded" chloroquines (BAILEY, 1969), 6,8-dichloro-4-[4-(di-diethylamino)-1-methybutylamino] quinoline, 7-chloro-4-(3-di-ethylamino-2-hydroxypropyl)-imino-1-methyl-1,4-dihydroquinoline, and 7-chloro-4-aminoquinoline. Thus reduction of the quinoline ring, presence of bulky side chains, and even removal of the side chain at position 4 does not destroy activity. At concentrations of 4 to 6 mM, quinine and quinidine, which lack an amino derivative at position 4, have activity. However, 7-chloro-4-hydroxyquinoline and primaquine are inactive at 1 and 2 mM, respectively. Quinacrine is inactive at a concentration of 1 mM. Higher concentrations of this and certain other drugs are difficult to test because of insolubility of the drug or of the drug-tRNA complex. Since quinacrine binds to DNA (LERMAN, 1964) and inhibits DNA and RNA polymerases [O'BRIEN et al., 1966 (2)] its lack of activity in conversion of $tRNA_i^{Trp}$ to $tRNA_a^{Trp}$ indicates specific structural requirements for the binding and/or conversion.

Another acridine derivative, proflavine, is of particular interest. The dye is a prototype for intercalating drugs (LERMAN, 1961; CAIRNS, 1962; WARING, 1968) and inhibits both DNA and RNA polymerases (HURWITZ, FURTH, MALAMY and ALEXANDER, 1962). A proflavine-tRNA complex is stable enough to be isolated by gel filtration, but is disrupted by Mg^{2+} (WERENNE, GROSJEAN and CHANTRENNE, 1966). Proflavine does not convert $tRNA_i^{Trp}$ into $tRNA_a^{Trp}$. Even at 2×10^{-4} M in the usual incubation mixture proflavine actually decreases the $tRNA_a^{Trp}$ detected, inhibiting the Mg^{2+}-dependent conversion of $tRNA_i^{Trp}$ into $tRNA_a^{Trp}$, which occurs slowly in the incubation mixture. Proflavine, then, binds to DNA and tRNA, thereby producing manv of the effects of chloroquine but not the specific effect on $tRNA^{Trp}$.

The action of chloroquine has been compared to that of polyamines (COHEN, MORGAN and STREIBEL, 1969). Spermine binds to DNA, increases its stability to thermal denaturation, and inhibits RNA polymerase [O'BRIEN et al., 1966 (2)]. Moreover, spermidine and putrescine have been found tightly bound to tRNA isolated from *E. coli* under conditions of low ionic strength, and have been postulated to play an important role in stabilization of tRNA structure *in vivo* (COHEN et al., 1969).

In the activation of yeast $tRNA^{Leu}$ at 60°, ethylenediamine, propanediamine, putrescine, spermidine, and histamine have some activity, but none are as effective as Mg^{2+} (FRESCO, ADAMS, ASCIONE, HENLEY and LINDAHL, 1966). Cadaverine, putrescine, propanediamine, and ethylenediamine, in order of increasing potency, activated $tRNA_i^{Trp}$ at 50°, as did Mg^{2+}, Ca^{2+}, Mn^{2+}, Ba^{2+}, Sr^{2+}, and Cu^{2+}, according to ISHIDA and SUEOKA (1968), who concluded that the activator must have a closely-spaced pair of positive charges. However, our results with 7-chloro-4-amino-quinoline show that a divalent cation is not essential, for this compound lacks the alkylamine side chain and is a monovalent cation at pH 7 (IRVIN and IRVIN, 1947). Moreover, even the most potent of the diamines at a concentration of 0.1 M gave incomplete conversion of $tRNA_i^{Trp}$ into $tRNA_a^{Trp}$ (ISHIDA and SUEOKA, 1968). Chloroquine is effective at concentrations 100 times lower.

When polyamines at a concentration of 2 mM are substituted for chloroquine under our standard assay conditions at 37°, putrescine and cadaverine have no effect on the relative levels of $tRNA_i^{Trp}$ and $tRNA_a^{Trp}$, and spermine and spermidine as proflavine, decrease the Mg^{2+}-dependent conversion of $tRNA_i^{Trp}$ into $tRNA_a^{Trp}$ in the assay medium. Therefore, the action of chloroquine clearly cannot be explained simply as that of a nonspecific polyamine.

We have briefly examined the hypothesis that a binding site for the indole ring of tryptophan may exist on $tRNA^{Trp}$, and that occupation of the site insures the active conformation of $tRNA^{Trp}$. Tryptamine and serotonin can substitute for chloroquine both in the conversion of $tRNA_i^{Trp}$ into $tRNA_a^{Trp}$ and in the conversion of Trp-$tRNA_i$ into Trp-$tRNA_a$. The concentration required, however, is 10 times greater than that of chloroquine for comparable conversion rates. The significance of this observation remains to be determined.

3. Binding Studies

Another approach to specificity is to compare pure $tRNA^{Trp}$ with whole tRNA in terms of chloroquine-binding properties. Least prevalent of the 20 specific tRNA's in *E. coli*, $tRNA^{Trp}$ constitutes less than 1% of unfractionated tRNA. However, because of its lipophilic nature $tRNA^{Trp}$ emerges later than most other tRNA's from benzoylated DEAE-cellulose columns (GILLAM, MILLWARD, BLEW, VON TIGERSTROM, WIMMER and TENER, 1967; ROY and SÖLL, 1968). A gradient of ethanol in 1.5 M NaCl, 0.05 M sodium acetate buffer, pH 5, elutes the $tRNA^{Trp}$. After this first chromatographic step the partially purified $tRNA^{Trp}$ is charged with tryptophan. The Trp-tRNA is even more lipophilic in its behavior on the benzoylated DEAE-cellulose-column, just as described for the Trp-tRNA from yeast (MAXWELL, WIMMER and TENER, 1968). From these two steps a product 75% pure is obtained. The final step is chromatography on hydroxylapatite at pH 5.8 [MUENCH and BERG, 1966 (1)]. The $tRNA^{Trp}$ recovered has a specific activity of 1800 pmoles/A_{260} unit, a value signifying complete purity. Purified $tRNA^{Trp}$ and whole tRNA have the same absorbance spectrum, including the broad peak at 335 nm characteristic of 4-thiouracil (LIPSETT and DOCTOR, 1967), and the same ultraviolet circular dichroism spectrum (A. H. BRADY and K. H. MUENCH, unpublished observations). Moreover, the difference between $tRNA_i^{Trp}$ and $tRNA_a^{Trp}$ cannot be detected by circular dichroism. Just as whole tRNA, purified $tRNA^{Trp}$ decreases the magnitude and causes a red shift in the absorbance spectrum of chloroquine [MUENCH, 1966, 1969 (2)].

The possibility that $tRNA^{Trp}$ has an unusual number of binding sites for chloroquine or has binding sites with an unusual association constant was investigated spectrophotometrically by observation of the change in chloroquine absorbance at 343 nm in the presence of $tRNA^{Trp}$. BLODGETT and YIELDING (1968) have shown that chloroquine bound to poly *A* absorbs light at 343 nm to a different extent than chloroquine bound to poly *G*. The absorbance of chloroquine bound to whole tRNA and to $tRNA^{Trp}$ is the same. As shown in Fig. 1 the Scatchard plots (SCATCHARD, 1949) obtained either by addition of chloroquine to a constant concentration of whole tRNA or $tRNA^{Trp}$ or by addition of whole tRNA or $tRNA^{Trp}$ to a constant chloroquine concentration differ only slightly. As expected, the plots are curved, a characteristic indicating presence of different binding sites or of interacting binding sites on each of the tRNA's (SCATCHARD, 1949). COHEN and YIELDING [1965 (2)] obtained similar results for chloroquine and DNA. If the maximum number of binding sites under these conditions is assumed to be 0.5 per nucleotidyl residue (PARKER and IRVIN, 1952), then the apparent association constants, actually expressing the average binding of chloroquine to different or interacting sites, are 7×10^3 for whole tRNA and 6×10^3 for $tRNA^{Trp}$.

Two major objections to such spectrophotometric studies exist: (1) The method detects only those chloroquine-tRNA interactions which result in changes in the absorbance of chloroquine, and these may not be the interactions producing the conformational change in $tRNA^{Trp}$; (2) The high molar absorbance of chloroquine prevents its use at the concentrations necessary to change the conformation of $tRNA^{Trp}$ at 37°.

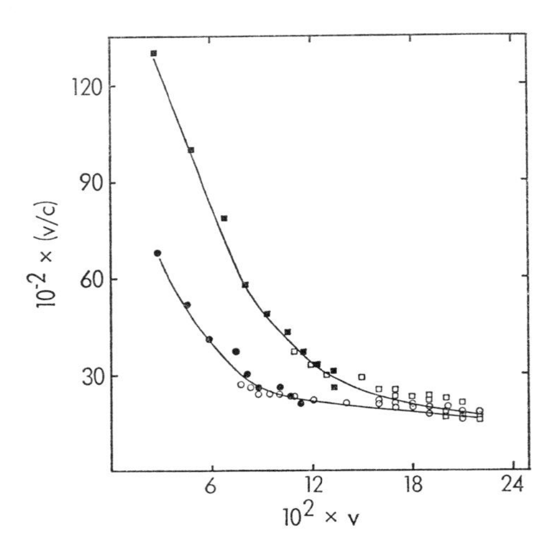

Fig. 1. Scatchard plots of the binding of chloroquine to unfractionated tRNA (squares) or to $tRNA^{Trp}$ (circles). In the first experiment (open symbols) with a constant concentration of chloroquine (1.45×10^{-4} M) increments of unfractionated tRNA or $tRNA^{Trp}$ were added to final concentrations of 1×10^{-3} M nucleotidyl residue. The A_{343} was determined initially and after each tRNA addition. After correction for the A_{343} contributed by the tRNA and for the instrumental deviation of chloroquine absorbance from Beer's law, the amount of chloroquine bound was determined as described [COHEN and YIELDING, 1965 (1)]. The value, v, equals moles of bound chloroquine per mole of nucleotidyl residue, and the quantity, c, equals the corresponding molar concentrations of unbound chloroquine. In the converse experiment (solid symbols) the concentration of tRNA or $tRNA^{Trp}$ was held constant at 2.6×10^{-4} M, and increments of chloroquine up to a final concentration of 0.9×10^{-4} M were added. The A_{343} of chloroquine totally bound to tRNA was determined by extrapolation of the plot of the reciprocal of tRNA concentration against the reciprocal of the decrease in A_{343} with each tRNA addition as described by COHEN and YIELDING [1965 (1)]

4. Leu-tRNA Synthesis

Perhaps unrelated to binding of tRNA, chloroquine changes the reaction rates of several aminoacyl-tRNA synthetases. Because of the experience with $tRNA^{Trp}$ such effects are easily assumed to result from the interaction of chloroquine with tRNA, with a consequent indirect effect on the enzyme. Such an assumption would seem to be supported by preliminary observations on the Leu-tRNA synthetase of *E. coli*. Whereas chloroquine has no effect on the leucine-dependent ATP-PP_i exchange, the drug markedly inhibits Leu-tRNA synthesis under similar assay conditions, as shown in Fig. 2. At lower Mg^{2+} concentrations chloroquine completely stops Leu-tRNA synthesis with no effect on leucine-dependent ATP-PP_i exchange. At higher Mg^{2+} concentrations the effect of chloroquine is decreased. These findings are consistent with a competition between Mg^{2+} and chloroquine for sites on $tRNA^{Trp}$. However, the kinetic analysis of the inhibition (Table 1) reveals chloroquine

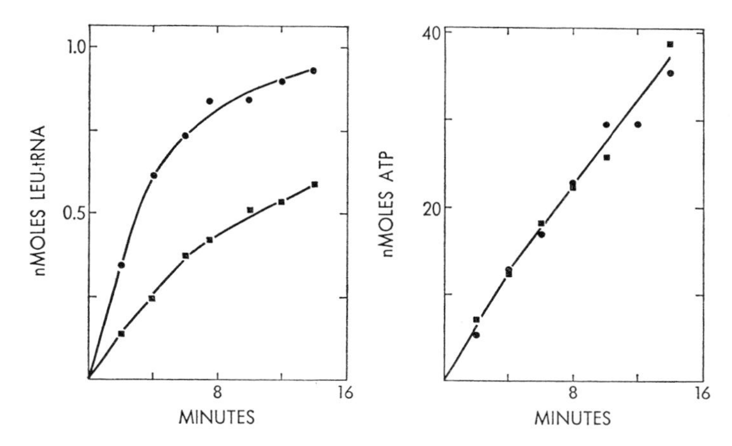

Fig. 2. Effect of chloroquine on Leu-tRNA synthesis and leucine-dependent ATP-PP_i exchange. In total reaction volumes of 3.75 ml, containing 28 μg of partially purified Leu-tRNA synthetase [MUENCH and BERG, 1966 (2)], 100 mM Tris · HCl buffer pH 8.0, 10 mM KCl, 5 mM $MgCl_2$, and 1 mM ATP, the formation of Leu-tRNA or the leucine-dependent incorporation of ^{32}P-PP_i into ATP was followed in 0.5 ml aliquots removed at the times shown during incubation at 30°. The reaction mixtures in the left panel were 0.08 mM in L-^{14}C-leucine, and contained 39 A_{260} units unfractionated tRNA/ml. The reaction mixtures in the right panel contained 0.08 mM L-leucine and 1 mM ^{32}P-PP_i. The mixtures contained no chloroquine (●) or 2.5 mM chloroquine (■). Measurement of radioactive ATP and Leu-tRNA was as indicated in section II

Table 1. *Chloroquine inhibition of Leu-tRNA synthetase*

Substrate	K_m [M]	Relative V_{max}	K_i [M]
Leucine	1.1×10^{-5}	17	
Leucine + 2.5 mM CQN	1.0×10^{-5}	7.0	1.4×10^{-4}
ATP	1.0×10^{-4}	17	
ATP + 2.5 mM CQN	0.8×10^{-4}	6.2	1.2×10^{-4}
$tRNA^{Leu}$	3.9×10^{-7}	17	
$tRNA^{Leu}$ + 2.5 mM CQN	2.0×10^{-7}	5.2	0.9×10^{-4}
Mg^{2+}	0.4×10^{-3}	17	
Mg^{2+} + 1.1 mM CQN	3.6×10^{-3}	16	1.2×10^{-4}

Highly purified Leu-tRNA synthetase (MUENCH, SAFILLE, LEE, JOSEPH, KESDEN and PITA, 1970) was assayed in the presence or absence of chloroquine by the usual assay of Leu-tRNA formation, but in an assay system containing Tris buffer, pH 8.0 (Fig. 2). Initial rates were determined, and a plot of rate against rate/substrate concentration (DOWD and RIGGS, 1965) was fit by the method of least squares to determine K_m and V_{max}. The K_i values were determined as described by WEBB (1963) for competitive inhibition with respect to an enzyme activator.

as a non-competitive inhibitor of Leu-tRNA synthetase with respect to leucine, ATP, and $tRNA^{Leu}$. Thus chloroquine cannot be acting in this case by binding to $tRNA^{Leu}$ and removing it as a substrate. That mechanism would present the picture of competitive inhibition with respect to $tRNA^{Leu}$, as with respect to DNA in the inhibition of DNA and RNA polymerases by proflavine (HURWITZ et al., 1962) or in the inhibition of DNA polymerase by chloroquine [COHEN and YIELDING, 1965 (2)]. Moreover, chloroquine is a competitive inhibitor of Leu-tRNA synthetase with respect to Mg^{2+}. Since Mg^{2+} is required for leucine-dependent ATP-PP_i exchange, and since chloroquine does not inhibit that exchange, there must be two binding sites on the enzyme for Mg^{2+}. Chloroquine competes for only one of these.

5. Conclusion

Thus the study of effects of chloroquine on aminoacyl-tRNA synthetases promises information on their active sites, as already provided for Leu-tRNA synthetase. The study of chloroquine and tRNA has provided new information on conformational states of $tRNA^{Trp}$. The basis for the specificity of the chloroquine-$tRNA^{Trp}$ interaction remains unknown, but the phenomenon gives rise to the hope that specific and controlled conformational changes in tRNA may be induced by drugs *in vivo.*

Acknowledgement

Able technical assistance was provided by Mrs. ALICIA SAFILLE.

References

ADAMS, A., LINDAHL, T., FRESCO, J. R.: Conformational differences between the biologically active and inactive forms of a transfer ribonucleic acid. Proc. nat. Acad. Sci. (Wash.) **57**, 1684—1691 (1967).

ALLISON, J. L., O'BRIEN, R. L., HAHN, F. E.: DNA: Reaction with chloroquine. Science **149**, 1111—1113 (1965).

BAILEY, D. M.: Quinoline antimalarials. Folded chloroquine. J. med. chem. **12**, 184—185 (1969).

BLODGETT, L. W., YIELDING, K. L.: Comparison of chloroquine binding to DNA, and polyadenylic and polyguanylic acids. Biochim. biophys. Acta (Amst.) **169**, 451—456 (1968).

CAIRNS, J.: The application of autoradiography to the study of DNA viruses. Cold Spr. Harb. Symp. quant. Biol. **27**, 311—318 (1962).

COHEN, S. N., YIELDING, K. L.: (1) Spectrophotometric studies of the interaction of chloroquine with deoxyribonucleic acid. J. biol. Chem. **240**, 3123—3131 (1965).

— — (2) Inhibition of DNA and RNA polymerase reactions by chloroquine. Proc. nat. Acad. Sci. (Wash.) **54**, 521—527 (1965).

COHEN, S. S., MORGAN, S., STREIBEL, E.: The polyamine content of the tRNA of *E. coli.* Proc. nat. Acad. Sci. (Wash.) **64**, 669—676 (1969).

DOWD, J. E., RIGGS, D. S.: A comparison of estimates of Michaelis-Menton kinetic constants from various linear transformations. J. biol. Chem. **240**, 863—869 (1965).

FRESCO, J. R., ADAMS, A., ASCIONE, R., HENLEY, D., LINDAHL, T.: Tertiary structure in transfer ribonucleic acids. Cold Spr. Harb. Symp. quant. Biol. **31**, 527—537 (1966).

GARTLAND, W. J., SUEOKA, N.: Two interconvertible forms of tryptophanyl sRNA in *E. coli.* Proc. nat. Acad. Sci. (Wash.) **55**, 948—956 (1966).

— ISHIDA, T., SUEOKA, N., NIRENBERG, M. W.: Coding properties of two conformations of tryptophanyl-tRNA in *Escherichia coli.* J. molec. Biol. **44**, 403—413 (1969).

GILLAM, I., MILLWARD, S., BLEW, D., VON TIGERSTROM, M., WIMMER, E., TENER, G. M.: the separation of soluble ribonucleic acids on benzoylated diethylaminoethylcellulose. Biochemistry **6**, 3043—3056 (1967).

HALL, R. H.: A general procedure for the isolation of "minor" nucleosides from ribonucleic acid hydrolysates. Biochemistry **4**, 661—670 (1965).

HURWITZ, J., FURTH, J. J., MALAMY, M., ALEXANDER, M.: The role of deoxyribonucleic acid in ribonucleic acid synthesis, III. The inhibition of the enzymatic synthesis of ribonucleic acid and deoxyribonucleic acid by actinomycin D and proflavin. Proc. nat. Acad. Sci. (Wash.) **48**, 1222—1230 (1962).

IRVIN, J. L., IRVIN, E. M.: Spectrophotometric and potentiometric evaluation of apparent acid dissociation exponents of various 4-aminoquinolines. J. Amer. chem. Soc. **69**, 1091—1099 (1947).

ISHIDA, T., SUEOKA, N.: Rearrangement of the secondary structure of tryptophan sRNA in *Escherichia coli*. Proc. nat. Acad. Sci. (Wash.) **58**, 1080—1087 (1967).

— — Effect of ambient conditions on conformations of tryptophan transfer ribonucleic acid of *Escherichia coli*. J. biol. Chem. **243**, 5329—5336 (1968).

LERMAN, L. S.: Structural considerations on the interaction of DNA and acridines. J. molec. Biol. **3**, 18—30 (1961).

— Acridine mutagens and DNA structure. J. cell. comp. Physiol. **64**, (Suppl. 1), 1—18 (1964).

LEVITT, M.: Detailed molecular model for transfer ribonucleic acid. Nature (Lond.) **224**, 759—763 (1969).

LINDAHL, T., ADAMS, A., FRESCO, J. R.: Renaturation of transfer ribonucleic acids through site binding of magnesium. Proc. nat. Acad. Sci. (Wash.) **55**, 941—948 (1966).

— — GEROCH, M., FRESCO, J. R.: Selective recognition of the native conformation of transfer ribonucleic acids by enzymes. Proc. nat. Acad. Sci. (Wash.) **57**, 178—185 (1967).

LIPSETT, M. N., DOCTOR, B. P.: Studies on tyrosine transfer ribonucleic acid, a sulfur-rich species from *Escherichia coli*. J. biol. Chem. **242**, 4072—4077 (1967).

MAXWELL, I. H., WIMMER, E., TENER, G. M.: The isolation of yeast tyrosine and tryptophan transfer ribonucleic acids. Biochemistry **7**, 2629—2634 (1968).

MUENCH, K. H.: Chloroquine-mediated conversion of transfer ribonucleic acid of *Escherichia coli* from an inactive to an active state. Cold Spr. Harb. Symp. quant. Biol. **31**, 539—542 (1966).

— BERG, P.: (1) Resolution of aminoacyl transfer ribonucleic acid by hydroxylapatite chromatography. Biochemistry **5**, 982—987 (1966).

— — (2) Preparation of aminoacyl ribonucleic acid synthetases from *Escherichia coli*. In: Procedures in nucleic acid research, p. 375. New York: Harper and Row 1966.

— SAFILLE, P. A.: Transfer ribonucleic acids in *Escherichia coli*. Multiplicity and variation. Biochemistry **7**, 2799—2808 (1968).

— (1) Chloroquine and synthesis of aminoacyl transfer ribonucleic acids. Tryptophanyl transfer ribonucleic acid synthetase of *Escherichia coli* and tryptophanyladenosine triphosphate formation. Biochemistry **8**, 4872—4879 (1969).

— (2) Chloroquine and synthesis of transfer ribonucleic acids. Conformational changes in tryptophanyl and tryptophan transfer ribonucleic acids. Biochemistry **8**, 4880—4888 (1969).

— SAFILLE, A., LEE, M., JOSEPH, D. R., KESDEN, D., PITA, J. C., Jr.: Aminoacyl-tRNA synthetases of *Escherichia coli* and of man. In: Miami Winter Symposia, Vol. I. Amsterdam: North Holland 1970.

O'BRIEN, R. L., ALLISON, J. L., HAHN, F. E.: Evidence for intercalation of chloroquine into DNA. Biochim. biophys. Acta (Amst.) **129**, 622—624 (1966).

— OLENICK, J. G., HAHN, F. E.: (2) Reactions of quinine, chloroquine, and quinacrine with DNA and their effects on the DNA and RNA polymerase reactions. Proc. nat. Acad. Sci. (Wash.) **55**, 1511—1517 (1966).

PARKER, F. S., IRVIN, J. L.: The interaction of chloroquine with nucleic acids and nucleoproteins. J. biol. Chem. **199**, 897—909 (1952).

ROY, K. L., SÖLL, D.: Fractionation of *Escherichia coli* transfer RNA on benzoylated DEAE-cellulose. Biochim. biophys. Acta (Amst.) **161**, 572—574 (1968).

SCATCHARD, G.: The attractions of proteins for small molecules and ions. Ann. N.Y. Acad. Sci. **51**, 660—672 (1949).

SUEOKA, N., HARDY, J.: Deproteinization of cell extract with silicic acid. Arch. Biochem. **125**, 558—566 (1968).

SUTHERLAND, J. C., SUTHERLAND, B. M.: Energy transfer in the DNA-chloroquine complex. Biochim. biophys. Acta (Amst.) **190**, 545—548 (1969).

WARING, M. J.: Drugs which affect the structure and function of DNA. Nature (Lond.) **219**, 1320—1325 (1968).

WEBB, J. L.: Enzyme and metabolic inhibitors, Vol. I, p. 168. New York: Academic Press 1963.

WERENNE, J., GROSJEAN, H., CHANTRENNE, H.: Effect of proflavine on the binding of isoleucine to transfer RNA. Biochim. biophys. Acta (Amst.) **129**, 585—593 (1966).

Interaction of Antimalarial Aminoquinolines (Primaquine, Pentaquine, and Chloroquine) with Nucleic Acids, and Effects on Various Enzymatic Reactions *in Vitro*

DAVID J. HOLBROOK, JR., LEONA P. WHICHARD, CARL R. MORRIS and LIDIA A. WHITE

I. Introduction

Two classes of aminoquinolines are effective as antimalarial drugs. The 4-aminoquinolines, including the prototype chloroquine, are active against the erythrocytic stage of Plasmodia. In contrast, the 8-aminoquinolines, including as representatives primaquine and pentaquine, are active against the tissue stages of the malaria parasite (POWELL, 1966). In addition, chloroquine is also used for several other parasitic diseases and for rheumatoid arthritis.

Since the initial studies by IRVIN and co-workers [IRVIN, IRVIN and PARKER, 1949; PARKER and IRVIN, 1952 (1); IRVIN and IRVIN, 1954, 1954 (1)], the *in vitro* interaction of the antimalarials chloroquine, quinacrine and quinine with nucleic acids has been extensively studied. In most cases, it is found that drug binding occurs to both DNA and RNA, as well as to their related polynucleotides and nucleoside phosphates [KURNICK and RADCLIFFE, 1962; STOLLAR and LEVINE, 1963; COHEN and YIELDING, 1965; ALLISON, O'BRIEN and HAHN, 1965; HAHN, O'BRIEN, CIAK, ALLISON and OLENICK, 1966; O'BRIEN, ALLISON and HAHN, 1966; MUENCH, 1966, 1969 (1); WHICHARD, MORRIS, SMITH and HOLBROOK, 1968; BLODGETT and YIELDING, 1968; ESTENSEN, KREY and HAHN, 1969; STERNGLANZ, YIELDING and PRUITT, 1969; SUTHERLAND and SUTHERLAND, 1969; MORRIS, ANDREW, WHICHARD and HOLBROOK, 1970].

It has been proposed that the antimalarial activity (against the erythrocytic stage) of chloroquine and quinine is due to their interaction with the parasitic nucleic acid and the inhibition of DNA (or RNA) function [PARKER and IRVIN, 1952 (1); COHEN and YIELDING, 1965 (1); ESTENSEN et al., 1969]. Although a different stage in the life cycle of the malaria parasite is sensitive to the antimalarial 8-aminoquinolines, a number of these compounds — including primaquine and pentaquine — also bind to DNA, RNA and various polynucleotides (WHICHARD et al., 1968; MORRIS et al., 1970). Our laboratory has been interested in the study of reactions in nucleic acid metabolism which may be involved in the observed biological activities of the aminoquinolines, namely (a) the antimalarial activity, and (b) some of the toxic effects of these drugs in animals. The emphasis generally has been placed on the 8-aminoquinolines — pri-

Abbreviations: DNA-P, DNA-phosphorus, equivalent to DNA-nucleotide; PN-P, polynucleotide-phosphorus; DNase, deoxyribonuclease; RNase, ribonuclease. In figures: AQ, aminoquinoline; PR, primaquine; PE, pentaquine; CQ, chloroquine; PL, plasmocid; PAM, pamaquine; DHPE, dihydroxypentaquine.

maquine and pentaquine, although chloroquine has also been studied as a representative of the 4-aminoquinolines.

II. Results and Discussion

Each of the antimalarial aminoquinolines contains a 1,4- or 1,5-diaminopentyl side chain at the 4- or 8-position of the quinoline ring. The terminal nitrogen, with a pK of 10 to 10.5 (IRVIN and IRVIN, 1947), was protonated at the pH of these studies. The pK_2 of chloroquine is about 7.8 (PARKER and IRVIN, 1952); thus, in the binding studies conducted at pH 6 or 7, chloroquine occurs predominantly as a divalent cation.

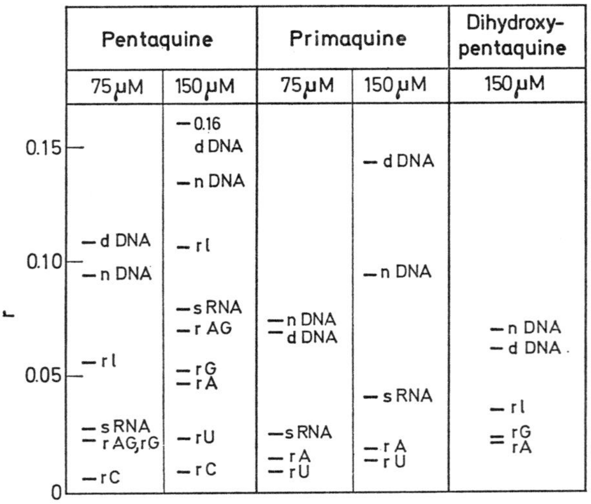

r	Pentaquine		Primaquine		Dihydroxy-pentaquine
	75 µM	150 µM	75 µM	150 µM	150 µM
0.15		– 0.16 d DNA – n DNA		– d DNA	
0.10	– d DNA – n DNA	– rI		– n DNA	
0.05	– rI	– sRNA – r AG – rG – rA	– n DNA – d DNA	– s RNA	– n DNA – d DNA – rI
0	– sRNA – rAG,rG – rC	– rU – rC	– sRNA – rA – rU	– rA – rU	– rG – rA

Fig. 1. The binding of 8-aminoquinolines to DNA, RNA, and various polynucleotides. The ratio *r* (= moles of aminoquinoline bound per mole of PN-P), as measured by equilibrium dialysis, is given at a concentration of free, unbound aminoquinoline of 75 and 150 µM. DNA-P or PN-P concentrations = 1.0 to 5.5 mM

The antimalarial 8-aminoquinolines, including primaquine and pentaquine, generally have a pK_2 of 3.0 to 3.2 (WHICHARD et al., 1968; MORRIS et al., 1970) and, therefore, exist as monoprotonated cations at pH values of 6 and above.

A common metabolic reaction of the 8-aminoquinolines in various species (SMITH, 1956) is an 0-demethylation of the 6-methoxy group, followed by subsequent oxidation to form the 5,6-dihydroxy derivative. One of the potential metabolites of the 8-aminoquinolines will be referred to as "dihydroxypentaquine", that is, the 5,6-dihydroxy aminoquinoline which has the same diamino side chain as pentaquine in the 8-position.

The binding of the 8-aminoquinolines to DNA, RNA, various polynucleotides and chromatin has been studied by equilibrium dialysis and by direct spectrophotometry. The results from the earlier studies (WHICHARD et al., 1968) on binding to DNA may be summarized as follows.

a) The binding of the 8-aminoquinolines to DNA is accompanied by a decrease in the absorbancy of the aminoquinoline. In addition to the decrease in the absorbancy, the wavelength of the maximum absorbance undergoes a slight bathochromic shift (to longer wavelengths) of 2 to 6 nm for the 8-aminoquinolines and 4 to 8 nm for chloroquine.

b) The binding of the 8-aminoquinolines (and chloroquine) to DNA is decreased by an increase in the ionic strength of the medium, and the binding is decreased by Mg^{2+} ions to a greater extent than would be expected from ionic strength effects alone.

c) At a DNA-nucleotide to aminoquinoline molar ratio of 10 and an ionic strength of 0.012 (at pH 6), the percentages of the aminoquinolines bound to native calf

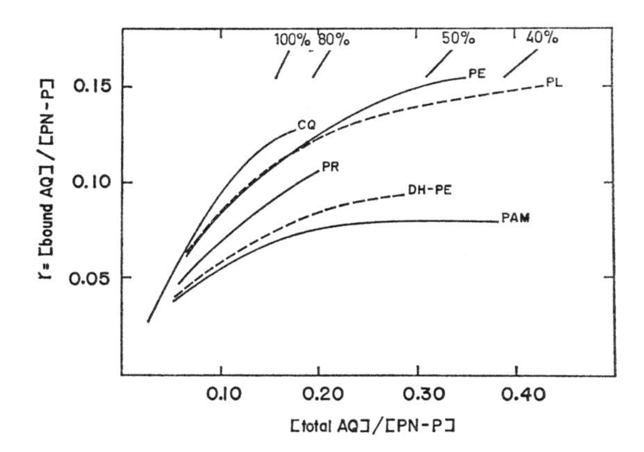

Fig. 2. The binding of the 8-aminoquinolines and chloroquine to native calf thymus DNA as a function of the ratio of the total aminoquinoline per polynucleotidephosphorus. Vertical axis: r = moles of aminoquinoline bound per mole of PN-P (= DNA-P); horizontal axis: molar ratio of total (bound plus free) aminoquinoline per mole of PN-P. DNA-P concentrations = 1.1 – 4.4 mM. At the top of the figure are markers which indicate the position which a binding curve would occupy if various percentages of the total aminoquinoline were bound to the DNA. From these markers (and their extrapolation to the origin of the figure), the percentages of the total aminoquinoline bound to the DNA at various molar ratios of total aminoquinoline to PN-P can be evaluated. Of the 100 experimental points used to construct this figure, only four points varied from the line drawn by more than $r = \pm 0.01$

thymus DNA are in the following order: chloroquine, a divalent cation, followed in decreasing order by the monovalent 8-aminoquinolines, namely, pentaquine, plasmocid, primaquine, and pamaquine.

Results, obtained by equilibrium dialysis, for the binding of pentaquine, primaquine, and dihydroxypentaquine to DNA, RNA and various polyribonucleotides are given in Fig. 1 at two concentrations (75 and 150 μM) of free, unbound aminoquinolines. The greatest extent of binding occurs with DNA; the data given are for calf thymus DNA (56% A+ T base composition) but similar or somewhat lower numbers are obtained for DNA from *Clostridium perfringens* (73% A+ T) and *Micrococcus lysodeikticus* (29% A+ T). The extent of binding to polyribonucleotides falls below that of binding to various DNAs in the following order: purine-containing polyribonucleotides (poly *rI*, poly *rG* ,poly *rA*) and yeast soluble RNA, followed by poly *rU*. The lowest level of binding of the 8-aminoquinolines occurs to poly *rC*. Under these same experimental conditions (0.01 M potassium phosphate, pH 6; ionic strength of 0.012), it is not possible to distinguish a selectivity of chloroquine binding to the various polyribonucleotides since more than 90% of the chloroquine present is bound to each polynucleotide.

A comparison of the binding curves of chloroquine and five of the 8-aminoquinolines obtained with native calf thymus DNA is presented in Fig. 2. At the top of Fig. 2 (and Fig. 3) are lines (along with the respective extrapolations to the origin of the figure) denoting the position that a binding curve would occupy if 100, 80, 50, or 40%, respectively, of the total ligand were bound to the polymer. Thus, the percentages of the respective aminoquinolines bound to the native DNA can be

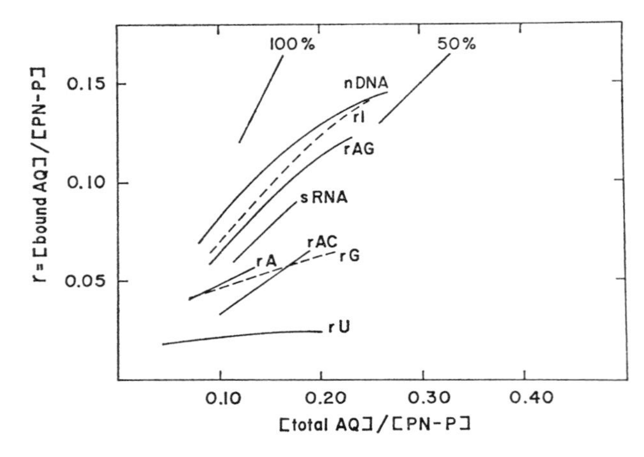

Fig. 3. The binding of pentaquine to native calf thymus DNA, yeast soluble RNA, and various polyribonucleotides. Vertical axis, horizontal axis, and markers: same as Fig. 2

estimated at various molar ratios of total aminoquinoline to DNA-nucleotide. As the relative amount of the aminoquinolines is increased and a greater proportion of the binding sites on the DNA are occupied, the percentage of the total aminoquinoline bound to the DNA decreases. At a molar ratio of total aminoquinoline to DNA-P (DNA-nucleotide) of 0.10, approximately 95% of the chloroquine is bound to the DNA, 80 to 85% of the pentaquine and plasmocid, 68% of the primaquine, and 55 to 60% of the dihydroxypentaquine and pamaquine. For the 8-aminoquinolines, this is equivalent to 5.5 to 8.5 bound aminoquinolines per 100 DNA-nucleotides (in a medium containing 10 aminoquinolines per 100 DNA-nucleotides). At higher ionic strengths (0.15 to 0.2) the percentage of binding of plasmocid and pentaquine are approximately equal to the binding of chloroquine to native DNA (at 20 to 25% binding).

In Fig. 3 is presented a comparison of the binding of a single 8-aminoquinoline (pentaquine) with native calf thymus DNA, yeast soluble RNA, and a series of polyribonucleotides, as a function of the molar ratio of the total aminoquinoline to polynucleotide-phosphorus. Two polyribocopolymers (poly *rAG* and poly *rAC*) were studied in this series. A markedly greater extent of binding occurs to poly *rAG* than to poly *rA* or to poly *rG*, although the secondary structure of the latter polymer may affect the total binding observed by equilibrium dialysis.

The binding parameters K (= the association constants, expressed in M^{-1}) and n (= the maximum number of binding sites per PN-P) have also been determined. Generally, n is approximately 0.12 to 0.20 for all of the polymers examined (WHICHARD et al., 1968; MORRIS et al., 1970).

In 0.01 M potassium phosphate, pH 6 (ionic strength, 0.012), the association constants for the binding of primaquine and pentaquine to native DNA are 10^4 to 1.2×10^4 M^{-1}, although the K for plasmocid binding to native DNA is $> 2 \times 10^4$ M^{-1} (MORRIS et al., 1970). An increase in the ionic strength from 0.012 to 0.025 decreases the association constants of primaquine and pentaquine for native calf thymus DNA but does not appreciably change the maximum number of binding sites per DNA-P for the respective aminoquinolines (MORRIS et al., 1970).

A number of agents which bind to DNA cause an increase in the transition temperature (T_m) for the conversion from the helical to coil structures of DNA. HAHN and co-workers (ALLISON et al., 1965; ESTENSEN et al., 1969) and COHEN and YIELDING (1965) have reported that the antimalarials chloroquine and quinine raise the T_m of DNA. In this laboratory, in a medium of 0.01 M potassium phosphate-0.001 M EDTA (pH 6.9), it was observed that the presence of chloroquine (4×10^{-5} M; molar ratio of DNA-P/chloroquine = 4.5) causes an increase in the T_m of calf thymus DNA of approximately 10°. The effect of the 8-aminoquinolines (primaquine and pentaquine) on the T_m of calf thymus DNA has also been examined in the same medium. The results (Fig. 4) show that neither pentaquine nor primaquine (even at molar ratios of DNA-P/aminoquinoline of 3.0 and 4.5, respectively) caused an appreciable change in the T_m of calf thymus DNA. Under these conditions (including ionic strength) appreciable binding of the 8-aminoquinolines to DNA would be expected, but the binding does not appear to inhibit strand separation of native DNA.

The mode of binding of the 8-aminoquinolines to DNA is still unknown. As with chloroquine (COHEN and YIELDING, 1965), there is evidence from direct spectrophotometry that binding occurs to at least two sites or by two modes (WHICHARD et al., 1968). In addition to a lack of effect of primaquine and pentaquine on the T_m of DNA strand separation, primaquine also does not cause a change in at least one of the hydrodynamic properties of native DNA. The binding of a compound to DNA by intercalation results in a more rigid structure of DNA and brings about an increase in the viscosity of the DNA (LERMAN, 1961). In agreement with an earlier report by O'BRIEN et al. (1966), chloroquine produced an increase in the viscosity of native calf thymus DNA (MORRIS et al., 1970). Primaquine, however, did not induce an increase in the viscosity of native DNA (MORRIS et al., 1970). Although this does not preclude that a minor portion of the binding of primaquine and the other 8-aminoquinolines to native DNA occurs by intercalation, it is apparent that any possible intercalation by primaquine would occur to such a small extent that it is not readily detectable by an increase in the viscosity of the DNA.

The binding of antimalarial 8-aminoquinolines and chloroquine to DNA, RNA and various polynucleotides has been measured by equlibrium dialysis and by direct spectrophotometry. The binding measured by these physical techniques furnishes only a *potential* mechanism for some of the biological activities of these compounds, including the antiplasmodial action and the toxic effects of these drugs in animals. The binding of 8-aminoquinolines and chloroquine to various polynucleotides may cause alterations in the function of various RNA species involved in protein synthesis, in addition to the previously proposed mechanism for chloroquine and quinacrine, namely, the inhibition of DNA polymerase and RNA polymerase by binding to template DNA [COHEN and YIELDING, 1965 (1); O'BRIEN, OLENICK and HAHN, 1966].

As a consequence of the observation of binding of the 8-aminoquinolines (and chloroquine) to various nucleic acids and polynucleotides, it was of interest to

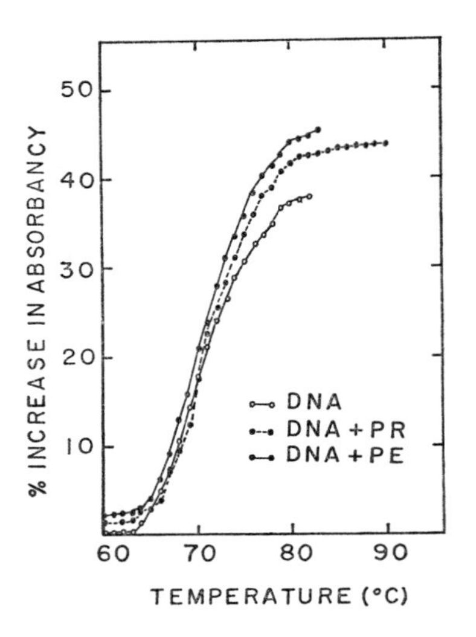

Fig. 4. Lack of effect of pentaquine and primaquine on the transition temperature (T_m) of calf thymus DNA. Melting curves were determined in a Gilford model 2400 spectrophotometer. Native calf thymus DNA and the respective aminoquinolines were dissolved in 0.01 M potassium phosphate-0.001 M EDTA (pH 6.94 at 25°); final concentrations: 80 μg DNA/ml; primaquine, 4×10^{-5} M; pentaquine, 6×10^{-5} M. Molar ratios of DNA-P/aminoquinolines were 4.5 and 3.0 for primaquine and pentaquine, respectively

examine the possible effects that such interactions might have on enzymatic reactions of the nucleic acids studied in vitro. The binding of drugs to DNA has been commonly observed by the inhibition of nuclease activity, presumably due to the formation of a DNA-drug complex which is less sensitive to nuclease activity than the free DNA. The activity of a wide range of deoxyribonucleases (DNases) and phosphodiesterases upon DNA is inhibited by interaction of DNA with actinomycin D, ethidium, acridine orange and other compounds which interact with DNA (LEITH, 1963; ERON and MCAUSLAN, 1966; SARKAR, 1967; ZELEZNICK and SWEENEY, 1967; LAGOWSKI and FORREST, 1967; SULKOWSKI and LASKOWSKI, 1968; BEHR, HONIKEL and HARTMANN, 1969; BATES, KUENZIG and WATSON, 1969). One report of special interest is that by KURNICK and RADCLIFFE (1962) who demonstrated that the antimalarials chloroquine and quinacrine inhibit the hydrolysis of DNA by pancreatic DNase and a rabbit serum DNase.

In agreement with KURNICK and RADCLIFFE, it was found that chloroquine inhibited the hydrolysis of native calf thymus DNA by 25 to 60% at a molar ratio of DNA-P/chloroquine of 5 (Table 1)*. Micrococcal nuclease is a phosphodiesterase which is capable of hydrolysis of either DNA or RNA. Ca^{2+} ions are required for its activity and are included in the assay system. Chloroquine inhibited the hydrolysis by micrococcal nuclease of native or denatured DNA. It was observed that primaquine and pentaquine were also capable of inhibiting the hydrolysis of denatured DNA by micrococcal nuclease (Table 1). Such an inhibition might have been expected from the pattern of other nucleases and phosphodiesterases which are inhibited by DNA-drug interactions.

Table 1. *Effect of chloroquine and 8-aminoquinolines on enzymatic hydrolysis of DNA*

Enzyme	Substrate[a]	Amino-quinoline[a]	Activity	Hydrolysis (control = 100)
Pancreatic DNase	*n* DNA	chloroquine	inhibition	39—73[b]
Micrococcal nuclease	*n* DNA	chloroquine	inhibition	63—67[c]
	d DNA	chloroquine	inhibition	83 ± 6 (10)[c, d]
		primaquine	inhibition	85 ± 6 (10)[c, d]
		pentaquine	inhibition	71 ± 7 (10)[c, d]

[a] In each assay, the concentrations of DNA-P and the aminoquinolines were 0.5 to 1.1 mM and 0.1 to 0.25 mM, respectively.
[b] DNA-P/CQ molar ratio = 5; 2 mM Mg^{2+}; 0.01 M Tris, pH 6.7.
[c] DNA-P/AQ molar ratio = 3.6 to 4.4; 0.2 mM Ca^{2+}; 0.02 M Tris, pH 7.7.
[d] Data from 5 experiments; if one experiment excluded: CQ, 75 ± 3 (8); PR, 77 ± 2 (8); PE, 61 ± 4 (8).
± = S.E.M.

Chloroquine and the 8-aminoquinolines bind to RNA, including soluble RNA, and various polyribonucleotides [PARKER and IRVIN, 1952 (1); MUENCH, 1966; BLODGETT and YIELDING, 1968; MORRIS et al., 1970]. Consequently, it was of interest to examine the effect of these aminoquinolines on the hydrolysis of soluble RNA and various polyribonucleotides. In contrast to our expectations, chloroquine, primaquine, and pentaquine caused a stimulation in the hydrolysis of yeast soluble (transfer) RNA by pancreatic RNase (Table 2). Transfer RNA consists of intramolecular double-standed regions with single-stranded loops. Since the stimulation of the hydrolysis of soluble RNA might have been due to an aminoquinoline-induced conformational change in the structure of the RNA [MUENCH, 1969 (1)], the hydrolysis of single-stranded poly *rAC* and poly *rAU* was also examined. The similar aminoquinoline-induced stimulation of RNase activity against the latter two polyribonucleotides (data for poly *rAC* in Table 2) implies that the stimulation of RNase activity is not caused by a specific change in the tertiary structure of soluble (transfer) RNA but instead

* The extent of hydrolysis of DNA and RNA was measured by the formation of material (soluble in cold 0.5 N $HClO_4$ plus bovine serum albumin coprecipitant) which gave a positive reaction with diphenylamine and orcinol, respectively.

is due to a more general change in the (perhaps secondary) structure of polyribonucleotides.

Micrococcal nuclease activity against soluble RNA and poly *rA* was also stimulated by chloroquine (against sRNA and poly *rA*) and by primaquine and pentaquine (against poly *rA*) (Table 2) although the activity of the same enzyme against DNA is inhibited by the aminoquinolines. The contrasting nature of the aminoquinoline effects on micrococcal nuclease (depending on the ribo- or deoxyribo-character of the polymer), and the stimulation of RNA hydrolysis by both micrococcal nuclease and by pancreatic RNase tend to obviate the possibility that the stimulation of RNA hydrolysis is due to an aminoquinoline-enzyme complex rather than an aminoquinoline-RNA complex.

Table 2. *Effect of chloroquine and 8-aminoquinolines on enzymatic hydrolysis of RNA*

Enzyme	Substrate[a]	Amino-quinoline[a]	Activity	Hydrolysis (control = 100)
Pancreatic RNase	*s* RNA	chloroquine	stimulation	150—200[b]
		primaquine	stimulation	124 ± 5 (10)[c]
		pentaquine	stimulation	136 ± 6 (10)[c]
	poly *AC*	chloroquine	stimulation	136 ± 4 (7)[c]
		primaquine	stimulation	117 ± 3 (7)[c]
		pentaquine	stimulation	122 ± 3 (7)[c]
Micrococcal nuclease	*s* RNA	chloroquine	stimulation	131 ± 4 (4)[d]
	poly *A*	chloroquine	stimulation	138 ± 12 (6)[d]
		primaquine	stimulation	149 ± 19 (4)[d]
		pentaquine	stimulation	134 ± 14 (5)[d]
Spleen phospho-diesterase	*s* RNA	chloroquine	inhibition	65—72[e]
	poly *A*	chloroquine	inhibition	73—80[e]

[a] In each assay, the concentrations of RNA-P (or PN-P) and the aminoquinolines were 0.3 to 1.2 mM and 0.06 to 0.24 mM, respectively.
[b] When 50% or less of total sRNA is hydrolyzed; 0.02 M Tris-HCl, pH 6.7.
[c] PN-P/AQ molar ratio = 3.6 to 4.0; 0.02 M Tris-HCl, pH 6.7.
[d] PN-P/AQ molar ratio = 3.8 to 4.1; 0.2 mM Ca^{2+}; 0.02 M Tris-HCl, pH 7.7.
[e] PN-P/CQ molar ratio = 3.9 to 4.2; 0.02 M Tris-HCl, pH 6.8.

Pancreatic DNase, micrococcal nuclease, and pancreatic RNase are (predominantly) endonucleases. In contrast, spleen phosphodiesterase is an exonuclease which removes sequentially one terminal nucleotide per hydrolytic reaction. The activity of the latter enzyme against soluble RNA or poly *rA* was inhibited by chloroquine (Table 2).

The interaction of a number of drugs with DNA commonly can be detected by an inhibition of DNA function in the DNA polymerase or RNA polymerase reactions (Waring, 1965). *E. coli* DNA polymerase and RNA polymerase are inhibited by various antimalarials, including chloroquine, quinacrine and quinine, which bind to the template DNA [Cohen and Yielding, 1965 (1); O'Brien, Olenick and Hahn, 1966]. In addition, these drugs markedly decrease the nucleic acid synthesis in the intact malaria parasite (Schellenberg and Coatney, 1961; Polet and Barr, 1968; Van Dyke, Szustkiewicz, Lantz and Saxe, 1969).

The effect of primaquine and pentaquine on the activity of *E. coli* RNA polymerase was examined with native calf thymus DNA template in the presence of 1.0 mM Mn^{2+} ions. For comparison, data on the effect of chloroquine are included and generally agree with previously published reports [Cohen and Yielding, 1965 (1); O'Brien, Olenick and Hahn, 1966]. The stated concentrations of the aminoquinolines reflect the total amount of the aminoquinolines added to the incubation system. However, the RNA polymerase preparation (obtained from Biopolymers Laboratory) contains carrier bovine serum albumin which, from the work of Parker and Irvin (1952), has been demonstrated to bind chloroquine. Thus, the effective concentrations of the

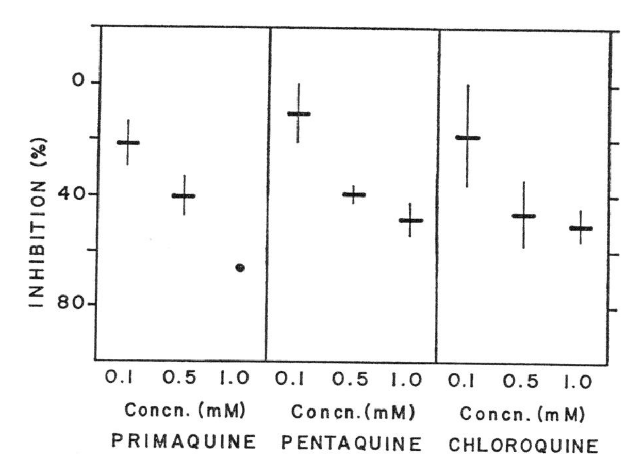

Fig. 5. Inhibition of *E. coli* RNA polymerase by primaquine, pentaquine, and chloroquine. Incubations were carried out for 45 to 60 min at 37° in the presence of 0.23 mM DNA-P and 1.0 mM Mn^{2+} ions. Three to five assays were conducted at each aminoquinoline concentration (except a single assay at 1 mM primaquine). The short horizontal bar indicates the average of the inhibition observed at a specific aminoquinoline concentration; the vertical bar indicates the mean deviation from the mean for the 3 to 5 determinations

aminoquinolines in the incubation system may have been less than stated owing to binding by albumin of a portion of the chloroquine and perhaps the 8-aminoquinolines.

Primaquine, pentaquine, and chloroquine, at concentrations of 5×10^{-4} M and 1×10^{-3} M, inhibited the RNA polymerase reaction by 40 to 50% upon incubation for 45 to 60 min (Fig. 5). Although there is a trend of inhibition by each aminoquinoline at a concentration of 1×10^{-4} M, the results were too scattered to establish definite inhibition. Upon incubation for 15 to 20 min, primaquine and pentaquine caused only a slight inhibition (11 and 22%, respectively, at 5×10^{-4} M) of the RNA polymerase. Considerable inhibition (38 to 45%) occurred in the presence of relatively high concentrations (1×10^{-3} M) of the 8-aminoquinolines.

The binding of chloroquine and other antimalarial drugs to RNA and polyribonucleotides suggests the possibility that such binding might alter the ability of the RNA to function in protein (or polypeptide) synthesis. At least four laboratories have examined the effect of chloroquine on the aminoacylation of transfer RNA. MUENCH (1966, 1969) has reported that chloroquine induces the conversion of an inactive form of *E. coli* tryptophan transfer RNA into an active form. In contrast, chloroquine is a specific inhibitor of leucine aminoacylation by direct action on the synthetase (MUENCH, 1969). ILAN and ILAN (1969) found that chloroquine (0.1 mM) did not appreciably affect the valine aminoacylation to rabbit liver or *E. coli* transfer RNA by synthetase from *Plasmodium berghei* in a medium containing 10 mM Mg^{2+} ions. LANDEZ, ROSKOSKI and COPPOC (1969), however, found that chloroquine (0.3 mM) decreased the valine aminoacylation in a rat liver system containing 2 mM Mg^{2+}. A concentration of 3 mM

Table 3. *Effect of primaquine on in vitro aminoacylation by mouse liver system*

Amino acid	Time of Incubation min	Primaquine concentration (relative incorporation with control = 100; ± mean deviation) 1.0 mM	0.1 mM	0.01 mM
Phenylalanine (3)	10—15	111 ± 10	119 ± 6	94 ± 2
	30	114 ± 15	116 ± 11	106 ± 3
	60	94 ± 12	100 ± 12	94 ± 3
Tryptophan (2)	10—15	68 ± 14	104 ± 36	118 ± 30
	30	91 ± 3	103 ± 10	109 ± 12
	60	106 ± 0	109 ± 18	136 ± 32
Leucine (3)	10	121 ± 20	108 ± 5	87 ± 10
	30	99 ± 19	110 ± 12	95 ± 13
	60	79 ± 16	112 ± 15	110 ± 25
Reconstituted algal hydrolysate (2)	10	105 ± 9	103 ± 3	95 ± 4
	30	93 ± 3	99 ± 0	87 ± 2
	60	87 ± 4	98 ± 9	88 ± 9

Incubation medium: Mg^{2+}, 5 mM; ATP, 10 mM; endogenous sRNA, 59 to 132 μg RNA/ml; Tris, 10 mM, pH 7.4; total volume 0.5 ml.

The number of enzyme preparations studied is denoted in parentheses; each preparation was assayed in triplicate.

chloroquine was necessary to inhibit the aminoacylation of phenylalanine or methionine. In the same in vitro assay system (rat liver; 2 mM Mg^{2+}), primaquine (1.0 mM) did not appreciably affect the aminoacylation of phenylalanine, valine, or methionine (LANDEZ et al., 1969).

The effect of chloroquine and primaquine on the aminoacylation of transfer RNA was studied also in a mouse liver system, containing 5 mM Mg^{2+} ions during assay. The lack of effect of primaquine on the in vitro aminoacylation in a mouse liver system is shown in Table 3. The aminoacylation of five amino acids (histidine and tyrosine, in addition to the three in Table 3) and a reconstituted amino acid mixture was not appreciably altered by primaquine at concentrations from 1.0 mM to 0.01 mM, when the aminoacylation was assayed in a medium containing 5 mM Mg^{2+} ions. It should be considered, however, that variations in the reported effects of the antimalarial aminoquinolines on aminoacylation measured in vitro may be due, in large part, to the conditions of the assay systems (especially Mg^{2+} ion concentration) rather than to apparent species differences.

III. Summary

Some of the properties of the interaction of antimalarial 8-aminoquinolines with DNA, RNA and various polynucleotides have been studied. Binding can be demonstrated by equilibrium dialysis and by direct spectrophotometry. Although appreciable binding of pentaquine and primaquine occurs, the binding does not cause a significant change in the transition temperature (T_m) or the viscosity of native DNA; the antimalarial 4-aminoquinoline, chloroquine, induces marked changes in both properties upon binding to native DNA. Pentaquine and primaquine, like chloroquine, inhibit the activity of *E. coli* RNA polymerase assayed with calf thymus native DNA template in a medium containing 1 mM Mn^{2+}. In agreement with reports of other DNA-drug interactions, the DNA-aminoquinoline complex is less sensitive to nuclease activity than free DNA. In contrast, the interaction of chloroquine, primaquine, or pentaquine with RNA results in an increased sensitivity of the RNA to enzymatic hydrolysis by several endonucleases. Primaquine has little effect on the aminoacylation of transfer RNA in a mouse liver system assayed in the presence of 5 mM Mg^{2+} ions.

On the basis of the interaction of the antimalarial aminoquinolines with nucleic acids and the consequent interference in nucleic acid synthesis and function, it is probable that such interactions are one mode of the antimalarial activity of these compounds and are responsible for some of the toxic reactions of these drugs in animals.

Acknowledgements

The authors wish to express their appreciation to Dr. THOMAS C. BUTLER for the encouragement and support of this program, and to Dr. J. LOGAN IRVIN for the initial discussions of the potential for this project. We gratefully acknowledge the excellent technical assistance of Mrs. MARY C. PARKER in this work.

This study was supported by the United States Public Health Service, National Institute of General Medical Sciences Grant 13606 (Dr. THOMAS C. BUTLER, Principal Investigator) and by the United States Army Medical Research and Development Command Contract DADA 17-69-C-9075; this is contribution No. 803 from the Army Malaria Research Program.

References

ALLISON, J. L., O'BRIEN, R. L., HAHN, F. E.: DNA: Reaction with chloroquine. Science **149**, 1111 (1965).

BATES, H. M., KUENZIG, W., WATSON, W. B.: Studies on the mechanism of action of anthramycin methyl ether, a new antitumor antibiotic. Cancer Res. **29**, 2195 (1969).

BEHR, W., HONIKEL, K., HARTMANN, G.: Interaction of the RNA polymerase inhibitor chromomycin with DNA. Europ. J. Biochem. **9**, 82 (1969).

BLODGETT, L. W., YIELDING, K. L.: Comparison of chloroquine binding to DNA, and polyadenylic and polyguanylic acids. Biochim. biophys. Acta (Amst.) **169**, 451 (1968).

COHEN, S. N., YIELDING, K. L.: Spectrophotometric studies of the interaction of chloroquine with deoxyribonucleic acid. J. biol. Chem. **240**, 3123 (1965).

— — (1) Inhibition of DNA and RNA polymerase reactions by chloroquine. Proc. nat. Acad. Sci. (Wash.) **54**, 521 (1965).

ERON, L. J., MCAUSLAN, B. R.: Inhibition of deoxyribonuclease action by actinomycin D and ethidium bromide. Biochim. biophys. Acta (Amst.) **114**, 633 (1966).

ESTENSEN, R. D., KREY, A. K., HAHN, F. E.: Studies on a deoxyribonucleic acid-quinine complex. Molec. Pharmacol. **5**, 532 (1969).

HAHN, F. E., O'BRIEN, R. L., CIAK, J., ALLISON, J. L., OLENICK, J. G.: Studies on modes of action of chloroquine, quinacrine, and quinine and on chloroquine resistance. Milit. Med. **131**, 1071 (1966).

ILAN, J., ILAN, J.: Aminoacyl transfer ribonucleic acid synthetases from cell-free extract of *Plasmoidum berghei*. Science **164**, 560 (1969).

IRVIN, J. L., IRVIN, E. M.: Spectrophotometric and potentiometric evaluation of apparent acid dissociation exponents of various 4-aminoquinolines. J. Amer. chem. Soc. **69**, 1091 (1947).

— — The interaction of a 9-aminoacridine derivative with nucleic acids and nucleoproteins. J. biol. Chem. **206**, 39 (1954).

— — (1) The interaction of quinacrine with adenine nucleotides. J. biol. Chem. **210**, 45 (1954).

— — PARKER, F. S.: The interaction of antimalarials with nucleic acids. I. Acridines. II. Quinolines. Science **110**, 426 (1949).

KURNICK, N. B., RADCLIFFE, I. E.: Reaction between DNA and quinacrine and other antimalarials. J. Lab. clin. Med. **60**, 669 (1962).

LAGOWSKI, J. M., FORREST, H. S.: Interaction in vitro between isoxanthopterin and DNA. Proc. nat. Acad. Sci. (Wash.) **58**, 1541 (1967).

LANDEZ, J. H., ROSKOSKI, R., Jr., COPPOC, G. L.: Ethidium bromide and chloroquine inhibition of rat liver cell-free aminoacylation. Biochim. biophys. Acta (Amst.) **195**, 276 (1969).

LEITH, J. D., Jr.: Acridine orange and acriflavin inhibit deoxyribonuclease action. Biochim. biophys. Acta (Amst.) **72**, 643 (1963).

LERMAN, L. S.: Structural considerations in the interaction of DNA and acridines. J. molec. Biol. **3**, 18 (1961).

MORRIS, C. R., ANDREW, L. V., WHICHARD, L. P., HOLBROOK, D. J., Jr.: The binding of antimalarial aminoquinolines to nucleic acids and polynucleotides. Molec. Pharmacol. **6**, 240 (1970).

MUENCH, K. H.: Chloroquine-mediated conversion of transfer ribonucleic acid of *Escherichia coli* from an inactive to an active state. Cold Spr. Harb. Symp. quant. Biol. **31**, 539 (1966).

— Chloroquine and synthesis of aminoacyl transfer ribonucleic acids. Tryptophanyl transfer ribonucleic acid synthetase of *Escherichia coli* and tryptophanyladenosine triphosphate formation. Biochemistry **8**, 4872 (1969).

— (1) Chloroquine and synthesis of aminoacyl transfer ribonucleic acids. Conformational changes in tryptophanyl and tryptophan transfer ribonucleic acids. Biochemistry **8**, 4880 (1969).

O'BRIEN, R. L., ALLISON, J. L., HAHN, F. E.: Evidence for intercalation of chloroquine into DNA. Biochim. biophys. Acta (Amst.) **129**, 622 (1966).

— OLENICK, J. G., HAHN, F. E.: Reactions of quinine, chloroquine, and quinacrine with DNA and their effects on the DNA and RNA polymerase reactions. Proc. nat. Acad. Sci. (Wash.) **55**, 1511 (1966).

PARKER, F. S., IRVIN, J. L.: The interaction of chloroquine with the albumin of bovine plasma. J. biol. Chem. **199**, 889 (1952).

— — (1) The interaction of chloroquine with nucleic acids and nucleoproteins. J. biol. Chem. **199**, 897 (1952).

POLET, H., BARR, C. F.: Chloroquine and dihydroquinine. In vitro studies of their antimalarial effect upon *Plasmodium knowlesi*. J. Pharmacol. exp. Ther. **164**, 380 (1968).

POWELL, R. D.: The chemotherapy of malaria. Clin. Pharmacol. Ther. **7**, 48 (1966).

SARKAR, N. K.: Effects of actinomycin D and mitomycin C on the degradation of deoxyribonucleic acid and polydeoxyribonucleotide by deoxyribonucleases and venom phosphodiesterase. Biochim. biophys. Acta (Amst.) **145**, 174 (1967).

SCHELLENBERG, K. A., COATNEY, G. R.: The influence of antimalarial drugs on nucleic acid synthesis in *Plasmodium gallinaceum* and *Plasmodium berghei*. Biochem. Pharmacol. **6**, 143 (1961).

SMITH, C. C.: Metabolism of pentaquine in the rhesus monkey. J. Pharmacol. exp. Ther. **116**, 67 (1956).

STERNGLANZ, H., YIELDING, K. L., PRUITT, K. M.: Nuclear magnetic resonance studies of the interaction of chloroquine diphosphate with adenosine 5′-phosphate and other nucleotides. Molec. Pharmacol. **5**, 376 (1969).

STOLLAR, D., LEVINE, L.: Antibodies to denatured deoxyribonucleic acid in lupus erythematosus serum. V. Mechanism of DNA-antiDNA inhibition by chloroquine. Arch. Biochem. **101**, 335 (1963).

SULKOWSKI, E., LASKOWSKI, M., Sr.: Degradation of thymus DNA and crab poly *d(A-T)* by micrococcal nuclease in the presence of actinomycin D. Biochim. biophys. Acta (Amst.) **157**, 207 (1968).

SUTHERLAND, J. C., SUTHERLAND, B. M.: Energy transfer in the DNA-chloroquine complex. Biochim. biophys. Acta (Amst.) **190**, 545 (1969).

VAN DYKE, K., SZUSTKIEWICZ, C., LANTZ, C. H., SAXE, L. H.: Studies concerning the mechanism of action of antimalarial drugs. Inhibition of the incorporation of adenosine-8-^{3}H into nucleic acids of *Plasmodium berghei*. Biochem. Pharmacol. **18**, 1417 (1969).

WARING, M. J.: The effects of antimicrobial agents on ribonucleic acid polymerase. Molec. Pharmacol. **1**, 1 (1965).

WHICHARD, L. P., MORRIS, C. R., SMITH, J. M., HOLBROOK, D. J., Jr.: The binding of primaquine, pentaquine, pamaquine, and plasmocid to deoxyribonucleic acid. Molec. Pharmacol. **4**, 630 (1968).

ZELEZNICK, L. D., SWEENEY, C. M.: Inhibition of deoxyribonuclease action by nogalamycin and U-12241 by their interaction with DNA. Arch. Biochem. **120**, 292 (1967).

Interpretation of Antimalarial Activity in Terms of Regression Analyses, Molecular Orbital Calculations, and Theory of DNA-Drug Binding

GEORGE E. BASS, DONNA R. HUDSON, JANE E. PARKER and WILLIAM P. PURCELL

I. Introduction

In 1965 a model to account for the antimalarial activity of chloroquine and its congeners was proposed by O'BRIEN and HAHN (O'BRIEN and HAHN, 1965) (see Fig. 1). In particular, they suggested that:

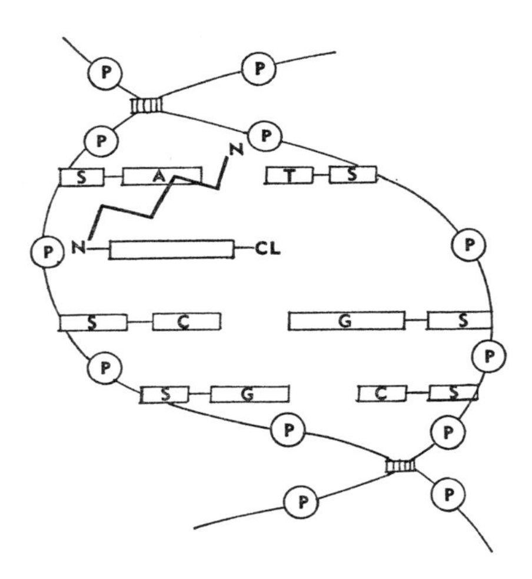

Fig. 1. Proposed structure of DNA-chloroquine complex consistent with model of R. L. O'BRIEN and F. E. HAHN (O'BRIEN and HAHN, 1965)

1. These compounds exert their antimalarial effect by intercalation with the parasite DNA, and that the activity of a given compound depends on the stability of its complex with DNA.

2. High activity requires an electronegative group attached to position 7 of the quinoline ring.

3. The diamino side chain bridges the two DNA strands by electrostatic interactions between the diamino nitrogens and the DNA phosphate groups.

To support this model, considerable evidence derived from *in vitro* intercalation studies (O'BRIEN, ALLISON and HAHN, 1966), *in vivo* bacteriological studies (HAHN, O'BRIEN, CIAK, ALLISON and OLENICK, 1966), and antimalarial activity data (O'BRIEN and HAHN, 1965) selected from the literature (COATNEY, COOPER, EDDY and GREENBERG, 1953; WISELOGLE, 1946) was offered. The first of these studies demonstrates that chloroquine can intercalate with DNA and the second shows that, at least for the strain of *Bacillus megaterium* studied, chloroquine inhibits DNA replication with subsequent death of the affected cell.

It is from the third study, the antimalarial activities of chloroquine and its congeners, that O'BRIEN and HAHN deduced the roles to be played by the substituents attached to position 7 of the ring and of the diamino side chain attached to position 4. Their observations, based on general trends in the activity data, were qualitative in nature. The results presented here represent an attempt to evaluate quantitatively these trends and to test some of the features of O'BRIEN and HAHN's model.

II. Methods

In the search for quantitative correlation, a free energy related expression, Eq. 1, which is patterned after the HANSCH model (HANSCH and FUJITA, 1964) was used.

$$\mathrm{Log}\frac{1}{C} = a\pi^2 + b\pi + cQ_x + dQ_{N1} + eQ_{N2} + f \qquad (1)$$

The drug concentration required to produce a specified response is designated C, π is a substituent constant derived from octanol/water partition coefficients, Q_x represents the net charge calculated for the substituent attached to position 7, Q_{N1} is the net charge of the nitrogen attached to position 4, and Q_{N2} is the net charge of the terminal nitrogen of the diamino side chain. The model of O'BRIEN and HAHN is based on electrostatic interactions involving these three atoms. The HANSCH parameter, π, has been introduced in the hopes of uncovering any apparent activity dependence on the transport or hydrophobic properties of these molecules. In practice, the relative activities reported by O'BRIEN and HAHN (O'BRIEN and HAHN, 1965) were used in place of 1/C. These quantities are proportional and, thus, the results are the same within an additive constant.

As a measure of the "degree of fit" for a particular equation, the square of the correlation coefficient, R^2, the overall F ratio of the regression and corresponding significance level, and the amount of explained variance (SNEDECOR and COCHRAN, 1967) were calculated. The net charges were obtained by combining the results of HÜCKEL pi electron calculations (PULLMAN and PULLMAN, 1963) and DEL RE sigma electron calculations (DEL RE, 1958). In both cases, parameter values recommended to reproduce dipole moments (BERTHOD, GIESSNER-PRETTRE and PULLMAN, 1967) were used. Values for the substituent constant π were gleaned from the publications of HANSCH and co-workers (HANSCH and FUJITA, 1964; FUJITA, IWASA and HANSCH, 1964; IWASA, FUJITA and HANSCH, 1965; HANSCH, QUINLAN and LAWRENCE, 1968; HANSCH and HELMERS, 1968; HELMER, KIEHS and HANSCH, 1968; LEO, HANSCH and CHURCH, 1969). The antimalarial activity data relate to the survival of baby chicks infected with *Plasmodium gallinaceum*.

In these analyses only compounds selected from the series listed by O'BRIEN and HAHN (O'BRIEN and HAHN, 1965) have been considered. These can be divided into three series:

1. Compounds which differ from chloroquine only in the substituent attached to position 7 of the ring, i.e.,

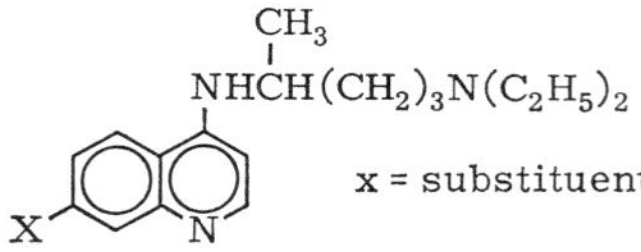

CH_3
$NHCH(CH_2)_3N(C_2H_5)_2$
X N
x = substituent group

Series I

2. Compounds which differ from chloroquine in the substituents attached to position 7 and one other position but not at position 4, i.e.,

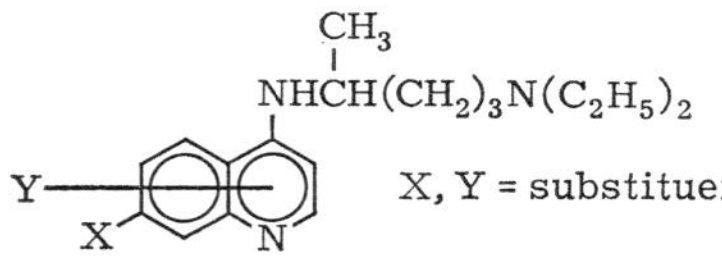

CH_3
$NHCH(CH_2)_3N(C_2H_5)_2$
Y X N
X, Y = substituent groups

Series II

3. Compounds which differ from chloroquine only in the diamino side chain attached at position 4, i.e.,

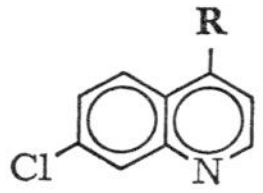

R
Cl N
R = diamino side chain

Series III

The molecules included in each of these series and corresponding parameter values are given in Tables 1 to 3. The results to be summarized here are the end product of some 20 HÜCKEL pi electron calculations, 59 DEL RE sigma electron calculations and 68 regression analyses.

III. Results

1. Analyses Involving Series I, II, and III

When the 33 compounds listed in Series I, II, and III were combined and subjected to a series of regression analyses, the essential results obtained are:

	R^2	Significance levels (%)	Explained variance
$\text{Log}\,(1/C) = a\pi^2 + b\pi + cQ_x + dQ_{N1} + eQ_{N2} + f$	0.26	85—90	0.13
$\text{Log}\,(1/C) = a\pi^2 + b\pi + cQ_x + d$	0.12	<75	0.03
$\text{Log}\,(1/C) = a\pi^2 + b\pi + c$	0.11	75—80	0.05

The very low correlation and explained variance for each equation indicate that antimalarial activity is too complex to be explained by such a general, simple model. To investigate the role of a particular segment of the molecule, it will be necessary to consider series of compounds in which only that segment varies.

Table 1. *Antimalarial activities and parameter values for series I*

CH_3
$NHCH(CH_2)_3N(C_2H_5)_2$
X N

X	Activity[a]	$\Sigma\pi$[b]	ϱ_X	ϱ_{N1}	ϱ_{N2}
Cl	100	4.20	−0.124	−0.264	−0.205
I	67	4.71	−0.101	−0.264	−0.205
Br	50	4.35	−0.109	−0.264	−0.205
F	50	3.63	−0.176	−0.264	−0.205
CF_3	50	4.65	−0.042	−0.264	−0.205
OCH_3	14	3.47	−0.071	−0.264	−0.205
CH_3	7	3.99	−0.026	−0.264	−0.205
H	7	3.49	+0.053	−0.264	−0.205

[a] Activities are relative to chloroquine which is 100 (O'BRIEN and HAHN, 1965).
[b] $\Sigma\pi$ = Sum of Hansch π values of all segments (including the quinoline ring) of the molecule.

Table 2. *Antimalarial activities and parameter values for series II*

CH_3
$NHCH(CH_2)_3N(C_2H_5)_2$
Y
X N

X	*Y*	Activity[a]	$\Sigma\pi$[b]	ϱ_X	ϱ_{N1}	ϱ_{N2}
Cl	H	100	4.20	−0.124	−0.264	−0.205
H	6-Cl	100	4.20	+0.055	−0.264	−0.205
H	5-Cl	3	4.20	+0.053	−0.264	−0.205
H	8-Cl	3	4.20	+0.055	−0.264	−0.205
H	3-Br	3	4.35	+0.053	−0.261	−0.205
H	6-OCH_3	10	3.47	−0.213	−0.266	−0.205
Cl	6-CH_3	25	4.70	−0.125	−0.264	−0.205
Cl	3-CH_3	15	4.70	−0.124	−0.264	−0.205
Cl	2-CH_3	10	4.70	−0.124	−0.264	−0.205
Cl	2-Φ	2	6.33	−0.124	−0.265	−0.205
Cl	3-Br	6	5.06	−0.124	−0.262	−0.205
Cl	5-Br	5	5.06	−0.120	−0.269	−0.205
Cl	8-NH_2	2	2.97	−0.127	−0.265	−0.205

[a] Activities are relative to chloroquine which is 100 (O'BRIEN and HAHN, 1965).
[b] $\Sigma\pi$ = Sum of Hansch π values of all segments (including the quinoline ring) of the molecule.

2. Analyses Involving Series I

Series I is comprised of 8 molecules which vary only in the substituent attached at position 7 of the quinoline ring. The overall results of the regression analyses on

Table 3. *Antimalarial activities and parameter values for series III*

(Structure: 7-chloroquinoline with R at position 4)

R	Activity[a]	$\Sigma\pi$[b]	Q_X	Q_{N1}	Q_{N2}	Q_{N2}^+
$-NHCH(CH_3)(CH_2)_3N(C_2H_5)_2$	100	4.20	−0.124	−0.264	−0.205	+0.525
-NH-[ring]-$N(C_2H_5)_2$	100	4.18	−0.124	−0.264	−0.208	+0.524
-NH-[ring]-$N(C_4H_9)_2$	25	6.26	−0.124	−0.264	−0.208	+0.524
-NH-[ring]-NHC_2H_5	100	3.25	−0.124	−0.264	−0.359	+0.346
-NH-[ring]-$NHCH(CH_3)CH_3$	50	3.64	−0.124	−0.264	−0.361	+0.344
-NH-[ring]-NH-[ring] Cis	50	4.66	−0.124	−0.264	−0.361	+0.348
-NH-[ring]-NH-[ring] Trans	25	4.54	−0.124	−0.264	−0.361	+0.348
$-NHCH(CH_3)(CH_2)_3NHCH_3$	100	2.75	−0.124	−0.264	−0.355	+0.349
$-NHCH(CH_3)(CH_2)_3NHC_2H_5$	50	3.27	−0.124	−0.264	−0.357	+0.347
$-NH(CH_2)_3N(C_2H_5)_2$	80	3.28	−0.124	−0.262	−0.209	+0.526
$-NH(CH_2)_3N(CH_2CH_2OH)_2$	8	−0.32	−0.124	−0.262	−0.205	+0.527
$-NH(CH_2)_3N(C_6H_{13})_2$	6	7.44	−0.124	−0.262	−0.210	+0.525
$-NH(CH_2)_3N(C_8H_{17})_2$	5	9.52	−0.124	−0.262	−2.210	+0.525
$-NHCH_2CH(OH)CH_2N(C_2H_5)_2$	100	0.96	−0.124	−0.261	−0.208	+0.527

[a] Activities are relative to chloroquine which is 100 (O'BRIEN and HAHN, 1965).
[b] $\Sigma\pi$ = Sum of Hansch π values of all segements (including the quinoline ring) of the molecule

these compounds indicate that the Hansch parameter, π, plays a secondary role in this series, and that, in keeping with the model of O'BRIEN and HAHN, the net charge on the substituent, Q_x, is of considerable importance. By carrying out successive analyses of Log $1/C$ *vs.* Q_x, starting with the substituents H, F, and Cl and expanding to include all the substituents of Series I, the results presented in Table 4 were obtained. These data demonstrate the manner in which correlation and explained variance change as larger and larger substituents are included. It is seen that optimum results are obtained when the largest group is bromine, and that incorporation of still larger groups into the series steadily decreases both correlation and explained variance. Thus, it is found that the ability of the parameter Q_x alone to explain the variation in activity begins to decrease at the very point where size of the largest substituent would tend to inhibit intercalation. This is in complete agreement with the model of O'BRIEN and HAHN.

3. Analyses Involving Series III

Series III consists of 14 molecules which differ from one another only at the 4-diamino side chain. In the model of O'BRIEN and HAHN, this side chain is depicted as being electrostatically bound through the diamino nitrogens to phosphate groups on opposite DNA strands. The DNA geometry led these authors to predict that the effectiveness of this side group should depend on the number of carbon atoms separating the two nitrogens; the optimal number for this separation is four. Nine of the compounds in Series III have the two diamino nitrogens separated by four carbons while for the remaining five compounds, the separation is three carbons.

When all of the molecules in Series III were considered together, the only meaningful correlation was obtained with the Hansch parameter, π. The charges on the nitrogens, Q_{N1} and Q_{N2}, taken alone, could account for essentially none of the variation in biological activity. Furthermore, it was found that inclusion of these parameters after π and/or π^2 was statistically unjustified. These results are summarized in Table 5.

To investigate the possible influence of the separation of the diamino nitrogens, the subgroups with three and four carbon separations were analyzed separately. In both examples, it was again found that the charges on the nitrogens could account for essentially none of the variance, while the Hansch parameter did seem important. When the regression analyses were carried out using both the Hansch parameter (as π and/or π^2) and Q_{N2}, however, very substantial increases in correlation and explained variance were obtained, suggesting a cooperative effect (see Table 5). Thus, these results indicate that, consistent with O'BRIEN and HAHN's model, antimalarial activity does depend on nitrogen separation. When this effect is taken into consideration, the charge on the terminal nitrogen can be correlated with antimalarial activity if it is considered along with the Hansch parameter. The calculated results do imply, however, that the role played by the charge is subordinate to that of the Hansch parameter. Along these lines, it should not be overlooked that the variation in π parallels that of the size of the groups attached to the terminal nitrogen. Thus, it is not immediately apparent whether the dependence of antimalarial activity on π can be related to the hydrophobic-transport properties of the molecule or to steric requirements of the receptor site (presumably, DNA).

Table 4. *Regression analyses on series I: Log* $(1/C) = AQ_X + B$

CH_3
$NHCH(CH_2)_3N(C_2H_5)_2$
X N

Substituents included	R^2	Significance level (%)	Explained variance
H, F, Cl	0.79	<75	0.59
H, F, Cl, Br	0.80	85—90	0.70
H, F, Cl, Br, I	0.78	93—95	0.69
H, F, Cl, Br, CH_3	0.75	90—95	0.67
H, F, Cl, Br, I, CH_3	0.73	95—97	0.66
H, F, Cl, Br, I, CH_3, CF_3	0.63	95—97	0.56
H, F, Cl, Br, I, CH_3, CF_3, OCH_3	0.58	95—97	0.52

Table 5. *Regression analyses on Series III*

R
Cl N

Regression equation	Number of compounds	R^2	Significance level (%)	Explained varianice
$\text{Log}\,(1/C) = A\pi + B$	14	0.24	90—95	0.17
$\text{Log}\,(1/C) = A\pi^2 + B\pi + C$	14	0.69	> 99	0.63
$\text{Log}\,(1/C) = AQ_{N2} + B$	14	0.09	< 75	0.02
$\text{Log}\,(1/C) = A\pi^2 + B\pi + CQ_{N2} + D$	14	0.70	> 99	0.60
$\text{Log}\,(1/C) = A\pi^2 + B\pi + CQ_{N2} + DQ_{N1} + E$	14	0.70	97—99	0.57
$\text{Log}\,(1/C) = A\pi + B$	9	0.44	90—95	0.36
$\text{Log}\,(1/C) = A\pi^2 + B\pi + C$	9	0.44	80—85	0.26
$\text{Log}\,(1/C) = AQ_{N2} + B$	9	0.01	< < 75	0.00
$\text{Log}\,(1/C) = A\pi + BQ_{N2} + C$	9	0.78	97—99	0.71
$\text{Log}\,(1/C) = A\pi^2 + B\pi + CQ_{N2} + D$	9	0.79	95—97	0.66
$\text{Log}\,(1/C) = A\pi + B$	5	0.30	< 75	0.07
$\text{Log}\,(1/C) = A\pi^2 + B\pi + C$	5	0.64	< 75	0.28
$\text{Log}\,(1/C) = AQ_{N2} + B$	5	0.01	< < 75	0.00
$\text{Log}\,(1/C) = A\pi + BQ_{N2} + C$	5	0.94	90—95	0.88
$\text{Log}\,(1/C) = A\pi^2 + B\pi + CO_{N2} + D$	5	0.95	< 75	0.80

IV. Conclusions

In these attempts to correlate the antimalarial activities of chloroquine and its congeners with their physical and molecular orbital properties, it has been found that the charge (and by inference, the size) of the substituent attached to position 7 of the quinoline ring is important. With regard to the 4-diamino side chain, the results tend to indicate that the hydrophobic-transport (π-dependent) properties and/or size of this side group should be considered along with the charges on the nitrogens. The net results of these findings are consistent with the model proposed by O'Brien and Hahn.

Acknowledgements

This research is being supported by the U. S. Army Medical Research and Development Command (DA-49-193-MD-2779), the Cotton Producers Institute (through the National Cotton Council of America), the National Science Foundation (GB-7383), and a grant from Eli Lilly and Company. This paper is contribution number 801 from the Army Research Program on Malaria. Computer facilities were provided through Grant HE-09495 from the National Institutes of Health.

References

Berthod, H., Giessner-Prettre, Cl., Pullman, A.: Sur les rôles respectifs des électrons σ et π dans les propriétés des dérivés halogénés des molécules conjuguées. Application à l'étude de l'uracile et du fluorouracile. Theoret. Chim. Acta **8**, 212—222 (1967).

Coatney, G. R., Cooper, W. C., Eddy, N. B., Greenberg, J.: Survey of antimalarial agents. Chemotherapy of *Plasmodium gallinaceum* infections; toxicity; correlation of structure and action. Publ. Hlth. Monogr. **1953**, 9.

del Re, G.: A simple MO-LCAO method for the calculation of charge distributions in saturated organic molecules. J. chem. Soc. **1958**, 4031.

Fujita, T., Iwasa, J., Hansch, C.: A new substituent constant, π, derived from partition coefficients. J. Amer. chem. Soc. **86**, 5175—5180 (1964).

Hahn, F. E., O'Brien, R. L., Ciak, J., Allison, J. L., Olenick, J. G.: Studies on modes of action of chloroquine, quinacrine, and quinine and on chloroquine resistance. Milit. Med. **131**, 1071—1089 (1966).

Hansch, C., Fujita, T.: ϱ-σ-π Analysis. A method for the correlation of biological activity and chemical structure. J. Amer. chem. Soc. **86**, 1616—1626 (1964).

— Helmer, F.: Extrathermodynamic approach to the study of the adsorption of organic compounds by macromolecules. J. Polymer Sci., Part A 1 **6**, 3295—3302 (1968).

— Quinlan, J. E., Lawrence, G. L.: The linear free-energy relationship between partition coefficients and the aqueous solubility of organic liquids. J. organ. Chem. **33**, 347—350 (1968).

Helmer, F., Kiehs, K., Hansch, C.: The linear free-energy relationship between partition coefficients and the binding and conformational perturbation of macromolecules by small organic compounds. Biochemistry **7**, 2858—2863 (1968).

Iwasa, J., Fujita, T., Hansch, C.: Substituent constants for aliphatic functions obtained from partition coefficients. J. med. Chem. **8**, 150—153 (1965).

Leo, A., Hansch, C., Church, C.: Comparison of parameters currently used in the study of structure-activity relationships. J. med. Chem. **12**, 766—771 (1969).

O'Brien, R. L., Allison, J. L., Hahn, F. E.: Evidence for intercalation of chloroquine into DNA. Biochim. biophys. Acta (Amst.) **129**, 622—624 (1966).

— Hahn, F. E.: Chloroquine structural requirements for binding to deoxyribonucleic acid and antimalarial activity. Antimicrob. Agents and Chemother. **1965**, 315.

Pullman, B., Pullman, A.: Quantum biochemistry, p. 67. New York: Interscience Publ. 1963.

Snedecor, G. W., Cochran, W. G.: Statistical methods, p. 386, 402. Ames, Iowa: Iowa State University Press 1967.

Wiselogle, F. Y. (Ed.): A survey of antimalarial drugs 1941—1945. Ann Arbor, Michigan: J. W. Edwards 1946.

Interactions of Alkaloids with DNA

FRED E. HAHN and ANNE K. KREY

I. Introduction

Alkaloids are nitrogen-containing bases of complex structures produced by higher plants. Although many of these substances show physiological activity and approximately 1000 alkaloids are known, only some 20 of them are used in practical medicine.

Owing to their cationic properties, alkaloids can be *a priori* expected to bind to polyanions, such as DNA, by electrostatic attraction. However, only in a limited number of instances has the interaction of alkaloids with DNA been investigated.

Extensive studies on their interactions with nucleic acids have been carried out with steroidal diamines such as cyclobuxin from *Buxus sempervirens* (MAHLER and DUTTON, 1964), irehdiamine from *Funtumia elastica* (MAHLER, GOUTAREL, KHUONG-HUU and HO, 1966; LEFRESNE, SAUCIER and PAOLETTI, 1967; SAUCIER, LEFRESNE and PAOLETTI, 1968; MAHLER, GREEN, GOUTAREL and KHUONG-HUU, 1968; WARING, this volume) and structurally related substances (MAHLER et al., 1966, 1968). These alkaloids permit optical studies of changes which they induce in the conformation of DNA because their absorption maxima do not occlude the major absorption band of nucleic acids around 260 nm. Irehdiamine and malouetine are effective inhibitors of bacterial growth and of bacteriophage replication, and also are mutagens (MAHLER and BAYLOR, 1967); these properties have been attributed to the formation of DNA-alkaloid complexes *in vivo*.

In the work to be presented, we have investigated the complex formation with DNA of three alkaloids, quinine, colchicine and berberine, with proved or presumed antiplasmodial activity. We also have studied the DNA complex of one synthetic quinine analog. This work has advanced the elucidation of the mechanism of action of quinine.

II. Colchicine

The antimitotic alkaloid, colchicine, has been shown to cure, synergistically with quinine, a number of cases of falciparum malaria caused by chloroquine-resistant strains of *Plasmodium falciparum* (REBA and SHEEHY, 1967). Colchicine has come under consideration as a possible DNA-complexing substance. The optical rotations at 589 nm of DNA-colchicine mixtures have been reported to be markedly different from the arithmetic sum of the optical rotations of the two substances measured individually (ILAN and QUASTEL, 1966). We have reproduced this result. However, in our hands, calf-thymus DNA did not alter the ultraviolet absorption or optical rotatory dispersion spectra of colchicine, and the alkaloid did not influence the thermal denaturation profile of DNA. Since colchicine owes its antimitotic action to its ability to combine with a component protein of the microtubules of the mitotic apparatus (BORISY and TAYLOR, 1967) it is doubtful that a putative ability of the alkaloid to complex with DNA is of wider biological importance.

In contrast, the antibiotic alkaloid anthramycin (HORWITZ, this volume) whose structure is reminiscent of that of colchicine binds to native double-stranded DNA and inhibits the DNA-dependent RNA polymerase reaction when double-stranded DNA is used as a template. Anthramycin binds covalently to duplex DNA (KOHN and SPEARS, 1970).

III. The DNA-Quinine Complex

An alkaloid of major medical importance is quinine which has been used as an antimalarial drug for centuries and is the first chemotherapeutic substance in history for which selective toxicity for an invasive pathogenic organism has been demonstrated (ROMANOVSKY, 1891). Quinine is selectively toxic for plasmodia because it inhibits plasmodial DNA biosynthesis, i.e. it inhibits the incorporation of building blocks into plasmodial DNA (SCHELLENBERG and COATNEY, 1961; POLET and BARR, 1968). Synthetic antimalarial drugs, foremost chloroquine (ANDERSAG, BREITNER and JUNG, 1939), have been developed through molecular modification, retaining as a key component the quinoline ring system of quinine. After it was found (HAHN, O'BRIEN, CIAK, ALLISON and OLENICK, 1966; YIELDING, this volume) that chloroquine forms a molecular complex with DNA and that quinine, like chloroquine, inhibits the DNA-dependent DNA polymerase reaction *in vitro* (O'BRIEN, OLENICK and HAHN, 1966), detailed studies on the interaction of quinine and DNA were undertaken (ESTENSEN, KREY and HAHN, 1969) with the following results.

1. Qualitative Optical Observations

Calf-thymus DNA altered the absorption spectrum of quinine (O'BRIEN, OLENICK and HAHN, 1966). Fig. 1 shows a difference spectrum quinine + DNA minus quinine alone; the figure also shows that this difference became insignificant when measured in the presence of 6 M urea. It was concluded that hydrogen bonding is involved in the formation of the DNA quinine complex. However, NaCl and $MgCl_2$ (Figs. 2 and 3) abolished the influence of DNA on quinine's absorption spectrum, suggesting that, in addition to hydrogen bond formation, electrostatic attractions are also involved in the formation of the complex of quinine with DNA. Denatured DNA, prepared by heating and rapid cooling or single-stranded DNA prepared by heating in the presence of 3.3×10^{-2} M formaldehyde (STOLLAR and GROSSMAN, 1962), had no influence on quinine's absorption spectrum, while unheated native DNA, treated with formaldehyde, produced the same effect as DNA which had not been so treated. Double-strandedness of DNA is, therefore, an essential precondition for observable changes in the absorption spectrum of quinine.

The intensity of absorption bands of other drugs which bind preferentially to double-stranded DNA or to double-helical DNA-like polymers such as actinomycin (GOLDBERG, RABINOWITZ and REICH, 1962) or chloroquine (O'BRIEN, OLENICK and HAHN, 1966), is strongly decreased by polymers containing guanine. This was also found to be the case for quinine. Fig. 4 shows a family of difference spectra of the

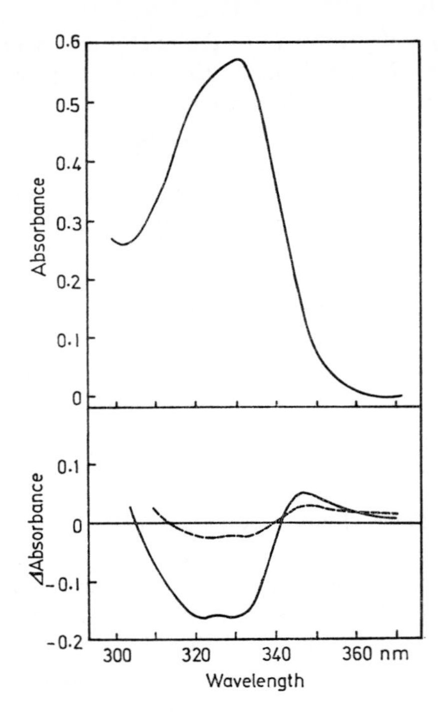

Fig. 1. Influence of calf thymus DNA (10^{-3} M component nucleotides) on the light absorption of quinine (10^{-4} M). Upper frame: absorption spectrum of quinine in 5×10^{-3} M Tris HCl at pH 7. 5. Lower frame: ——— quinine-DNA mixture minus quinine alone; ----- the same in the presence of 6 M urea

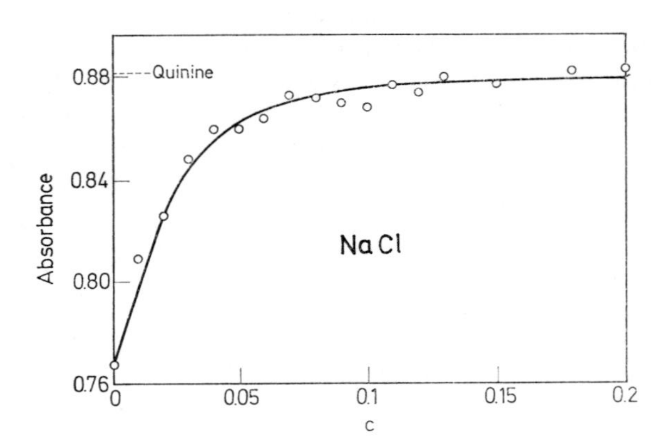

Fig. 2. Influence of NaCl on the light absorption of a mixture of 250 μg/ml of DNA and 2×10^{-4} M quinine. The ordinate represents optical density at 331 nm and the abscissa molar concentrations of NaCl

complexes of quinine with DNA, poly *dG:dC*, poly *dI:dC* and poly *dA:dT*. The greatest difference of a complex spectrum from that of quinine alone was seen for DNA, followed by poly *dG:dC*. In contrast, poly *dI:dC* and poly *dA:dT* produced only a slight nonspecific effect on quinine's spectrum with no difference detected at 331 nm, the absorption maximum of the alkaloid's spectrum.

The effect of graded concentration of single-stranded ribopolynucleotides on quinine's peak absorption at 331 nm was also measured. The only marked depression in quinine's absorption was caused by poly *G* at high concentrations.

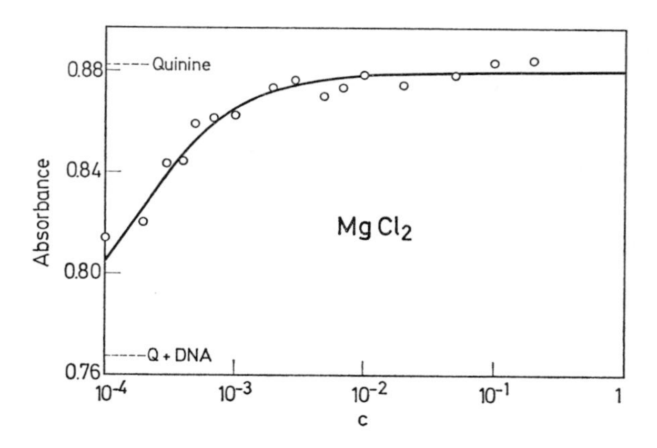

Fig. 3. Influence of $MgCl_2$ on the light absorption of a mixture of DNA and quinine. Conditions as in Fig. 2 except that the abscissa represents molar concentrations of $MgCl_2$

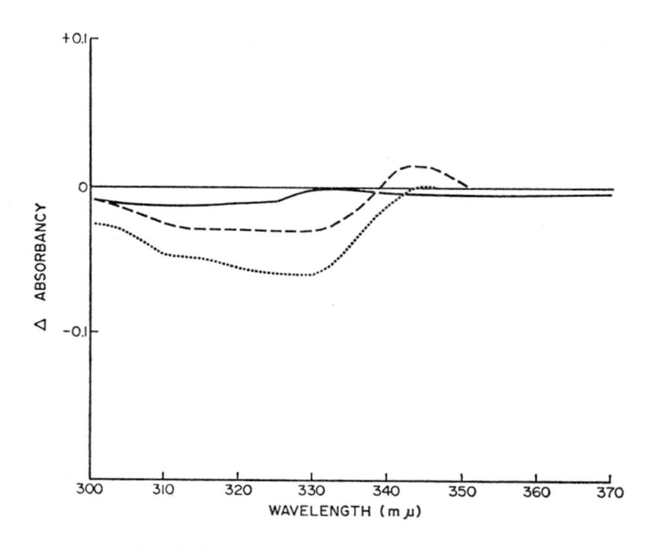

Fig. 4. Difference spectra of polydeoxyribotide or of DNA complexes with quinine (polymer-quinine complex minus quinine alone). Quinine concentration was 2×10^{-4} M. Polymer concentration was 6×10^{-4} M with respect to component nucleotides. ......: DNA; ------: poly *dG:dC*; ———: poly *dA:dT* or poly *dI:dC*

2. Spectrophotometric Titration of Quinine with DNA

The ability of DNA to decrease the light absorption of quinine was used to carry out spectrophotometric titrations of a constant concentration of quinine with graded concentrations of calf-thymus DNA according to principles summarized by Blake and Peacocke (1968). The results of a typical experiment transformed into an adsorption isotherm, are shown in Fig. 5. Some problems in the interpretation of such curves are discussed by Crothers (this volume). It is customary to consider a curved isotherm as an indication of the existence of more than one class of binding sites to which a drug attaches itself by more than one binding process.

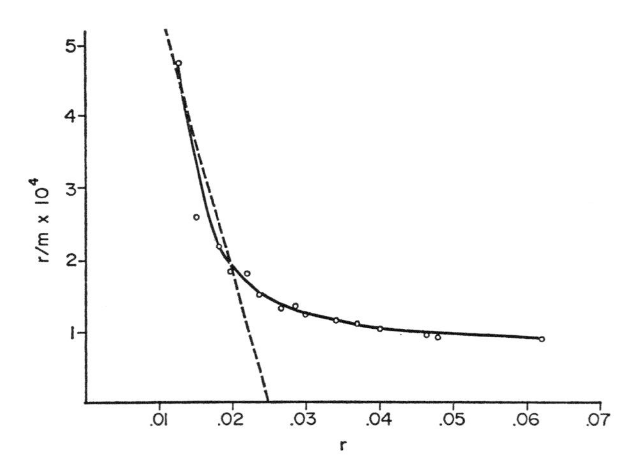

Fig. 5. Isotherm for the binding of quinine to calf thymus DNA. *r* number of bound quinine molecules per base pair; *m* = concentration of unbound quinine. ------: Regression line computed from the first five points

A tangent to the isotherm at the side of low values of *r*, as shown in Fig. 5, yielded for the strongest binding process a stoichiometry of 0.025 quinine molecules per base pair, i.e. one molecule of quinine bound per 40 base pairs with an apparent association constant of 3.6×10^6 M^{-1}. A corresponding tangent to the side of large values of *r* (not shown) suggested a stoichiometry of 0.125 molecules of quinine per base pair, i.e. one molecule of quinine bound to eight or nine base pairs with an apparent association constant of 1.5×10^5 M^{-1}. Additional binding of quinine to DNA without spectrophotometric manifestations was demonstrated by equilibrium dialysis which determines total binding and showed that one molecule of quinine was bound to approximately four or five base pairs of DNA. At quinine concentrations $> 10^{-3}$ M, DNA was aggregated or precipitated. We do not understand the conditions underlying the quantitative limitations in the strong binding of quinine to DNA and can only speculate as to the existence of some exclusion principle comparable to that which limits the number of molecules of actinomycin which can bind strongly to DNA. In view of the studies of Wells (this volume) it is unlikely that the presence of guanine in DNA is alone responsible for such selectivity of strong binding; one will have to assume that a specific arrangement of several bases jointly constitutes one strong attachment site.

3. Hydrodynamic Experiments with the DNA-quinine Complex

Since the quinoline ring of chloroquine intercalates into DNA (O'Brien, Allison and Hahn, 1966) it was speculated that the quinoline moiety of quinine too might be intercalated; this was also suggested by the preference of the drug for double-

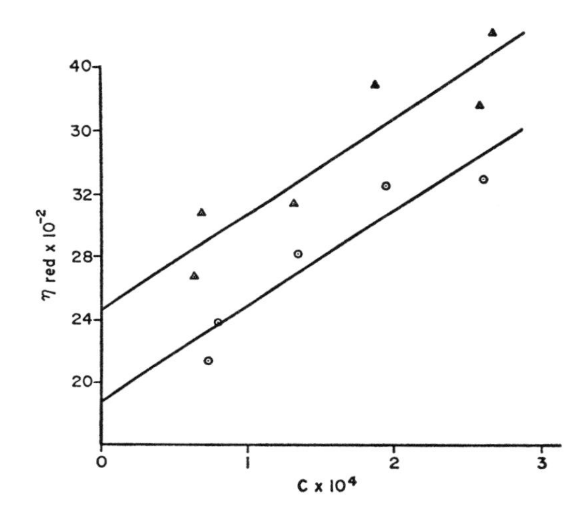

Fig. 6. Enhancement of DNA's viscosity by quinine. Upper line: DNA-quinine mixtures of a DNA phosphorus to quinine ratio of 10:1. Lower line: DNA alone. DNA concentrations are represented on the abscissa. The ordinate represents reduced viscosities, i.e. specific viscosity divided by DNA concentration

helical DNA. Among the criteria for intercalation binding are requirements that the binding substance increases the intrinsic viscosity of DNA but decreases the sedimentation constant of the polymer. Fig. 6 shows such an increase in the viscosity of a DNA preparation upon complexing with quinine, and Fig. 7 shows the decrease in the sedimentation constant when quinine binds to DNA. Although it would be desirable to add support to the quinine intercalation hypothesis by furnishing additional experimental criteria such as measurement of flow dichroism, radioautography of DNA complexes with quinine, light scattering and low angle *X*-ray scattering, or steric hindrance to chemical reactivity of the intercalated drug, such experiments are either unpromising on statistical grounds since one molecule of quinine is strongly bound per only 40 base pairs or are physically not feasible because quinine at $> 10^{-3}$ M precipitates DNA.

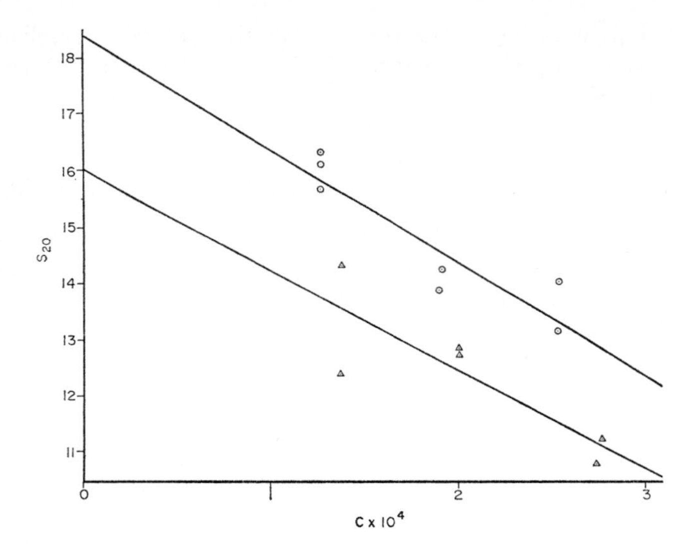

Fig. 7. Diminution of DNA's sedimentation coefficient by quinine. Lower line: DNA-quinine mixture; upper line: DNA alone. Concentrations as in Fig. 6

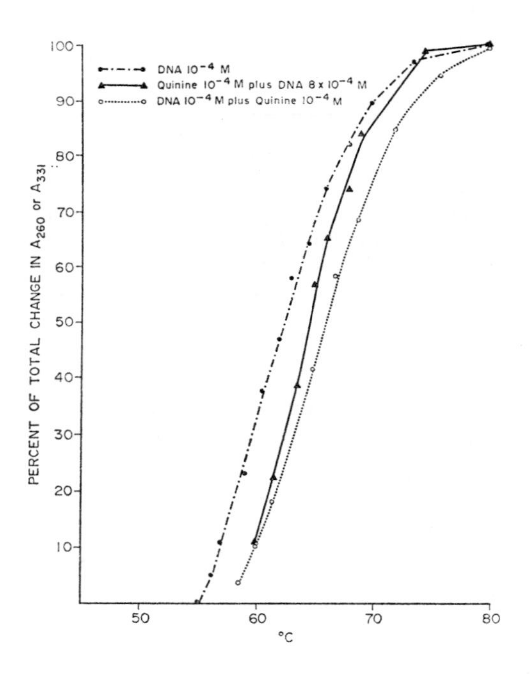

Fig. 8. Thermal denaturation profiles of DNA (25 μg/ml) in the absence ----- and in the presence · · · · of 2 × 10^{-4} M quinine and thermal reversal of the effect of 250 μg/ml of DNA on the absorption of quinine at 331 nm ———

4. Stabilization of DNA by Quinine

Quinine stabilized DNA to thermal denaturation as shown in Fig. 8. The increase in T_m was smaller than that produced by chloroquine (HAHN et al., 1966). Fig. 8 also shows that quinine became dissociated from DNA upon heating to the extent to which the hydrogen-bonded structure of the double helix was melting out. The melting of the DNA-quinine complex did not show evidence of greater cooperativity than that of DNA alone: quinine once dissociated from melting DNA did not become re-associated with as yet undenatured stretches of the double helix.

It has been postulated (O'BRIEN, OLENICK and HAHN, 1966) that substances which interfere with DNA's strand separation will inhibit DNA replication *in vitro*, a process

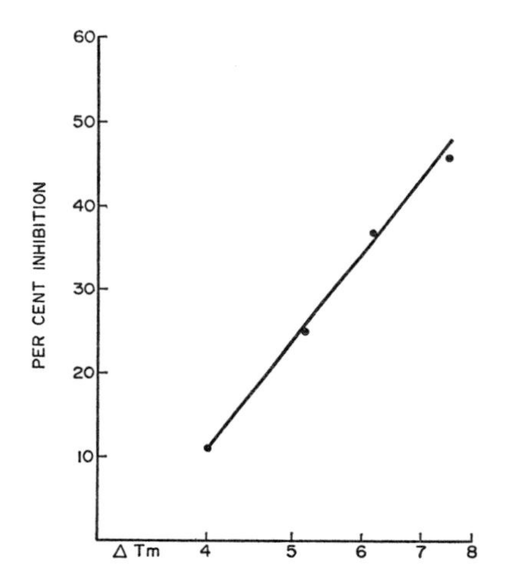

Fig. 9. Correlation between the stabilization of DNA to heat by quinine (ΔT_m) and the inhibition by quinine of a cell-free DNA polymerase reaction. The points represent quinine concentrations of 2.25, 5.5, 7.75 and 10 × 10^{-4} M from left to right

for which it is necessary that the two companion strands of the parental DNA molecule become separated. The same authors also presented a dosage response curve for the inhibition of the DNA polymerase reaction by quinine. When ΔT_m, the increase in median strand separation temperature of complexes of DNA with graded concentrations of quinine (2.25×10^{-4} M to 10^{-3} M), was determined and the per cent inhibition of the DNA polymerase reactions produced by the same quinine concentrations was plotted as a function of log ΔT_m, the linear correlation of Fig. 9 was obtained which suggests that over the concentration range which can be studied, the inhibition of the DNA polymerase reaction is, indeed, a function of the stabilization of the double helix by quinine.

5. Hypothetical Structure of the DNA-Quinine Complex and Structure-Activity Rules

A structural model of a DNA-quinine complex must account for the following findings: (1) DNA must be double-stranded in order to alter significantly quinine's absorption spectrum. (2) Quinine enhances DNA's viscosity and diminishes its sedimentation coefficient. (3) DNA's effect upon the absorption spectrum of quinine is reversed by urea as well as by inorganic ions. It has been proposed (ESTENSEN et al., 1969) that quinine forms by a strong binding process an intercalation complex with DNA in which the quinoline ring is inserted from the minor groove between base pairs into DNA, the alcoholic hydroxyl group of the drug molecule engages in hydrogen bond formation, perhaps with the 2-amino group of guanine, and the quinuclidine moiety protrudes into the minor groove of the double helix and is electrostatically attracted with its tertiary alicyclic amino group to phosphate groups of deoxyribonucleotides. Model building experiments have shown that this proposed structure is possible.

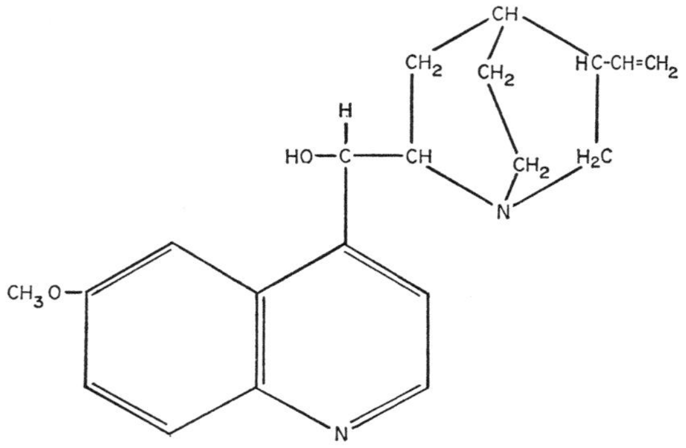

Fig. 10. Chemical structure of quinine

Fig. 10 shows the structure of quinine. Provided that inhibition of plasmodial DNA biosynthesis is the mode of action of the drug and is based upon the formation of a DNA complex with the structure suggested above, it should be possible to rationalize certain empirical structure-activity rules known to apply to the antimalarial potency of quinine derivatives. (1) The model predicts that elimination of the alcoholic hydroxyl group of quinine, essential for hydrogen bonding to DNA, should abolish antimalarial activity. This is, indeed, the case (COHEN and KING, 1938). Similarly, steric inversion of this hydroxyl group at its asymmetric carbon atom, yielding epiquinine, may introduce steric hindrance to hydrogen bond formation and does, in fact, result in marked reduction in antimalarial potency (COHEN and KING, 1938). (2) Changes in the distance and spatial relationship between the alcoholic hydroxyl group and the alicyclic amino group should interfere with the bimodal binding of quinine to DNA. Indeed, eliminating the bond between the amino nitrogen and carbon atom number 8 destroys antimalarial activity (AINLEY and KING, 1938), while substitution of the entire quinuclidine moiety by -CH_2NR_2 which does maintain a tertiary aliphatic amino group in the correct distance from the hydroxyl group, yields active compounds (KING and WORK, 1942).

6. Complex Formation of a Synthetic Quinoline Methanol with DNA

When this critical distance is maintained by substituting piperidine for the quinuclidine ring system of quinine, an antimalarial quinoline methanol is obtained (AINLEY and KING, 1938). From the more than 200 synthetic quinoline methanols, synthesized as promising antimalarials on this basis, one compound, Fig. 11, was selected for study of possible complex formation with DNA (HAHN and FEAN, 1970). Experiments were rendered difficult by the fact that this substance, more strongly than quinine, aggregated or precipitated DNA.

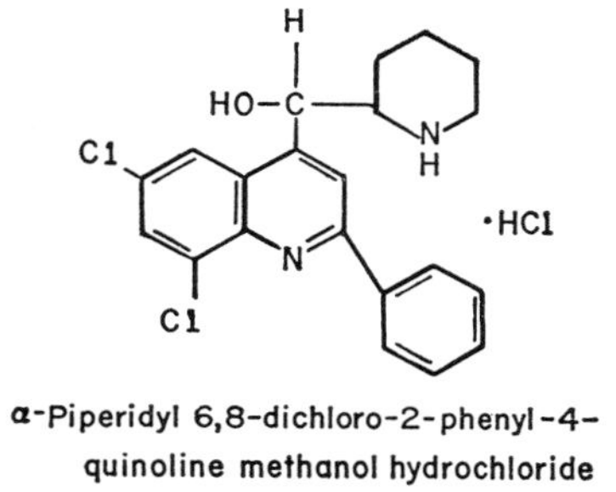

HO–C(H)– · HCl

α-Piperidyl 6,8-dichloro-2-phenyl-4-quinoline methanol hydrochloride

Fig. 11. Chemical structure of the synthetic quinoline methanol studied

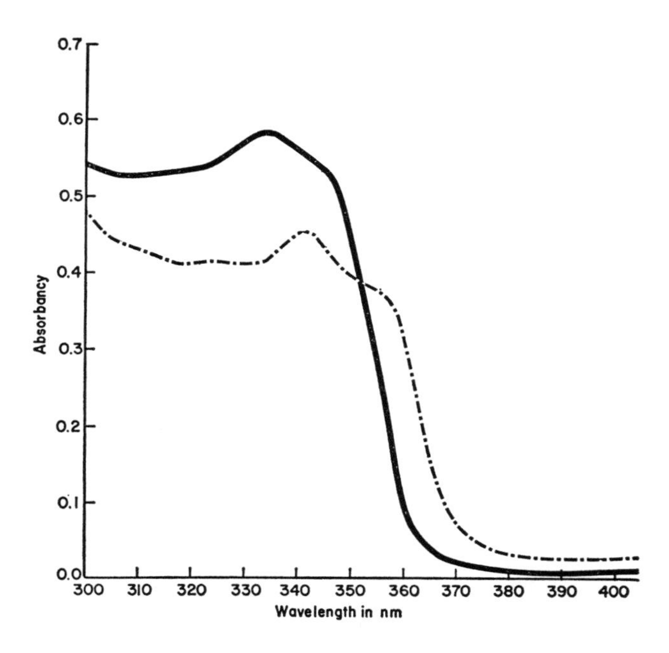

Fig. 12. Influence of DNA (6.9×10^{-5} M component nucleotides) upon the absorption spectrum of the quinoline methanol in Fig. 11 (6.2×10^{-6} M). ———: Drug alone; -·-·-·: drug-DNA complex

Fig. 12 shows that DNA decreased the intensity of the drug's absorption band and shifted it to longer wavelengths. This effect was observed at low concentrations of the drug (of the order of 10^{-6} M) and of DNA (of the order of 10^{-5} M component nucleotide).

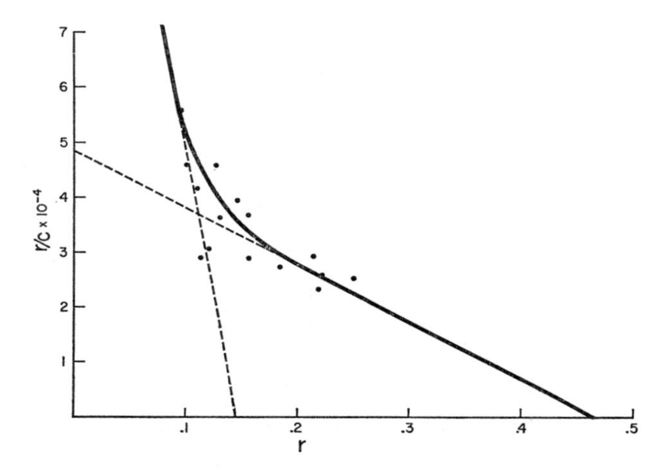

Fig. 13. Isotherm for the binding of the quinoline methanol, Fig. 11, to DNA. r = number of bound drug molecules per nucleotide; c = concentration of unbound drug

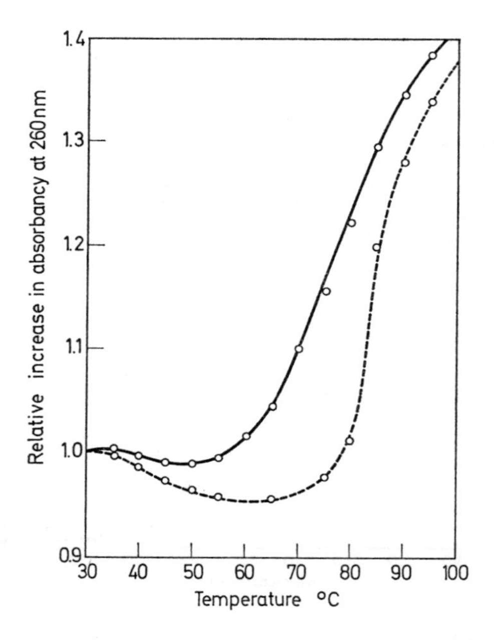

Fig. 14. Thermal denaturation profiles of DNA (5.2×10^{-5} M component nucleotides) in the absence ———, or in the presence ----- of quinoline methanol (1.16×10^{-5} M)

Like for quinine, a spectrophotometric titration of a constant concentration of the quinoline methanol with graded concentrations of DNA was carried out and the results transformed into the adsorption isotherm, Fig. 13. Again, a curved plot suggested the existence of more than one binding site in DNA and of more than one binding process. The apparent association constants for the assumed strong and weak binding processes were of the same magnitude as those estimated for quinine but the number of quinoline molecules bound by either process to DNA was approximately ten times greater than the corresponding numbers of quinine molecules. This may well account for the comparatively greater antimalarial potency of this synthetic quinoline methanol. The quinoline methanol shifted the thermal denaturation profile of DNA to higher temperatures as shown in Fig. 14. The drug was much more active in this respect than quinine.

The important point is that one quinoline methanol, typical for many such potent antimalarial compounds and embodying those structural features which according to our ideas are responsible for the formation of the DNA-quinine complex, did, indeed, form a complex with DNA albeit difficult to study in detail owing to the poor solubility of the drug and its property of precipitating DNA.

IV. The DNA-Berberine Complex

Berberine is a yellow, optically inactive, alkaloid which occurs in a number of unrelated plant species, in North America foremost in *Hydrastis canadensis*. Its salts are formed with loss of one molecule of water and are, therefore, properly designated

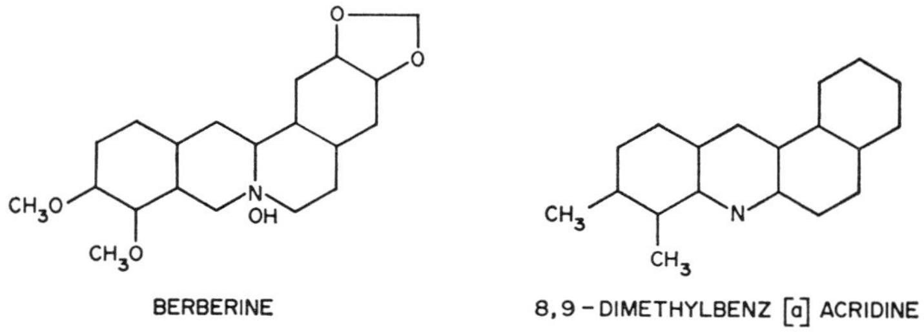

BERBERINE

8,9-DIMETHYLBENZ [a] ACRIDINE

Fig. 15. Structure of berberine and of a similar benzacridine

as berberinium compounds. The alkaloid has long been known to have antibacterial properties (STICKL, 1928), is useful in the treatment of cutaneous leishmaniasis (VARMA, 1927) and has been often claimed to be an antimalarial drug (reviewed by SHIDEMAN, 1950) without conclusive experimental evidence to that effect.

DNA and, to a lesser extent, RNA alter the absorption spectrum of berberinium sulfate (MORTHLAND, DE BRUYN and SMITH, 1954; KLÍMEK and HNILICA, 1958), and DNA markedly increases the intensity of the fluorescence of the alkaloid (YAMAGISHI, 1962). Another suggestion that berberine interacts with DNA comes from observations (MEISEL and SOKOLOVA, 1960) that the alkaloid is a mitochondrial mutagen in yeast.

The structure of berberine, Fig. 15, is reminiscent of that of 8,9-dimethylbenz[a]-acridine, Fig. 15, which has been reported to enhance the viscosity of DNA owing to intercalation of the ring system between base pairs (LERMAN, 1964). We have reported (KREY and HAHN, 1969) and expanded studies which show that berberine forms a complex with DNA and suggest that the alkaloid is intercalated.

1. Qualitative Optical Observations

Visually observed, the yellow berberine sedimented with DNA in the analytical ultracentrifuge and accumulated below the schlieren boundary of DNA. This constitutes the most direct and general type of proof of the existence of a DNA complex

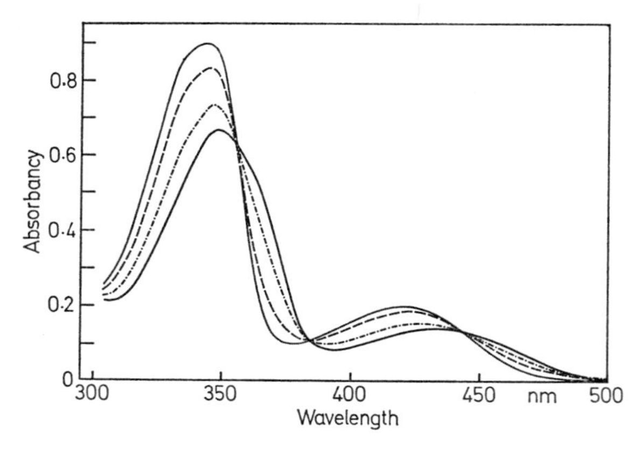

Fig. 16. Influence of DNA on the absorption spectrum of berberine (4×10^{-5} M) in Tris HCl at pH 7.5 ····: Berberine alone; -----: berberine plus 1.14×10^{-4} M component bases DNA; -·-·-·-·: plus 2.27×10^{-4} M component bases DNA; ———: plus 4.5×10^{-4} M component bases DNA

with a chromophoric substance. The absorption spectrum of berberine was altered by DNA in two respects, Fig. 16. The intensity of the alkaloid's absorption bands was progressively decreased by increasing concentrations of DNA and the absorption maxima were shifted progressively to longer wavelengths. This suggests that single berberine molecules are bound to DNA and that the alkaloid in the bound condition exhibits a true "monomer" spectrum (MICHAELIS, 1947). Solutions of berberinium sulfate alone exhibited a comparable red shift upon progressive dilution. It appears, therefore, that berberine has a tendency to self-aggregation.

The changes which DNA produces in the absorption spectrum of berberine were partly reversed by 6 M urea and completely reversed by 3×10^{-1} M CsCl. This suggests that the alkaloid binds to DNA through formation of hydrogen bonds as well as by electrostatic attraction. The spectrum of berberine was also changed by transfer RNA and by DNA which had been rendered single-stranded by heating in the presence of 3.3×10^{-2} M formaldehyde (STOLLAR and GROSSMAN, 1962). The binding of berberine to nucleic acids, therefore, does not depend upon double-helicity of the polymers.

2. Titrations of Berberine with DNA

The reduction by DNA of the absorption band of berberine at 344 nm and the strong enhancement of the alkaloid's fluorescence at 520 nm, when excited by light of 350 nm, were used to carry out spectrophotometric and spectrophotofluoremetric titrations of a constant concentration of berberine (5×10^{-5} M) with graded concentrations of DNA. The results of both titrations could be converted into curved adsorption isotherms similar to those in Figs. 5 and 13. The binding parameters for one strong and one weak binding process were derived from these curves and are tabulated in Table 1. Both titrations determined a stoichiometry of strong binding of one molecule of berberine per 5 base pairs and a weaker binding of the alkaloid per 2 base pairs. Either of the two binding frequencies is within the theoretically postulated limit of intercalation binding of one ligand molecule per 2 base pairs (CAIRNS, 1962).

Table 1. *Parameters of binding of berberine to DNA*

Binding process	Association constants		Number of base pairs binding one berberine molecule	
	Spectrophotometric	Fluorometric	Spectrophoto-metric	Fluorometric
Strong	4.8×10^5 M^{-1}	21.2×10^5 M^{-1}	5	5
Weak	0.9×10^5 M^{-1}	1.2×10^5 M^{-1}	2	2

3. Hydrodynamic Properties of the DNA-Berberine Complex

Intercalation binding was, in fact, suggested by the results of flow dichroism experiments on the DNA-berberine complex and by comparing the intrinsic viscosities of this complex to that of the uncomplexed DNA preparation. Flow dichroism was measured as before (O'BRIEN, ALLISON and HAHN, 1966), based on the technique and theoretical considerations developed by LERMAN (1963) in his studies of the acridine, quinacrine. Table 2 shows that the magnitudes and signs of the flow dichroism of the DNA bases, measured at 259 nm, and of berberine in the DNA complex, measured at 350 nm, were the same. Extending LERMAN's argument (1963) concerning the direction of the transition moment of the long wavelength absorption band in quinacrine, the data in Table 2 suggest that in the complex with DNA, the ring system of berberine lies parallel to the planes of the component base pairs of the double helix. This is one of the self-evident characteristics of intercalation binding.

Fig. 17 shows the results of viscosimetric measurements of a concentration series of the DNA-berberine complex and of the corresponding series of DNA concentrations determined as described previously (O'BRIEN, ALLISON and HAHN, 1966). Extrapolation to concentration naught yielded on the ordinate values of the intrinsic viscosities. The intrinsic viscosity of the DNA-berberine complex was significantly

10*

greater than that of the DNA preparation alone. Such a viscosity enhancement represents one hydrodynamic criterion for intercalation binding (LERMAN, 1964; O'BRIEN, ALLISON and HAHN, 1966).

Table 2. *Flow dichroism of the DNA-berberine complex*

Dichroism	259 nm	350 nm
At 0°	− 0.22	− 0.18
At 90°	+ .09	+ .08

This method has been described by LERMAN (1963) and O'BRIEN, ALLISON and HAHN (1966). Dichroism is expressed as fractional change in absorbancy of polarized light when the complex is flow-oriented. 0°, plane of polarized light parallel to axis of flow; 90°, plane of polarized light perpendicular to axis of flow.

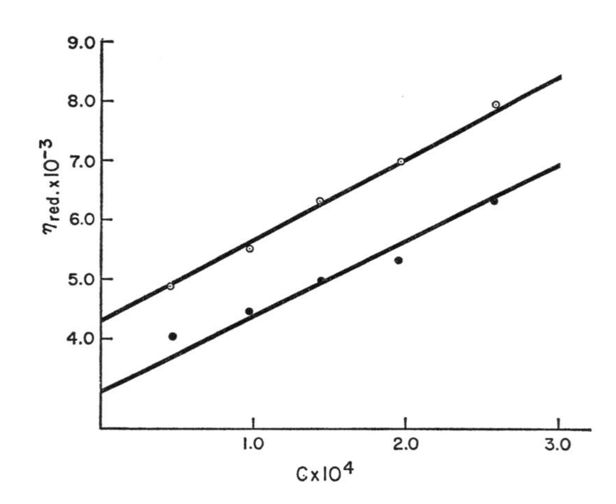

Fig. 17. Enhancement of DNA's viscosity by berberine. Experimental conditions and parameters the same as in Fig. 6

4. Stabilization of DNA by Berberine

Berberine stabilized DNA to thermal heat denaturation as shown in Fig. 18. The melting curve of the DNA-berberine complex was steeper than that of DNA alone, indicating a more cooperative type of melting. This is usually observed when a DNA-complexing molecule, separating from heated DNA, reassociates with an as yet undenatured part of the double helix.

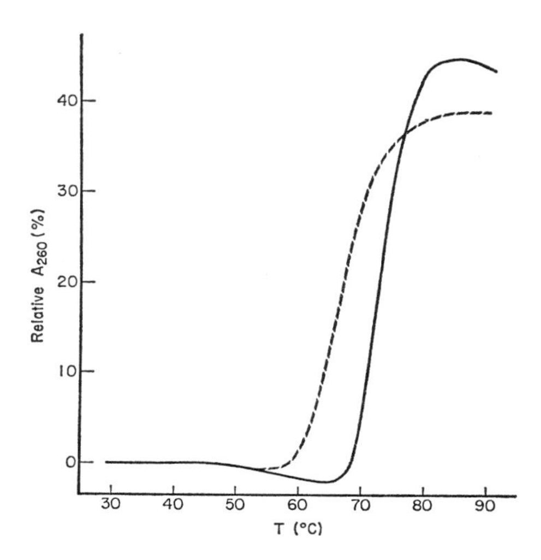

Fig. 18. Thermal denaturation of 20 μg of calf thymus DNA in the presence of 5×10^{-5} M berberine ——— and in its absence ----- (KREY and HAHN, 1969)

V. Summary

Quinine, colchicine and berberine, three alkaloids with proved or presumed anti-plasmodial activity have been studied as to their ability to form complexes with DNA.

1. It was confirmed that the optical rotations of colchicine-DNA mixtures differ markedly from the arithmetic sum of the rotations of colchicine and of DNA measured separately, but no other indications of the formation of a DNA-colchicine complex were detected.

2. Berberine forms a complex with DNA and binds maximally to the extent of one alkaloid molecule per two base pairs. Hydrodynamic criteria for intercalation binding (flow-dichroism, viscosity enhancement) have been satisfied for the DNA-berberine complex.

3. Quinine forms a complex specifically with native, double-helical DNA. The extent to which the alkaloid inhibits the cell-free polymerization of DNA is a function of the stabilization of the double helix to forces which bring about strand separation. In order for DNA to be replicated, the two companion strands of duplex DNA must undergo separation. An hypothetical model of the structure of the DNA-quinine intercalation complex serves to rationalize the known empirical structure-(anti-malarial)activity rules for quinine derivatives. One synthetic quinoline methanol of high antimalarial activity which conforms to these rules also forms a complex with DNA. Since quinine is known to inhibit plasmodial DNA synthesis *in vitro* and *in vivo*, the formation of the DNA-quinine complex probably represents the mechanism of action of quinine.

References

AINLEY, A. D., KING, H.: Antiplasmodial action and chemical constitution. II. Some simple synthetic analogs of quinine and cinchonine. Proc. roy. Soc. B **125**, 60 (1938).

ANDERSAG, H., BREITNER, S., JUNG, H.: Verfahren zur Darstellung von in 4-Stellung basisch substituierte Aminogruppen enthaltenden Chinolinverbindungen. German Pat. 683692 (1939).

BLAKE, A., PEACOCKE, A. R.: The interaction of aminoacridines with nucleic acids. Biopolymers **6**, 1225 (1968).

BORISY, G. G., TAYLOR, E. W.: The mechanism of action of colchicine. J. Cell Biol. **34**, 535 (1967).

CAIRNS, J.: The application of autoradiography to the study of DNA viruses. Cold Spr. Harb. Symp. quant. Biol. **27**, 311 (1962).

COHEN, A., KING, H.: Antiplasmodial action and chemical constitution. I. Cinchona alkaloidal derivatives and allied substances. Proc. roy. Soc. B **125**, 49 (1938).

ESTENSEN, R. D., KREY, A. K., HAHN, F. E.: Studies on a deoxyribonucleic acid-quinine complex. Molec. Pharmacol. **5**, 532 (1969).

GOLDBERG, I. H., RABINOWITZ, M., REICH, E.: Basis of actinomycin action. I. DNA binding and inhibition of RNA polymerase synthetic reactions by actinomycin. Proc. nat. Acad. Sci. (Wash.) **48**, 2094 (1962).

HAHN, F. E., FEAN, C. F.: Spectrophotometric studies of the interaction of an antimalarial quinoline methanol with deoxyribonucleic acid. Antimicrobial Agents and Chemother. **1969**, 63 (1970).

— O'BRIEN, R. L., CIAK, J., ALLISON, J. L., OLENICK, J. G.: Studies on modes of action of chloroquine, quinacrine and quinine and on chloroquine resistance. Milit. Med. **131**, 1071 (1966).

ILAN, J., QUASTEL, J. H.: Effects of colchicine on nucleic acid metabolism during metamorphosis of *Tenebrio molitor* L. and in some mammalian tissues. Biochem. J. **100**, 448 (1966).

KING, H., WORK, T. S.: Antiplasmodial action and chemical constitution. V. Carbinolamines derived from 6-methoxyquinoline. J. chem. Soc. **1942**, 401.

KLÍMEK, M., HNILICA, L.: The influence of deoxyribonucleic acid on ultraviolet and visible light absorption of berberine. Arch. Biochem. **81**, 105 (1959).

KOHN, K. W., SPEARS, C. L.: Reaction of anthramycin with deoxyribonucleic acid. J. molec. Biol. **51**, 551 (1970).

KREY, A., HAHN, F. E.: Berberine: Complex with DNA. Science **166**, 757 (1969).

LEFRESNE, P., SAUCIER, J. M., PAOLETTI, C.: Structural changes of polyriboinosinic acid induced by a steroidal diamine, irehdiamine A. Biochem. biophys. Res. Commun. **29**, 216 (1967).

LERMAN, L. S.: The structure of the DNA-acridine complex. Proc. nat. Acad. Sci. (Wash.) **49**, 94 (1963).

— Acridine mutagens and DNA structure. J. cell. comp. Physiol. **64**, Suppl. **1**, 1, (1964).

MAHLER, H. R., BAYLOR, M. B.: Effects of steroidal diamines on DNA duplication and mutagenesis. Proc. nat. Acad. Sci. (Wash.) **58**, 256 (1967).

— DUTTON, G.: Nucleic acid interactions. V. Effects of cyclobuxine. J. molec. Biol. **10**, 157 (1964).

— GOUTAREL, R., KHUONG-HUU, Q., TRUONG HO, M.: Nucleic acid interactions. VI. Effects of steroidal diamines. Biochemistry **5**, 2177 (1966).

— GREEN, G., GOUTAREL, R., KHUONG-HUU, Q.: Nucleic acid-small molecule interactions. VII. Further characterization of deoxyribonucleic acid-diamino steroid complexes. Biochemistry **7**, 1568 (1968).

MEISEL, M. N., SOKOLOVA, T. S.: Inherited cytoplasmic changes induced in yeast by acriflavine and berberine. Dokl. Akad. Nauk SSSR **131**, 436 (1959).

MICHAELIS, L.: The nature of the interaction of nucleic acids and nuclei with basic dye stuffs. Cold Spr. Harb. Symp. quant. Biol. **12**, 131 (1947).

MORTHLAND, F. W., DE BRUYN, P. P. H., SMITH, N. H.: Spectrophotometric studies on the interaction of nucleic acids with aminoacridines and other basic dyes. Exp. Cell Res. **7**, 201 (1954).

O'BRIEN, R. L., ALLISON, J. L., HAHN, F. E.: Evidence for intercalation of chloroquine into DNA. Biochim. biophys. Acta (Amst.) **129**, 622 (1966).

— Olenick, J. G., Hahn, F. E.: Reactions of quinine, chloroquine and quinacrine with DNA and their effects on the DNA and RNA polymerase reaction. Proc. nat. Acad. Sci. (Wash.) **55**, 1511 (1966).

Polet, H., Barr, C. F.: Chloroquine and dihydroquinine. *In vitro* studies of their antimalarial effect upon *Plasmodium knowlesi*. J. Pharmacol. exp. Ther. **164**, 380 (1968).

Reba, R. C., Sheehy, T. W.: Colchicine-quinine therapy for acute falciparum malaria acquired in Vietnam. J. Amer. med. Ass. **201**, 143 (1967).

Romanovsky, D. L.: Specific action of quinine in malaria. Vrach (St. Petersb.) **12**, 438 (1891).

Saucier, J. M., Lefresne, P., Paoletti, C.: Etudes sur l'interaction d'un steroïde diamine, l'irehdiamine A, avec l'acide desoxyribonucléique. C. R. Acad. Sci. Paris **266**, 731 (1968).

Schellenberg, K. A., Coatney, G. R.: The influence of antimalarial drugs on nucleic acid synthesis in *Plasmodium gallinaceum* and *Plasmodium berghei*. Biochem. Pharmacol. **6**, 143 (1961).

Shideman, F. E.: A review of the pharmacology and therapeutics of hydrastis and its alkaloids, hydrastine, berberine and canadine. Bull. nat. Formulary Comm. **18**, 3 (1950).

Stickl, O.: Die bactericide Wirkung der Extrakte und Alkaloide des Schöllkrautes (Chelidonium maius) auf grampositive pathogene Mikroorganismen. Z. ges. Hyg. **108**, 567 (1928).

Stollar, D., Grossman, L.: The reaction of formaldehyde with denatured DNA: Spectrophotometric, immunologic, and enzymic studies. J. molec. Biol. **4**, 31 (1962).

Varma, R. L.: Berberine sulphate in oriental sore. Indian. med. Gaz. **62**, 84 (1927).

Yamagishi, H.: Interaction between nucleic acids and berberine sulfate. J. Cell Biol. **15**, 589 (1962).

Physical Studies on the Interaction of Lysergic Acid Diethylamide and Trypanocidal Dyes with DNA and DNA-Containing Genetic Material

THOMAS E. WAGNER

I. Introduction

Both lysergic acid diethylamide (LSD-25) and the trypanocidal dyes ethidium bromide and isometamidium methanesulfonate have been shown to affect genetic structures. The induction of chromosomal damage in peripheral lymphocytes by the lysergic acid derivative LSD-25 has been clearly demonstrated in both *in vitro* and *in vivo* experiments (COHEN, 1969). The majority of the chromosomal abnormalities apparently induced by LSD-25 are simple chromatid and isochromatid breaks, although several chromatid exchanges in dicentric chromosomes have been observed (IRVIN and EGOZCUE, 1967; COHEN, MORINELLO and BACK, 1967). The observation of a specific interaction between LSD-25 and DNA (WAGNER, 1969; YIELDING and STERNGLANZ, 1968; SMYTHIES and ANTUN, 1969) suggests that LSD-25 may function to damage chromosomal material by interacting directly with the DNA component of the chromosome.

Similarly, the trypanocidal dye ethidium bromide displays several interesting biological properties related to DNA-containing biological structures. This dye has been found to have anti-viral properties (DICKINSON, CHANTRILL, INKLEY and THOMPSON, 1953). It interferes with nucleic acid synthesis *in vivo* (NEWTON, 1957; HENDERSON, 1963) and it inhibits both DNA polymerase and the DNA-dependent RNA polymerase by binding to the template DNA (ELLIOTT, 1963; WARING, 1964). As in the case of LSD-25, the direct interaction of ethidium bromide with DNA has been investigated (LEPECQ, YOT and PAOLETTI, 1964; WARING, 1965). The preferential binding of ethidium bromide to double stranded helical DNA, the occurrence of energy transfer from DNA to dye, and the absence of energy transfer from dye to dye (LEPECQ and PAOLETTI, 1967) suggest that the mode of binding of ethidium bromide to DNA is intercalation. The X-ray diffraction data from the ethidium bromide-DNA complex is consistent with this suggestion (FULLER and WARING, 1964).

Isometamidium methanesulfonate, like ethidium bromide, has marked antitumor effects which appear to be related to the inhibition of nucleic acid synthesis (HENDERSON). In addition to these effects isometamidium methanesulfonate has been shown to be an effective precipitant of DNA (PHILIPS, STERNBERG, CRONIN, SADERGREN and VIDAL, 1967). Recent studies of the interaction between isometamidium methanesulfonate and DNA carried out in our laboratory will be discussed in detail in a later section of this paper.

Because each of the three small molecules already discussed interact in a specific manner with DNA and because the structural characteristics of these molecules are similar (i.e., a planer aromatic arrangement of atoms containing a single positive ammonium ion oriented in the plane of the aromatic portion of the molecule) it is of interest to compare the physical parameters of the complexes formed between each of these molecules and DNA. Such a comparison should be useful both in determining the nature of binding in the case of each molecule and in determining the effects of the binding of these molecules on the overall structure of DNA.

II. The Effects of LSD-25 on the CD Spectrum of Isolated Calf Thymus Nuclei

Reproducible circular dichroism (CD) spectra of intact nuclei may be measured in dilute nuclear suspensions containing a low concentration of calcium ions to assure

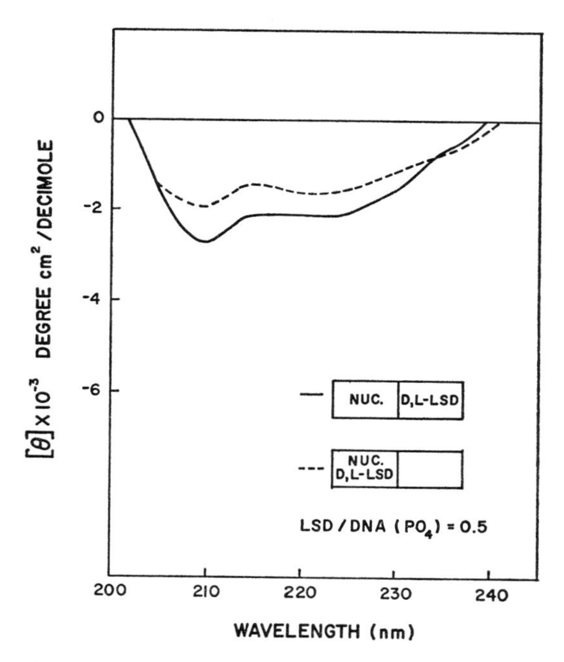

Fig. 1. The circular dichroism spectra of isolated calf thymus nuclei in the presence and abscence of LSD-25. Conditions: 0.01 M Tris buffer (pH 8) containing 0.0033 M $CaCl_2$

the rigidity of the nuclear membrane (WAGNER, 1970). The region of these spectra below 240 nm is characterized by two negative bands at 222.5 nm, with $[\theta]_{222.5} - 2{,}150$ and at 208.5 nm with $[\theta]_{208.5} - 2{,}650$. This region of the nuclear spectrum, where optical activity from both the DNA and protein components of nuclei contribute to the CD spectrum, appears to be very similar to a simple helical protein spectrum in terms of the spectral shape and the positions of the ellipticity bands although the magnitude of the ellipticity of these bands is much reduced from the expected ellipticity of a helical protein. The reduced ellipticity may be due to contributions from the DNA component of nuclei in the low wavelength region of the spectrum or from other effects.

The CD spectrum of calf thymus nuclei prepared by the procedure of ALLFREY, MIRSKY and OSAWA (1957) and further purified by the method KODAMA and TEDESCHI (1963) shows marked changes in the presence of LSD-25 (Fig. 1). These changes are characterized by a decrease in the ellipticity of the two helical protein bands in the CD spectrum of the treated nuclei. This result may indicate a decrease in the protein helical content of LSD-25 treated nuclei. Ionic interactions between cationic LSD molecules and the anionic phosphate sites of the DNA component of nuclei would be expected to effect the partial dissociation of histones ionically bound to DNA in the nucleus. Although CD studies of the interaction of lysine-rich histones with DNA (OLINS, 1969) do not indicate an induced helical conformation for the lysine-rich histones in the presence of DNA, recent studies in our laboratory (WAGNER, 1970) on the f_2a_1 arginine-rich histone indicate that this histone assumes a highly helical conformation upon interaction with DNA. In the case of the f_2a_1 histone (and probably in the case of many arginine and slightly arginine-rich histones) dissociation from DNA would result in a marked decrease in helical content. It has been suggested (JI and URRY, 1969; URRY and JI, 1968) that "Distortions in circular dichroism patterns of particulate systems" may be the result of light scattering changes and absorption flattening effects in the system. It is unlikely that the effects of LSD-25 on the CD spectrum of calf thymus nuclei are the result of such changes since no changes in light scattering as reflected in phase microscopy, absorption at 700 nm, or spectroscopy from 700 nm to 350 nm are observed upon mixing LSD-25 with the nuclear preparation investigated.

III. The Effects of LSD-25 and Ethidium Bromide on the CD Spectrum of Isolated Calf Thymus Interphase Chromosomes

Optical rotatory dispersion spectra have been recorded from solutions of calf thymus interphase chromosomes (chromatin) (ORIEL, 1966) and (TUAN and BONNER, 1969). We have recently recorded the CD spectrum of calf thymus chromatin in our laboratory. The chromatin CD spectrum (Fig. 2) displays two bands above 240 nm; a positive band at 275 nm, with $[\theta]_{275} + 5{,}400$ and a negative band at 244 nm, with $[\theta]_{244} + 9{,}000$. Below 240 nm, the chromatin spectrum is characterized by two negative bands at 222.5 nm with $[\theta]_{222.5} - 7{,}200$ and at 208 nm, with $[\theta]_{208}$ nm $- 8{,}500$. The crossover point in the circular dichroism spectrum of isolated chromatin occurs at 257 nm.

The CD spectrum of calf thymus chromatin prepared by the method of MARUSHIGE and BONNER (1966) was altered by the presence of either LSD-25 or ethidium bromide (Fig. 2). Lysergic acid diethylamide caused a slight decrease in the 278 nm band and a large increase in the 247 nm trough without significantly altering the protein region of the spectrum. The mixture of ethidium bromide and chromatin yielded a more complicated CD spectrum. Two new bands at 310 nm and 296 nm appeared along with a marked increase in negative ellipticity from 280 nm to approximately 215 nm. The observed alterations in the CD spectrum of calf thymus chromatin in the presence of optically inactive LSD-25 (*d, l* mixture) or of ethidium bromide are markedly similar to the changes observed in the CD spectrum of purified DNA upon interaction with these molecules. Interpretation of these changes is more suited to the simpler DNA system and will be discussed in the following section.

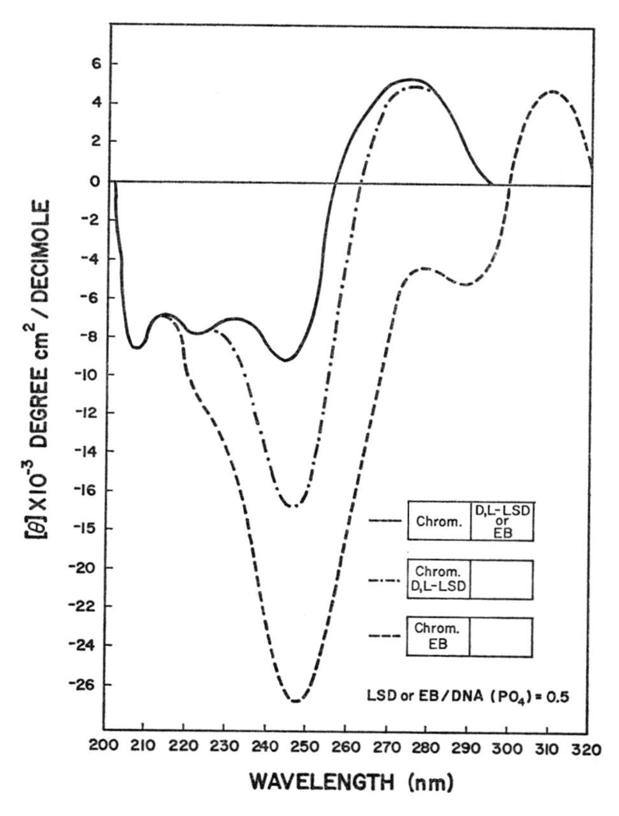

Fig. 2. The circular dichroism spectra of isolated interphase calf thymus chromosomes in the presence and absence of LSD-25 and ethidium bromide. Conditions: 0.01 M Tris buffer

IV. Interactions of Ethidium Bromide and LSD with Calf Thymus DNA

1. The Ethidium Bromide — DNA Interaction

The formation of a complex between ethidium bromide and DNA has been suggested by a number of independent observations. In the presence of DNA the absorption maximum of the dye shifts from 479 nm to somewhere between 516 nm and 518 nm depending upon the source of the DNA. Utilizing this characteristic of the DNA-ethidium bromide complex, dissociation constants for the binding of the dye to DNA from several sources were calculated. The strength of binding was found to be independent of the $G+C$ content of the DNA and saturation was reached at a ratio of one dye molecule per 2 to 3 base-pairs. Secondary binding was observed after saturation, as was the binding of dye to denatured DNA and RNA (WARING, 1965). X-ray diffraction studies of the ethidium bromide-DNA complex indicated that the helical structure of the complex was significantly different from the B-form of native helical DNA particularly in the spacing between base pairs. These X-ray diffraction studies were considered compatible with a model for the complex in which ethidium bromide is intercalated between the base pairs of DNA (FULLER and WARING, 1964). Upon binding to DNA the quantum efficiency of the ethidium bromide fluorescence increases dramatically. This effect is specific for ethidium bromide bound to native double-stranded DNA (LEPECQ and PAOLETTI, 1967). Intercalation of ethidium bromide between the base planes of native DNA would place the dye molecules in an environment of considerably lower dielectric constant than the medium of the unbound dye. The increased quantum efficiency of the bound dye most probably results from this change in environment. Flow dichroism studies of the DNA-dye complex indicate that the planes of the bound dye molecules are perpendicular to the axis of the DNA helix (LEPECQ and PAOLETTI, 1967), a finding which is consistent with the intercalation model for the binding of ethidium bromide to native DNA. In addition to the above mentioned effects of complex formation on the dye, binding of ethidium bromide has been shown to decrease the buoyant density of DNA (LEPECQ and PAOLETTI, 1967; RADLOFF, BAUER and VINOGRAD, 1967).

The circular dichroism spectrum of the complex formed between DNA and optically inactive ethidium bromide is markedly different from that of native DNA (WAGNER, 1969) (Fig. 3 A). In the region of the spectrum above 290 nm two new circular dichroism bands are observed upon interaction of DNA with ethidium bromide. These bands, a peak at 310 nm and a trough at 296 nm, appear in a region where ethidium bromide has two absorption maxima (289 nm and 301 nm) in the presence of DNA (i.e., measured from the difference absorption spectrum of the ethidium bromide-DNA complex minus DNA). It seems likely that these bands are the result of rotatory strengths arising from electronic transitions in the helically arranged dye molecules within the complex. In addition to the appearance of these new CD bands, the CD spectrum of the complex differs from native DNA in the width and intensity of both the 278 nm and 247 nm CD bands. The narrowing and increased magnitude in the 278 nm band is difficult to interpret due to the overlapping effects of the new 296 nm band. Conversely, the changes in the CD spectrum associated with the 247 nm band appear to be clearly the result of changes in the conformation of the DNA component of the complex since they are far removed from the dye absorption bands. This suggested alteration in the helical structure of DNA upon interaction with ethidium bromide is completely consistent with the intercalation model suggested by FULLER and WARING (1964) and is supported by the findings of LEPECQ and PAOLETTI (1967) for the DNA-ethidium bromide complex.

2. The (LSD-25)-DNA Interaction

A direct interaction between LSD-25 and DNA is indicated by three observations. Addition of DNA to a solution of LSD-25 results in a decrease in the extinction coefficient of the drug at 310 nm and a decrease in the fluorescence intensity of the drug at 450 nm (Ex. 310 nm) (YIELDING and STERNGLANZ, 1968). The effects of DNA

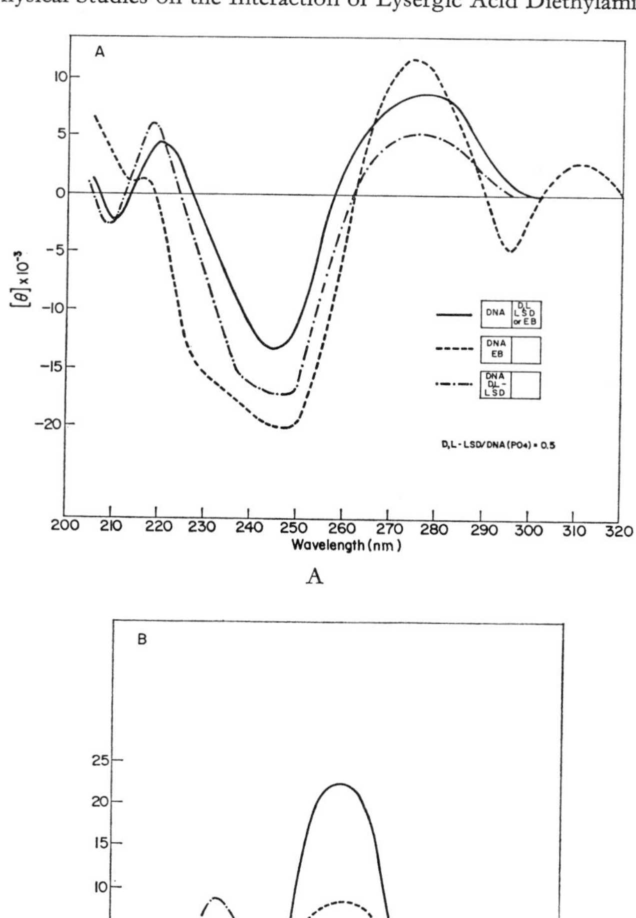

A

B

Fig. 3. (A) The circular dichroism spectra of calf thymus DNA in the presence and absence of LSD-25 and ethidium bromide. Conditions: 0.1 M NaCl (pH 7.2). — (B) The circular dichroism spectra of calf thymus DNA, *l*-LSD and a mixture of *l*-LSD and DNA. Conditions: 0.1 M NaCl (pH 7.2) on the fluorescence intensity of LSD-25 have been further substantiated by the work of SMYTHIES and ANTUN (1969). In addition to these changes in the absorption and fluorescence character of the drug upon interaction with DNA, LSD-25 has been shown to alter the circular dichroism spectrum of DNA (Fig. 3) (WAGNER, 1969).

The effect of DNA on the absorption and fluorescence characteristics of LSD-25 appears to be a related phenomenon. Decreased absorption at the excitation wavelength (310 nm) for fluorescence would *a priori* result in decreased fluorescence emission. The quantum efficiency of fluorescence would appear to remain unchanged for LSD-25 in the presence of DNA. Therefore, the significant observation made by YIELDING and STERNGLANZ (1968) is the hypochromic effect of DNA-binding on LSD-25. It is pertinent that this effect was only observed by these workers in the presence of native helical DNA and appears to be relatively unrelated to the ionic character of the drug.

The changes observed in the CD spectrum of calf thymus DNA in the presence of LSD-25 (Fig. 3 A) are characterized by a small decrease in the ellipticity of the 278 nm band and a large increase in the ellipticity and width of the 247 nm band. The changes in the 247 nm band are remarkably similar to the changes in this band resulting from the interaction of ethidium bromide with DNA. Since neither the dye nor the drug have absorption bands in this region, it would appear that these changes are the result of similar change in the DNA helical structure caused independently by both LSD-25 and ethidium bromide. This suggests that the mode of LSD-25 binding to DNA is similar to the "intercalative" binding of ethidium bromide by DNA. The observed hypochromicity of the bound LSD-25 could result from the stacking of drug molecules between the DNA base-pairs in a polymeric array (TINOCO, 1960; 1961). It is disappointing that an increase in the fluorescence quantum efficiency of LSD-25 bound to DNA has not been observed. Intercalation of drug molecules between the base-pair of DNA would be expected to decrease the dielectric constant in the neighborhood of the drug resulting in increased fluorescence quantum efficiency.

3. The (1-LSD)-DNA Interaction

Attempts by BRADY (1970) to reproduce the effects of LSD on the CD spectrum of DNA observed in our laboratory with LSD-25 (WAGNER, 1969) using *d*-LSD were totally unsuccessful. The spectrum of *d*-LSD and DNA mixed and separated are identical. Since the LSD-25 used in our preliminary experiments is a racemic mixture of the drug, the results of BRADY suggest that the *l*-component of this mixture may be solely responsible for our observations. The effects of *l*-LSD on the CD spectrum of DNA (Fig. 3 B) observed both in our laboratory and by BRADY support this suggestion. These effects are characterized by large increase in ellipticity in both the 278 nm and 247 nm bands of the DNA spectrum and are remarkably similar to the effects of ethidium bromide and isometamidium methanesulfonate (see next section) on the CD spectrum of DNA.

V. The Interaction of Isometamidium Methanesulfonate with DNA

1. Fluorescence Studies

We have recently investigated the binding of isometamidium methanesulfonate (IMS) to calf thymus DNA by difference fluorescence spectroscopy. Difference spectrofluorometric measurements were recorded using a prototype Cary difference spectroflurometer. The addition of DNA to a solution of IMS results in a dramatic increase in the fluorescence of the dye (Fig. 4). The increase in the fluorescence quantum efficiency of IMS in the presence of DNA suggests that the interaction of IMS with DNA is similar to the binding of ethidium bromide to DNA.

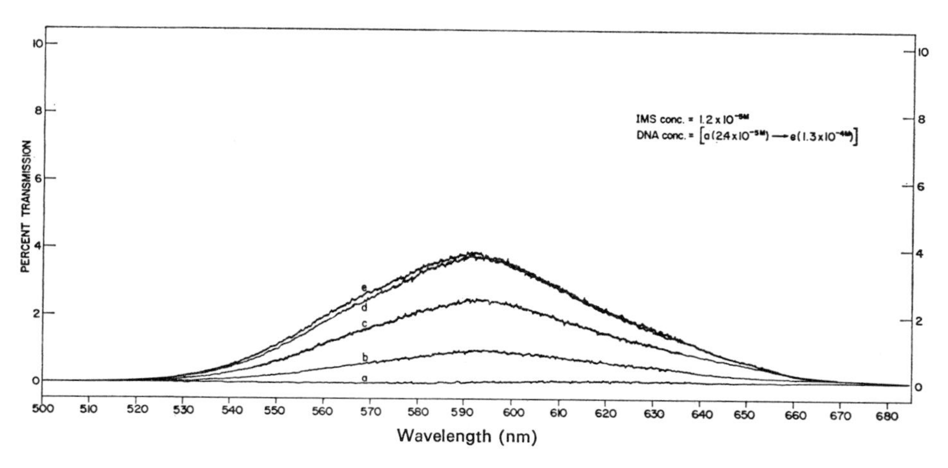

Fig. 4. The difference fluorescence spectra of calf thymus DNA and IMS minus IMS at varying DNA concentration. Conditions: 0.1 M NaCl (pH 7.2)

2. Circular Dichroism Studies

The incremental addition of IMS to a solution of DNA results in a most interesting change in the CD spectrum of DNA (Fig. 5). This change is characterized by a total increase in ellipticity between 300 nm and 220 nm, reaching a maximum at a ratio of dye to DNA phosphate of about 0.1. The IMS-induced changes in ellipticity are totally symmetrical with the CD spectrum of pure DNA, and result in a maximum increase in the 278 nm band of 100% and in the 247 nm band of 50%. Since IMS has no observable absorption bands between 300 nm and 220 nm, and since the changes caused in the DNA CD spectrum are symmetrical with the CD spectrum of DNA alone, the changes in the DNA CD spectrum in the presence of IMS are assumed to be the result of changes in the structural parameters of the DNA helix due to the interaction with IMS.

VI. Conclusions

1. The Relation between the Structure of Dye and Drug Complexes with DNA and Their CD Spectrum

In the present model (Lerman, 1961; Fuller and Waring, 1964) for intercalation with DNA each base-pair in the DNA is a potential binding site for an intercalating molecule. No evidence has been found for any base-pair specificity in DNA. It is therefore surprising that complexes containing one intercalated small molecule per

DNA base-pair are not observed. Complexes containing a maximum of one intercalated molecule per 2 to 5 DNA base-pairs are usually observed (LERMAN, 1961; WARING, 1965). One explaination for these observations may involve a local conformational change in the DNA structure induced by an intercalated molecule. The altered DNA structure in the region close to the bound molecule may render base-pairs in this region unsuitable for interaction with molecules which otherwise would undergo intercalative binding to native regions of DNA.

The *l*-form of LSD and IMS both affect the CD spectrum of DNA in the same manner. The molecules cause large, symetrical, increases in the ellipticity of both

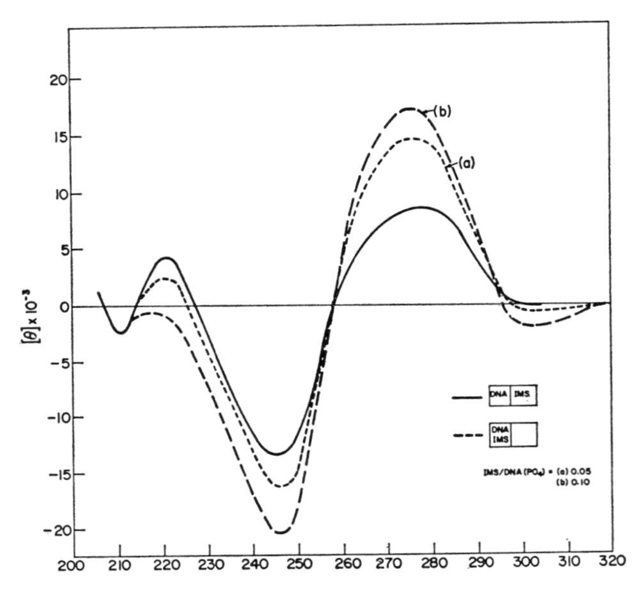

Fig. 5. The circular dichroism spectra of calf thymus DNA in the presence and absence of IMS. Conditions: 0.1 M NaCl (pH 7.2)

major bands in the CD spectrum of DNA. Because the altered spectra are remarkably similar in shape to the native DNA spectrum, and because neither of the bound molecules has major absorption bands in the region where the alterations are observed, we have assumed that these changes are not the result of rotatory strengths induced in the small bound molecules. Conversly, these changes appear to be the result of alterations in the conformation of the DNA itself and may be relevant to the neighbor exclusion shown by intercalated molecules.

Although structural interpretation of the observed changes in the CD spectrum of DNA in the presence of molecules such as LSD and IMS is premature at this point, several indications of the nature of the structural alterations are apparent. These indications are most apparent in the IMS-DNA complex (Fig. 5). Addition of one IMS molecule per 10 DNA base-pairs resulted in a 70% increase in the ellipticity of the positive DNA CD band while the negative band underwent only a 20% increase.

The resulting spectrum begins to take on the characteristics of a non-conservative spectrum. This change as well as the appearance of a small negative band at about 296 nm in the CD spectrum of the IMS-DNA complex suggests that the DNA molecule may be undergoing regional changes in conformation from the B-form of DNA to some other form closely related to either the A-form of DNA or helical RNA (YANG and SAMEJIMA, 1969). Such a change in conformation would involve base tilting and may partially explain the inability of molecules like IMS to intercalate near already bound molecules.

2. The Specificity of *l*-LSD for DNA Binding Sites

Lysergic acid diethylamide in the *d*-Form does not undergo interaction with DNA resulting in changes in the DNA CD spectrum (BRADY, 1970). In addition, it has been recently demonstrated in our laboratory that *d*-LSD does not alter the CD spectrum of calf thymus nuclei or chromatin. Conversely, *l*-LSD interacts with DNA to form a complex which is characterized by its unusually large rotatory strength. The mode of binding to DNA shown by *l*-LSD appears to be intercalation. Since *d* and *l*-LSD are mirror images of each other it is most difficult to imagine what the selectivity for binding to DNA is based upon.

3. General

The propensity of the *l*-form of LSD to interact with DNA and DNA containing genetic structures, and the similarity of interaction with DNA between *l*-LSD and potent mutagens such as the acridine dyes (LERMAN, 1961) suggests the real danger of this form of the drug to humans.

Acknowledgements

The author is grateful for valuable discussions with Dr. A. H. BRADY, Dr. F. LANDSBERGER, and Dr. M. SONENBERG during the preparation of this manuscript. The research carried out by the author was supported by the National Cancer Institute Grant 08748.

References

ALLFREY, V. G., MIRSKY, A. E., OSAWA, S.: Protein synthesis in isolated cell nuclei. J. gen. Physiol. **40**, 451—490 (1957).

BRADY, A. H.: Personal communication.

COHEN, M. M.: The interaction of various drugs with human chromosomes. Canad. J. Genet. Cytol. **11**, 1—24 (1969).

— MARINELLO, M. J., BACK, N.: Chromosomal damage in human leukocytes induced by lysergic acid diethylamide. Science **155**, 1417—1419 (1967).

DICKINSON, L., CHANTRILL, B. H., INKLEY, G. W., THOMPSON, M. J.: Antiviral action of phenanthridinium compounds. Brit. J. Pharmacol. **8**, 139—142 (1953).

ELLIOTT, W. H.: The effects of antimicrobial agents on deoxyribonucleic acid polymerase. Biochem. J. **86**, 562—567 (1963).

FULLER, W., WARING, M. J.: A molecular model for the interaction of ethidium bromide with deoxyribonucleic acid. Ber. Bunseges. Physik. Chem. **68**, 805—808 (1964).

HENDERSON, J. F.: Inhibition of purine metabolism in Ehrlich ascites carcinoma cells by phenanthridinium compounds related to ethidium bromide. Cancer Res. **23**, 491—495 (1963).

HOFFER, A., OSMOND, H.: The hallucinogens. New York: Academic Press 1967.
IRWIN, S., EGOZCUE, J.: Chromosomal abnormalities in leukocytes from LSD-25 users. Science **157**, 313—314 (1967).
JI, T. H., URRY, D. W.: Correlation of light scattering and absorption flattening effects with distortions in the circular dichroism patterns of mitochondrial membrane fragments. Biochem. biophys. Res. Commun. **34**, 404—411 (1969).
KODAMA, R., TEDESCHI, H.: An electron microscope study of calf thymus nuclear preparations isolated in surcrose solutions. J. Cell Biol. **18**, 541—553 (1963).
LEPECQ, J.-B., PAOLETTI, C.: A fluorescent complex between ethidium bromide and nucleic acids. J. molec. Biol. **27**, 87—106 (1967).
— YOT, P., PAOLETTI, C.: Interaction du bromhydrate d'ethidium (BET) avec les acides nucléiques (A.N.) Etude spectrofluorimétrique. C. R. Acad. Sci. Paris **259**, 1786—1789 (1964).
LERMAN, L. S.: Structural considerations in the interaction of DNA and acridenes. J. molec. Biol. **3**, 18—30 (1961).
MARUSHIGE, K., BONNER, J.: Template properties of liver chromatin. J. molec. Biol. **15**, 160—174 (1966).
NEWTON, B. A.: The mode of action of phenanthridines: The effect of ethidium bromide on cell division and nucleic acid synthesis. J. gen. Microbiol. **17**, 718—730 (1957).
OLINS, D. E.: Interaction of lysine-rich histones and DNA. J. molec. Biol. **43**, 439—460 (1969).
PHILIPS, F. S., STERNBERG, S. S., CRONIN, A. P., SODERGREN, J. E., VIDAL, P. M.: Physiologic disposition and intracellular localization of isometamidium. Cancer Res. **27**, 333—349 (1967).
RADLOFF, R., BAUER, W., VINOGRAD, J.: A dye-buoyant-density method for the detection and isolation of closed circular duplex DNA: The closed circular DNA in hela cells. Proc. nat. Acad. Sci. (Wash.) **57**, 1514—1521 (1967).
SMYTHIES, J. R., ANTUN, F.: Binding of tryptamine and allied compounds to nucleic acids. Nature (Lond.) **223**, 1061—1063 (1969).
TINOCO, I.: Optical and other electronic properties of polymers. J. chem. Phys. **33**, 1332 to 1338 (1960).
— Erratum: Optical and other electronic properties of polymers. J. chem. Phys. **34**, 1067—1069 (1961).
TUNIS, M. J., HEARST, J. E.: On the hydration of DNA. I. Preferential hydration and stability of DNA in concentrated trifluoroacetate solution. Biopolymers **6**, 1218—1223 (1968).
URRY, D. W., JI, T. H.: Distortions in circular dichroism of particulate (or membranous) systems. Arch. Biochem. **128**, 802—807 (1962).
WAGNER, T. E.: In vitro interaction of LSD with purified calf thymus DNA. Nature (Lond.) **222**, 1170—1172 (1969).
— A trypsin sensitive site for the action of hydrocortisone on calf thymus nuclei. Biochem. biophys. Res. Commun. (1970) (in press).
— A circular dichroism study of the f_2a_1 histone in the presence of polyvinylphosphate and DNA. Nature (Lond.) (1970) (in press).
WARING, M. J.: Complex formation with DNA and inhibition of Escherichia coli RNA polymerase by ethidium bromide. Biochem. biophys. Acta (Amst.) **87**, 358—361 (1964).
— Complex formation between ethidium bromide and nucleic acids. J. molec. Biol. **13**, 269—282 (1965).
YANG, J. T., SAMEJIMA, T.: Optical rotatory dispersion and circular dichroism of nucleic acids. Progress in Nucleic Acid Research and Molecular Biology, Vol. 9. New York: Academic Press 1969.
YIELDING, K. L., STERNGLANZ, H.: Lysergic acid diethylamide (LSD) binding to deoxyribnucleic acid (DNA). Proc. Soc. exp. Biol. (N. Y.) **128**, 1096—1098 (1968)

Comments on the Interaction of LSD with DNA

K. Lemone Yielding and Helene Sternglanz

The actions of drugs on the central nervous system have attracted wide interest in recent years. To the scientist, drug studies provide a means for probing the biochemical processes of learning and behaviour, and for understanding ultimately the basis for abnormal function. In most instances, of course, a drug will not be expected to exert its influence exclusively on the central nervous system and conversely, many drugs known for their systemic effects also influence the nervous system.

Of particular concern has been the use by the public of potent central nervous system agents such as the hallucinogens, with their attendent side effects, both on the individual and society. Recently, LSD has been singled out for concern because of the reports that its use resulted in chromosomal breaks and therefore could possibly lead to birth defects (Cohen et al., 1967, 1968; Irvin and Ecozcue, 1967). These reports prompted us to seek evidence for interaction of the drug with DNA. These findings that a complex can indeed be formed between LSD and helical DNA have appeared in a preliminary communication (Yielding and Sternglanz, 1968). These findings were confirmed by Wagner (1969) and by Smythies and Antun (1969). Wagner conducted a more detailed study of the nature of the binding process which is reported in the present volume.

Such experiments as these, while they bear directly on the basis for chromosomal damage, also may be relevant to the question of whether nucleic acids are involved in information processing in the central nervous system.

Binding of LSD to DNA

LSD has an absorption in the near UV which may be used for spectrophotometric examination of its interactions. Fig. 1 shows that consecutive additions of DNA to D-LSD influenced this absorption spectrum, demonstrating that an interaction occurred. From a series of such experiments using a variety of LSD concentrations the extent of drug binding to DNA was determined. Analysis of data and use of a standard Scatchard plot resulted in a dissociation constant $Kd = 5 \times 10^{-4}$ M and the number of binding sites, n, of approximately 1 drug molecule per DNA base.

Essentially identical results were obtained with L-LSD, and with the biologically ineffective analogue D, L-2 Bromo-LSD. Fluorescence spectroscopy was also used to detect the complex and permitted the use of considerably lower concentrations of LSD (Fig. 2).

In addition, an attempt was made to detect the presence of a limited number of stronger bindings sites on native DNA for LSD by performing equilibrium dialysis with large excess of DNA where binding of 1 drug/1000 bases could have been detected. No evidence for strong binding was obtained. It appeared that binding was favored by helical structure since it did not occur at pH 3, nor to DNA which had been heated to 100° and cooled suddenly. Binding was independent of pH between 6.5 and 8.5, however, indicating that the drug does not have to be completely protonated to bind. $MgCl_2$ at 3.4×10^{-3} M, limited binding as shown spectrophoto-

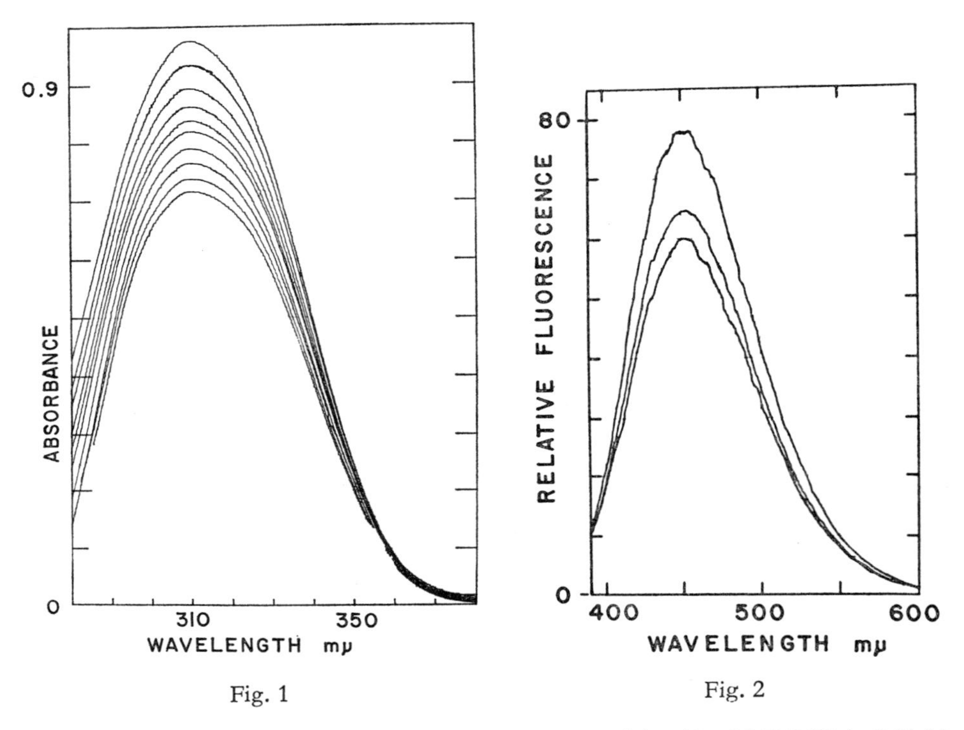

Fig. 1

Fig. 2

Fig. 1. Depression by DNA of the absorption spectrum of 1×10^{-4} M D-LSD in 0.01 M phosphate buffer pH 6.1. DNA additions ranged from 0.86×10^{-4} M (2nd curve) to 7.9×10^{-4} M (bottom curve). Reprinted from YIELDING and STERNGLANZ (1968)

Fig. 2. Effect of DNA on the fluorescence emission of 1×10^{-6} M LSD in 0.01 M PO_4 buffer pH 6.1: Top curve = control; middle and bottom curves resulted from addition of 1.76 and 8.80×10^{-4} M DNA

metrically. The spectrum of 1×10^{-4} M LSD was not altered by 3.4×10^{-4} M RNA, indicating specificity for DNA. Likewise, poly *U* and poly *A* showed only weak interaction with the drug.

Our present results have shown that LSD interacts with DNA and suggests that it might interfere with DNA functions, thus providing a molecular basis for the observed chromosomal damage. Certain drugs, interacting with DNA, interfere with the DNA repair mechanism. If such were the case for LSD it would explain why the drug can cause accumulation of chromosomal damage (WILKIN, 1961; YIELDING et al., 1969; CLEAVER, 1969). Our studies have not determined the precise nature of the DNA-LSD interaction but are, nevertheless, of interest as one additional example of drug binding to the double helix which is conceivably related to a biological effect.

It is questionable whether interaction of LSD with DNA has a relationship to the hallucinogenic action of the drug. It should be noted, however, that several other drugs which bind to DNA are known to be neurotoxic, for example, quinacrine, miracil and chloroquine. SMYTHIES and collaborators (1970) have recently proposed a model for the neuroreceptor which consists of RNA complexed with prostaglandins. Therefore, studies such as these with LSD, may also serve a useful role in ultimately understanding the CNS effects of various drugs.

References

CLEAVER, J. E.: Repair replication of mammalian cell DNA: Effects of compounds that inhibit DNA synthesis on dark repair. Radiat. Res. **37**, 334—348 (1969).

COHEN, M. M., MARINELLO, M. J., BACK, N.: Chromosomal damage in human leukocytes induced by LSD. Science **155**, 1417—1419 (1967).

— HORSCHHORN, K., VERBO, S., FROSCH, W. A., GROESCHEL, M. M.: The effect of LSD-25 on the chromosomes of children exposed *in utero*. Radiat. Res. **2**, 486—492 (1968).

IRWIN, S., EGOZCUE, J.: Chromosomal abnormalities in leukocytes from LSD-25 users. Science **157**, 313—314 (1967).

SMYTHIES, J. R., ANTUM, F.: Binding of tryptamine and allied compounds to nucleic acids. Nature (Lond.) **223**, 1061—1063 (1969).

—, BENNINGTON, F., MORIN, R. D.: The mechanism of action of hallucinogenic drugs on a possible serotonin receptor in the brain. Int. Rev. Neurobiol. **12**, 207—233 (1970).

WAGNER, T. E.: *In vitro* interaction of LSD with purified calf thymus DNA. Nature (Lond.) **222**, 1170—1172 (1969).

WILKIN, E. M.: Modification of mutagenesis initiated by UV light through posttreatment of bacteria with basic dyes. J. cell. comp. Physiol. **58**, 135 (1961).

YIELDING, K. L., STERNGLANZ, H.: LSD binding to DNA. Proc. Soc. exp. Biol. (N. Y.) **128**, 1096—1098 (1968).

— YIELDING, L., GAUDIN, D.: Inhibition by chloroquine of UV repair in E. coli B. Proc. Soc. exp. Biol. (N. Y.) **133**, 999—1001 (1969).

The Interaction of Polyamines with Nucleic Acids

EDWARD J. HERBST and ROBERT B. TANGUAY

The strong interaction of the polyamines spermine and spermidine with ordered structures of nucleic acids is attributed to the molecular dimensions and the imino and amino groups of these organic bases.

$$NH_2CH_2CH_2CH_2NHCH_2CH_2CH_2CH_2NH_2$$
Spermidine

$$NH_2CH_2CH_2CH_2NHCH_2CH_2CH_2CH_2NHCH_2CH_2CH_2NH_2$$
Spermine

LIQUORI, CONSTANTINO, CRESCENZI, ELIA, GIGLIO, PULITI, DE SANTIS SAVINO and VITAGLIANO (1967) showed that the conformation of these molecules, as established

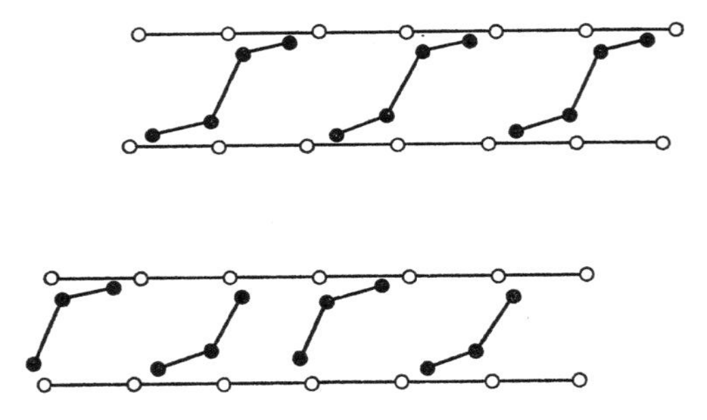

Fig. 1. Schematic models showing spermidine and spermine molecules in the narrow grooves of DNA. Models as presented by LIQUORI et al. Spermidine (●—●—●); Spermine (●—●—●—●); DNA phosphates (○—○)

by *X*-ray analysis of the crystal structures of the hydrochlorides, permits a specific interaction with the DNA double helix, the polyamine bridging the narrow groove of the macromolecule. Their models of the electrostatic interaction between spermine or spermidine protonated imino and amino groups and the ionized phosphate groups of pairs of nucleotide units across the narrow groove of DNA are illustrated schematically in Fig. 1.

The neutralization of the charged phosphate "backbone" and the interstrand bridging by the tetramethylene unit of the polyamines stabilizes helical nucleic acids against thermal denaturation. FELSENFELD and HUANG (1960) first cited the strong interaction between spermine and ribohomopolymers and reported thermal stabilization of *rA:rU* helices by the polyamine. MAHLER and his coworkers (MAHLER, MEHROTRA and SHARP, 1961; MAHLER and MEHROTRA, 1963; MEHROTRA and MAHLER, 1964) performed extensive studies on the effect of aliphatic diamines on thermal denaturation of DNA and concluded that optimal stabilization was obtained with C_3 to C_5 diamines. These structures can be accomodated in the model of LIQUORI et al. and the length of the carbon chain specificity is probably related to favorable interstrand interaction of the diamines and DNA.

TABOR (1962) demonstrated that spermine and spermidine increase the *Tm* of solutions of DNA and that the polyamines are effective at low concentrations (10^{-4} to 10^{-5} M) by comparison to those of diamines or Mg^{++} (10^{-2} M) which produced comparable thermal stabilization. More recently, SZER [SZER, 1966 (1, 2)] critically examined diamine and polyamine interactions with synthetic polyribonucleotides and the effect of these interactions on the thermal transitions of ordered structures. He concluded that diamines are less effective than divalent metal cations in the stabilization of ordered structures of ribohomopolymers while spermidine and spermine increase the *Tm* of all ribohomopolymers more effectively than divalent metal ions.

I. Biological Effects of Polyamines

It is very likely that the polyamine-nucleic interactions manifested in stabilized ordered structures of nucleic acids cause the diverse biological activities of polyamines. Spermine and spermidine are growth factors (HERBST and SNELL, 1949; SNEATH, 1955; HAM, 1964; BERTOSSI, BAGNI, MORUZZI and CALDARERA, 1965; DAVIS, 1966) and the cellular levels of these compounds are dramatically increased during rapid growth of regenerating liver (DYKSTRA and HERBST, 1965; RAINA, JÄNNE and SIIMES, 1966) and in specific stages of the growth of embryos (RAINA, 1963; CALDARERA, BARBIROLI and MORUZZI, 1965; DION and HERBST, 1967; RUSSELL, SNYDER and MEDINA, 1969).

Observations of increased polyamine levels during rapid growth are indicative of the participation of the organic cations in protein and nucleic acid synthesis and there is considerable evidence for the stimulation of the *in vitro* incorporation of amino acids into proteins and of ribonucleotides into RNA by spermine and spermidine.

Effects of Polyamines on Protein Synthesis — The stimulation of the *in vitro* incorporation of amino acids into protein by spermine was first reported by HERSHKO, AMOZ and MAGER, 1961 in a study with rat liver microsomes. The polyamine replaced about 80% of the Mg^{++} requirement of the protein-synthesizing system and a combination of spermine and suboptimal amounts of Mg^{++} resulted in an increase of incorporation above the maximum level attainable with optimal concentrations of Mg^{++} alone. MARTIN and AMES (1962) also reported a reduction of the Mg^{++} requirement by polyamines for poly *U*-directed phenylalanine synthesis by the amino acid incorporating system from *Salmonella typhimurium*. Combinations of Mg^{++} (12 mM), spermidine (5 mM) and putrescine (25 mM) were superior to optimal Mg^{++} (17 mM) for phenylalanine incorporation by this system. TAKEDA (1969) performed similar experiments with a cell-free amino acid incorporating system from *Escherichia coli* and established polyamine stimulation of both poly *U*-directed and MS2 RNA-directed polypeptide synthesis. Spermidine also stimulates amino acid incorporation by cell-free extracts of yeast (BRETTHAUER, MARCUS, CHALOUPKA, HALVORSON and BOCK, 1963) and combinations of spermidine (2.5 mM) and putrescine (12.5 mM) increase the amino acid incorporating activity of a cell-free system derived from *Helianthus tuberosus* tissue (COCUCCI and BAGNI, 1968).

The specific effect of the polyamines in the complex protein-synthesizing system has not been explained but several RNA-polyamine interactions which could affect the rate of peptide bond formation have been identified: (1) Functional ribosomes are stabilized by divalent cations and polyamines (COHEN and LICHTENSTEIN, 1960; COLBOURN, WITHERSPOON and HERBST, 1961; MARTIN and AMES); (2) The binding of aminoacyl-tRNA to ribosomes is facilitated by polyamines (TANNER, 1967; TAKEDA, 1969); (3) Aminoacyl-tRNA formation by purified aminoacyl-tRNA synthetases from *E. coli* (TAKEDA and IGARASHI, 1970) is stimulated by polyamines; (4) Polyamines bind to ribosomes and rRNA (STEVENS, 1969) and to tRNA (COHEN, MORGAN and STREIBEL, 1969; IGARASHI and TAKEDA, 1970); (5) Biologically-active conformations of tRNA are stabilized by polyamines (FRESCO, ADAMS, ASCIONE, HENLEY and LINDAHL, 1966).

The relative importance of these polyamine-RNA interactions in protein biosynthesis is difficult to assess. There is evidence, however, that spermidine is essential for the binding of aminoacyl-tRNA to yeast ribosomes (TANNER) and it appears likely that the polyamine has a function in protein synthesis in yeast which cannot be fulfilled by Mg^{++}.

Effects of Polyamines on RNA Synthesis — Polyamines stimulate the incorporation of labeled uracil into the RNA of bacteria (COHEN and RAINA, 1966) and into early gastrula stages of the sea urchin (BARROS and GUIDICE, 1968). Incorporation of labeled orotic acid or formate into RNA of tenth day chick embryos (MORUZZI, BARBIROLI and CALDARERA, 1968) or chick embryo brain (CALDARERA, MORUZZI, ROSSONI and BARBIROLI, 1969) is also increased by injections of spermidine or spermine into the air sac of the embryos. RAINA and JÄNNE (1970) have performed experiments with Ehrlich ascites cells and with rat liver nuclei or nucleoli which indicate that polyamines can increase RNA synthesis and decrease turnover. DION and HERBST (1967) also demonstrated a pronounced increase of uptake of labeled uridine by nuclei of excised salivary glands of *Drosophila melanogaster* which had been incubated in media containing spermidine.

These observatious of the involvement of polyamines in RNA synthesis by intact cells or isolated nuclei are supported by numerous reports of the stimulation of the incorporation of ribonucleotides into RNA by purified RNA polymerases. A study by FOX and WEISS (1964) with the *Micrococcus lysodeikticus* (ML) RNA polymerase demonstrated a two-fold increase in RNA synthesis when the assay containing optimal levels of all components was supplemented with 1 to 2 mM spermidine or 0.2 mM spermine. The increased rates of RNA synthesis were observed when the reactions were primed by a variety of native DNA templates but the addition of polyamines did not increase RNA synthesis with single-stranded DNA templates and polyamines inhibited RNA-directed RNA synthesis (FOX, ROBINSON, HASELKORN and WEISS, 1964). KRAKOW (1963) obtained similar results with the purified *Azotobacter vinelandii* RNA polymerase and FUCHS, MILLETTE, ZILLIG and WALTER (1967) showed that the

E. coli RNA polymerase is also stimulated by spermidine. The greatest effect of spermidine was obtained when Mg^{++} concentrations in the assay were sub-optimal but the polyamine could not entirely replace Mg^{++}. A yeast RNA polymerase, recently purified by FREDERICK, MAITRA and HURWITZ (1969), is stimulated by spermine (2.4 fold) and by spermidine (2.1 fold) only when native DNA is provided as a template. Of great interest is the preferential utilization of denatured DNA templates by this enzyme in the absence of polyamines. It is possible that polyamines improve the interaction between the polymerase and native DNA since a deficiency in the binding of native templates relative to the binding of denatured templates was demonstrated. Polyamines have also been reported to stimulate a partially purified testicular RNA polymerase (BALLARD and WILLIAMS-ASHMAN, 1966) and in this instance the effect of spermine and spermidine can be demonstrated with denatured as well as native DNA templates.

The mechanism of the stimulation of bacterial RNA polymerases by polyamines is of great interest because of the possible relationship to the regulation of RNA synthesis *in vivo*. FOX, GUMPORT and WEISS (1965) showed that the inhibition of the ML RNA polymerase by RNA could be partially reversed by 2 mM spermidine and KRAKOW (1963) reported similar results with the *A. vinelandii* enzyme. The conclusion that polyamines could reverse product inhibition was reached by both laboratories and this interpretation was strengthened by the observation by FUCHS, MILLETTE, ZILLIG and WALTER (1967) that spermidine can re-start the *E. coli* RNA polymerase reaction after a plateau in the rate of RNA synthesis occurred. The reactivation of the "stalled" reaction was attributed to the elimination of the product RNA from a complex of enzyme · DNA · RNA which had been demonstrated by BREMER and KONRAD (1964). KRAKOW (1966) proved that product inhibition was an important factor when he eliminated the plateau in the release of γ^{32}P-labeled pyrophosphate from ribonucleoside triphosphate substrates by the addition of RNases to the reaction to remove the product RNA. He attempted to demonstrate the release of product RNA from the complex with the enzyme and template DNA by adding spermidine and filtering the reaction mixture through membrane filters. The product formed in the reaction was retained by the filter in both spermidine-supplemented and unsupplemented reactions and a direct effect of the polyamine in product release could not be inferred from these results.

ABRAHAM (1968) concluded that spermidine prevented non-specific binding of enzyme molecules to RNA thus making more enzyme available to chain initiation sites. This interpretation, based on experiments performed with *E. coli* RNA polymerase, is strengthened by results of FOX et al. (1965) on the interaction of RNA or DNA with the ML RNA polymerase in the presence of spermidine. Complexes of enzyme and RNA were dissociated by spermidine while enzyme · DNA complexes were not disrupted when sedimented in a glycerol gradiant containing 2 mM spermidine.

Additional studies of the *E. coli* RNA polymerase assay by PETERSEN, KRÖGER and HAGEN (1968) revealed that the polyamine increases both chain initiation and chain elongation since both the incorporation of γ^{32}P-labeled ATP and ^{14}C-labeled AMP were similarly increased by spermidine. Very recently, GUMPORT and WEISS (1969) reported that spermidine improved the asymmetrical transcription of the replicative form DNA of bacteriophage ϕX 174 by ML RNA polymerase. Similar results were obtained by substituting Mg^{++} for Mn^{++} or by combinations of Mg^{++} and spermidine. The mechanism by which strand selection is modified by the polyamine is unknown.

We have attempted to gain additional information concerning the mechanism by which polyamines stimulate RNA synthesis. Triple-strand "hybrid" homopolymers have been utilized as models of DNA · RNA complexes. (The extensive literature supporting the occurrence of "hybrid pairs" of ribo-deoxyribohomopolymer triple stranded structures was reviewed recently by FELSENFELD and TODD, 1967). We selected *(dA:dT)rU* as an appropriate model to study the effect of polyamines on the formation and release of "RNA product" from a simulated DNA · RNA complex in a stalled RNA polymerase reaction. STRAAT and Ts'o (1969) utilized *(dA:dT)rU* as a template for the incorporation of AMP and UMP by ML RNA polymerase. UMP incorporation was severely inhibited while AMP incorporation was equivalent to that observed with a *dA:dT* template. MORGAN and WELLS (1968) reported similar result with *(dA:dT)rU* as a template for UMP incorporation by *E. coli* RNA polymerase. Polyamines were not utilized in these investigations but GLASER and GABBAY (1968) reported that spermine or spermidine promote the formation of *(rA:rU)rU* triple stranded structures and previously SZER [1966 (2)] briefly noted a similar activity of the polyamines in the formation of ribohomopolymer triple strands. The work to be reported is, to our knowledge, the first study of the effect of polyamines on mixed or "hybrid" triple strands of deoxyribo- and ribohomopolymers.

II. Effect of Spermidine on Thermal Denaturation of (dA:dT)rU

RILEY, MALING and CHAMBERLIN (1966) prepared *(dA:dT)rU* and studied the thermal denaturation of the complex in phosphate buffer containing variable concentrations of NaCl. Although *Tm* data were not presented, they stated: "The mixed complex *(dA:dT)rU* is less stable than *dA:dT* at all sodium ion concentrations tested and invariably gives a characteristic two-step melting profile in which the second step is identical to that found with the two-stranded pair indicating that it is the *rU* strand which is melted off first." STRAAT and Ts'o (1969) also observed the formation of *(dA:dT)rU* in 0.1 M Tris buffer, pH 7.5 and 0.001 M $MnCl_2$. All *Tm* data to be reported in this study were obtained on solutions of *(dA:dT)rU* in 0.1 M Tris buffer, pH 7.5 since we sought to provide conditions comparable to those utilized to demonstrate the stimulatory effect of polyamines in the ML RNA polymerase assay.

Fig. 2 illustrates the melting of *(dA:dT)rU* in this buffer and in Tris containing 2.5×10^{-4} M spermidine (SD). A single thermal transition is obtained in 0.1 M Tris at 260 nm and at 280 nm. The 280 nm hyperchromicity during the melting of triple strand homopolymers has been characterized by STEVENS and FELSENFELD (1964) as indicative of the release of the third strand from the structure. The *Tm* data for the 0.1 M Tris control suggest a simultaneous "melting" of the third strand, *rU*, and of the double stranded *dA:dT*. In the presence of 2.5×10^{-4} M spermidine, a two-step thermal transition was observed indicative of a low temperature (approximately 50°) release of *rU* while *dA:dT* is stabilized in the presence of the polyamine to a *Tm* 10° higher than the 0.1 M Tris control. This two-step thermal transition is dependent upon a critical concentration of the polyamine. In Fig. 3 the *Tm* data for higher

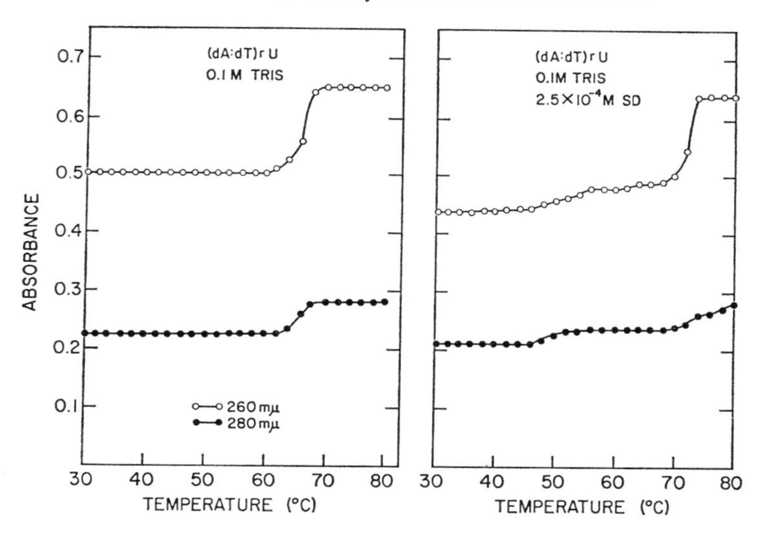

Fig. 2. The effect of spermidine on the thermal denaturation of *(dA:dT)rU*. Optical density measurements were made with the cuvettes in a electrically-heated cell block of a Beckman DB-G spectrophotometer. Heating at the rate of 1 °C per minute and recording of cuvette temperature was performed with a Beckman *Tm* Analyzer and Temperature Programmer

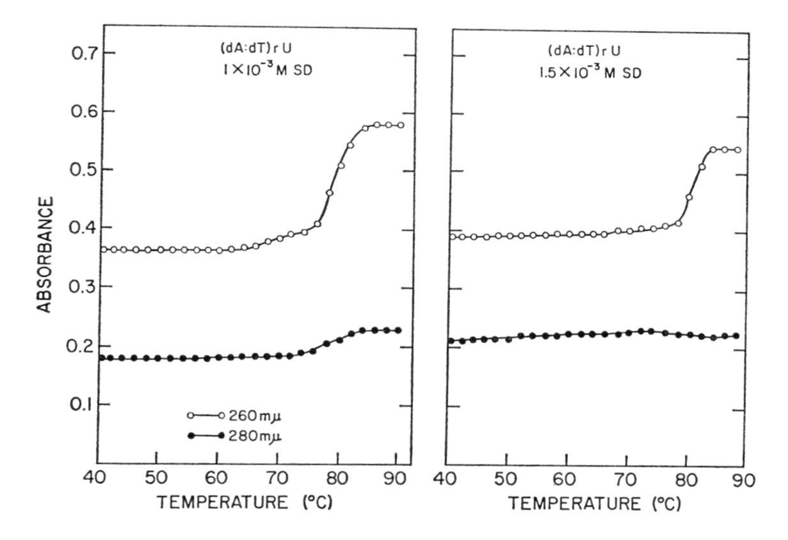

Fig. 3. Modification of the thermal transition of *(dA:dT)rU* at 280 mμ by spermidine

concentrations of spermidine are illustrated and it is apparent that the low temperature release of homopolymer from a triple stranded complex does not occur. On the other hand, the *Tm* of the simultaneous double strand and third strand melting is stabilized at a higher temperature when spermidine is increased to 1×10^{-3} M. At yet higher concentrations of polyamine (1.5×10^{-3} M), a single thermal transition is

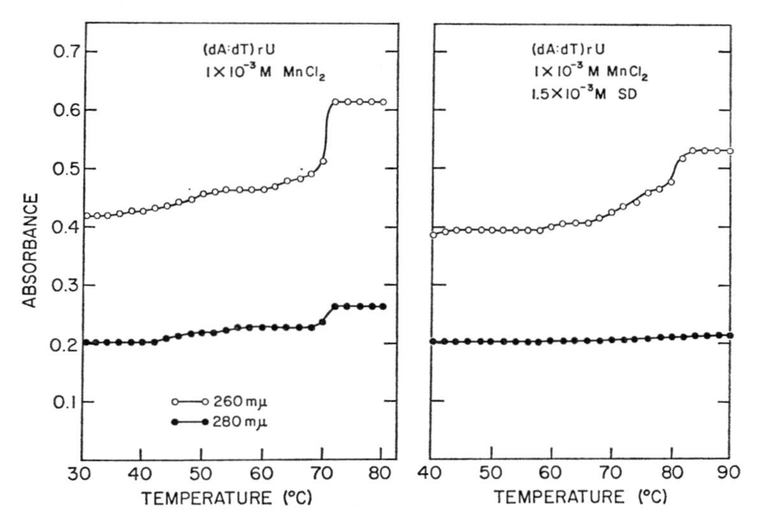

Fig. 4. Apparent elimination of the two-step melting of *(dA:dT)rU* by spermidine

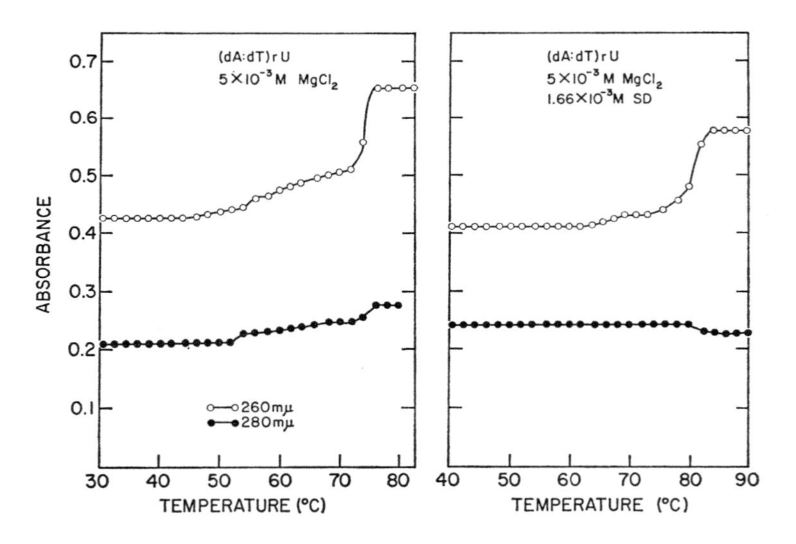

Fig. 5. Apparent elimination of the two-step melting of *(dA:dT)rU* by spermidine

obtained at 260 nm while the 280 nm hyperchromicity is eliminated. The experiment suggests that triple strand hybrid homopolymers are very sensitive to polyamine counterions and that the addition of *rU* to the *dA:dT* helix can be prevented by 1.5×10^{-3} M spermidine.

Similar results were obtained with spermine and with the diamine, putrescine. As has been generally observed in comparisons of the polyamines and diamines, spermine is effective at lower concentrations (10^{-5} M) and putrescine at higher concentrations (2.5×10^{-3} M) than spermidine (2.5×10^{-4} M) in the stabilization of *dA:dT*. With 1×10^{-4} M spermine, the thermal denaturation of *(dA:dT)rU* is characteristic of the data obtained with 1.5×10^{-3} M spermidine. A single thermal transition occurs at 260 nm indicative of the absence of a stable triple stranded complex.

The effect of Mn^{++} and Mg^{++} or of polyamines and divalent cations combined on the thermal denaturation of *(dA:dT)rU* is of interest since these inorganic cations are present in RNA polymerase assays and during transcription in cells. The addition of appropriate concentrations of Mg^{++} or Mn^{++} to *(dA:dT)rU* in 0.1 M Tris buffer induces a two-step melting transition of the complex. In Fig. 4 the melting curves obtained with 1×10^{-3} M Mn^{++} are illustrated. The *Tm* data are entirely comparable to those obtained with 2.5×10^{-4} M spermidine. Combinations of low concentrations of spermidine (10^{-4} M) and Mn^{++} (10^{-4} to 10^{-3} M) yield similar two-step melting curves and hyperchromic shifts are demonstrable at 260 nm and 280 nm. However, 1.5×10^{-3} M spermidine eliminates the 280 nm hyperchromic shift, attributed to the release of *rU*, when combined with 1×10^{-3} M Mn^{++}. Comparable results have been obtained in *Tm* studies which involved Mg^{++} or Mg^{++} and spermidine (Fig. 5). Mg^{++} (5×10^{-3} M) permits a two-step melting of *(dA:dT)rU* and, as previously observed with Mn^{++} and spermidine combined, the combination of Mg^{++} and 1.66×10^{-3} M spermidine produces a single 260 nm thermal transition.

III. Triple Stranded Complexes as Templates for Transcription

The *Tm* data suggested that spermidine might influence the release of RNA from the complex of RNA polymerase · DNA · product RNA which has been demonstrated during transcription *in vitro* (Bremer et al.). Experiments were undertaken to test this possibility with ML RNA polymerase purified and assayed according to Weiss (1968). Straat and Ts'o (1969) reported that *(dA:dT)rU* is inert as a template for the incorporation of UMP by the ML RNA polymerase when UTP is the only substrate supplied in the assay. Furthermore, *dA:dT* is a poor template for the incorporation of UMP but serves as an excellent template when ATP is supplied as the only substrate. This latter observation suggested that the formation of *(dA:dT)rU* during UMP incorporation inhibited transcription due to the failure te release the product, *rU*, from the triple stranded complex. When ATP is supplied as the only substrate, "product" inhibition does not occur since the triple stranded complex *(dA:dT)rA* does not form under assay conditions and, in fact, has never been observed (Felsenfeld and Miles, 1967).

The failure of *(dA:dT)rU* to serve as a template for UMP incorporation by RNA polymerase suggests that the triple stranded complex is stabilized under the assay conditions and that the third strand cannot be removed by interaction with the enzyme. Our *Tm* data indicated that combinations of spermidine and Mn^{++} or Mg^{++} might labilize the association of the third strand with *dA : dT* and thereby activate *(dA:dT)rU* as a template for the transcription reaction. The presence of spermidine in the assay might also reduce the association of the nascent *rU* synthesized on a *dA:dT* template with the enzyme · template complex and thereby reduce "product" inhibition of transcription.

These ideas have been tested and the results do not support the hypothesis of the participation of spermidine in the RNA polymerase reaction. Table 1 summarizes experiments performed with *(dA:dT)rU* as a template. Incorporation of UMP is minimal in the standard assay in the absence of spermidine and addition of spermidine does not activate the triple stranded homopolymer as a template. A number of variations in the assay conditions have been tested in an attempt to improve the response to spermidine. No effect of the polyamine was observed when the temperature of the assay was increased to 37°, when Mg^{++} was substituted for Mn^{++} or when the time of incubation was extended to 3 h. FUCHS et al. demonstrated the activation of a "stalled" *E. coli* RNA polymerase reaction by the addition of either

Table 1. *Effect of spermidine on the ML RNA polymerase reaction; Attempts to activate (dA:dT)rU as a template for UMP incorporation*

Additions	Assay 1	Assay 2
	mμmoles ^{14}C-UMP	incorporated
0 spermidine	3.3	0.45
2 mM spermidine	4.0	0.38
4 mM spermidine	6.7	0.43
8 mM spermidine	8.2	0.50

The complete reaction mixture (Assay 1) contained 0.1 M Tris, pH 7.5, 0.4 mM each of ATP, GTP, CTP and ^{14}C-UTP (100 c.p.m. per μmole), 2.5 mM $MnCl_2$, 30 μg of rat liver DNA, 30 μg of enzyme protein and spermidine as indicated in a final volume of 0.2 ml. In Assay 2, *(dA:dT)rU* (100 mμmoles of each homopolymer) replaced DNA as a template and ^{14}C-UTP was provided as a single substrate. The reactions were incubated for 30 min at 30° and assayed for ^{14}C-UMP incorporation by precipitation with 10% trichloroacetic acid on SCHLEICHER and SCHUELL B-6 Membrane Filters which were washed and counted in PPO-POPOP-toluene by the liquid scintillation method.

8 mM spermidine, 70 mM Mg^{++} or 0.15 m NH_4Cl to the assay. The re-starting of the stalled reaction was attributed to the activation of the enzyme, possibly by the elimination of product inhibition. However, in our experiments the addition of NH_4Cl or Mg^{++} or a combination of the inorganic salts and the polyamine did not increase incorporation of UMP in assays containing *(dA:dT)rU* as a template.

The polyamine also failed to stimulate the incorporation of UMP when *dA:dT* was utilized as a template. A wide range of concentrations of spermidine was tested but no increased incorporation of UMP was obtained. Inhibition of UMP incorporation occurred when spermidine was included in assays in which both UTP and ATP were provided as substrates. The inability of spermidine to release the inhibition of the RNA polymerase reaction which occurs when *(dA:dT)rU* is provided as a template is surprising in view of the results obtained in the thermal denaturation studies. The 280 nm thermal transition was not detected in the melting curves at critical concentrations of spermidine and the polyamine was present in the assays in the range of these cation concentrations. The combined results do not support the assumption based upon the *Tm* studies that spermidine can labilize the ribohomopolymer strand from a *(dA:dT)rU* complex with RNA polymerase. The effect of the polyamine is apparently modified in the presence of the enzyme perhaps because the triple strand is stabilized by interaction with the template binding site of the protein.

IV. Inhibition of RNA Polymerase by RNA

The addition of RNA to the ML RNA polymerase reaction produces an inhibition of RNA synthesis which is reduced but not eliminated by spermidine (Fox, Gumport and Weiss, 1965). We have studied the effect of spermidine on the reversal of the

Table 2. *Stimulation of RNA synthesis and reversal of RNA inhibition by spermidine*

Additions	0 SD	4 mM SD
	mμmoles ^{14}C-UMP	incorporated
None	3.8	9.2
5 μg "pH 8.3" RNA	2.2	7.6
15 μg "pH 8.3" RNA	1.5	7.2
5 μg *rU*	3.8	5.9
15 μg *rU*	2.9	3.4
5 μg 16 S RNA	3.6	6.9
15 μg 16 S RNA	1.4	7.0
5 μg tRNA	2.9	6.7
15 μg tRNA	2.7	7.0

The assay was performed as described in Table 1 with rat liver DNA as template and with additions of RNA and spermidine (SD) as indicated.

inhibition of the polymerase reaction by several RNA preparations as well as by the ribohomopolymer, *rU*. Drews and Brawerman (1967) described the preparation of a rat liver nuclear RNA designated "pH 8.3" RNA. This nuclear RNA was characterized as "DNA-like" by base analysis and by hybridization with rat liver DNA. We have prepared this RNA as well as rat liver DNA by the procedures described by Drews and Brawerman (1967). The effect of spermidine on transcription of the liver DNA in the presence of the "pH 8.3" RNA was studied to attempt to gain evidence in support of a polyamine role in specific DNA · RNA interactions.

The effect of spermidine on the reversal of "RNA inhibition" of the RNA polymerase reaction with a rat liver DNA template is summarized in Table 2. The nuclear "pH 8.3" RNA inhibits the reaction and spermidine can reverse the inhibition (and simultaneously stimulate transcription) at several levels of inhibitor. Neither the inhibition nor the reversal by spermidine is indicative of a species-*specific* effect of polyamine on the interaction between a homologous template and inhibitor, since bacterial RNA and, to some extent, *rU* inhibition and reversal by spermidine were entirely comparable when assayed with liver DNA as template. Table 3 summarizes similar results obtained with calf thymus DNA as a template. Spermidine produces a substantial increase in the incorporation of ^{14}C-UMP and there is a partial reversal of the inhibition of the reaction by "pH 8.3" RNA and by *rU*.

Finally, we have sought evidence for the polyamine-facilitated release of product RNA from the DNA · RNA polymerase complex. An RNA polymerase reaction mixture in which RNA synthesis was stimulated by spermidine was applied to a sucrose gradient and the sedimentation of the labeled RNA was compared with the sedimentation of RNA from an assay mixture lacking spermidine. The results (Fig. 6) suggest that spermidine *increases* the retention of product RNA by the DNA · RNA polymerase complex. The labeled products sediment more rapidly than "free" RNA

Table 3. *Stimulation of RNA synthesis and reversal of RNA inhibition by spermidine*

Additions	^{14}C-UMP incorporated mμmoles
None	7.1
4 mM spermidine	11.5
8 mM spermidine	12.7
10 μg "pH 8.3" RNA	3.0
"pH 8.3" RNA + 4 mM spermidine	6.8
"pH 8.3" RNA + 8 mM spermidine	8.1
10 μg *rU*	3.3
rU + 4 mM spermidine	6.7
rU + 8 mM spermidine	6.8

The assay was performed as described in Table 1 except for the substitution of calf thymus DNA (30 μg) for rat liver DNA template.

released by treatment with a detergent by the procedure utilized by BREMER and KONRAD (1964) to illustrate the formation of a DNA · RNA polymerase · product RNA complex. The results are in agreement with the previous observation of KRAKOW that spermidine does not release product RNA from the DNA · RNA polymerase complex which is retained by a membrane filter.

Our results agree with the well documented evidence of a polyamine effect in RNA synthesis but the site of polyamine action does not appear to be at the level of DNA · RNA complexes. An alternative explanation of the role of polyamines in the RNA polymerase reaction can be derived from the results obtained by Fox et al. which demonstrated that spermidine reduced the binding of RNA to the enzyme but did not release DNA from the initiation complex with the polymerase. Fox et al. rejected this possible mechanism of polyamine labilization of enzyme · RNA complexes since the irreversibility of the DNA · RNA polymerase interaction seemed to eliminate any significant RNA binding to the enzyme. The association of product RNA with a binding site on the enzyme cannot be excluded, however, and might limit transcription throughout the synthetic reaction. Such an RNA binding site on the enzyme which is sensitive to polyamines might be labilized by critical concentrations of the organic cations without a physical release of the product from the complex. We are investigating this potentially important function of the polyamines.

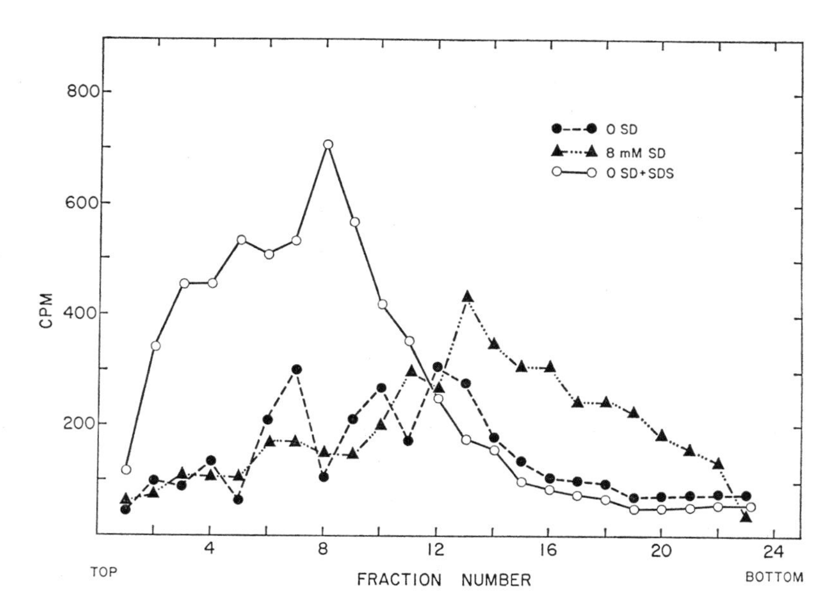

Fig. 6. Failure of spermidine to promote release of product RNA from the DNA · RNA polymerase complex. Duplicate reaction mixtures as described in Table 1 (Assay 1) containing either zero or 8 mM spermidine were incubated for 30 min. One-tenth ml was removed for assay of ^{14}C-UMP incorporation with the following results: zero spermidine; 11.6 and 10.2 mμmoles; 8 mM spermidine; 18.6 and 17.6 mμmoles. Three-tenths ml of each reaction mixture was diluted with chilled buffer to 0.9 ml and applied to 4.4 ml of a 4 to 20% sucrose gradiant containing 0.1 M Tris, pH 7.5 and 2.5 mM $MnCl_2$. A duplicate zero spermidine reaction mixture was supplemented with sodium dodecyl sulfate (SDS) to 0.4% and applied to the third gradiant. The gradiants were centrifuged for 5.5 h at 30,000 rpm in the SW 50 Spinco swinging bucket rotor at 5 °C. Fractions were collected with an ISCO Density Gradiant Fractionator, precipitated with 10% trichloroacetic acid in the presence of 200 μg carrier DNA and dried on SCHLEICHER and SCHUELL B-6 membrane filters which were assayed by liquid scintillation counting

Acknowledgements

The authors wish to acknowledge the generous gift of the deoxyribohomopolymers from Dr. F. J. BOLLUM, University of Kentucky Medical School, Lexington, Ky.

One of us (E. J. H.) received important assistance from Dr. PATRICIA A. STRAAT and most generous use of facilities and instruments in the laboratories of Dr. ROLAND BEERS and Dr. PAUL TS'O of the Department of Radiological Sciences, Johns Hopkins University School of Hygiene and Public Health, Baltimore, Md.

The research was supported by NSF Traineeship GZ-528 to R. B. T. and U. S. Public Health Service Special Research Fellowship, 1 F03 GM 44165-01 to E. J. H. and by Hatch Project No. 170 of the New Hampshire Agricultural Experiment Station, Durham, N. H.

References

Abraham, K. A.: Studies on DNA-dependent RNA polymerase from *Escherichia coli:* 1. The mechanism of polyamine induced stimulation of enzyme activity. Europ. J. Biochem. **5**, 143 (1968).

Ballard, P. L., Williams-Ashman, H. G.: Isolation and properties of a testicular ribonucleic acid polymerase. J. biol. Chem. **241**, 1602 (1966).

Barros, C., Guidice, G.: Effect of polyamines on ribosomal RNA synthesis during sea urchin development. Exp. Cell Res. **50**, 671 (1968).

Bertossi, F., Bagni, N., Moruzzi, G., Caldarera, C. M.: Spermine as a new growth-promoting substance for *Helianthus tuberosus* (Jerusalem artichoke) *in vitro*. Experientia (Basel) **21**, 80 (1965).

Bretthauer, R. K., Marcus, L., Chaloupka, J., Halvorson, H. O., Bock, R. M.: Amino acid incorporation by cell-free extracts of yeast. Biochemistry **2**, 1079 (1963).

Bremer, H., Konrad, M. W.: A complex of enzymatically synthesized RNA and template DNA. Proc. nat. Acad. Sci. (Wash.) **51**, 801 (1964).

Caldarera, C. M., Barbiroli, B., Moruzzi, G.: Polyamines and nucleic acids during development of the chick embryo. Biochem. J. **97**, 84 (1965).

— Moruzzi, M. S., Rossoni, C., Barbiroli, B.: Polyamines and nucleic acid metabolism during development of chick embryo brain. J. Neurochem. **16**, 309 (1969).

Cocucci, S., Bagni, N.: Polyamine induced activation of protein synthesis in ribosomal preparation from *Helianthus tuberosus* tissue. Life Sci. **7**, 113 (1968).

Cohen, S. S., Lichtenstein, J.: Polyamines and ribosome structure. J. biol. Chem. **235**, 2122 (1960).

— Morgan, S., Streibel, E.: The polyamine content of the tRNA of *E. coli*. Proc. nat. Acad. Sci. (Wash.) **64**, 669 (1969).

Colbourn, J. L., Witherspoon, B. H., Herbst, E. J.: Effect of intracellular spermine on ribosomes of *Escherichia coli*. Biochim. biophys. Acta (Amst.) **49**, 422 (1961).

Davis, G. R. F.: Putrescine and spermidine as growth-promoting substances for the saw-toothed grain beetle, *Oryzaephilus surinamensis*. Comp. Biochem. Physiol. **19**, 619 (1966).

Dion, A. S., Herbst, E. J.: The localization of spermidine in salivary gland cells of *Drosophila melanogaster* and its effect on H^3-uridine incorporation. Proc. nat. Acad. Sci. (Wash.) **58**, 2367 (1967).

Drews, J., Brawerman, G.: Alterations in the nature of ribonucleic acid synthesized in rat liver during regeneration and after cortisol administration. J. biol. Chem. **242**, 801 (1967).

Dykstra, W. G., Herbst, E. J.: Spermidine in regenerating liver: Relation to rapid synthesis of ribonucleic acid. Science **149**, 428 (1965).

Felsenfeld, G., Huang, S.: The interaction of polynucleotides with metal ions, amino acids and polyamines. Biochim. biophys. Acta (Amst.) **37**, 425 (1960).

— Miles, H. T.: The physical and chemical properties of nucleic acids. Ann. Rev. Biochem. **36**, 407 (1967).

Fox, D. F., Gumport, R. I., Weiss, S. B.: The enzymatic synthesis of ribonucleic acid. V. The interaction of ribonucleic acid polymerase with nucleic acids. J. biol. Chem. **240**, 2101 (1965).

Fox, C. F., Weiss, S. B.: Enzymatic synthesis of ribonucleic acid. II. Properties of the deoxyribonucleic acid-primed reaction with *Micrococcus lysodeikticus* ribonucleic acid polymerase. J. biol. Chem. **239**, 175 (1964).

— Robinson, W. S., Haselkorn, R., Weiss, S. B.: Enzymatic synthesis of ribonucleic acid with *Micrococcus lysodeikticus* ribonucleic acid polymerase. J. biol. Chem. **239**, 186 (1964).

FREDERICK, W. W., MAITRA, U., HURWITZ, J.: The role of DNA in ribonucleic acid synthesis. XVI. The purification and properties of ribonucleic acid polymerase from yeast: Preferential utilization of denatured deoxyribonucleic acid as template. J. biol. Chem. **244**, 413 (1969).

FRESCO, J. R., ADAMS, A., ASCIONE, R., HENLEY, D., LINDAHL, T.: Tertiary structure of transfer ribonucleic acids. Cold Spr. Harb. Symp. quant. Biol. **31**, 527 (1966).

FUCHS, E., MILLETTE, R. L., ZILLIG, W., WALTER, G.: Influence of salts on RNA synthesis by DNA-dependent RNA-polymerase from *Escherichia coli*. Europ. J. Biochem. **3**, 183 (1967).

GLASER, R., GABBAY, E. J.: Topography of nucleic acid helixes in solutions. III. Interactions of spermine and spermidine derivatives with polyadenylic-polyuridylic and polyinosinic-polycytidylic acid helices. Biopolymers **6**, 243 (1968).

GUMPORT, R. F., WEISS, S. B.: Transcriptional and sedimentation properties of ribonucleic acid polymerase from *Micrococcus lysodeikticus*. Biochemistry **8**, 3618 (1969).

HAM, R. G.: Putrescine and related amines as growth factors for a mammalian cell line. Biochem. biophys. Res. Commun. **14**, 34 (1964).

HERBST, E. J., SNELL, E. E.: Putrescine and related compounds as growth factors for *Hemophilus parainfluenzae* 7901. J. biol. Chem. **181**, 47 (1949).

HERSHKO, A., AMOZ, S., MAGER, J.: Effect of polyamines and divalent metals on *in vitro* incorporation of amino acids into ribonucleoprotein particles. Biochem. biophys. Res. Commun. **5**, 46 (1961).

IGARASKI, K., TAKEDA, Y.: Polyamines and protein synthesis. V. Interaction of polyamines with transfer RNA. Biochim. biophys. Acta (Amst.) (in press).

KRAKOW, J. S.: Ribonucleic acid polymerase of *Azotobacter vinelandii*. III. Effect of polyamines. Biochim. biophys. Acta (Amst.) **72**, 566 (1963).

— *Azotobacter vinelandii* ribonucleic acid polymerase. II. Effect of ribonuclease on polymerase activity. J. biol. Chem. **241**, 1830 (1966).

LIQUORI, A. M., CONSTANTINO, L., CRESCENZI, V., ELIA, V., GIGLIO, E., PULITI, R., DE SANTIS SAVINO, M., VITAGLIANO, V.: Complexes between DNA and polyamines: a molecular model. J. molec. Biol. **24**, 113 (1967).

MAHLER, H. R., MEHROTRA, B. D.: The interaction of nucleic acids with diamines. Biochim. biophys. Acta (Amst.) **68**, 211 (1963).

— — SHARP, C. W.: Effect of diamines on the thermal transition of DNA. Biochem. biophys. Res. Commun. **4**, 79 (1961).

MARTIN, R. G., AMES, B. N.: The effect of polyamines and of poly U size on phenylalanine incorporation. Proc. nat. Acad. Sci. (Wash.) **48**, 2171 (1962).

MEHROTRA, B. D., MAHLER, H. R.: Studies on polynucleotide-small molecule interactions. III. The effect of diamines on helical polynucleotides. Biochim. biophys. Acta (Amst.) **91**, 78 (1964).

MORGAN, A. R., WELLS, R. D.: Specificity of the three-stranded complex formation between double-stranded DNA and single-stranded RNA containing repeating nucleotide sequences. J. molec. Biol. **37**, 63 (1968).

MORUZZI, G., BARBIROLI, B., CALDARERA, C. M.: Polyamines and nucleic acid metabolism in chick embryo. Incorporation of labeled precursors into nucleic acids of subcellular fractions and polyribosomal patterns. Biochem. J. **107**, 609 (1968).

PETERSEN, E. E., KRÖGER, H., HAGEN, U.: The influence of spermidine on the reaction of RNA nucleotidyltransferase. Biochim. biophys. Acta (Amst.) **161**, 325 (1968).

RAINA, A.: Studies on the determination of spermidine and spermine and their metabolism in the developing chick embryo. Acta physiol. scand. **60**, 7 (1963).

— COHEN, S. S.: Polyamines and RNA synthesis in a polyauxotrophic strain of *E. coli*. Proc. nat. Acad. Sci. (Wash.) **55**, 1587 (1966).

— JÄNNE, J.: Polyamines and the accumulation of RNA in mammalian systems. Fed. Proc. **29**, 1568 (1970).

— — SIIMES, M.: Stimulation of polyamine synthesis in relation to nucleic acids in regenerating rat liver. Biochim. biophys. Acta (Amst.) **123**, 197 (1966).

RILEY, M., MALING, B., CHAMBERLIN, M. J.: Physical and chemical characterization of two- and three-stranded adenine-thymine and adenine-uracil homopolymer complexes. J. molec. Biol. **20**, 359 (1966).

Russell, D., Snyder, S. H.: Amine synthesis in rapidly growing tissues: Ornithine decarboxylase activity in regenerating rat liver, chick embryo, and various tumors. Proc. nat. Acad. Sci. (Wash.) **60**, 1420 (1968).

Russell, D. H., Snyder, S. H., Medina, V. J.: Presence and biosynthesis of putrescine and polyamines in amphibians. Life Sci. **8**, 1247 (1969).

Sneath, P. H. A.: Putrescine as an essential growth factor for a mutant of *Aspergillus nidulans*. Nature (Lond.) **175**, 818 (1955).

Stevens, C. L., Felsenfeld, G.: The conversion of two-stranded poly (A + U) to three-stranded poly (A + 2U) and poly A by heat. Biopolymers **2**, 293 (1964).

Stevens, L.: The binding of spermine to the ribosomes and ribosomal ribonucleic acid from *Bacillus stearothermophilus*. Biochem. J. **113**, 117 (1969).

Straat, P. A., Ts'o, P. O. P.: Ribonucleic acid polymerase from *Micrococcus lysodeikticus*. II. Studies on double stranded homopolynucleotide templates. J. biol. Chem. **244**, 391 (1969).

Szer, W.: (1) Effect of di- and polyamines on the thermal transition of synthetic polyribonucleotides. Biochem. biophys. Res. Commun. **22**, 559 (1966).

— (2) Interaction of polyribothymidylic acid with metal ions and aliphatic amines. Acta biochim. pol. **10**, 251 (1966).

Tabor, H.: The protective effect of spermine and other polyamines against heat denaturation of deoxyribonucleic acid. Biochemistry **1**, 496 (1962).

Takeda, Y.: Polyamines and protein synthesis. III. The effect of polyamines on the binding of phenylalanyl transfer RNA to ribosomes. Biochim. biophys. Acta (Amst.) **182**, 258 (1969).

— Igaraski, K.: Polyamines and protein synthesis. IV. Stimulation of aminoacyl transfer RNA formation by polyamines. Biochem. biophys. Res. Commun. **37**, 917 (1969).

Tanner, M. J. A.: The effect of polyamines on the binding of aminoacyl transfer ribonucleic acid to ribosomes in a yeast system. Biochemistry **6**, 2686 (1967).

Weiss, S. B.: In: Grossman, L., Moldave, K. (Eds.): Methods in enzymology, Vol. 12, part b, p. 559. New York: Academic Press 1968.

The Use of Intercalative Dyes in the Study of Closed Circular DNA

WILLIAM BAUER and JEROME VINOGRAD

I. Introduction

1. The Occurrence of Closed Circular DNA

A wide variety of naturally occurring duplex DNA's are now known to possess the dual characteristics of closed circularity and superhelicity. These species of DNA may be broadly classified according to biological origin as arising from animal and bacterial viruses, animal mitochondria and extra-mitochondrial cytoplasm, protozoa, and bacteria. This latter category contains several types, including viral intracellular forms, plasmids, and episomes. A summary of the known closed circular duplex DNA's (DNA I) is presented in Table 1.

The DNA molecules described here are all relatively small, ranging in size from the smallest circles isolated from HeLa cells, molecular weight 0.4×10^6 daltons (RADLOFF, BAUER and VINOGRAD, 1967), to the bacterial sex factor *FColVColBtrycys* (HICKSON, ROTH and HELSINKI, 1967), molecular weight 107×10^6 daltons. Much larger DNA molecules are known to be circular, including the chromosomal DNA of *E. coli* (CAIRNS, 1963). The determination of whether or not these larger forms are ever closed is, however, extremely difficult due to the relative ease of introduction of a chain scission with consequent abolition of closure (VINOGRAD, LEBOWITZ, RADLOFF, WATSON and LAIPIS, 1965). Certain fractions of the mitochondrial DNA from yeast have been shown to be circular by electron microscopy (HOLLENBERG, BORST, THURING and VAN BRUGGEN, 1969; AVERS, BILLHEIMER, HOFFMAN and PAULI, 1968; BILLHEIMER and AVERS, 1968) but up to the present time the preparation of linear molecules observed in a given sample is usually high and variable (AVERS, 1967; SHAPIRO, GROSSMAN, MARMUR and KLEINSCHMIDT, 1968). If the mature yeast mitochondrial DNA is closed circular, it would appear to be extremely sensitive to shear degradation (AVERS et al., 1968).

2. The Properties of Closed Circular DNA Elucidated Prior to the Use of Dyes

a) Experimental Results

Several properties of closed circular DNA were described between the time of its discovery in extracts of polyoma virus (WEIL and VINOGRAD, 1963; DULBECCO and VOGT, 1963) and the first experimental studies of its interaction with intercalating dyes (CRAWFORD and WARING, 1967; RADLOFF et al., 1967). These include elevated sedimentation coefficients in neutral and alkaline solvents, elevated buoyant density in

Table 1. *Occurrence of closed circular DNA*

Source of DNA	Reference
Animal viruses	
Polyoma	Weil and Vinograd (1963); Dulbecco and Vogt (1963)
SV40	Crawford and Black (1964)
Rabbit papilloma	Crawford (1964)
Human papilloma	Crawford (1965)
Bovine papilloma	Bujard (1967)
Animal mitochondria	
Sheep heart	Kroon et al. (1966)
Beef heart	Borst and Ruttenberg (1966)
Mouse liver	Borst and Ruttenberg (1966); Sinclair and Stevens (1966)
Chicken liver	Borst and Ruttenberg (1966); Borst et al. [1967 (1), 1967 (2)]
Human leukocytes	Clayton and Vinograd (1967)[c]
HeLa cells	Radloff et al. (1967)[c]
Unfertilized sea urchin eggs	Pikó et al. (1967, 1968)[c]
Monkey liver	Suyama and Muira (1968)
Frog oöcytes	Dawid and Wolstenholme (1967); Wolstenholme and Dawid (1967)[a]
Frog liver	Wolstenholme and Dawid (1967)
Mouse L-cells	Nass [1966[a], 1969 (1)[c]]
Rabbit brain, kidney, liver and bone marrow	Clayton et al. (1968)[c]; Hudson et al. (1968)[c]
Guinea pig brain and liver	Clayton et al. (1968)[c]
Mouse embryo	Clayton et al. (1968)[c]; Hudson et al. (1968)
Animal cell cytoplasm	
HeLa cells	Radloff et al. (1967)[c]
Boar sperm	Hotta and Bassel (1965)[b]
Protozoa	
Kinetoplast of *T. cruzi*	Riou and Paoletti (1967)[a]: Riou and Delain (1969)[c]
Mature bacteriophage DNA	
PM-2	Espejo et al. (1969)
Bacteriophage intracellular forms	
ϕX174	Kleinschmidt et al. 1963)
M13	Ray et al. (1966)
λ	Young and Sinsheimer (1964); Bode and Kaiser (1965)
λdv	Mackinlay and Kaiser (1969)
P22	Rhodes and Thomas (1968)
N1	Lee et al. (1968)[c]
fd	Sugiura et al. (1969)
Bacterial sex factors	
F	Freifelder (1968)
F'2	Freifelder (1968)
F'lac	Freifelder (1968)
F'gal	Freifelder (1968)
FColVColBtrycys	Hickson et al. (1967)[c]

Table 1 (continued)

Source of DNA	Reference
Bacterial plasmids	
E. coli 15, minute	COZZARELLI et al. (1968)[c]
Penicillinase plasmid from *S. Aureus*	RUSH et al. (1969)[c]
Col. E1	ROTH and HELSINKI (1967)
Shigella disenteriae Y6R	RUSH et al. [1969 (2)][c]
Shigella paradysenteriae	JANSZ et al. (1969)[c]
Bacillus megaterium	CARLTON and HELINSKI (1969)[c]
Col. E2	BAZARAL and HELINSKI (1968)[c]
Col. E3	BAZARAL and HELINSKI (1968)[c]
Micrococcus lysodeikticus	LEE and DAVIDSON (1968)[c]

[a] Based on circular appearance in electron micrographs only.
[b] Based on non-specific electron microscopy. Found circles ranging 0.5 μ to 16.8 μ in length, some of which could have been mitochondrial in origin.
[c] Ethidium bromide used in isolation procedure.

alkaline CsCl (VINOGRAD and LEBOWITZ, 1966; VINOGRAD, LEBOWITZ and WATSON, 1968), resistance to thermal and alkaline denaturation (WEIL and VINOGRAD, 1963; DULBECCO and VOGT, 1963), a dip in the sedimentation velocity-pH titration curve with both increasing and decreasing pH (VINOGRAD and LEBOWITZ, 1966), and the appearance of circularity (twisted or untwisted) in electron micrographs (STOECKENIUS in WEIL and VINOGRAD, 1963; VINOGRAD et al., 1965). As pointed out by VINOGRAD and LEBOWITZ (1966), this latter property is a necessary but not sufficient indication of closed circularity. In addition to the above properties, conversion to a nicked circle (DNA II) was shown to take place by the introduction of at least one chain scission or nick. This may be done, for example, by thermal hydrolysis, acid depurination, reaction with reducing agents, or treatment with endonucleases. The conversion is normally irreversible, and the nicked molecule fails to exhibit the characteristic properties cited above. A more detailed discussion of these effects may be found in the review by VINOGRAD and LEBOWITZ (1966).

b) A Structural and Topological Model

The special properties of closed circular DNA were satisfactorily explained by the hypothesis that the total number of rotations of one strand about the other is invariant (VINOGRAD et al., 1965). The closed molecule is therefore subject to a *topological restraint*. The consequences of this restraint have been discussed in detail elsewhere [BAUER and VINOGRAD, 1968; WANG, 1969 (1); BAUER and VINOGRAD, 1970 (1)] and will be only briefly summarized here. The net amount of winding of the duplex strands about each other is indicated by the topological winding number, α. This quantity is constant in the absence of a strand scission, and it may be conceptually determined by counting the number of complete revolutions made by either strand about the duplex axis constrained to lie in a plane.

It is convenient to regard α as being composed of two winding numbers, β and τ, although higher order windings are also possible. The *duplex winding number*, β, is defined as the number of revolutions made by either strand about the duplex axis in the unconstrained molecule. The *superhelix winding number*, τ, is a tertiary winding number and is equal to the number of revolutions made by the duplex axis about a three-dimensional space axis, the superhelix axis, again in the unconstrained molecule. The winding numbers are related by the equation (VINOGRAD and LEBOWITZ, 1966; GLAUBIGER and HEARST, 1967)

$$\alpha = \beta + \tau \,. \tag{1}$$

Although the sum of β and τ is necessarily constant, they are each functions of both intra- and intermolecular interactions; these interactions might be subject to variation with, for example, ionic strength and temperature (WANG, BAUMGARTEN and OLIVERA, 1968), pH (VINOGRAD and LEBOWITZ, 1966; VINOGRAD et al., 1968), heating in formaldehyde (CRAWFORD and BLACK, 1964) and the addition of nonaqueous solvents (HERSKOVITS, 1962). These two winding numbers might also vary with the bending and torsional stresses upon the duplex which are induced by the superhelical winding itself (WANG et al., 1967). The above factors are discussed in Appendix I to BAUER and VINOGRAD [1970 (1)]. In the absence of these complications the duplex winding number β is assumed to be a constant and equal to $\beta°$, which represents one-tenth the number of base pairs in a DNA molecule of the Watson-Crick B form. The superhelix winding number (number of superhelical turns), τ, is maximal in absolute value when $\beta = \beta°$, excluding the experimentally unexplored possibility of 0 verwinding the duplex in solution. Since α is a constant, changes in τ and in β are related by

$$\Delta\tau = -\Delta\beta \,. \tag{2}$$

Eq. (2) is of general utility in studies of the interaction of intercalating dyes with closed circular DNA, as discussed in detail in the sections to follow.

The Watson-Crick duplex is right-handed and we correspondingly define β_0, the duplex winding number in the native molecule, to have a positive sign. The results of the alkaline titration of closed circular polyoma DNA (VINOGRAD and LEBOWITZ, 1966; VINOGRAD et al., 1968) show that the native polyoma DNA molecule behaves as though the duplex strands were closed before the completion of the duplex base pairing. The sign of τ_0, the superhelix winding number of the native molecule, is therefore negative. All naturally occurring and enzymatically prepared closed circular DNA's known to the present time are characterized by $\tau_0 < 0$.

The handedness of the native superhelix is, however, not so readily determined. Two different superhelical structures have been described (VINOGRAD et al., 1965; VINOGRAD and LEBOWITZ, 1966; BAUER and VINOGRAD, 1968), the toroidal superhelix and the interwound superhelix. These two forms are illustrated diagramatically in Fig. 1. The duplex is bent in both models, as indicated in the figure, but the pure interwound form can be constructed in principle without the additional stress of torsion on the duplex, neglecting end effects. For this reason the interwound model, Fig. 1 a, is generally considered to be more nearly representative of closed circular DNA as it occurs in solution. Requirements of duplex continuity dictate, however, that even a preponderately interwound superhelix contain at least two toroidal regions. It should also be emphasized that nothing is yet known concerning the extent of regularity of the native superhelical structure. The conformation of the molecule is expected to resemble a random coil at a low density of tertiary turns.

3. Complex Forms of Closed Circular DNA

Recent investigations into the occurrence of closed circular DNA have revealed the presence of a variety of complex multimeric forms. The naturally occurring complex oligomers are of two general types, catenated oligomers (HUDSON and VINOGRAD, 1967) and circular oligomers (CLAYTON and VINOGRAD, 1967). The catenated molecules are characterized by the presence of independently closed circles joined by

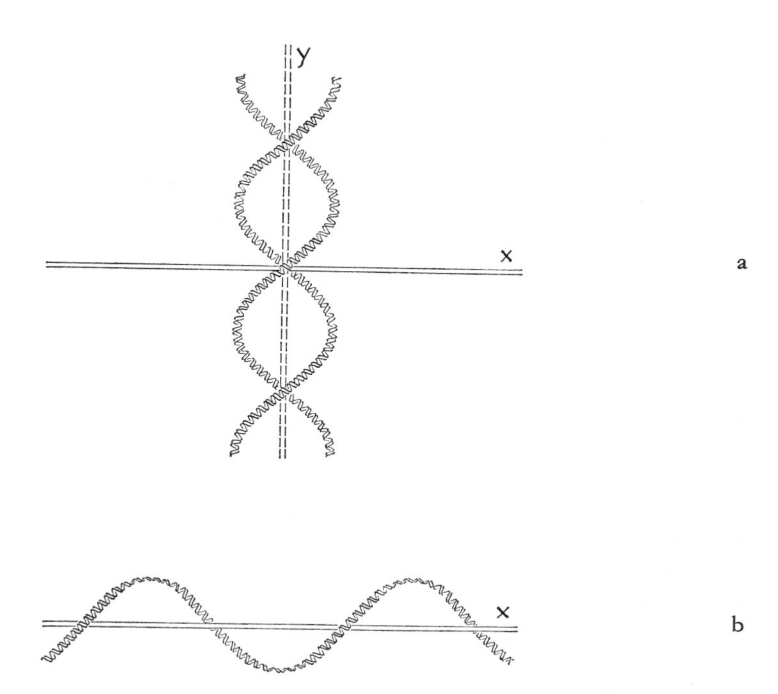

Fig. 1 a and b. Sections of the two possible first-order superhelical models for closed circular DNA. Either of these models, or a combination of them, formally represents the closed circular molecule which contains a negative superhelix density, $\sigma < 0$. — a A section of the toroidal model shown about its superhelical axis, x. The winding is left-handed about this axis. — b A section of the interwound model shown twisted about its superhelical axis, y. The winding is right-handed about this axis, but remains left-handed about the axis x. Both models contain the correct number of duplex turns per superhelical turn for native SV40 DNA (BAUER and VINOGRAD, 1968)

tertiary topological bonds only. The circular oligomers, on the other hand, contain only one duplex axis but are present in multiples of a monomeric length. Table 2 summarizes the present state of knowledge concerning the distribution of such molecules.

The catenated and circular oligomers are further to be distinguished on the basis of their natural distributions, as indicated in Table 2 and discussed by CLAYTON et al. (1970). Catenated oligomers are commonly found in normal animal mitochondrial extracts, although only to the extent of a few percent. Circular oligomers have not yet been reported to occur as components of normal animal cells. The only unambiguous report of a naturally occurring circular oligomer in animals (CLAYTON et al., 1970) has been in studies of malignant human cells, the leukemic human leukocyte. Circular dimers and higher multimeric forms of mitochondrial DNA have also been found in mouse L-cells grown in tissue culture [NASS, 1969 (1); JORDAN, VAN DER EB and

Table 2. *Complex forms of closed circular DNA*

Source of DNA	Reference
Animal viruses	
Polyoma, in 3T3 cells transformed by *Ts-a*	CUZIN et al. (1970)[b]
Animal mitochondria	
HeLa cells	RADLOFF et al. (1967); HUDSON and VINOGRAD (1967)
Unfertilized sea urchin eggs	PIKÓ et al. (1967)
Human leukemic leucocytes	CLAYTON and VINOGRAD (1967)[b]; CLAYTON et al. (1970)[c]
Normal human leucocytes	CLAYTON and VINOGRAD (1967)
Rabbit brain, kidney, liver and bone marrow	CLAYTON et al. (1968); HUDSON et al. (1968)
Guinea pig brain and liver	CLAYTON et al. (1968)
Mouse embryos	CLAYTON et al. (1968); HUDSON et al. (1968)
Human tumors	C. A. SMITH (private communication)
Mouse L-cells	NASS [1969 (1, 2)][b]; JORDAN, VAN DER EB and VINOGRAD (1970)[b]
Protozoa	
Kinetoplast of *T. cruzi*	RIOU and DELAIN (1969)
Bacteriophage intercellular	
P22	RHOADES and THOMAS (1968)
ϕX174	RUSH et al. (1967)[b]
Bacterial plasmids	
Col. E1 (*Proteus mirabilis*)	ROTH and HELINSKI (1967); BAZARAL and HELINSKI (1968)[b]
Shigella paradysenteriae	JANSZ et al. (1969)[b]
Penicillinase plasmid, *S. aurens*	RUSH et al. (1969)[d]

[a] Catenated oligomers (HUDSON and VINOGRAD, 1967) unless otherwise indicated.
[b] Circular oligomers (CLAYTON and VINOGRAD, 1967) as shown by multiple length relationship.
[c] Circular oligomers as shown by reannealing studies (CLAYTON et al., 1970).
[d] Both catenated and circular oligomers found by electron microscopy.

VINOGRAD, 1971]. The circular oligomers in this latter case [NASS, 1969 (1)] could be reversibly induced by treatment with cycloheximide, by amino acid starvation and by keeping the cells in the stationary phase of growth [NASS, 1969 (2)]. Circular oligomers of polyoma DNA have also been reported in large amounts (up to 40%) in intracellular extracts of mouse 3T3 cells transformed with a thermosensitive mutant of polyoma (*Ts-a*) and thermally induced (CUZIN, VOGT, DIECKMANN and

BERG, 1970). Catenated oligomers have been discovered in one protozoan, *Trypanosoma cruzi* (RIOU and DELAIN, 1968). Circular oligomers may also be induced in the kinetoplast DNA of this organism upon treatment with ethidium bromide [RIOU and DELAIN, 1969 (2)], as discussed in Section II. 7 below.

Altough circular oligomers have been reported from prokaryotic organisms as well (see Table 2), these reports are based primarily upon measurements of electron micrographs indicating the presence of DNA molecules of lengths which are integral multiples of an identified unit circular DNA. An explanation of the biological meaning of the circular catenated and multimeric forms is yet to be forthcoming.

II. Intercalators as Probes for the Structure and Properties of Closed Circular DNA

1. The Intercalation of Dyes with DNA

The concept of intercalation as applied to the interaction of certain ligands with nucleic acids originated in the investigations of LERMAN and coworkers into the

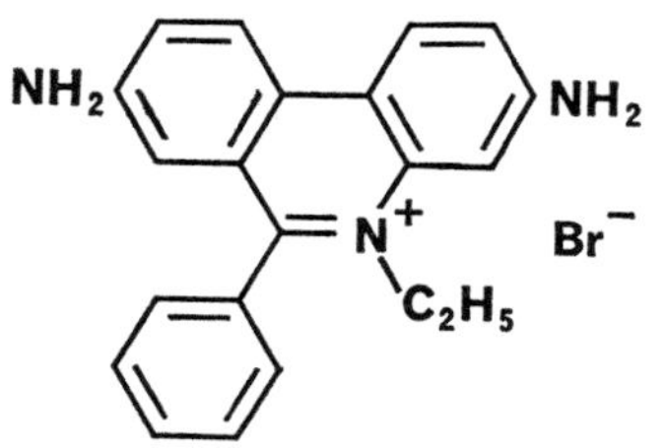

NH_2 NH_2 N^+ Br^- C_2H_5

Fig. 2. Structure of ethidium bromide; 3,8-diamino-6-phenyl-5-ethylphenanthridium bromide

binding of the aminoacridines. These compounds were found to induce an enhanced viscosity and diminished sedimentation coefficient (LERMAN, 1961; COHEN and EISENBERG, 1969), contrary to the expected behavior on the basis of simple hydrodynamic theory. Later studies revealed a diminished mass per unit length upon binding of proflavine to DNA (LUZZATI, MASSON and LERMAN, 1961), based upon low angle *X*-ray scattering experiments. Flow dichroism (LERMAN, 1963) and measurements of the effect of complexing on the rates of diazotization of amino acridines (LERMAN, 1964) lent further support to the intercalation hypothesis (LERMAN, 1961). Fiber *X*-ray diffraction studies (NEVILLE and DAVIS, 1966) are also consistent with the intercalation hypothesis, for the binding of both acridine orange and proflavine.

The basic notion involved in intercalation is that the planar dye molecules bind by physical insertion between adjoining base pairs, which are themselves in a parallel configuration. For space filling reasons, this insertion may take place only with an accompanying axial separation between the base pairs involved, a process which requires a local unwinding of the duplex. The rotational angle between adjoining base pairs in the B form of DNA is 36.0° (LANGRIDGE, SEEDS, WILSON, HOOPER, WILKINS and HAMILTON, 1957). LERMAN (1961) calculated that the DNA-phosphate backbone can be assembled with reasonable connections if the maximum change from the B form is $-45°$, allowing more than enough room for an intercalated dye molecule. In the case of the trypanocidal dye ethidium bromide (EB), Fig. 2, FULLER and WARING (1964) calculate by means of molecular models that an unwinding of only 12° is necessary to account for the space filling requirements of intercalation. Such a value for the unwinding angle is compatible with the *X*-ray fiber diffraction results (FULLER and WARING, 1964) but is not demanded by them. LI and CROTHERS (1969) have recently found that the circular dichroism of an isolated intercalated proflavine dye molecule depends strongly on salt concentration over the range 0.001 M to 1.0 M. This indicates that the value of the unwinding angle, ϕ, might depend upon the solution environment even in the absence of interaction among bound dye molecules. The magnitude of this effect, if it exists, is unknown at present.

At least two types of binding other than intercalation are possible between planar dye molecules and DNA; these include simple ion pair formation with the DNA phosphates and a stacked configuration external to the double helix (PEACOCKE and SKERRETT, 1956; DRUMMOND, SIMPSON-GILDEMEISTER and PEACOCKE, 1965). These mechanisms are expected to become progressively less important as the ionic strength is raised, although they are possibly never eliminated completely (LI and CROTHERS, 1968). Neither of these mechanisms has been implicated in an unwinding to the DNA duplex (change in ϕ), but such an effect should not be excluded from consideration.

2. Topological Relationships Involved in the Binding of Intercalators to Closed Circular DNA

The binding of one molecule of an intercalating dye to the DNA duplex brings about an unwinding of ϕ radians, as discussed in the preceding section. When such an unwinding takes place in a closed circular DNA, the additional condition expressed by Eq. (1) must be satisfied and the topological winding number, α, remains constant. When there are no second order complications (WANG, 1969), Eq. (2) holds and $\Delta\tau = -\Delta\beta$. That is, the duplex winding alteration is accompanied by an equal and opposite superhelix winding alteration.

In the absence of dye $\tau = \tau_0$ and $\beta = \beta_0$, and Eq. (1) becomes

$$\alpha = \beta_0 + \tau_0 \,. \tag{3}$$

If v moles dye are bound per mole DNA nucleotide, the duplex angular winding alteration is given by (BAUER and VINOGRAD, 1968)

$$\Delta\beta = -\left(\frac{\phi}{2\pi}\right)(10\,\beta°)\,(2\,v)\,, \tag{4}$$

where ϕ is positive and is expressed in radians. We combine Eq. (1), (3) and (4) with the relationship $\Delta\tau = \tau - \tau_0$ to obtain

$$\tau = \tau_0 + \left(\frac{\phi}{2\pi}\right)(10\,\beta°)\,(2\,v)\,. \tag{5a}$$

Eq. (5a) may be expressed more generally in terms of the *superhelix density*, σ, the number of turns per 10 base pairs, defined as $\sigma = \tau/\beta°$.

$$\sigma = \sigma_0 + \frac{10\phi}{\pi}\,v\,. \tag{5b}$$

Here σ_0 is the superhelix density in the absence of dye. If we accept the value of 12° unwinding per intercalated dye molecule (FULLER and WARING, 1964), $\phi = \pi/15$ and

$$\tau = \tau_0 + 0.67\,\beta^\circ\, v \tag{6a}$$

$$\sigma = \sigma_0 + 0.67\, v\,. \tag{6b}$$

Results involving closed circular DNA molecules of different lengths are most conveniently expressed in terms of σ, since this quantity removes any linear dependence of τ upon molecular weight. In the present communication we discuss experimental results exclusively in terms of the superhelix density and employ τ only for special cases.

As pointed out in Section I, the native duplex winding number β_0 is taken to be positive, corresponding to the right-handed winding of the Watson-Crick duplex. The experimental results cited above then require that $\sigma_0 < 0$ for all naturally occurring DNA's discovered to date (right-handed interwound superhelix). The binding of an intercalative dye then results in a decreasing absolute value of σ until the value $\sigma = 0$ is reached and all superhelical turns are removed. Further addition of dye causes σ to again increase, although in this region with a positive sign (left-handed interwound superhelix).

3. Variation of the Sedimentation Coefficient of Closed Circular DNA with the Binding of Intercalative Dyes

a) Changes in s with Dye Added

The binding of small molecules to DNA is often accompanied by changes in sedimentation coefficient and, for a wormlike chain model, s is expected to increase with the square root of the mass per unit length (HEARST and STOCKMAYER, 1962). The effect of taking into account considerations of excluded volume is to reduce the value of the exponent in this relationship slightly (GRAY, BLOOMFIELD and HEARST, 1967), but the general conclusion concerning the increase of s with the mass/length remains unchanged. Such an effect is observed with, for example, the binding of pinacyanol (LERMAN, 1965) and of magnesium (BAUER, WATSON and VINOGRAD, unpublished) to DNA. Both of these ligands therefore increase the mass/length of the DNA molecule, as is to be expected in the absence of an elongation of the DNA structure.

The binding of intercalating dyes, on the other hand, is generally accompanied by a small decrease in s (LERMAN, 1961; COHEN and EISENBERG, 1969). This observation is generally interpreted in terms of the increased length which results from the unwinding of the DNA duplex with the insertion of a dye molecule.

The binding of intercalative dyes to closed circular DNA results in sedimentation coefficient changes of a dramatically different variety. Fig. 3 presents the results of studies of the variation of sedimentation coefficient with ethidium bromide addition for both polyoma (CRAWFORD and WARING, 1967) and SV40 (BAUER and VINOGRAD, 1968) DNA's. These results were obtained with the technique of boundary sedimentation velocity at 20° in the analytical ultracentrifuge. The salt concentrations were 0.025 M and 1.0 M, respectively. The sedimentation data in Fig. 3a are plotted as a function of an apparent binding ratio determined spectrophotometrically with a mixture of DNA's I and II; in Fig. 3b the corresponding abscissa is ν_{II}, moles EB bound per mole nucleotide residue for the nicked molecule, calculated according to the equation of SCATCHARD (1949) with constants determined by LEPECQ (1965) and by LEPECQ and PAOLETTI (1967).

The experimental data for these two cases can be classified into three regions. In the initial phase of binding, extending from $\nu_{II} = 0$ to $\nu_{II} = 0.01$ in Fig. 3b, the

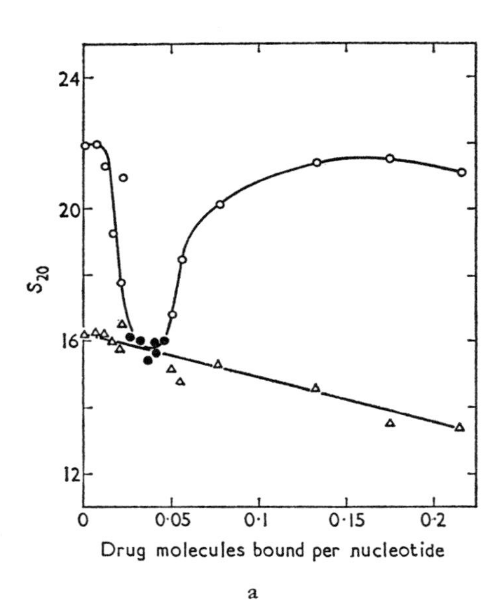

a

Fig. 3. a Effect of ethidium bromide on the sedimentation velocity of polyoma virus DNA. Samples of polyoma virus DNA, containing both fast (supercoiled) and slow (open ring) forms of the DNA, were exposed to a range of concentrations of ethidium bromide. Each sample contained 9 μg of DNA, plus ethidium bromide to give the average binding ratio shown, in 0.5 ml. Tris buffer (0.05 M, pH 8.0). The drug and the DNA were mixed and left at room temperature in the dark for 10 min before being centrifuged at 31,410 rev./min, 20 °C. The abscissa represents the results of spectrophotometric binding measurements in which it was assumed that binding takes place equally to both DNA components, an assumption now known to be incorrect (BAUER and VINOGRAD, 1968). —○—○—, Supercoiled molecules; —△—△—, open ring (unsupercoiled) molecules; —●—●—, only a single boundary was formed (CRAWFORD and WARING, 1967). — b Sedimentation velocity of SV40 DNA in 1.0 M NaCl in various concentrations of ethidium bromide. The symbols (○) and (●) refer to results from resolved boundaries of intact SV40 I and nicked SV40 II, respectively. The weight-average sedimentation coefficients from non-resolved boundaries are indicated by the symbol (◑). The error bars on the data points represent the 95% confidence limits in the sedimentation coefficients. When not shown the error was smaller than ± 0.155. The sedimentation coefficients are presented as a function of ν_{II}, the molar binding ratio of EB to the nicked DNA molecule. The dye concentrations used ranged from 0.0 to 12.0 μg/ml. The total DNA concentration was approximately 20 μg/ml and the ratio of I/II was either 0.5 or 2.0 in the six series of experiments performed. In the experiment at $\nu_{II} = 0.045$ the initial concentrations of I, II, total dye and free dye were 7.0, 13.0, 4.3, 3.2 μg/ml, respectively. The experiments were performed at 44 krev./min and at 20 °C. The measured sedimentation coefficients were corrected for the effects of NaCl on the viscosity and density of the solution (BAUER and VINOGRAD, 1968) change in s with ν_{II} (or with $\bar{\nu}$) is small and negative. The central phase involves rapid changes in s (the "dip" region): first a decrease, then equality or indistinguishability with s_{II}, then an increase of rapidity and magnitude comparable to the preceding decrease. The final phase, shown more clearly in Fig. 3a, consists of a gradual decrease of s as dye is added, with a slope comparable to the decrease evidenced by DNA II.

The theoretical interpretation of the above transition involves consideration of three distinct processes. These are (1) the variation of ν with c, (2) the variation of σ

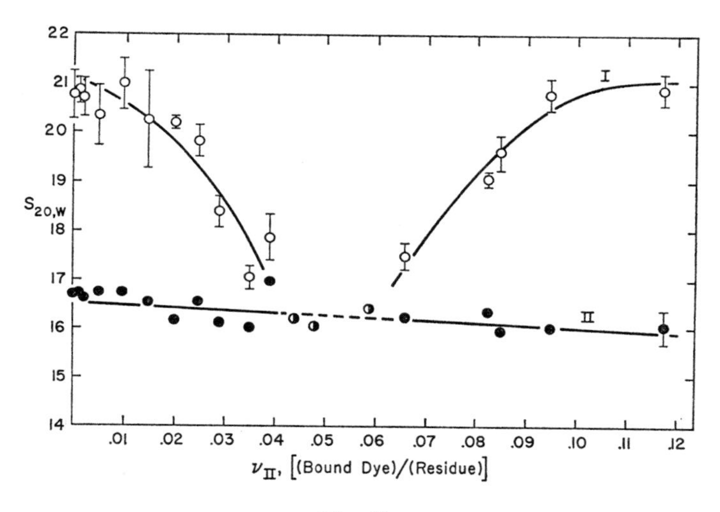

Fig. 3b

with ν, and (3) the variation of s with both ν and σ. This latter process is complicated by the fact that s depends upon ν, as demonstrated by the results with DNA II, even in the absence of changes in σ. This dependence may be used to account for the gradual decrease of s_I with ν_I in the initial and final regions of Fig. 3, and we will not discuss it further here. There remain two distinctive features possessed by DNA I: ν_I varies differently with c than does ν_{II} (BAUER and VINOGRAD, 1968) and s depends upon σ. In general, we expect s to decrease as $|\sigma|$ decreases. This is consistent with an initial decrease in $|\sigma|$ (with negative values) at the left of the dip region, passage through the state $\sigma = 0$ at the minimum, and increase in σ (with positive values) to the right of the minimum. This description is generally supported by the theoretical calculations of GRAY (1967) on the dependence of s_I on σ for rigid and regular superhelical molecules.

Similar titrations, employing ethidium bromide and comparable to those presented in Fig. 3, have since been performed for closed circular DNA's from a variety of natural sources and, in addition, for closed DNA's prepared *in vitro* by means of the polynucleotide ligase (see below). The list of naturally occurring DNA's which have been so studied now includes those from bovine papilloma (BUJARD, 1967), chick liver mitochondria (RUTTENBERG, SMIT, BORST and VAN BRUGGEN, 1968), bacteriophage

PM-2 (GRAY et al., 1971), the intracellular form of bacteriophages $\lambda b_2 b_5 c$ and $\lambda c I_{857}$ [WANG, 1969 (2)], and the plasmid DNA from *E. coli* 15 [WANG, 1969 (2)]. Ethidium bromide-sedimentation velocity titrations have also been carried out with several enzymatically closed DNA's from the SV40, polyoma (GRAY et al., 1971) and λ [WANG et al., 1968; WANG, 1969 (1)] series.

In addition to ethidium bromide, the dye 9-aminoacridine has been shown to unwind DNA with a corresponding dip in the sedimentation velocity-dye addition curve (BAUER and VINOGRAD, 1968). The unwinding angle in this case appeared to be the same as for the intercalation of ethidium into DNA, but the experimental precision was too low ($\pm$ 25%) to allow any firm conclusion to be reached.

b) The Calculation of Superhelix Density from the Sedimentation Velocity-dye Titration

The preceding section has described the nature of the variation in the sedimentation coefficient of closed circular DNA with a dye concentration variable such as the free dye concentration or the amount of dye bound to the nicked circular molecule. Each of these curves is characterized by the presence of a minimum which occurs at the critical dye concentration. The sedimentation coefficient of DNA I with all superhelical turns removed at this critical dye concentration may be expected to be substantially the same as the sedimentation coefficient of the corresponding nicked circular DNA. In addition, the binding affinity of DNA I for dye is equal to that of DNA II at the critical dye concentration. The known values of the binding parameters for nicked circular DNA may therefore be used to calculate v_c, the critical molar binding ratio to DNA I at this dye concentration. For this purpose Eq. (6b) may be used with $v = v_c$. At the critical dye concentration $\sigma = 0$, with the result that

$$\sigma_0 = -0.67\, v_c\,. \tag{7}$$

The sedimentation velocity-dye titration method for the determination of superhelix densities has been applied to a variety of both naturally occurring and enzymatically prepared closed circular DNA's. A list of such DNA's, along with the corresponding values of the superhelix density, the salt type and concentration, and the type of run technique employed are listed in Table 3. No significant difference between the band and boundary methods is apparent from the results, although there is some suggestion in the data of a dependence upon the salt concentration. A detailed discussion of the effect of salt concentration as well as a possible effect of sedimentation velocity run temperature upon observed superhelix density is discussed in the paper by WANG [1969 (1)].

The two series of enzymatically prepared closed circular DNA's, those with SV40 and with $\lambda b_2 b_5 c$, were prepared by the addition of varying amounts of ethidium bromide to the reaction mixture and later removal of the ethidium before the sedimentation velocity titration. The use of ethidium bromide in these preparative procedures is discussed in greater detail in Section IV below. The superhelix densities of the naturally occurring closed DNA are seen to vary under identical sedimentation conditions from -0.026 in the case of $\lambda b_2 b_5 c$ to -0.052 for the bacteriophage DNA from PM-2. These two DNA's differ in molecular weight by a factor of approximately 4, hence the number of superhelical turns in comparing these two species is clearly not proportional to the ratio of molecular weights. Nonetheless, it would appear that there is a significant difference in σ_0 between the DNA's from these two sources. No naturally occurring closed circular DNA has yet been discovered with a native superhelix density more positive than -0.025.

Table 3. *Superhelix densities determined with the sedimentation velocity-dye titration*

DNA Source	$-\sigma_0$	(Salt), *M*	Type
Polyoma[1]	0.025 ± 0.004[a]	0.05 Tris	Boundary
SV40[1]	0.033 ± 0.004	1.0 NaCl	Boundary
Chick liver mitochondria[3]	0.026 ± 0.004[a]	0.05 Tris	Boundary
Bovine papilloma[4]	0.025 ± 0.004[a]	0.05 Tris	Boundary
SV40 I[5]	0.038 ± 0.002	2.8 M CsCl	Band
SV40 ligase series[5]	0.0 to 0.083	2.8 M CsCl	Band
$\lambda b_2 b_5 c$ ligase series[5,6]	0.0 to 0.065	2.8 to 3.0 M CsCl	Band
PM-2[5]	0.052 ± 0.003	2.8 M CsCl	Band
Polyoma[5]	0.032 ± 0.002	2.8 M CsCl	Band
$\lambda b_2 b_5 c$[7]	0.026	3.0 M CsCl	Band
E. coli 15 plasmid[7]	0.039	3.0 M CsCl	Band
$\phi X174$ *RF*[7]	0.037	3.0 M CsCl	Band
$\lambda c I_{857}$ (sensitive cells)[7]	0.027	3.0 M CsCl	Band
$\lambda c I_{857}$ (superinfected immune cells)[7]	0.031	3.0 M CsCl	Band

[1] Crawford and Waring, 1967.
[2] Bauer and Vinograd, 1968.
[3] Ruttenberg et al., 1968.
[4] Bujard, 1968.
[5] Gray et al., 1970.
[6] Wang, 1969 (1).
[7] Wang, 1969 (2).
[a] Standard deviations estimated from published curves.

c) The Variation in Sedimentation Coefficient with Superhelix Density

The previous sections have discussed the variation of the sedimentation coefficient of closed circular DNA with the total formal concentration of ethidium bromide in the presence of the sedimenting species. This variation in sedimentation coefficient has been attributed to the increase in compactness of the molecule which results from the presence of the tertiary structure (Vinograd et al., 1965). It is therefore of interest to calculate the dependence of the sedimentation coefficient upon the superhelix density of the molecule as σ is varied by the addition of dye.

The molar binding ratio v_I is linearly related to σ in the presence of dye, as indicated by Eq. (6b). The variation of v_I with free dye concentration is complex, however as discussed in detail in Section II. 5 below. Based upon the measured binding isotherms of EB to SV40 DNA I, Bauer and Vinograd (1968) calculated experimental values of σ as a function of the EB concentration in μg/ml, c, over the range covered by the corresponding measurements of s_I *versus* c. These relationships were then combined to permit the calculation of the uncorrected sedimentation coefficient for the DNA-dye complex as a function of the superhelix density in the presence of dye, σ. The results are reproduced in Fig. 4. The effect of correcting for the variation of the binding constant with the superhelix density has been to considerably sharpen and narrow the dip region. Otherwise the general features of the transition are much like those already encountered in Fig. 3. Similar results have been obtained for lambda DNA by WANG [1969 (1)].

An additional problem is presented by the calculation of $s_{20,w}$ for closed circular DNA in the presence of v moles dye bound per mole nucleotide. If the sedimentation coefficient measurements are made at low salt concentrations, the partial specific volume of the DNA-dye complex at the particular level of occupancy chosen for the experiment must be known. If, on the other hand, the velocity experiments are

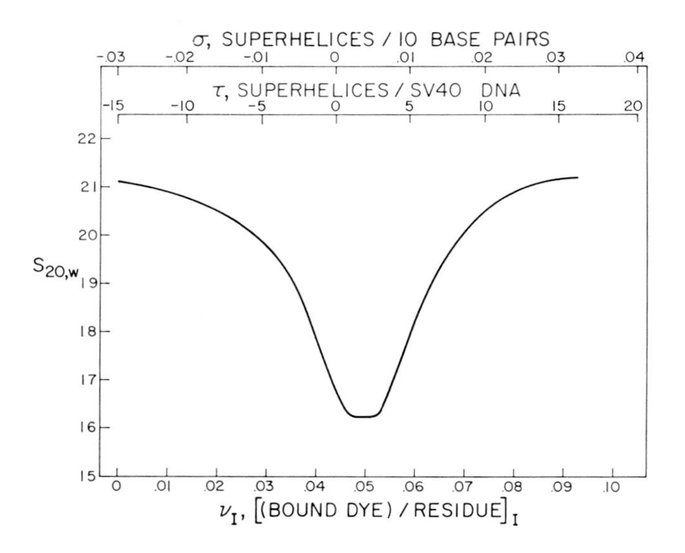

Fig. 4. The dependence of the sedimentation coefficient of SV40 DNA I on ν_I, the moles of ethidium bromide bound per mole nucleotide residue. The upper abscissa shows τ, the number of superhelical turns per molecule, and σ, the superhelix density, as calculated with Eqs. (3) and (4) of BAUER and VINOGRAD (1968)

performed in concentrated CsCl solutions, as is usually the case in the method of band sedimentation velocity, the value of the buoyant density of the DNA-dye complex, again at the particular level of occupancy chosen for the experiment, must be known. These quantities are experimentally available for SV40 DNA, and with the use of the appropriate corrections [BAUER and VINOGRAD, 1970 (2)] may be extended to closed circular DNA's of different base compositions and superhelix densities.

The quantity $s_{20,w}$ has been measured experimentally as a function of σ, the superhelix density in the presence of dye, for a series of enzymatically prepared SV40 DNA's by UPHOLT et al. (1971). The results of these experiments are presented in Fig. 5. The complexity of the variation of the sedimentation coefficient of closed DNA with superhelix density when the results are extended to superhelix densities beyond those for native SV40 DNA is readily apparent from inspection of the figure. The physical basis for this complexity is not yet fully understood. It may also be concluded from these experiments that the general variation of sedimentation coefficient with σ is largely independent of the amount of dye bound to the DNA I molecule. Thus, the curve for the ligase-generated SV40 $L3$, with $\sigma_0 = -0.083$, is qualitatively similar to that for the ligase-generated SV40 $L1$, $\sigma_0 = -0.010$, over the range of common values of σ.

The variation of sedimentation coefficient with the superhelix density of the native molecule, σ_0, has also been studied by UPHOLT et al. (1971). Fig. 6 presents the variation of $s_{20,w}$ with σ_0, again for a series of enzymatically-prepared SV40 DNA's. The same qualitative features are apparent in these results as were observed in Fig. 5,

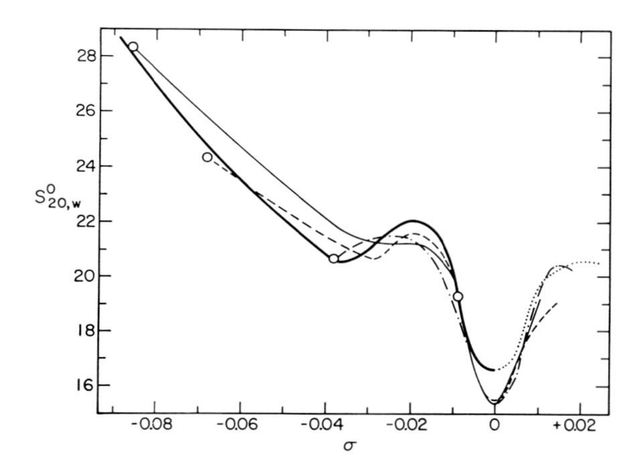

Fig. 5. The variation of the fully-corrected sedimentation coefficient, $s^0_{20,w}$, for a series of enzymatically-closed SV40 DNA's with the superhelix density in the presence of dye, σ. The sedimentation coefficients were measured by the method of band sedimentation velocity (VINOGRAD, BRUNER, KENT and WEIGLE, 1963) and were corrected to standard conditions with the aid of the known relationship between buoyant density and v (BAUER and VINOGRAD, 1968) and the binding coefficients calculated by BAUER and VINOGRAD [1970 (1)]. (UPHOLT et al., 1971)

and full account has been taken of the contribution of the dye to the sedimentation velocity behavior of the closed circular DNA. The results for the variation of $s_{20,w}$ with both σ and σ_0 are characterized by an initial diminution in the sedimentation coefficient, followed by the appearance of a minimum, a rise to a local maximum, and a subsequent decrease to the dip region which was earlier observed with the naturally occurring SV40 DNA I. The enzymatically prepared SV40 DNA's used in the above studies were prepared by closure in varying concentrations of ethidium bromide as discussed in Section IV below. This intercalative dye has therefore been of great utility for the study of the dependence of the sedimentation coefficient upon the superhelix density.

Several attempts have been made to explain the theoretical basis for a decrease in sedimentation coefficient with decreasing superhelix density. BLOOMFIELD (1966) considered the hydrodynamic behavior of a model consisting of adhering flexible loops of beads of equal size, according to the fundamental equations developed by KIRKWOOD (1954). KURATA (1966), using a basically similar approach, took into account the stiffness term expressed by the constant parameter in the equation of HEARST and STOCKMAYER (1962) for the variation of the DNA sedimentation coefficient with molecular weight. Neither of these two approaches was able to quantitatively account for the experimental variation of s_I with σ over the range encountered in the naturally occurring DNA's, in which s_I decreases as σ decreases. GRAY (1968) used the KIRKWOOD (1964) equation to calculate the sedimentation coefficient of a

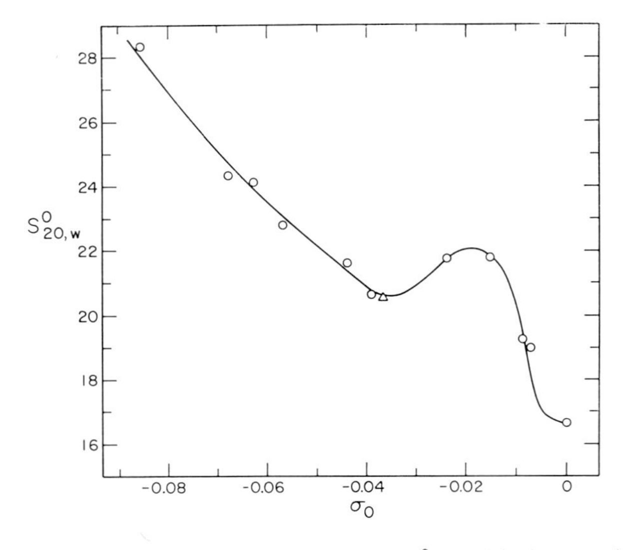

Fig. 6. The variation of the sedimentations coefficient, $s^0_{20,w}$, with the superhelix density of the dye-free molecule, σ_0, for a series of enzymatically-prepared SV40 DNA's (UPHOLT et al., 1971)

rigid superhelix. Again, his results require a uniform decrease in s_I with $|\sigma|$. It seems unlikely that the complex behavior presented in Figs. 5 and 6 can be accounted for in terms of a simple one-parameter variation such as has been attempted in terms of σ.

4. The Measurement of Binding Isotherms with Closed Circular DNA

A variety of experimental methods is commonly used for the study of the interaction of small molecules with DNA. These include changes in the ligand absorption spectrum, changes in the fluorescence efficiency of bound versus free ligand, equilibrium dialysis, radioactive labeling of ligand followed by pelleting of the complex in the preparative ultracentrifuge, and quantitation of the shift in buoyant density which occurs upon the addition of relatively light ligands, such as ethidium bromide, to DNA. The only method which has so far been applied to closed circular DNA is measurement of the buoyant shift (BAUER and VINOGRAD, 1968). This technique has the advantage of requiring only microgram quantities of DNA and of providing a coordinate assay for the integrity of the closed molecule. This latter characteristic is especially useful, considering the relative ease of conversion of the closed to the nicked circular species (VINOGRAD et al., 1965). In addition, the buoyant shift method measures the total amount of dye preferentially associated with the DNA on a mass basis. The results do not depend upon the details of any spectral changes which may occur upon complex formation.

The principal limitation of this method is that the solution environment is determined by the requirements of buoyancy and is therefore not an independent experimental variable. The most usual experimental conditions in the case of ethidium bromide binding are 25° and 4.5 to 5.5 M CsCl, pH 7 to 8. The method is therefore inapplicable to dyes, such as the aminoacridines, which fail to bind to DNA at high salt concentrations. A second disadvantage of the method, for some applications, is its failure to discriminate between binding at internal and external sites (intercalative and non-intercalative, for example). In the presence of high salt concentrations secondary or outside binding is, however, generally considered to be unimportant in the case of ethidium.

The theoretical derivation of the mathematical details of the buoyant shift method has been presented elsewhere (BAUER and VINOGRAD, 1969) and will not be repeated here. The general result relating v to $\Delta\theta = \theta - \theta_0$ is

$$v = \left(\frac{M_3}{M_4}\right)[(1 + \Gamma'_0)(\Delta\theta/\theta_0) - \Delta\Gamma'(\bar{v}_1\theta - 1)]/(\bar{v}_4\theta - 1), \quad (8)$$

where M_3 and M_4 are the molecular weights of neutral DNA and neutral dye, Γ'_0 is the preferential hydration of the DNA in the absence of dye, $\bar{v}_1$ and $\bar{v}_4$ are the partial specific volumes of water and dye, and $\Delta\Gamma'$ is the change in preferential hydration of the complex in moving to a region of higher water activity. This latter quantity is given by BAUER and VINOGRAD (1968)

$$\Delta\Gamma'(\theta) = 0.04092 - 0.7282\eta + 4.592\eta^2 - 16.08\eta^3 + 20.56\eta^4 \quad (9a)$$

$$\Delta\Gamma' = \Delta\Gamma'(\theta) - \Delta\Gamma'(\theta_0), \quad (9b)$$

where $\eta = \theta - 1.5$.

a) The Variation of σ with v for Closed Circular DNA

The buoyant shift method has been applied to the determination of the binding isotherms of ethidium bromide with SV40 DNA's I and II, as shown in Fig. 7. The data taken from BAUER and VINOGRAD (1968) are presented as a plot of v versus log $(1 + c)$, where c is the free EB concentration at band center in μg/ml. The curve may be naturally divided into two regions: (1) to the left of the critical dye concentration, c', the association constant ratio $k_I/k_{II} > 1$ and $v_I > v_{II}$; (2) to the right of c', $k_I/k_{II} < 1$ and $v_{II} > v_I$. At the critical dye concentration (5.4 μg/ml for SV40 DNA I), $k_I = k_{II}$ and $v_I = v_{II}$. This corresponds to the midpoint of the dip in the sedimentation velocity-dye titration, Fig. 3, and all net superhelical structure is removed. Fig. 7 indicates that the binding ratio difference $\Delta v = v_{II} - v_I$ changes only slowly over the dye concentration range $c > 100$ μg/ml. The significance of this behavior will be discussed in greater detail in Sections II. 6 and III, below.

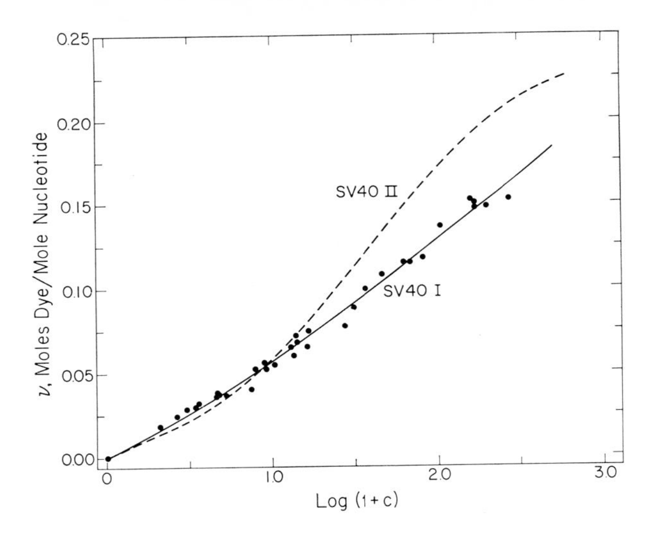

Fig. 7. The molar binding ratio, v, plotted as a function of log $(1 + c)$ for SV40 DNA's I and II. The free dye concentration, c, is in units of μg/ml and the experimental data were taken from BAUER and VINOGRAD (1968)

b) The Use of Binding Isotherms to Calculate σ_0

At the critical dye concentration indicated by the intersection of the curves for v_{I} and v_{II} in Fig. 7, the closed circular molecule is fully relaxed and $\sigma = 0$. Eq. (7) may then be used to calculate σ_0, the superhelix density of the dye-free molecule at the CsCl concentration corresponding to $c = c'$. This calculation has been performed for native *SV*40 DNA, with the result that $\sigma_0 = -0.030 \pm 0.003$ in 5.5 M CsCl at 25°. This is to be compared with the values $\sigma_0 = -0.038 \pm 0.002$ obtained by GRAY et al. (1971) in 2.8 M CsCl at 20° and with the value $\sigma_0 = -0.033 \pm 0.004$ obtained by BAUER and VINOGRAD in 1.0 M NaCl at 20°. These latter two values were obtained by means of the band and boundary sedimentation velocity-dye titrations, respectively. It is as yet uncertain whether this variation reflects a real dependence of σ_0 upon ionic strength and temperature, or whether it is merely a reflection of the difference in experimental techniques employed.

5. The Use of Ethidium Bromide Binding Isotherms for the Calculation of the Free Energy Associated with Superhelix Formation

The results presented in Sections II. 3 and II. 4 above have clearly shown that the net binding affinity of closed circular DNA for the intercalative dye ethidium bromide depends strongly upon the number of superhelical turns present in the molecule. For example, naturally occurring closed circular DNA contains negative superhelical turns and the binding of ethidium requires a concommitant reduction in the amount of superhelical winding. In this region of the binding isotherm the apparent binding constant of DNA I is greater than that of DNA II and decreases with the amount of dye bound. Conversely, to the right of the critical dye concentration as shown in

Figs. 4 and 7 binding of dye requires an accompanying increase in the superhelix density. Associated with this is a corresponding diminution in the binding affinity of closed relative to the nicked circular molecule. The extent of this relative diminution in the net binding affinity increases as dye is added to the right of the critical dye concentration. A positive free energy is therefore associated with the formation of superhelical turns. It has been shown by BAUER and VINOGRAD [1970 (1)] that this free energy may be quantitatively calculated from the observed binding isotherms of DNA I and II with ethidium bromide.

In general, the free energy required to form a superhelix of σ turns per 10 base pairs may be written as a Taylor series in the superhelix density σ.

$$\frac{\Delta G_\sigma}{RT} = \sum a_i \sigma^i \,. \tag{8a}$$

The boundary condition that ΔG_σ be everywhere greater than 0 requires that a_0 and a_1 be equal to 0. The next two higher order terms in the Taylor series are then selected as a first approximation, resulting in the relationship between ΔG_σ and the superhelix density, Eq. (8b),

$$\frac{\Delta G_\sigma}{RT} = a\sigma^2 + b\sigma^3 \,. \tag{8b}$$

Eq. (8b) is combined with more general considerations of the binding of small molecules to macromolecules [BAUER and VINOGRAD, 1970 (1); MÜLLER and CROTHERS, 1968] to obtain Eq. (9).

$$v/c = \left(\frac{k'}{2}\right) \frac{(1-4v)^2}{(1-2v)} \,. \tag{9}$$

The quantity k′ depends upon the type of DNA under consideration and differs from closed to nicked circular DNA. The value of this apparent binding constant is shown in Eqs. (10a) and (10b).

$$k'_{\mathrm{II}} = k_0 \tag{10a}$$

$$k'_{\mathrm{I}} = k_0 \exp(-\alpha\sigma - \beta\sigma^2) \,. \tag{10b}$$

In the case of nicked circular DNA, k'_{II} is equal to the intrinsic binding constant, k_0. In the case of closed circular DNA, however, k'_{I} is equal to the same intrinsic binding constant as obtained in Eq. (10a) but modulated by an exponential factor which takes into account the fact that the apparent binding constant for closed circular DNA depends also upon the superhelix density.

The above equations, which were derived in general terms, have been applied specifically to the case of the binding of ethidium bromide to closed circular SV40 DNA. The coefficients α and β in Eq. (10b), corresponding to the coefficients a and b in Eq. (8b), characterize the free energy required to form superhelical turns in closed circular DNA. The coefficients α and β were determined by requiring the best least squares fit to the experimental binding isotherms for SV40 DNA's I and II. Fig. 8a, b shows these isotherms and the resulting fitted curves for SV40 DNA's I and II, respectively. Both isotherms are well described by the formalism of Eqs. (9) and (10), with the values $k_0 = 7.3 \times 10^3 \ \mathrm{l \cdot mol^{-1}}$, $\alpha = 28.1$ and $\beta = -86.4$. The free energy of

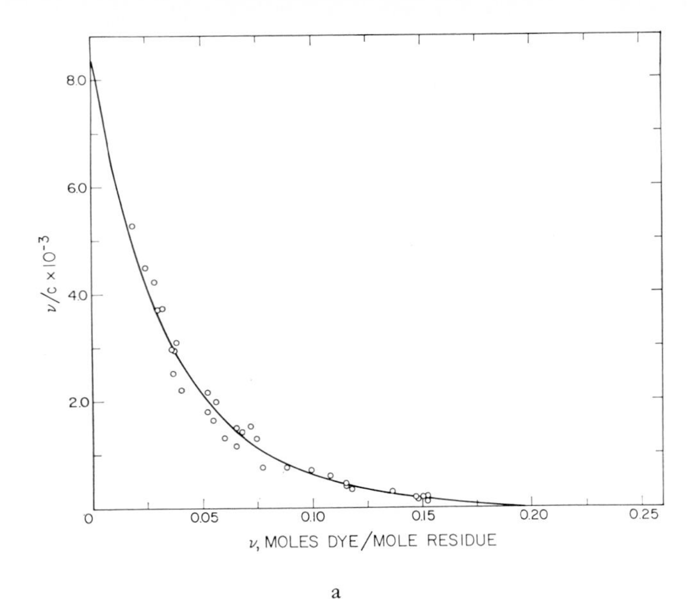

a

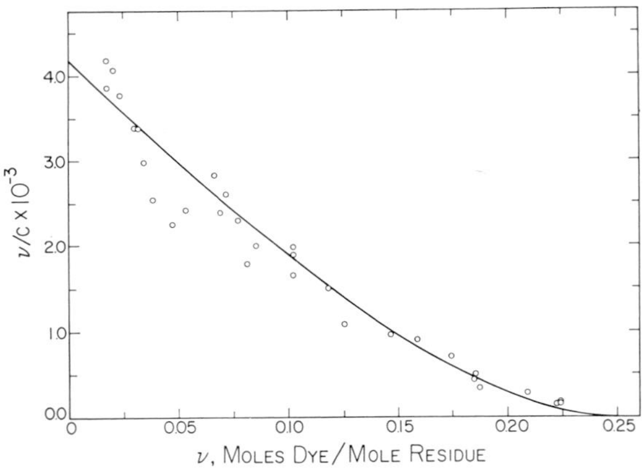

b

Fig. 8. a The quotient v/c plotted as a function of v for the binding of ethidium bromide to closed circular SV40 DNA in 5 M CsCl. The data points were obtained as described in BAUER and VINOGRAD (1968). The solid curve was calculated with the aid of the coefficients describing the free energy of superhelix formation as a function of σ [BAUER and VINOGRAD, 1970 (1)]. — b The quotient v/c plotted as a function of the binding ratio, v, in moles dye per mole nucleotide for the binding of ethidium bromide to nicked circular SV40 DNA in 5 M CsCl. The experimental data were obtained as described in BAUER and VINOGRAD (1968), and the solid line represents a plot of the right hand side of Eq. (9) versus v, with k' given by Eq. (10a) superhelix formation for SV40 DNA may then be expressed by the use of these constants and Eq. (8b).

$$\frac{\Delta G_\sigma}{RT} = 20.3 \times 10^4 \sigma^2 (1 - 1.6\,\sigma)\,. \tag{11}$$

For native SV40 DNA with $\sigma_0 = -0.030$, the free energy associated with the native superhelix is $\Delta G_0 = 100$ kcal/mole DNA or 21 cal/mole base pair.

6. The Induction of a Buoyant Separation between Closed and Nicked Circular DNA's by Intercalative Dyes

Although the introduction of a single-strand scission produces no change in the buoyant density of a native closed circular DNA, such is no longer the case in the presence of an intercalating (or other unwinding) dye of low density. The only exception to this observation is at the critical dye concentration at which $v_{\mathrm{I}} = v_{\mathrm{II}}$, as pointed out in Section II (4) above. In the region to the right of the buoyant crossover, $v_{\mathrm{II}} > v_{\mathrm{I}}$ and consequently $\theta_{\mathrm{I}} > \theta_{\mathrm{II}}$ (BAUER and VINOGRAD, 1968). The magnitude of the buoyant density difference, $\Delta\theta = \theta_{\mathrm{I}} - \theta_{\mathrm{II}}$, has been quantitatively related to the value of the native superhelix density, σ_0, both by theoretical [BAUER and VINOGRAD, 1970 (2)] and experimental (GRAY et al., 1971) studies.

The theoretical treatment begins with the expression for the free energy of superhelix formation, Eqs. (9) and (10b). Combination of Eq. (10b) with Eq. (6b) leads to the result

$$\ln k_{\mathrm{I}}' = \ln k_0 - \alpha(\sigma_0 + 0.67v) - \beta(\sigma_0 + 0.67v)^2\,. \tag{12}$$

Combination of the above equations makes possible the calculation of the variation of $\Delta v = v_{\mathrm{I}} - v_{\mathrm{II}}$ with σ_0 for SV40 DNA. This variation was found to depend only slightly upon the product k_0C, where C is the molar free EB concentration, over the range of EB concentration 100 to 450 µg/ml.

The dependence of θ upon v has also been described in Section II. 4 above. When account is taken of Eq. (8), and with the aid of certain approximations [BAUER and VINOGRAD, 1970 (2)], the relationship between the buoyant separation $\Delta\theta$ and the initial superhelix density σ_0 may be calculated. These quantities are most conveniently expressed with reference to a standard pair, taken to be SV40 DNA's I and II and denoted by an asterisk. BAUER and VINOGRAD [1970 (2)] found that, at $k_0C = 6.3$, the relationship is

$$\frac{\Delta\theta}{\Delta\theta^*} = \left(\frac{\theta_{01}\bar{v} - 1}{\theta_0^*\bar{v}_1 - 1}\right)^2 [1 + 9.61\,\Delta\sigma_0\,(1 + 0.66\,\Delta\sigma_0)]\,, \tag{13}$$

where θ_0 and θ^*_0 are the values of the buoyant densities in the absence of dye and $\Delta\sigma_0 = \sigma_0 - \sigma^*_0$. In the event that $\theta_0 = \theta^*_0$, Eq. (13) may be written to a good approximation in terms of the distance separations measured in the preparative ultracentrifuge

$$\Delta\sigma_0 = 0.104\left[\frac{\bar{r}}{\bar{r}^*}\,\frac{\Delta r}{\Delta r^*} - 1\right] \tag{14}$$

where $\bar{r}$ and $\bar{r}^*$ are the distances from the center of rotation to the average band position for the unknown and reference DNA pairs, and Δr and Δr^* are the corresponding distances between band centers.

The numerical coefficient in Eq. (14), 0.104, refers to 25°. When corrected to 20° it becomes 0.109, compared to the experimentally determined value of 0.113 (GRAY et al., 1971). The experimental determinations were made by measuring the superhelix density in 2.85 M CsCl, 20° by means of the sedimentation velocity-dye titration (Section II. 3), above, using enzymatically prepared samples of both SV40 and λ DNA's (see Part IV, below). The results were then plotted as a function of the experimentally measured buoyant separation at 4.5 M CsCl, 20° (see Part III, below). The close agreement between the theoretical and experimental values of the slope in Eq. (14) indicates that the free energy coefficients of Eq. (10) are largely independent of the amount of dye bound.

The method has been experimentally applied to the determination of the superhelix densities of a series of enzymatically prepared SV40 DNA's (GRAY et al., 1971) and of the plasmid from *E. coli* strain 15 (LEE and DAVIDSON, 1970). In addition, the buoyant separation relative to that between SV40 DNA's I and II has been measured for a variety of mitochondrial DNA's, including those from HeLa, *L. pictus*, rabbit liver, chicken liver and rat liver (HUDSON, UPHOLT, DEVINNY and VINOGRAD, 1969). These latter results indicated the presence of considerable variation in the superhelix density of mitochondrial DNA's from different sources.

The buoyant separation method is the most rapid and convenient method presently available for the measurement of σ_0. It requires but a single measurement which does not involve the acquisition of sophisticated detection and recording devices (WATSON, BAUER and VINOGRAD, in preparation). The method is a natural adjunct of the detection, separation and isolation method described in Section III. The buoyant conditions (4.5 M CsCl) to which the method is restricted (generally also at 330 μg/ml EB, 20°) might well prove convenient as a standard superhelix density solvent and can, in any case, be related to σ_0 as determined by other methods if the appropriate calibration experiments are performed.

7. The Selective Alteration *in vivo* of Closed Circular DNA Functions by Ethidium Bromide

The ability of ethidium bromide to alter certain functions associated with closed circular DNA has been examined in two different biological systems. These include preferential inhibition of the synthesis of mitochondria-associated 12 S and 21 S (ZYLBER, VESCO and PENMAN, 1969) and 4 S (ZYLBER and PENMAN, 1969; KNIGHT, 1969) RNA's in HeLa cells, as well as the induction of abnormal circular DNA molecules in the kinetoplast of *Trypanosoma cruzi* (RIOU and DELAIN, 1969). These latter DNA molecules appear to be similar in many respects to the multimeric forms observed in extracts of the mitochondrial DNA from leukemic human leukocytes (CLAYTON and VINOGRAD, 1969), in that both contain circular oligomers not found in normal cells.

Several species of RNA appear to be associated with animal mitochondrial *in vivo*, among which are the 12 S and 21 S fractions. Both of these RNA fractions have been shown to be homologous to the mitochondrial DNA in *Xenopus laevis* (DAWID, 1969). The effect of ethidium bromide upon *in vivo* synthesis of these two RNA components was studied by ZYLBER et al. (1969), using EB concentrations of 1.0 μg/ml or less.

It was found that synthesis of the 12 S and 21 S species of mitochondrial RNA from HeLa cells is completely inhibited at EB concentrations above 0.2 μg/ml, whereas the non-mitochondrial RNA fractions were unaffected. Their results also showed, however, that some EB-resistant RNA remains in the mitochondrial fraction. The previously formed 12 S and 21 S RNA's decay after EB addition with a half life of approximately 3 h. The mechanism of this inhibition could not be established, although these concentrations of EB were found not to inhibit the synthesis of cytoplasmic DNA. EB did not appear to act as a general metabolic inhibitor or as an inhibitor of protein synthesis. Similar results have also been obtained from studies upon the mitochondria-associated 4 S RNA fraction of HeLa cells (KNIGHT, 1969; ZYLBER and PENMAN, 1969) and of HeLa cells arrested in metaphase (FAN and PENMAN, 1970).

The mode of action of EB in specifically inhibiting the ribosomal RNA's of the mitochondrial fraction is not known, although ZYLBER et al. (1969) have suggested a possible link to changes in the tertiary structure of the closed circular mitochondrial DNA template. An alternative possibility is based upon the alteration of the apparent binding affinity of the dye with closed circular DNA and is illustrated in Fig. 9. Fig. 9a is based upon binding isotherms calculated by BAUER and VINOGRAD [1970 (2)] and represents a plot of the logarithm of the ratio of Eqs. (10a) and (10b) as a function of the logarithm of (kc) for various values of σ_0. Fig. 9b presents a plot of the critical free EB concentration, c', at which $\sigma = 0$ as a function of $-\sigma_0$ for 4.5 M CsCl and $k_0 = 7.3 \times 10^3\ l \cdot \text{mol}^{-1}$. The magnitude of c' is inversely proportional to k_0, as indicated in Eq. (9). The binding affinity of closed relative to nicked circular or linear DNA is clearly greater than unity for naturally occurring mitochondrial DNA's, and the relative advantage increases with $-\sigma_0$. Thus, at low EB concentration, a correspondingly greater amount cf dye is bound to the closed than to nicked or linear molecules. Although this effect may be present in the *in vivo* systems as well, it is not known whether or not the preferential inhibition of mitochondrial RNA synthesis may be thereby explained. *In vitro* experiments have shown, however, that the DNA-dependent RNA synthesis of the *Escherischia coli* system may be completely inhibited by EB concentrations of 8 μg/ml or greater (WARING, 1964). In addition, higher concentrations of proflavine have been shown to inhibit the synthesis of cytoplasmic RNA as well (PENMAN in. ZYLBER et al., 1969).

A preferential interaction of EB with closed circular DNA *in vivo* has also been studied with the kinetoplast DNA of *Trypanosoma cruzi* (RIOU and DELAIN, 1969). Normal kinetoplast DNA consists of closed circular molecules of contour length 0.43 μ (RIOU and PAOLETTI, 1967), along with a small percentage (RIOU and DELAIN, 1968) of catenated oligomers (HUDSON and VINOGRAD, 1967). Upon growth in the presence of an ethidium bromide concentration of 1.5 μg/ml, up to 50% by number of the closed circular molecules examined were found to be present as circular oligomers, a situation which resembles that found in the mitochondrial DNA of untreated leukemic human leukocytes (CLAYTON and VINOGRAD, 1967). This phenomenon occurs at an EB concentration which is too low to result in inhibition of trypanosome growth, nor was there any evidence of any functional impairment. Although these experiments clearly demonstrate a preferential effect of EB upon closed circular DNA *in vivo*, it is not yet clear whether the action is direct on the DNA or indirect on an enzyme or other metabolic system.

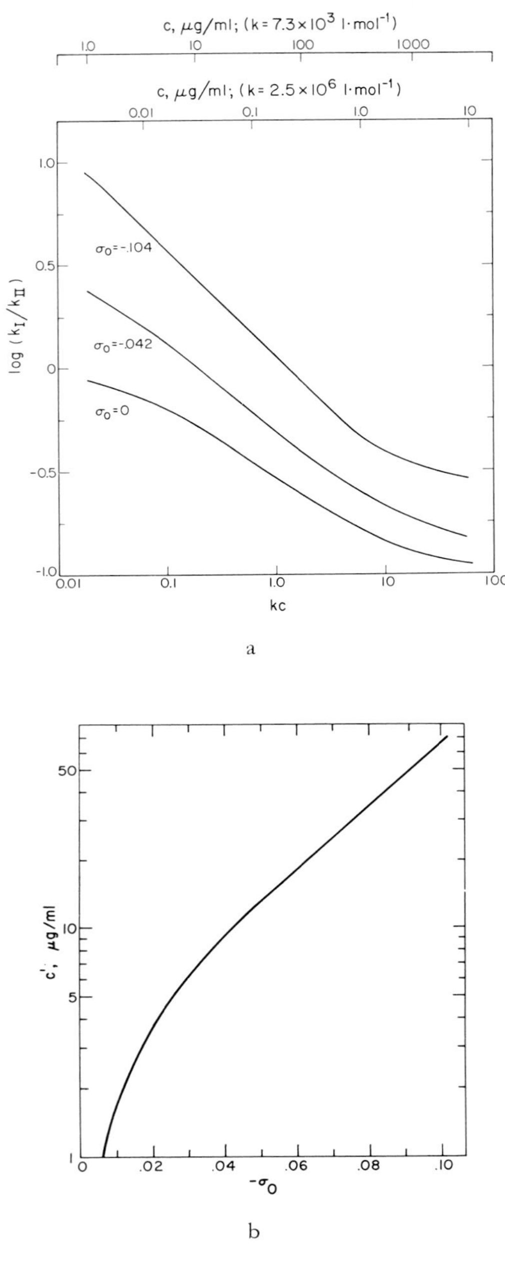

Fig. 9. a The logarithm of the binding constant ratio k_I/k_{II} plotted as a function of the log (kc) for a closed circular DNA similar to SV40 and with c expressed in μg/ml. The upper abscissae give the EB concentrations for the indicated values of k. — (b) The logarithm of the critical EB concentration, c', for which the superhelix density $\sigma = 0$, plotted as a function of $-\sigma_0$, for values of $k_0 = 7.3 \times 10^3\ l \cdot mol^{-1}$, 4.5 M CsCl and 25° [BAUER and VINOGRAD, 1970 (1)]

III. The Use of Intercalative Dyes for the Detection and Isolation of Closed Circular DNA

One of the fundamental problems to be found in the study of closed circular DNA is the separation in a pure form from an overwhelming preponderance of linear DNA's, generally arising from the chromosomes of the cell of interest. In the case of the extraction of mitochondrial DNA, it has been estimated that 99.8% of the starting mixture is linear (RADLOFF et al., 1967). The problem can be even more acute in the case of bacterial plasmids and bacteriophage intracellular forms. The plasmid DNA from *E. coli* 15, for example, is present only to the extent of 15 to 20 copies per cell (COZZARELLI et al., 1968). The intracellular form of the phage Nl from *micrococcus lysodeikticus* presents even greater difficulties, with only an average of one circular DNA molecule per cell (LEE, DAVIDSON and SCALETTI, 1968). Even in cases in which separation of the circular species from the bulk of host DNA is relatively easy, such as the *pseudomonas* phage PM-2 (ESPEJO, CANELO and SINSHEIMER, 1969) the purification problem remains.

1. Description of the Dye-Buoyant Density Method

The restricted binding of intercalating dyes to closed circular DNA at high dye concentrations has been discussed in Section II. 5 above. This relative binding restriction leads to a buoyant density decrement which is considerably smaller than that of the corresponding nicked circular DNA at the same free dye concentration. The buoyant density difference at high dye levels, $\Delta\theta$, may be quantitatively related to the native superhelix density, σ_0, as pointed out in Section II. 7.

The magnitude of $\Delta\theta$ is sufficiently large at EB concentrations in excess of approximately 100 μg/ml so that essentially complete physical separation may be obtained between buoyant nicked and closed circular DNA's. This effect was first used for preparative purposes by RADLOFF et al. (1967), who detected and isolated the closed circular DNA from HeLa cells. The difference in buoyant density between closed and nicked circular HeLa mitochondrial DNA was found to be 0.035 g/ml at an EB concentration of 100 μg/ml in a 3-ml column of CsCl, $\varrho^\circ = 1.550$ g/ml at 44,000 RPM in the SW50L rotor at 20°. The band maxima were separated under these conditions by approximately 12 fractions of four 7.5 μl drops per fraction.

Fig. 10a presents results with purified mixtures of nicked and closed circular polyoma DNA, detected in the dye-CsCl density gradient by scintillation counting of the ^{3}H-labelled DNA. These experiments demonstrated that no detectable single-strand scissions were introduced by ethidium bromide, and that the dye had only a small effect upon the relative counting efficiency. Comparable results with lambda DNA showed a normal titer with the protoplast assay (YOUNG and SINSHEINER, in RADLOFF et al., 1967) when samples of lambda DNA were tested after isolation with the dye-buoyant density method followed by 1000-fold dilution. Later experiments have demonstrated that single strand scissions are introduced at a slow rate upon allowing closed circular DNA to stand for prolonged periods in the presence of EB and intense visible light (CLAYTON, unpublished observations).

The DNA in a dye-CsCl density gradient may be detected by a variety of means including absorbance, fluorescence, and radioactivity, either directly or with the use of a linear marker DNA [Bauer and Vinograd, 1970 (1); Radloff et al., 1967). The absorbance and radioactivity counting methods are in general use with experiments conducted in the absence of dye and will not be discussed further here.

The fluorescence method of detection rests upon the observation that the molar fluorescence of bound EB is many times greater than that of free dye if the proper excitation wavelength is chosen (LePecq, 1965; LePecq and Paoletti, 1967). A convenient excitation wavelength for this purpose is 365 nm, with the corresponding fluorescence intensity maximum at 590 nm. The relative molar fluorescence of bound to unbound dye at these wavelengths is approximately 35 at 4.5 M CsCl and 20°

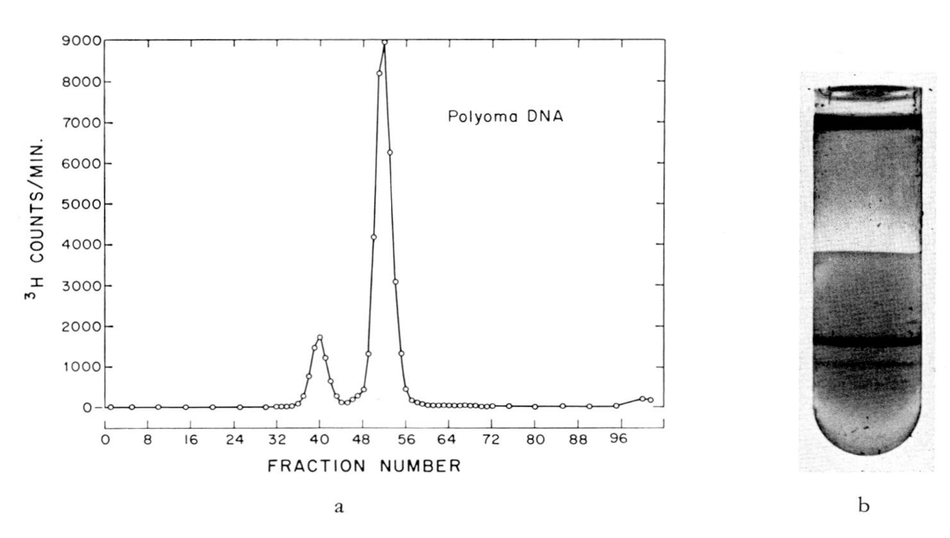

Fig. 10. A mixture of purified polyoma DNA's I and II, 3.00 ml, 1.558 gm/ml CsCl, 100 μg/ml ethidium bromide, 48 h at 43 krpm, 20°. — (a) The band maxima are separated by 12 fractions and 0.36 ml. The calculated buoyant densities are 1.592 and 1.556 gm/ml. — (b) A photograph of the centrifuge tube prior to drop collection. The tube contains a total of 4 μg of DNA, and 0.64 μg of component I (Radloff et al., 1967)

(Bauer, unpublished observations). Fig. 10b illustrates the corresponding ease of visual detection of fluorescent polyoma DNA buoyant bands under these conditions over the background of free EB at 100 μg/ml. Amounts of DNA of less than 0.2 μg can be readily detected and photographed by this method. In the event that the amount of closed circular DNA is below the detectable limit a marker of linear DNA may often be used to assist in its location. Further details of the method are discussed by Radloff et al. (1967).

2. The Use of Intercalators Other than Ethidium Bromide

The use of other intercalators as a substitute for ethidium bromide in the dye-buoyant density method has been investigated by Hudson, Upholt, DeVinny and Vinograd (1969). In general, the requirements set for a suitable dye are the following:

(1) sufficient unwinding of the DNA duplex; (2) a sufficient number of available binding sites; (3) the production of a shift in the buoyant density of DNA upon binding; (4) the retention of relatively strong binding at very high ionic strengths; and (5) a significant fluorescence enhancement upon binding. The first, second and third criteria are fundamental, in that the existence of the relative binding restrictions at high dye is predicated upon both an unwinding of the duplex and a number of available binding sites which is sufficient to generate a practical buoyant density difference. In the event that (1) is not satisfied, the closed DNA molecule will coband with the nicked; if (2) is neglected, both molecules will have already achieved their minimum density as the restriction in the closed form begins. Actinomycin D represents an example of an intercalator which does not satisfy condition (2) (Crothers and Müller, 1967). Condition (3) is not satisfied by, for example, the banding of DNA in NaI in the presence of EB. Here the thermodynamic components are ethidium

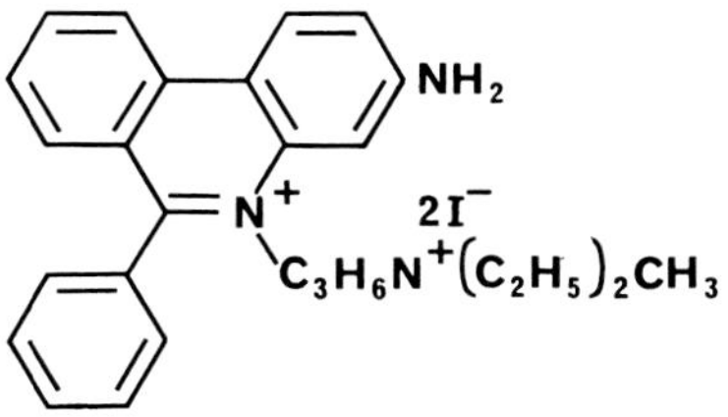

Fig. 11. The structure of propidium iodide

iodide and Na DNA and only a small buoyant shift results for either component (Anet and Strayer, 1969).

A variety of intercalators satisfy these three general requirements without being useful in the dye-buoyant density method. These include the aminoacridines and the quinine derivatives, both groups failing to evidence the strong binding at the high (4.5 M) CsCl concentrations herein required. Although condition (5), that of a fluorescence enhancement, is not strictly required, its absence clearly obviates one of the major advantages of the method.

Among the various ethidium analogs investigated by Hudson et al. (1969), only one was found to represent an improvement over ethidium bromide itself. This compound, called propidium diodide, is shown in Fig. 11. It was found to produce a buoyant density separation between DNA's I and II approximately 80% larger than that caused by ethidium, while retaining most of the fluorescence enhancement. It is anticipated that propidium will be of very great utility in the future use of the dye-buoyant density method. The reason for the increased buoyant separation has not yet been experimentally established.

3. Results Obtained with the Dye-Buoyant Density Method

The dye-buoyant density method has become widely used in the isolation and purification of closed circular DNA, as indicated in Table 1. Up to the present time some forty-six different closed circular DNA's have been extracted and studied, approximately half of them having involved the use of ethidium bromide in the buoyant system.

The dye-buoyant density method has also been instrumental in the discovery and investigation of complex forms of closed circular DNA. In the initial investigations of the closed circular DNA from HeLa cells (RADLOFF et al., 1967), three different classes of DNA were described. The major fraction consisted of circular molecules

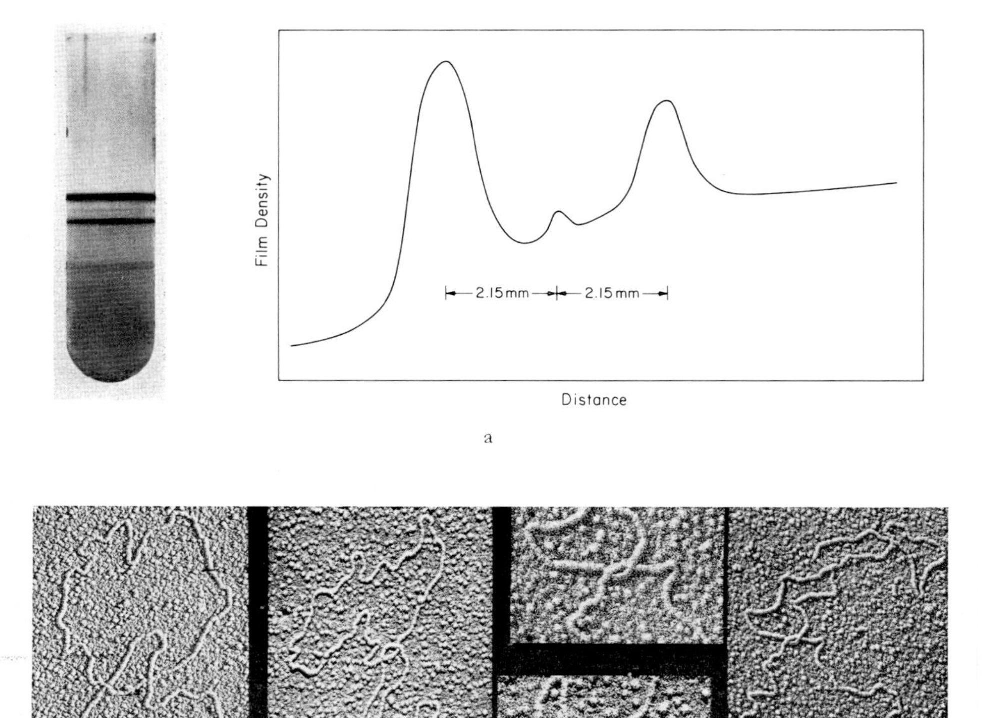

Fig. 12 a and b. Mitochondrial DNA from HeLa cells in a CsCl-EB density gradient. — a A photograph of a 5 ml cellulose nitrate tube illuminated with near ultraviolet light (365 nm). The upper band contains only nicked circular DNA. The lower band contains closed circular monomers and catenated oligomers. The middle band contains largely catenated dimers in which one submolecule is nicked. The centrifugation conditions were Spinco SW 50 rotor, 43 KRPM, 20°, 36 h. — b An electron micrograph showing the interlocking nature of the catenated dimers. The specimen grids were prepared as described by HUDSON and VINOGRAD (1967). The procedure involves rotary shadowing with platinum-palladium followed by shadowing from one direction in order to reveal the three-dimensional nature of the intersections. X 26,000 except third panel, which are X 49,000 (Figure taken from HUDSON and VINOGRAD, 1967)

of contour length 5 μ and were associated with the mitochondrion. A second pauci-disperse class consisted of integral multiples of the mitochondrial DNA size, and the third class was a polydisperse collection of closed circular molecules of contour length 0.2 to 3.5 μ. The origin and function of these small circles are yet unknown.

The origin of the second class, those closed circles which are even multiples of the mitochondrial DNA length, was later clarified by means of the dye-buoyant density method. Fig. 12a presents the results of a later isolation of the mitochondrial DNA from HeLa cells. The presence of a fluorescent band intermediate between the closed and nicked forms can be most readily accounted for in terms of the occurrence of catenated molecules in which a closed circular species is topologically bonded to a nicked circular molecule. The buoyant density for such a combination is expected to be approximately half-way between that of the closed and nicked forms band as separate species. The conclusions have since been confirmed by a combination of rotary and single-angle shadowing in the electron microscope, as shown in Fig. 12b. Ethidium bromide has been employed in all studies of complex forms of closed circular DNA, as indicated in Table 2.

IV. The Use of Intercalators in the Preparation of Closed Circular DNA

The ability to form closed circular DNA molecules from the corresponding nicked circular species involves at least two separate considerations: all phosphodiester bond scissions must be repaired, and account must be taken of the net interstrand winding at the time of closure.

The repair of single-strand phosphodiester bond scissions became possible through the discovery of the polynucleotide joining enzyme or ligase in extracts of *E. coli* [Gellert, 1967; Olivera and Lehman, 1967 (1); Gefter, Becker and Hurwitz, 1967]. This enzyme requires a divalent cation, such as Mg^{++}, and diphosphopyridine nucleotide as cofactors [Olivera and Lehman, 1967 (2); Zimmerman, Little, Oshinsky and Gellert, 1967]. A similar enzyme, but requiring ATP as cofactor, has been extracted from *T*4 infected *E. coli* (Weiss and Richardson, 1967). The ATP-dependent ligase has also been extracted from *E. coli* cells infected with phages *T*1, *T*2, *T*3, *T*6, and *T*7 (Becker, Lyn, Gefter and Hurwitz, 1967) and from phage λC_I-infected *E. coli* (Ando and Kosawa, 1970).

The topological considerations discussed in detail in Section I above indicate that the net winding of one DNA strand about the other is invariant after closure. If the topological winding number α is constrained to some value different from $\beta°$ at the time of closure [see Eq. (1)], and if solution conditions are later adjusted so that $\beta = \beta°$ is again attained, then a number of superhelical turns $\tau = \alpha - \beta°$ may thereby be generated. Such a procedure is possible in the event that a nicked circular DNA is closed with the ligase in the presence of an intercalating or other duplex-unwinding dye. Here $\alpha < \beta°$ at the time of scission repair. Upon later removal of dye a number of negative superhelical turns is generated. The quantitative relationships are determined by the molar unwinding angle of the dye, by the number of dye molecules bound at closure, and by the efficiency of the later dye removal.

The dye ethidium bromide has been used in conjunction with the polynucleotide ligase to induce superhelical turns into both lambda [Wang, 1969 (1)] and SV40

(HUDSON et al., 1970) DNA's. The enzyme reaction medium consists of typically 0.001 M EDTA, 0.002 M $MgCl_2$, 0.01 M Tris, pH 8. In such a medium the binding of ethidium is expected to be strong over the low binding region (WARING, 1965).

Removal of dye after completion of the closure reaction has been done by extraction with *i*-propanol (COZZARELLI et al., 1968) or *n*-butanol [WANG, 1969 (1)], by dialysis against 2 M urea, 0.2 M $MgSO_4$ and 0.1 M Tris, pH 8 [PAOLETTI, in WANG, 1969 (1)], and by passage through a Dowex-50 ion exchange column (RADLOFF et al., 1967). This latter method has been shown to result in the removal of EB to a binding ratio below that detectable by fluorescence (LEPECQ and PAOLETTI, 1967), a molar ratio of dye: nucleotide of 1:4000 (RADLOFF et al., 1967).

The above method has been used to generate closed circular DNA's of native superhelix density from approximately zero to -0.085 in both SV40 and lambda DNA's. The method is in principle applicable to DNA's of other molecular weights without regard to the ease of isolation or even the existence of a naturally occurring closed duplex. Any duplex DNA capable of forming hydrogen-bonded circles is a potential candidate for closure under the appropriate experimental conditions including, for example, the terminally redundant circularly permuted DNA molecule of bacteriophage *T4* (STREISINGER, EDGAR and DENHARDT, 1964). Closed circles of constant molecular weight and known σ_0 have already been used to study the details of the dependence of sedimentation coefficient upon superhelical structure (GRAY et al., 1971 and Section II above), and it is anticipated that a variety of other experimental uses will be soon forthcoming.

V. The Use of Closed Circular DNA to Assay the Unwinding of DNA Introduced by Low Molecular Weight Ligands

The mechanism of binding by intercalation, both as originally proposed by LERMAN (1961) and as subsequently developed by others (FULLER and WARING, 1964) requires that the DNA duplex be unwound as binding proceeds. It is therefore valuable in the study of the intercalation of a particular ligand with DNA to be able to determine whether or not unwinding accompanies the process; a negative result means clearly that intercalation does not take place under the experimental conditions chosen.

This determination may be made unambiguously by studying the effect of the ligand in question upon the sedimentation coefficient of closed circular DNA; since $\sigma_0 < 0$, as discussed above, a ligand may be adjudged to cause unwinding if and only if the characteristic dip appears in a plot of s *versus* dye bound. It should be emphasized that this procedure works only if there are a sufficient number of available binding sites as required by Eq. (6b). This requirement is that

$$v_{max} > -1.5\,\sigma_0\,, \tag{15}$$

where v_{max} is the maximum number of available binding sites. The method therefore requires the use of a DNA of previously determined σ_0, as well as a reliable method of measuring v_{max}. Within the above limitations, the sedimentation velocity-ligand titration is generally useful for detecting the absence of unwinding and therefore of intercalation. It should also be noted, of course, that the positive presence of a sedimentation velocity dip constitutes presumptive evidence in favor of unwinding. This is a necessary but not a sufficient condition for intercalation.

A second use of the ligand-sedimentation velocity titration assay is to estimate the magnitude of the unwinding angle ϕ for a given small molecular weight species which has been found to induce duplex unwinding. If the molar binding ratio of the ligand, v_c, can be measured at the minimum of the sedimentation velocity dip ($\sigma = 0$), comparison with a reference ligand (ethidium bromide) having a critical binding ratio of v_c^* allows calculation of the relative unwinding angles, ϕ and ϕ^*, according to Eq. (6b) with $\sigma = 0$.

$$\frac{\phi}{\phi^*} = \frac{v_c^*}{v_c} . \tag{16}$$

The binding ratios involved in Eq. (16) refer specifically to that mode of binding which brings about the unwinding. Especial care must be taken in measurements of v to exclude any other types of potential binding, such as outside stacking, from the quantitative results.

The above methods has been used for the sedimentation velocity titration of 9-aminoacridine with SV40 DNA in 0.02 M salt (BAUER and VINOGRAD, 1968). It was determined that $v_c = 0.043 \pm 0.02$, a number which is comparable to the results obtained with ethidium bromide at 1.0 M NaCl. WARING, as discussed elsewhere in this volume, has investigated the effect of a wide variety of dyes and antibiotics upon the sedimentation velocity behavior of the replicating form of ϕX-174 DNA.

An alternative method for the use of closed circular DNA in the unwinding assay is suggested by the behavior of the buoyant density in high EB concentrations (Section II. 6, above). The extent of the buoyant density separation between DNA's I and II in the presence of high dye is reciprocally related to ϕ [BAUER and VINOGRAD, 1970 (2)]. The buoyant separation between SV40 DNA's I and II at high concentrations of propidium iodide is, as noted above, approximately 80% greater than that with ethidium bromide under similar conditions (HUDSON, UPHOLT, DEVINNEY and VINOGRAD, 1969). The quantitative use of this method for the evaluation of ϕ in buoyant solvents also requires the determination of v_{max} for the ligand in question, and the method is applicable in practice only to those ligands which satisfy the conditions (1) through (5) specified in Section III. 2 above.

References

ANDO, T., KOSAWA, T.: Isolation of a DNA ligase from phage λC_1-infected *Escherichia coli*, Biochim. biophys. Acta (Amst.) **204**, 257 (1970).

AVERS, C. J.: Heterogeneous length distribution of circular DNA filaments from yeast mitochondria. Proc. nat. Acad. Sci. (Wash.) **58**, 620 (1967).

— BILLHEIMER, F. E., HOFFMAN, H.-P., PAULI, R. M.: Circularity of yeast mitochondrial DNA. Proc. nat. Acad. Sci. (Wash.) **61**, 90 (1968).

BAUER, W., VINOGRAD, J.: The interaction of closed circular DNA with intercalative dyes I. The superhelix density of SV40 DNA in the presence and absence of dye. J. molec. Biol. **33**, 141 (1968).

— — A thermodynamic theory for interactiong systems at equilibrium in a buoyant density gradient: The reaction between a small molecular species and a buoyant macromolecule. Ann. N. Y. Acad. Sci. **164**, Art. 1, 192 (1969).

— — (1) The interaction of closed circular DNA with intercalative dyes II. The free energy of superhelix formation in SV40 DNA. J. molec. Biol. **47**, 419 (1970).

— — (2) The interaction of closed circular DNA with intercalative dyes III. Dependence of the buoyant density upon superhelix density and base composition in the presence of dye. J. molec. Biol. **54,** 281 (1970).

BAZARAL, M., HELINSKI, D. R.: (1) Circular DNA forms of colincinogenic factors *E*1, *E*2, *E*3 from *Escherichia coli*. J. molec. Biol. **36**, 185 (1968).

— — (2) Characterization of multiple circular forms of colicinogenic factor E_1 from *Proteus mirabilis*. Biochemistry **7**, 3513 (1968).

BECKER, A., LYN, G., GEFTER, M., HURWITZ, J.: The enzymatic repair of DNA, II. Characterization of phage-induced sealase. Proc. nat. Acad. Sci. (Wash.) **58**, 1996 (1967).

BILLHEIMER, F. E., AVERS, C. J.: Nuclear and mitochondrial DNA from wild type and petite yeast: Circularity, length, and buoyant density. Proc. nat. Acad. Sci. (Wash.) **64**, 739 (1969).

BLOOMFIELD, V. A.: Twisted circular DNA: sedimentation coefficients and the number of twists. Proc. nat. Acad. Sci. (Wash.) **55**, 717 (1966).

BORST, P., RUTTENBERG, G. J. C. M.: Renaturation of mitochondrial DNA. Biochim. biophys. Acta (Amst.) **114**, 645 (1966).

— — KROON, A. M.: (1) Mitochondrial DNA I. Preparation and properties of mitochondrial DNA from chick liver. Biochim. biophys. Acta (Amst.) **149**, 140 (1967).

— VAN BRUGGEN, E. F. J., RUTTENBERG, G. J. C. M., KROON, A. M.: (2) Mitochondrial DNA II. Sedimentation analysis and electron microscopy of mitochondrial DNA from chick liver. Biochim. biophys. Acta (Amst.) **149**, 156 (1967).

BUJARD, H.: Isolation of bovine papilloma virus and characterization of its circular DNA. J. Virol. **1**, 1135 (1967).

— Studies on circular DNA II. Number of tertiary turns in papilloma DNA. J. molec. Biol. **33**, 503 (1968).

CARLTON, B. C., HELINSKI, D. R.: Heterogeneous circular DNA elements in vegetative cultures of *Bacillus megaterium*. Proc. nat. Acad. Sci. (Wash.) **64**, 592 (1969).

CLAYTON, D. A., DAVIS, R. W., VINOGRAD, J.: Homology and structural relationships between the dimeric and monomeric circular forms of mitochondrial DNA from human leukemic leucocytes. J. molec. Biol. **47**, 137 (1970).

— SMITH, C. A., JORDAN, J. M., TEPLITZ, M., VINOGRAD, J.: Occurrence of complex mitochondrial DNA in normal tissues. Nature (Lond.) **220**, 976 (1968).

— VINOGRAD, J.: Circular dimer and catenated forms of mitochondrial DNA in human leukemic leucocytes. Nature (Lond.) **216**, 652 (1967).

COHN, G., EISENBERG, H.: Viscosity and sedimentation study of sonicated DNA-proflavine complexes. Biopolymers **8**, 45 (1969).

COZZARELLI, N. R., KELLY, R. B., KORNBERG, A.: A minute circular DNA from *Escherichia coli* 15. Proc. nat. Acad. Sci. (Wash.) **60**, 992 (1968).

CRAWFORD, L. V.: A study of shope papilloma virus DNA. J. molec. Biol. **8**, 489 (1964).

— A study of human papilloma virus DNA. J. molec. Biol. **13**, 362 (1965).

— BLACK, P. H.: The nucleic acid of simian virus 40. Virology **24**, 388 (1964).

— WARING, M. J.: Supercoiling of polyoma virus DNA measured by its interaction with ethidium bromide. J. molec. Biol. **25**, 23 (1967).

CUZIN, F., VOGT, M., DRICKMANN, M., BERG, P.: Induction of virus multiplication in 3*T*3 cells transformed by thermosensitive mutant of polyoma virus. II. Formation of oligomeric polyoma DNA molecules. J. molec. Biol. **47**, 317 (1970).

DAWID, I. B.: Mitochondrial RNA in *Xenopus laevis*: Its homology to mitochondrial DNA. Fed. Proc. **28**, 349 (1969).

— WOLSTENHOLME, D. R.: Ultracentrifuge and electron microscope studies on the structure of mitochondrial DNA. J. molec. Biol. **28**, 233 (1967).

DULBECCO, R., VOGT, M.: Evidence for a ring structure of polyoma virus DNA. Proc. nat. Acad. Sci. (Wash.) **50**, 236 (1963).

DRUMMOND, D. S., SIMPSON-GILDEMEISTER, V. F. W., PEACOCKE, A. R.: Interaction of aminoacridines with DNA: Effects of ionic strength, denaturation, and structure. Biopolymers **3**, 135 (1965).

ESPEJO, R. T., CANELO, E. S., SINSHEIMER, R. L.: DNA of bacteriophage PM 2: A closed circular double-stranded molecule. Proc. nat. Acad. Sci. (Wash.) **63**, 1164 (1969).

FAN, H., PENMAN, S.: Mitochondrial RNA synthesis during meiosis. Science **168**, 135 (1970).

FREIFELDER, D.: Studies on *Escherichia coli* sex factors IV. Molecular weights of the DNA of several *F'* elements. J. molec. Biol. **35**, 95 (1968).

FULLER, W., WARING, M. T.: A molecular model for the interaction of ethidium bromide with deoxyribonucleic acid. Ber. der Bunsenges. **68**, 805 (1964).

GEFTER, M. L., BECHER, A., HURWITZ, J.: The enzymatic repair of DNA, I. Formation of circular λ DNA. Proc. nat. Acad. Sci. (Wash.) **58**, 240 (1967).

GELLERT, M.: Formation of covalent circles of lambda DNA by *E. coli* extracts. Proc. nat. Acad. Sci. (Wash.) **57**, 148 (1967).

GLAUBIGER, D., HEARST, J. E.: The effect of superhelical structure on the secondary structure of DNA rings. Biopolymers **8**, 692 (1967).

GRAY, H. B.: The sedimentation coefficient of polyoma virus DNA. Biopolymers **5**, 1009 (1967).

— BLOOMFIELD, V. A., HEARST, J. E.: Sedimentation coefficients of linear and cyclic wormlike coils with excluded-volume effects. J. chem. Phys. **46**, 1493 (1967).

— UPHOLT, W. D., VINOGRAD, J.: A buoyant method for the determination of superhelix density in closed circular DNA. Manuscript in preparation (1971).

HEARST, J. E., STOCKMAYER, W. H.: Sedimentation constants of broken chains and wormlike coils. J. chem. Phys. **37**, 1425 (1962).

HERSKOVITS, T. T.: Nonaqueous solutions of DNA: Factors determining the stability of the helical configuration insolution. Arch. Biochem. **97**, 474 (1962).

HICKSON, F. T., ROTH, T. F., HELINSKI, D. R.: Circular DNA forms of a bacterial sex factor. Proc. nat. Acad. Sci. (Wash.) **58**, 1731 (1967).

HOLLENBERG, C. P., BORST, P., THURING, R. W. J., VAN BRUGGEN, E. F. J.: Size, structure and genetic complexity of yeast mitochondrial DNA. Biochim. biophys. Acta (Amst.) **186**, 417 (1969).

HOTTA, Y., BASSEL, A.: Molecular size and circularity of DNA in cells of mammals and higher plants. Proc. nat. Acad. Sci. (Wash.) **53**, 356 (1965).

HUDSON, B., CLAYTON, D. A., VINOGRAD, J.: Complex mitochondrial DNA. Cold Spr. Harb. Symp. quant. Biol. **33**, 435 (1968).

— UPHOLT, W. B., DEVINNY, J., VINOGRAD, J.: The use of an ethidium analogue in the dye-buoyant density procedure for the isolation of closed circular DNA: The variation of the superhelix density of mitochondrial DNA. Proc. nat. Acad. Sci. (Wash.) **62**, 813 (1969).

— VINOGRAD, J.: Catenated circular DNA molecules in HeLa cell mitochondria. Nature (Lond.) **216**, 647 (1967).

JANSZ, H. S., ZANDBERG, J., VAN DE POL, J. H., VAN BRUGGEN, E. F. J.: Circular DNA from *Shigella paradysenteriae*. Europ. J. Biochem. **9**, 156 (1969).

JORDAN, J. M., VAN DER EB, A., VINOGRAD, J.: Complex mitochondiral DNA in cultured cells in an experimentally induced tumor. Manuscript in preparation (1970).

KIRKWOOD, J. G.: The general theory of irreversible processes in solutions of macromolecules. J. Polymer Sci. **12**, 1 (1954).

KNIGHT, E., JR.: Mitochondria-associated RNA of the HeLa cell. Effect of ethidium bromide on the synthesis of ribosomal and 4S RNA. Biochemistry **8**, 5089 (1969).

KROON, A. M., BORST, P., VAN BRUGGEN, E. F. J., RUTTENBERG, G. J. C. M.: Mitochondrial DNA from sheep heart. Proc. nat. Acad. Sci. (Wash.) **56**, 1836 (1966).

KURATA, M.: Hydrodynamic properties of dilute solutions of ring polymers, II. Twisted ring polymers. Bull. Inst. Chem. Res. Kyoto Univ. **44**, 150 (1966).

LANGRIDGE, R., SEEDS, W. E., WILSON, H. R., HOOPER, C. W., WILKINS, M. H. F., HAMILTON, L. D.: Molecular structure of DNA. J. biophys. biochem. Cytol. **3**, 767 (1957).

LEE, C. S., DAVIDSON, N.: Covalently closed minicircular DNA in *Micrococcus lysodeikticus*. Biochem. biophys. Res. Commun. **32**, 757 (1968).

— — Physicochemical studies on the minicircular DNA in *Escherichia coli* 15. Biochim. biophys. Acta (Amst.) **204**, 285 (1970).

— — SCALLETI, J. V.: Covalently closed circular DNA from *Micrococcus lysodeikticus* cells infected with phage Nl. Biochem. biophys. Res. Commun. **32**, 752 (1968).

LePecq, J. B.: Etude spectrofluorimetrique des modalités d'interaction du bromhydrate d'éthidium avec les acides nucleiques. Thèses, Faculté des Sciences de l'Université de Paris 1965.
— Paoletti, C.: A fluorescent complex between ethidium bromide and nucleic acids. J. molec. Biol. **27**, 87 (1967).
Lerman, L. S.: Structural considerations in the interaction of DNA and acridines. J. molec. Biol. **3**, 18 (1961).
— The structure of the DNA-acidine complex. Proc. nat. Acad. Sci. (Wash.) **49**, 94 (1963).
— Amino group reactivity in DNA-aminoacridine complexes. J. molec. Biol. **10**, 367 (1964).
Li, H. J., Crothers, D. M.: (1) Studies of the optical properties of the proflavine-DNA complex. Biopolymers **8**, 217 (1969).
— — (2) Relaxation studies of the proflavine-DNA complex: The kinetics of an intercalation reaction. J. molec. Biol. **39**, 461 (1969).
Luzzanti, V., Masson, F., Lerman, L. S.: Interaction of DNA and proflavine: A small-angle *X*-ray scattering study. J. molec. Biol. **3**, 634 (1961).
Mackinlay, A. G., Kaiser, A. D.: DNA replication in head mutants of bacteriophage λ. J. molec. Biol. **39**, 679 (1969).
Nass, M. M. K.: The circularity of mitochondrial DNA. Proc. nat. Acad. Sci. (Wash.) **56**, 1215 (1966).
— (1) Mitochondrial DNA II. Structure and physiochemical properties of isolated DNA. J. molec. Biol. **42**, 529 (1960).
— (2) Reversible generation of circular dimer and higher multiple forms of mitochondrial DNA. Nature (Lond.) **223**, 1124 (1969).
Neville, D. M., Davies, D. R.: The interaction of acridine dyes with DNA: An *X*-ray diffraction and optical investigation. J. molec. Biol. **17**, 57 (1966).
Olivera, B. M., Lehman, I. R.: (1) Linkage of polynucleotides through phosphodiester bonds by an enzyme from *Escherichia coli*. Proc. nat. Acad. Sci. (Wash.) **57**, 1426 (1967).
— — (2) Diphosphopyridine nucleotide: A cofactor for the polynucleotide — joining enzyme from *Escherichia coli*. Proc. nat. Acad. Sci. (Wash.) **57**, 1700 (1967).
Peacocke, A. R., Skerrett, J. N. H.: The interaction of aminoacridines with nucleic acids. Trans. Faraday Soc. **52**, 261 (1956).
Piko, L., Blair, D. G., Tyler, A., Vinograd, J.: Cytoplasmic DNA in the unfertilized sea urchin egg: Physical properties of circular mitochondrial DNA and the occurrence of catenated forms. Proc. nat. Acad. Sci. (Wash.) **59**, 838 (1968).
— Tyler, A., Vinograd, J.: Amount, location, priming capacity, circularity and other properties of cytoplasmic DNA in sea urchin eggs. Biol. Bull. **132**, 68 (1967).
Radloff, R., Bauer, W., Vinograd, J.: A dye-buoyant density method for the detection and isolation of closed circular duplex DNA: The closed circular DNA in HeLa cells. Proc. nat. Acad. Sci. (Wash.) **57**, 1514 (1967).
Rhoades, M., Thomas, C. A., Jr.: The 22P bacteriophage DNA molecule II. Circular intracellular forms. J. molec. Biol. **37**, 41 (1968).
Suguira, M., Okamoto, T., Takanimi, M.: Starting nucleotide sequences of RNA synthesized on the replicative form DNA of coliphage fd. J. molec. Biol. **43**, 299 (1969).
Suyama, Y., Muira, K.: Size and variations of mitochondrial DNA. Proc. nat. Acad. Sci. (Wash.) **60**, 235 (1968).
Upholt, W. D., Gray, H. B., Vinograd, J.: The sedimentation velocity properties of closed circular SV40 DNA as a function of superhelix density and ionic strength. Manuscript in preparation (1971).
Vinograd, J., Bruner, R., Kent, R., Weigle, J.: Band-centrifugation of macromolecules and viruses in self-generating density gradients. Proc. nat. Acad. Sci. (Wash.) **49**, 902 (1963).
— Lebowitz, J.: Physical and topological properties of circular DNA. J. gen. Physiol. **49**, 103 (1966).
— —, Radloff, R., Watson, R., Laipis, P.: The twisted circular form of polyoma viral DNA. Proc. nat. Acad. Sci. (Wash.) **53**, 1104 (1965).
— — Watson, R.: Early and late helix-coil transitions in closed circular DNA: The number of superhelical turns in polyoma DNA. J. molec. Biol. **33**, 173 (1968).

WANG, J. C.: (1) Variation of the average rotation angle of the DNA helix and the superhelical turns of covalently closed cyclic λ DNA. J. molec. Biol. **43**, 25 (1969).
— (2) Degree of superhelicity of covalently closed cyclic DNA's from *Escherichia coli.* J. molec. Biol. **43**, 263 (1969).
— BAUMGARTEN, D., OLIVERA, B. M.: On the origin of tertiary turns in covalently closed double-stranded cyclic DNA. Proc. nat. Acad. Sci. (Wash.) **58**, 1852 (1967).
WARING, M. J.: Complex formation with DNA and inhibition of *Escherichia coli* RNA polymerase by ethidium bromide. Biochim. biophys. Acta (Amst.) **87**, 358 (1964).
WEIL, R., VINOGRAD, J.: The cyclic helix and cyclic coil forms of polyoma viral DNA. Proc. nat. Acad. Sci. (Wash.) **50**, 730 (1963).
WEISS, B., RICHARDSON, C. C.: Enzymatic breakage and joining of DNA, I. Repair of single-strand breaks in DNA by an enzyme from *E. coli* infected with T4 bacteriophage. Proc. nat. Acad. Sci. (Wash.) **57**, 1021 (1967).
WOLSTENHOLME, D. R., DAWID, I. B.: Circular mitochondrial DNA from *Xenopus laevis* and *Rana pipiens.* Chromosoma (Berl.) **20**, 445 (1967).
ZIMMERMAN, S. B., LITTLE, J. W., OSHINSKY, C. K., GELLERT, M.: Enzymatic joining of DNA strands: A novel reaction of diphosphopyridine nucleotide. Proc. nat. Acad. Sci. (Wash.) **57**, 1841 (1967).
ZYBER, E., PENMAN, S.: Mitochondrial-associated 4S RNA synthesis inhibition by ethidium bromide. J. molec. Biol. **46**, 201 (1969).
— VESCO, C., PENMAN, S.: Selective inhibition of the synthesis of mitochondria-associated RNA by ethidium bromide. J. molec. Biol. **44**, 195 (1969).

Binding of Drugs to Supercoiled Circular DNA: Evidence for and against Intercalation

MICHAEL WARING

with the technical assistance of N. F. TOTTY

Outstanding among molecular models for interactions between drugs and DNA is the intercalation model, originally proposed by LERMAN (1961) for the binding of acridines to DNA. A primary objective of this model was to suggest a plausible molecular basis for the ability of acridine dyes, especially proflavine, to induce frame shift mutations in microorganisms. The phenomenon of frame shift mutagenesis by acridines is now firmly established (STREISINGER, OKADA, EMRICH, NEWTON, TSUGITA, TERZAGHI and INOUYE, 1966), and several suggestions have been made to account for such mutations in terms of intercalation (BRENNER, BARNETT, CRICK and ORGEL, 1961; LERMAN, 1963; STREISINGER et al., 1966). However, in recent years the intercalation model has become increasingly important for a different reason — namely, that it seems to account satisfactorily for the interaction between DNA and many other drugs, most of which are of interest for biological effects other than mutagenesis. These drugs include several chemotherapeutic agents, such as the trypanocide ethidium bromide (WARING, 1965, 1966), the schistosomicide Miracil D (LUCANTHONE) (HIRSCHBERG, WEINSTEIN, GERSTEN, MARNER, FINKELSTEIN and CARCHMAN, 1968), the antitumour antibiotics daunomycin and nogalamycin (WARD, REICH and GOLDBERG, 1965; KERSTEN, KERSTEN and SZYBALSKI, 1966), and the antimalarial drug chloroquine (O'BRIEN, ALLISON and HAHN, 1966). Intercalation is also a likely mode of action for two drugs important as biochemical tools, actinomycin D (MÜLLER and CROTHERS, 1968) and propidium iodide (HUDSON, UPHOLT, DEVINNY and VINOGRAD, 1969)*. Structural formulae of these compounds are included in Fig. 1. It will be noted that all posses a planar aromatic triple ring system, except for chloroquine which has only a two-ring chromophore. The flat ring systems are important, for it is an obvious requirement of the intercalation theory that the drug molecule must possess a reasonably large flat portion (but not too large or bulky) if it is to slip into a hydrophobic sandwich between the flat base-pairs of the DNA helix.

This paper is concerned with the relation between binding of drugs and the supercoiling of a closed circular duplex DNA, the replicative form of bacteriophage ϕX 174 DNA (ϕX 174 RF). The results provide new evidence that the drugs listed above bind to DNA by intercalation. They also contribute towards understanding the mechanisms of interaction between DNA and a number of non-intercalating substan-

* The references given are purely "key" references, and no attempt at a comprehensive list is attempted or implied. Many more references, and discussion of experimental results, will be found in other contributions to this volume.

ces. The principle of the test is simple: it is based on the fact that any intercalative process requires local uncoiling of the helix at the point of drug insertion, which alters the average pitch of the helix — more specifically, it causes an increase in the average number of base-pairs per turn (LERMAN, 1961, 1964; FULLER and WARING, 1964; FULLER, 1966). With closed circular DNAs, any change in the number of base-pairs per turn results in changes in the twisting (supercoiling) of the circles, which are observable experimentally by the consequent changes in sedimentation coefficient. In practical terms, when increasing amounts of an intercalating drug are bound to circular DNA there is first a decrease in the number of right-handed supercoils originally present in the circles, and their S_{20} falls. At a critical level of binding the accumulated drug-induced uncoiling has raised the average number of base-pairs per turn to a value where the circles are unstrained and have no supercoils at all. As more drug is added, the additional uncoiling forces the circles to form reversed (left-handed) supercoils and their sedimentation coefficient rises again. This sequence of events is readily demonstrable with the aid of molecular models, as illustrated in the plate. For a fuller description of theoretical and practical aspects of these phenomena see CRAWFORD and WARING [1967 (1)] and BAUER and VINOGRAD (1968).

Thus the test consists simply of measuring the S_{20} of closed circular DNA, in this case ϕX 174 RF, in the presence of increasing concentrations of drug. If the drug intercalates, the local uncoiling of the helix, necessitated by its intercalative mode of binding, should be revealed by characteristic changes in the S_{20}. The technique is already well investigated in the case of ethidium bromide, which has been shown to remove and reverse the supercoils of closed circular DNAs from a variety of sources [CRAWFORD and WARING, 1967 (1, 2); BAUER and VINOGRAD, 1968; other references collated in WARING, 1969]. Results with ethidium, proflavine, hycanthone, daunomycin, nogalamycin, actinomycin and propidium are collected in Fig. 2. Hycanthone is a hydroxylated metabolite of Miracil D which may be more active than the parent substance *in vivo* (ROSI, PERUZZOTTI, DENNIS, BERBERIAN, FREELE, TULLAR and ARCHER, 1967). Results with chloroquine are shown in Fig. 3a.

All eight drugs produce the same qualitative response in the S_{20} of the closed circles: a progressive fall followed by a rise. The ϕX 174 RF DNA preparations contained 25 to 35% nicked circles, and the S_{20} of this component is included in the Figures. With each drug, the minimum in the S_{20} curve for the closed circles coincides with the S_{20} of nicked circles (i.e. the two DNA components co-sedimented in the ultracentrifuge cell, yielding a single unresolved boundary). This reinforces the interpretation that at these levels of drug binding the closed circles have lost their supercoils completely, because nicked circles should not be supercoiled at any drug concentration since free rotation of one strand about the other is possible at the location of the nick(s) in the broken strand. In fact, the behaviour of the nicked circles in the DNA preparations provides a useful internal control which reveals effects on the S_{20} of the DNA caused by drug binding *per se*, i.e. effects other than changes in supercoiled tertiary structure. Such changes in S_{20} of the nicked circles as do occur are the same as those seen with ordinary (linear) DNA molecules; with seven of the eight drugs there is a small, effectively monotonic decrease in S_{20} or almost no effect at all. The small decrease in S_{20}, with linear DNA, is one of the features originally noted by LERMAN (1961) as suggestive of intercalation. The one drug which yields results deviating from this pattern is actinomycin D (Fig. 2g); here there is a sub-

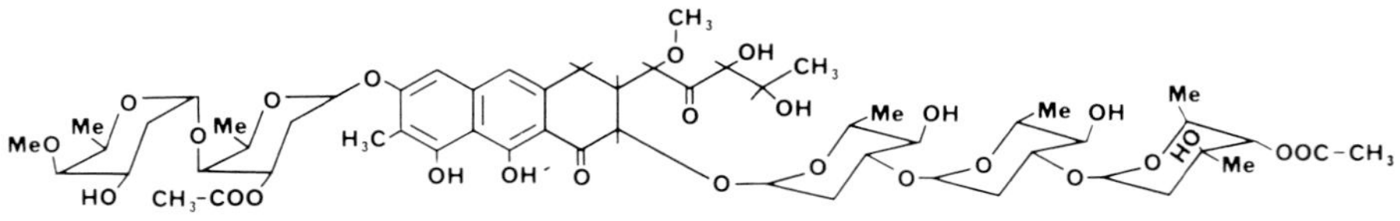

Chromomycin

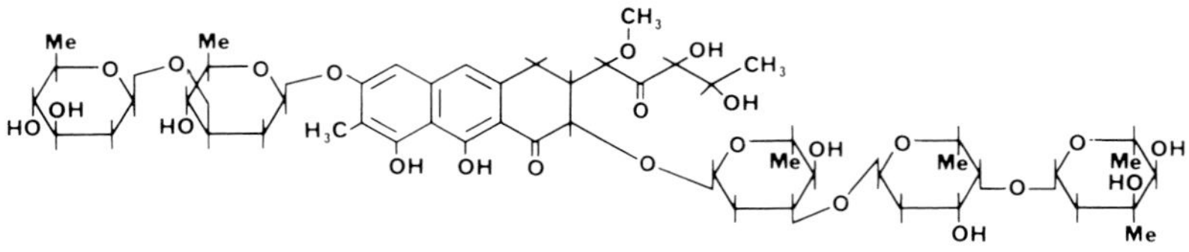

Mithramycin

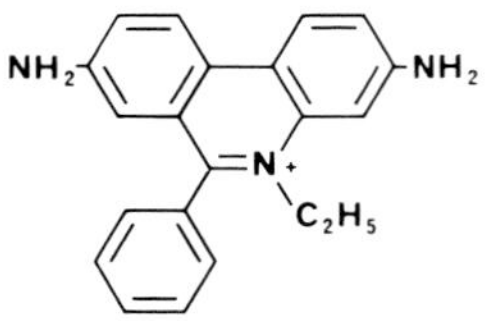

Ethidium

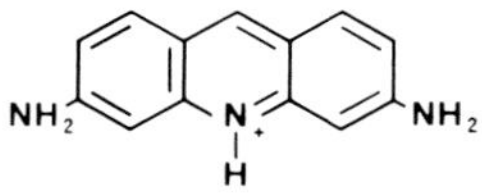

Proflavine

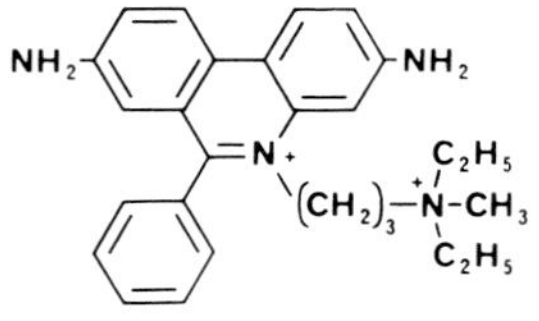

Propidium

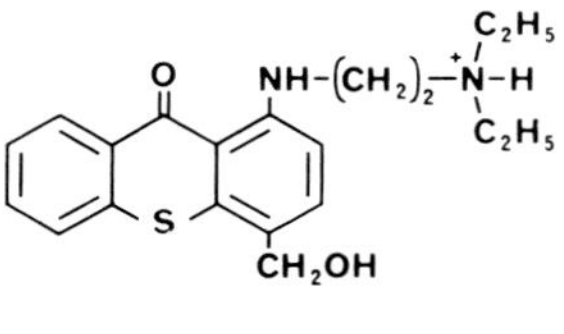

Hycanthone

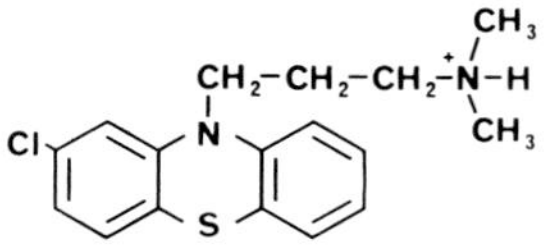

Chlorpromazine

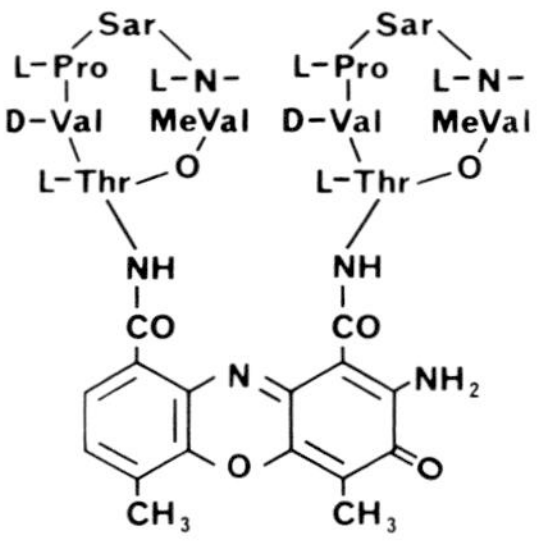

Actinomycin D

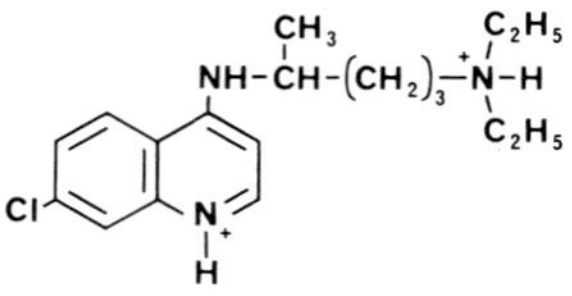

Chloroquine

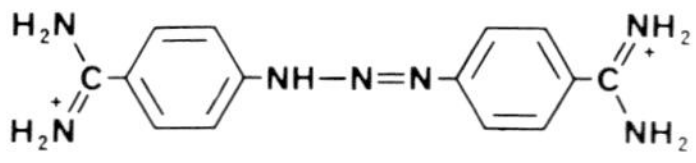

Berenil

Fig. 1. Structural formulae of compounds used in this work. The protonated forms expected to predominate at neutral pH are drawn

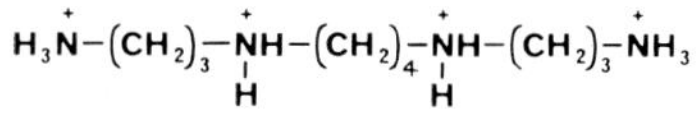

Spermine

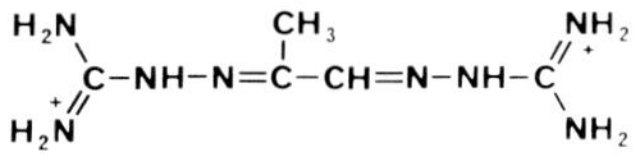

Methylglyoxal *-bis-* (guanylhydrazone) (Me GAG)

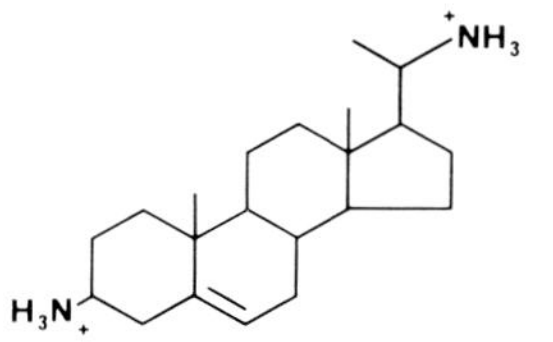

Irehdiamine A (IDA)

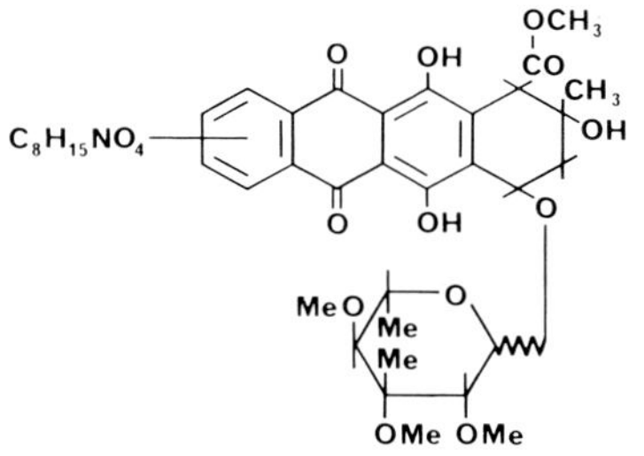

Nogalamycin

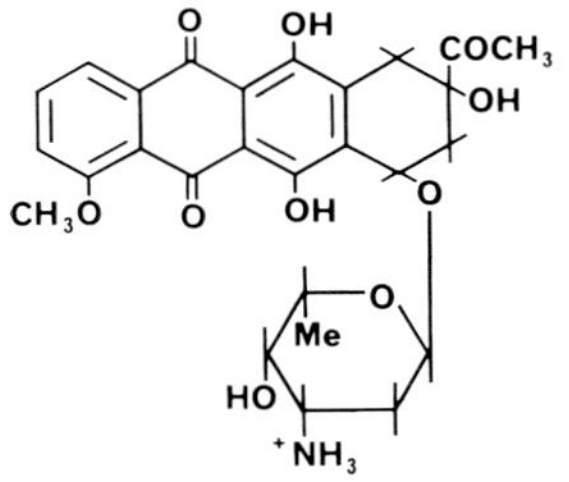

Daunomycin

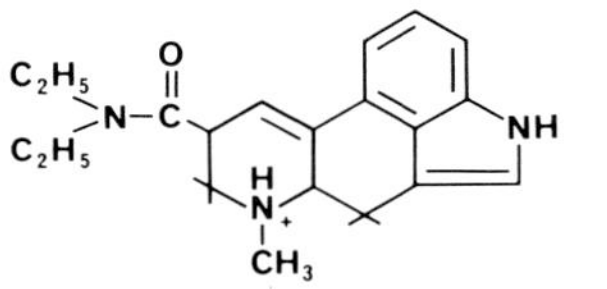

Lysergic acid diethylamide (LSD)

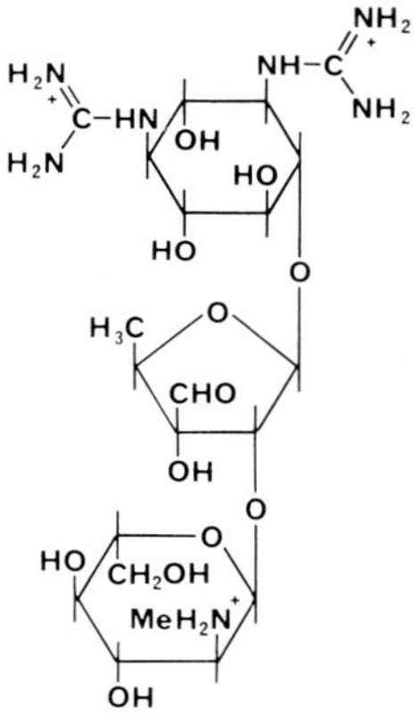

Streptomycin

Fig. 1 (Legend see p. 218)

stantial increase in the S_{20} of the nicked circles as the level of antibiotic binding rises. This difference from the behaviour of a "typical" intercalating drug, together with a related difference between the effects of actinomycin compared with other drugs on the viscosity of DNA, was an important factor leading to the rejection of intercalation and the development of an outside-binding model for the actinomycin-DNA complex by HAMILTON, FULLER and REICH (1963). However, MÜLLER and CROTHERS (1968) showed that under special circumstances this apparent anomaly disappears, and these authors proposed an intercalation model for actinomycin binding. When the anomalous effect of actinomycin on the S_{20} of the nicked circles is taken into account, its effect on the S_{20} of closed circles in Fig. 2g is plainly similar to the effects of the other seven drugs, which must be taken as strong support for the intercalation model

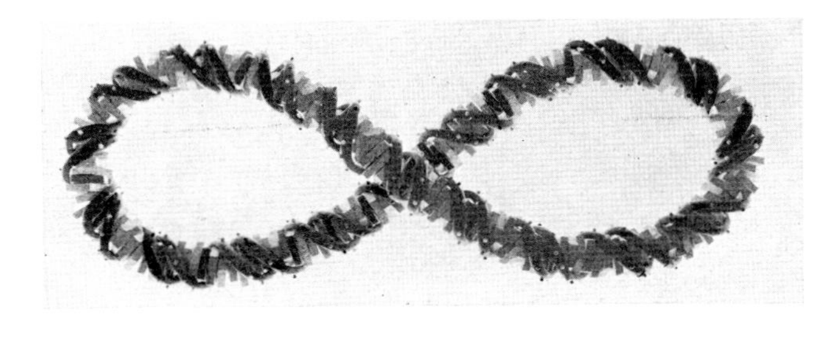

a

b

c

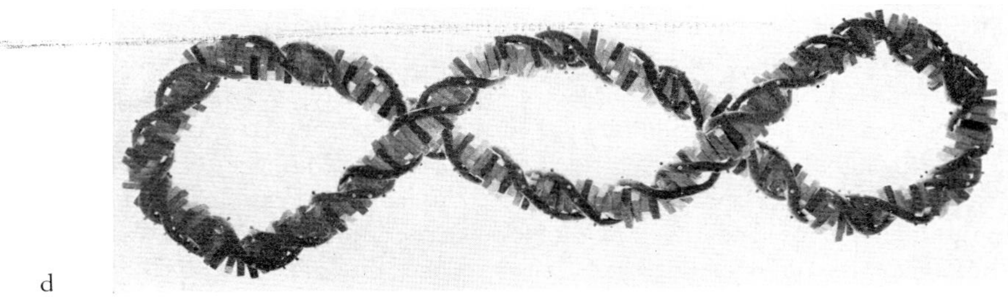

d

Plate

of MÜLLER and CROTHERS (1968). It remains to be seen whether refinement of the narrow groove-binding model of HAMILTON et al. (1963) would enable it to account for the apparent uncoiling of the helix associated with binding of actinomycin implied by the data in Fig. 2g.

The situation with the other seven drugs, including proflavine and ethidium, is less controversial and the results in Figs. 2 and 3a can be taken as confirmation of the conclusions that they intercalate. It is clear, however, that there are quantitative differences between the effects of different intercalating drugs on the supercoiling of the closed circles. Firstly, the curve for hycanthone (Fig. 2c) appears distinctly asymmetric compared to the curves for other drugs, and an even more marked asymmetry is implied in the curve for chloroquine (Fig. 3a) although the chloroquine data are plotted in terms of *added* drug/nucleotide ratio, not corrected to chloroquine molecules *bound* per nucleotide. This asymmetry is probably attributable to the lower binding constants of these two drugs (WARING, 1970). Secondly, the dip in the S_{20} curve for closed circles in the presence of different drugs occurs at different binding ratios. This observation indicates that differences exist in the angle of unwinding of the helix associated with binding of different drugs. This point will be taken up later.

Granted that the results in Figs. 2 and 3a confirm the expectations that the eight drugs studied intercalate, it is pertinent to ask whether the same sort of response in the S_{20} of closed circles may or may not be seen with drugs which do not intercalate,

---

Plate. Removal and reversal of the supercoils of closed circular duplex DNA by an intercalating drug.

The DNA model was constructed as follows. Balsa wood base-pairs (1 inch × $^3/_4$ inch × $^1/_4$ inch) are drilled with a hole off-set from the centre, and threaded on to a strand of solid rubber which acts as the axis of the helix. The rubber axis is under tension, as a result of which the base-pairs are held in face-to-face contact. Two lengths of rubber tubing simulate the phosphodiester backbones; each is pinned to each base-pair at a corner, the corners attaching the two tubes being those at the ends of a 1-inch side (the 1-inch side closest to the axis). The spacing of the pins along the rubber tubes is such as to create a stable helical screw of 36° per base-pair (i.e. ten base-pairs per turn). The ends of the model helix (4 to 5 feet long) are apposed, one end is rotated a few turns to introduce a deficiency of turns into the circle, and a neat join made. The resulting circular helix displays right-handed twisting, as shown (a). Intercalating drug molecules are represented by coloured pieces of balsa wood the same size as the base-pairs, with a *V*-shaped nick in the centre of one of the 1-inch sides; they can be pushed fully between the adjacent base-pairs with the rubber helix axis accommodated in the nick.

The model is not perfectly to scale, but the angle of uncoiling needed to permit insertion of the simulated intercalating drug molecule is fairly close to the desired angle of 12°. The plate shows, in the sequence of photographs (a) to (d), the changes in supercoiling of the circle which occur as progressively larger numbers of drug molecules are intercalated at random points around the circle. The restriction is applied that no insertions may occur into potential spaces adjacent to one already occupied.

As well as illustrating the effects of intercalation on the twisting of the circle, the model shows the increases in helix pitch and contour length of the molecule caused by the intercalation process. Each photograph is at the same magnification.

Photographs by courtesy of K. HARVEY, Medical Research Council Laboratory of Molecular Biology, Cambridge.

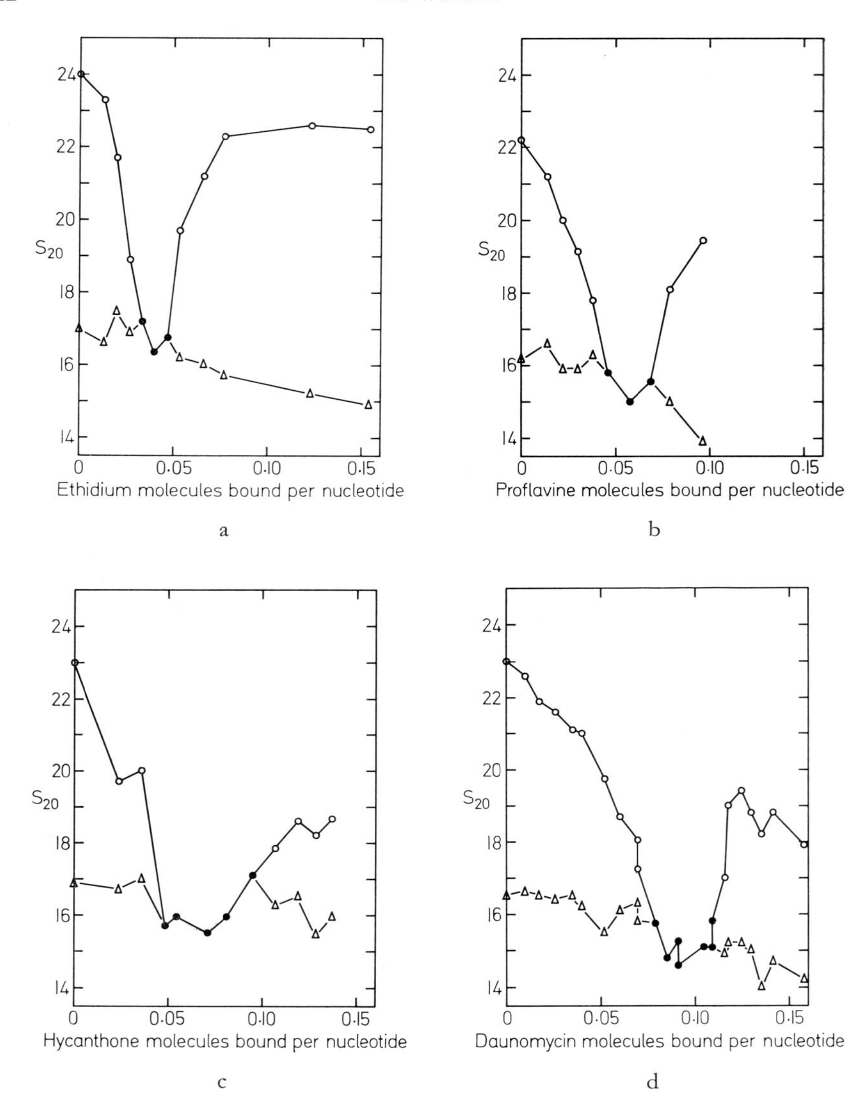

Fig. 2. a—g. Effects of intercalating drugs on the sedimentation coefficient of ϕX 174 RF. (a) Ethidium, (b) Proflavine, ()c Hycanthone, (d) Daunomycin, (e) Nogalamycin, (f) Propidium, (g) Actinomycin D. The buffer was 0.05 M tris-HCl, pH 7.9. For detailed experimental methods see WARING (1970). The abscissa shows the average level of drug binding to the two DNA components, closed and nicked circles, together. Symbols: 0, S_{20} of closed circles. △, S_{20} of nicked circles. ●, weight-average S_{20} from single boundary formed by closed and nicked circles sedimenting together

and thus whether or not this type of response may be taken as diagnostic of intercalation. Accordingly, the effects of nine further substances known to bind to DNA have been investigated. Four of them form a related group; they do not contain

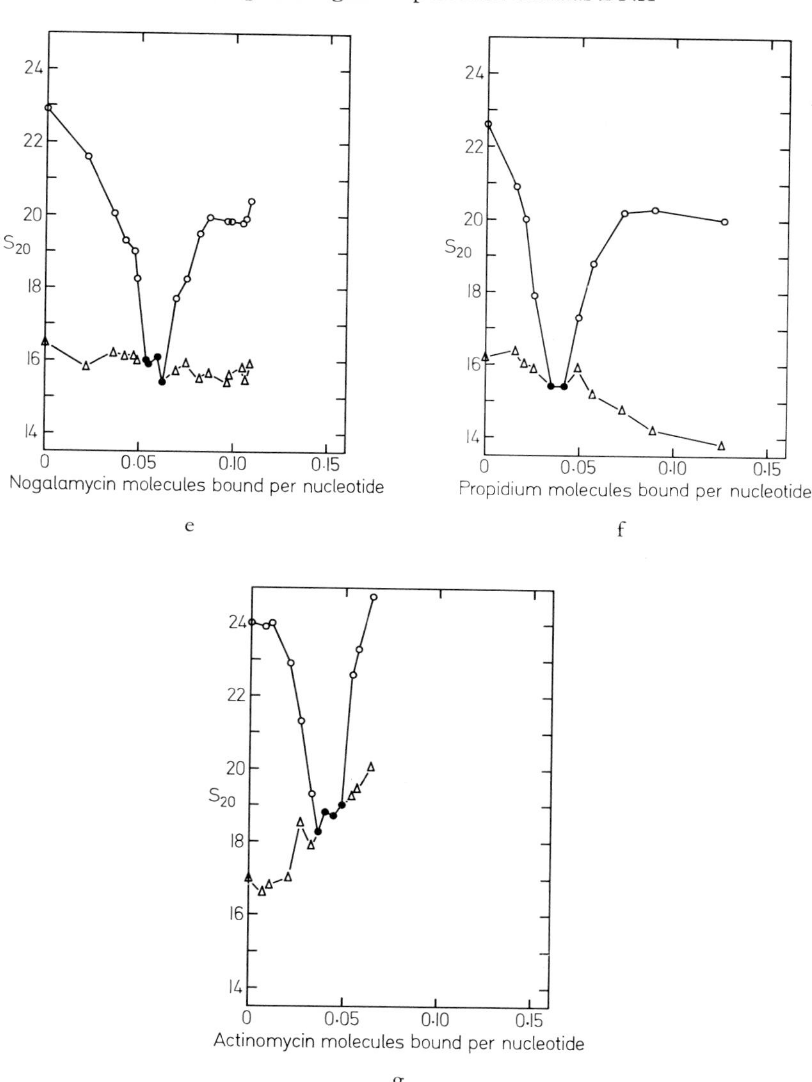

Fig. 2e—g

fused aromatic ring systems, but they carry substantial positive charges at neutral pH and most probably bind to DNA primarily via ionic forces — these are the polyamine spermine (HIRSCHMAN, LENG and FELSENFELD, 1967; SUWALSKY, TRAUB, SHMUELI and SUBIRANA, 1969), streptomycin (BROCK, 1966), the antineoplastic drug methylglyoxal-*bis*-(guanylhydrazone) (abbreviation: MeGAG; SARTORELLI, IANNOTTI, BOOTH, SCHNEIDER, BERTINO and JOHNS, 1965) and the trypanocide berenil (NEWTON, 1967). Their structures are included in Fig. 1 and their effects on the S_{20} of ϕX 174 RF are shown in Figs. 3 and 4. None of the four has much effect on the S_{20} of either closed

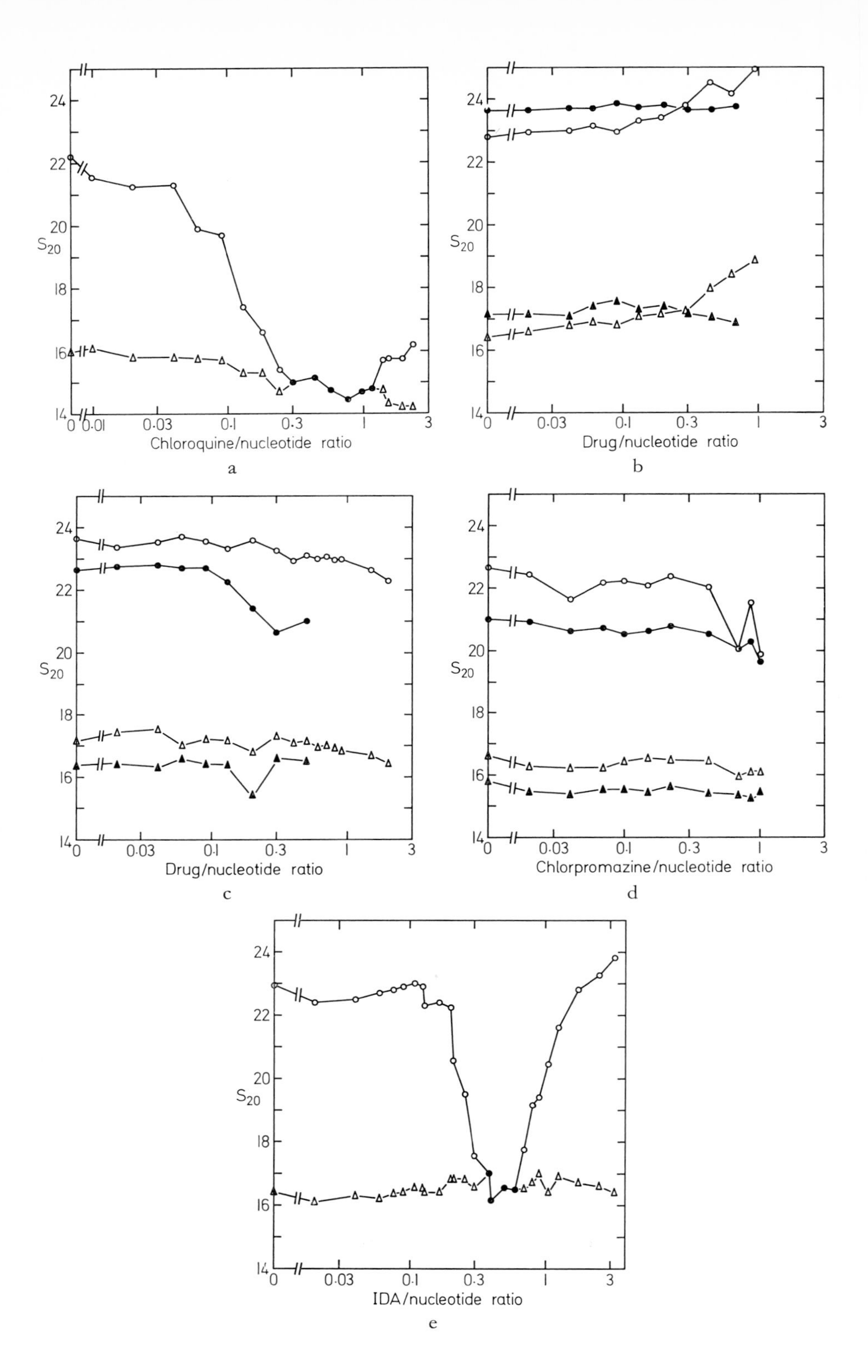
24
22
20
S20
18
16
14
0
0·01
0·03
0·1
0·3
1
3
Chloroquine/nucleotide ratio
a
Drug/nucleotide ratio
b
Drug/nucleotide ratio
c
Chlorpromazine/nucleotide ratio
d
IDA/nucleotide ratio
e

or nicked circles up to the highest concentrations tested. There is certainly no sign of the characteristic dip reflecting loss and reversal of supercoils produced by intercalating drugs, even though the drug/nucleotide ratios tested extend to values much higher than those used in Fig. 2. In the cases of spermine and berenil the highest concentrations which could be tested were limited by the onset of precipitation, which probably accounts for such changes in S_{20} as are apparent at the highest values plotted. The sample of berenil used was a commercial preparation containing 69% by weight of an antipyrine stabiliser; nevertheless this stabiliser does not prevent binding of the berenil to DNA as shown by the binding curve plotted in Fig. 4a, which climbs to a saturation plateau corresponding to approximately 0.13 berenil molecules bound per nucleotide. The effect of berenil on circular DNA has been checked using a pure drug preparation and closed circular DNA of bacteriophage PM2; again no effect on the S_{20} was found up to concentrations at which precipitation occurred (unpublished observations). In the case of this drug it may be safely concluded that there is no evidence for uncoiling of the helix associated with its strong primary interaction with DNA. This conclusion cannot be stated with equal certainty for spermine, streptomycin and Me GAG because there is no independent supporting evidence that the compounds were binding to the ϕX 174 RF DNA under the conditions of test; nevertheless the small but definite monotonic increase in S_{20} produced by streptomycin constitutes presumptive evidence that binding was taking place.

The antibiotics chromomycin A_3 and mithramycin (Fig. 1) are known to bind strongly to DNA in the presence of Mg^{++} ions (WARD, REICH and GOLDBERG, 1965; KERSTEN, KERSTEN and SZYBALSKI, 1966; BEHR, HONIKEL and HARTMANN, 1969). Their interaction with DNA differs significantly from the pattern typically seen with intercalating drugs, and the very bulky sugar substituents on their chromophore make an intercalative mode of binding seem unlikely. Their effects when added to ϕX 174 RF support the notion that they do not intercalate; in Fig. 4b and c the effects of adding these antibiotics up to quite high drug/nucleotide ratios are shown. The binding curves, obtained independently by a spectrophotometric method, show that strong binding well into the region of saturation occurred. The data are re-plotted in Fig. 4d in the form of S_{20} versus antibiotic molecules bound per nucleotide. It can be seen that both antibiotics caused a small monotonic increase in the S_{20} of the nicked circles, and the same is true of the effect of mithramycin on the S_{20} of closed circles, though higher binding ratios of chromomycin produced a definite fall in the S_{20} of the closed circles. Whether this effect of chromomycin reflects an incipient loss of the supercoils of closed circles it is not possible to say. Nevertheless, from the data in

---

Fig. 3. (a) Effect of chloroquine on the sedimentation coefficient of ϕX 174 RF. Symbols as in Fig. 2. (b) Effects of streptomycin (open symbols) and MeGAG (closed symbols). (c) Effects of spermine (closed symbols) and LSD (open symbols). (d) Effect of chlorpromazine in 0.05 M tris-HCl pH 7.9 (open symbols) and in 0.05 M sodium phosphate buffer pH 6.0 (closed symbols). (e) Effect of irehdiamine A (symbols as in Fig. 2). Unless otherwise stated the buffer was 0.05 M tris-HCl pH 7.9. Experimental methods as in WARING (1970). In each graph the abscissa shows, on a log scale, the molar ratio of *added* drug to DNA nucleotides

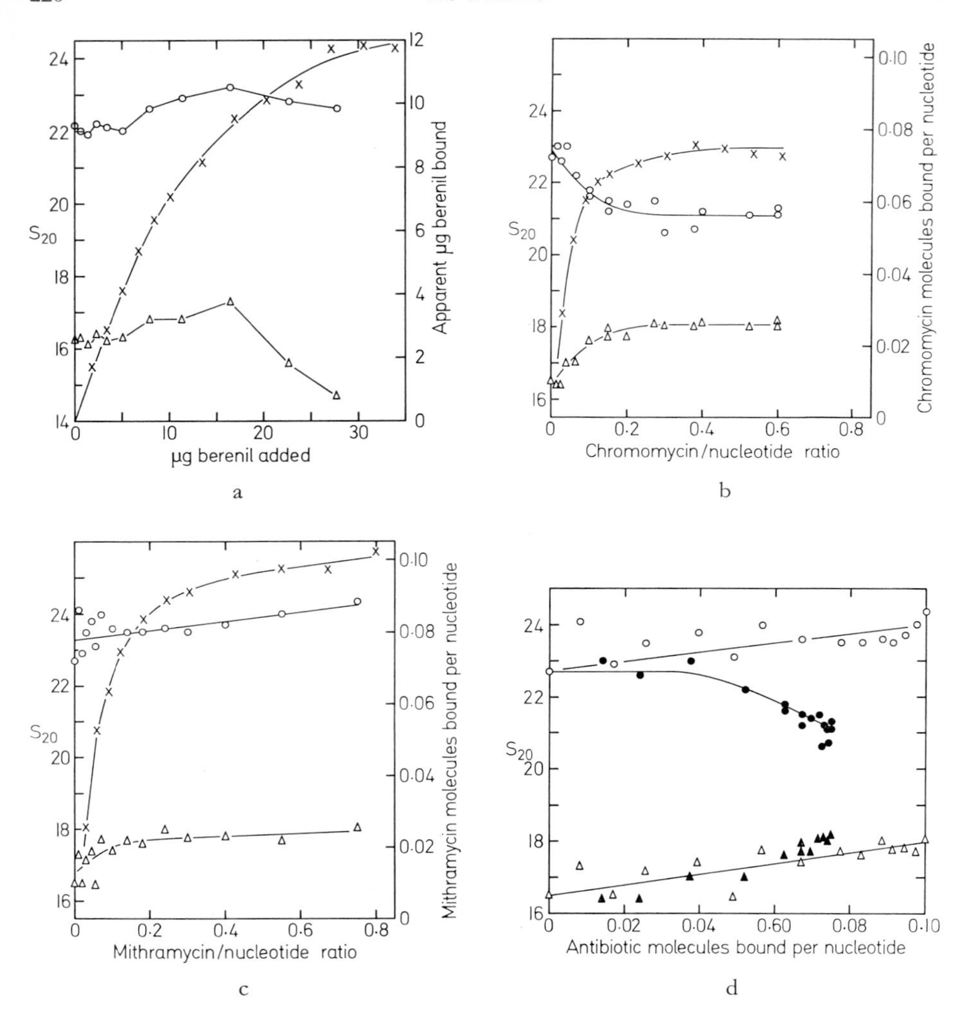

Fig. 4. (a) Effect of berenil on the sedimentation coefficient of ϕX 174 RF (left ordinate). Symbols: 0, S_{20} of closed circles. △, S_{20} of nicked circles. The right-hand ordinate shows a binding curve (symbols: X) calibrated in terms of the apparent weight of drug bound by 0.6 ml of ϕX 174 RF DNA having an absorbancy at 260 nm of 0.600, but it should be noted that the drug preparation contained a substantial weight fraction of material other than berenil (see text). The buffer was 0.05 M tris-HCl pH 7.9. (b) Effect of chromomycin A_3, and (c) effect of mithramycin. Symbols and ordinates as in (a). The buffer was 0.05 M tris-HCl pH 7.9 containing 10^{-4} M $MgCl_2$. The antibiotics were allowed to react with the DNA for at least 2 h before centrifugation. (d) Data of (b) and (c) re-plotted. Closed symbols, chromomycin. Open symbols, mithramycin. Further details of experimental methods in WARING (1970)

Fig. 4 there is little reason to suspect that binding of chromomycin involves an intercalative process, and none for mithramycin.

Two further drugs of interest are the hallucinogen lysergic acid diethylamide (LSD) and the tranquillizer chlorpromazine. LSD has been found to bind to DNA and some evidence suggestive of intercalation reported (WAGNER, 1969). The positive radical ion of chlorpromazine, prepared by persulphate oxidation, may also intercalate (OHNISHI and MCCONNELL, 1965). In Fig. 3 results with LSD and chlorpromazine itself are shown: in neither case is there evidence of the dip in S_{20} of the closed circles expected for intercalation, though again it must be pointed out that direct evidence of binding in these tests is lacking. The failure of LSD to affect the supercoiling of closed circular DNA has recently been confirmed using DNA of bacteriophage PM2 (unpublished observations). Experiments are currently in progress to try to test the effect of the chlorpromazine radical ion.

A most surprising result was obtained with the steroidal diamine irehdiamine A (abbreviation: IDA). This substance shows a primary and secondary interaction with DNA (MAHLER, GREEN, GOUTAREL and KHUONG-HUU, 1968). When added to ϕX 174 RF it produced a clear fall and rise in the S_{20} of the closed circles (Fig. 3e) reminiscent of the effect produced by intercalating drugs, yet IDA has no aromatic rings at all (its rings are puckered, not planar) and could not become intercalated in the usual sense. The drug/nucleotide ratios at which the dip in the S_{20} plot occurs are about ten-fold higher than those which produce similar effects with intercalating drugs, but binding data are not yet available and it may be that actual binding ratios are a good deal lower than the ratios of added IDA to nucleotides plotted in Fig. 3e. In any event, this result with IDA makes it seem highly probable that the dip in S_{20} of closed circular DNA, thus far seen only with intercalating drugs, can in fact arise in consequence of quite different modes of binding to DNA. It is not yet possible to correlate this apparent uncoiling effect of IDA with the primary or the secondary mode of interaction described by MAHLER et al. (1968), but in principle, such apparent uncoiling could be accounted for on the basis of either of the models proposed by these authors — conversion of DNA from a B-helical form to an A-type conformation, or local denaturation caused by binding of steroid micelles to the outside of the helix (WARING, 1970).

It remains to return to the observation that when different intercalating drugs bind to ϕX 174 RF the dip in the S_{20} plot for the closed circles occurs at different binding ratios. The position of the minimum in these plots, termed the equivalence point by CRAWFORD and WARING [1967 (1)], represents in effect the end-point of a titration; at this point the accumulated uncoiling of the helix due to drug binding just balances the initial number of right-handed superhelical turns in the circles. If the unwinding angle ϕ associated with the binding of the drug is known, the binding ratio at the equivalence point v_c may be used to calculate τ_0, the initial number of superhelical turns in the absence of drug, using the relation

$$\tau_0 = -\frac{N\phi v_c}{2\pi} \qquad (1)$$

where N is the number of nucleotides per DNA molecule (10^4 for ϕX 174 RF). The mathematics of supercoiling in circular DNA is discussed in BAUER and VINOGRAD (1968). It is generally agreed that the unwinding angle associated with intercalation of ethidium is 12° (that is $\phi = \pi/15$ radians) as proposed in the model of FULLER and WARING (1964). Using this value for ϕ, and $v_c = 0.04 \pm 0.008$ from Fig. 2a, it may be calculated that ϕX 174 RF circles have -13.3 ± 2.7 superhelical turns in the absence of drugs.

Since this same DNA yields different values of v_c with different drugs (Fig. 2) it would appear that there are differences in the unwinding angle per bound drug molecule for the various drugs. In general, for two drugs which uncoil the helix by angles of ϕ_1 and ϕ_2 per bound drug molecule respectively, it can be seen from Eq. (1) that

$$v_{c_1}\phi_1 = v_{c_2}\phi_2 \quad (2)$$

where v_{c_1} and v_{c_2} are the respective equivalence points. The data of Fig. 2 allow estimation of ϕ values for the various drugs based on the observed equivalence point for ethidium $v_{c_e} = 0.040$, the observed equivalence points for the other drugs, and the assumption that the unwinding angle per bound ethidium molecule $\phi_e = 12°$. The values of ϕ calculated for the drugs in Fig. 2 are shown in the Table. For propidium and actinomycin the calculated unwinding angle does not differ from that of

Table. *Binding of intercalating drugs to ϕX 174 RF DNA*

Drug	Equivalence binding ratio v_c	Unwinding angle ϕ	Fraction intercalated α
Ethidium bromide	0.040 ± 0.008	(12°)	(1.00)
Proflavine	0.057 ± 0.012	8.4°	0.70
Hycanthone	0.071 ± 0.024	6.8°	0.56
Daunomycin	0.092 ± 0.017	5.2°	0.44
Nogalamycin	0.059 ± 0.007	8.1°	0.68
Propidium iodide	0.040 ± 0.008	12.0°	1.00
Actinomycin D	0.042 ± 0.007	11.4°	0.95

ethidium, but for proflavine and nogalamycin it is significantly lower. Hycanthone and daunomycin yield clearly lower values. Of course, the estimates of ϕ in the Table are wholly dependent upon the value of 12° assumed for ethidium, but there is independent evidence that this value must be about correct if only because of the agreement between estimates of the numbers of superhelical turns in various closed circular DNAs obtained using ethidium (assuming 12° uncoiling) and estimates obtained by other methods not dependent upon the use of drugs (WARING, 1969, 1970; WANG, 1969). There seems now to be little justification for basing the calculations on the unwinding angle of 36° proposed by LERMAN (1964) for proflavine, and it is clear from Figs. 2a and b that the true unwinding angles per bound ethidium or proflavine molecule cannot differ by a factor of three. However, the ϕ values collected in the Table bring to light a paradox: where the calculated unwinding angles differ from that of ethidium they are lower, whereas the 12° for ethidium is supposed to represent the *minimum* uncoiling needed to accommodate a planar aromatic ring system between the stacked base-pairs (FULLER and WARING, 1964; for discussion see WARING, 1970). The simplest solution to this discrepancy is to postulate that not all the *bound* drug molecules may be in an intercalated state at the same time; for example an equilibrium between outside-bound and intercalated states may exist, with only a certain fraction (α) of the molecules actually intercalated at any given instant. This suggestion is not the only possible explanation (WARING, 1970), but it is not new (NEVILLE and DAVIES, 1966; LI and CROTHERS, 1969). If we assume that no uncoiling is associated with binding in the non-intercalated state we may re-write Eq. (2) in the form

$$\alpha_i \, v_{c_i} \, \phi_i' = \alpha_e \, v_{c_e} \phi_e' \tag{3}$$

where, for drug i and ethidium indicated by subscripts i and e respectively, α is the fraction of drug molecules in the intercalated state at the observed equivalence binding ration v_c, and ϕ' is the unwinding angle per *intercalated* drug molecule. Dr. D. M. CROTHERS (personal communication) finds that the kinetics of ethidium-DNA interaction suggest the existence of only a single bound form of the drug; attributing this to the intercalated state sets $\alpha_e = 1.00$ and $\phi_e = \phi_e' = 12°$. A further assumption, that the unwinding angle required to accommodate any aromatic chromophore is constant, i.e. that ϕ' for any other intercalated ring system is the same as that for ethidium ϕ_e', simplifies Eq. (3) to

$$\alpha_i \, v_{c_i} = v_{c_e} \, . \tag{4}$$

Values of α for the drugs studied in Fig. 2 are included in the Table. According to this reasoning, virtually all bound propidium and actinomycin molecules are intercalated, whereas in the series proflavine, nogalamycin, hycanthone and daunomycin an increasing proportion of the bound molecules are in a non-intercalated state or states. It is well to emphasise that these conclusions involve the assumptions stated above, and in addition to point to their dependence upon the reliability of the estimates of binding ratios at equivalence. In the experiments of Fig. 2 these binding ratios were determined independently of the ultracentrifugation experiments, using a spectrophotometric method. This method is potentially inapplicable in cases where more than a single bound form of the drug exists, i.e. in just the situation which is being postulated here to explain the results. Although there are no specific indications that the determinations of binding ratios were severely in error (WARING, 1970), it would be reassuring to have independent confirmation of the conclusions. For proflavine such confirmation is available in the results of LI and CROTHERS (1969), whose kinetic studies revealed that at an ionic strength comparable to that used here about 30% of DNA-bound proflavine molecules appeared to exist in an outside-bound form (cf. α for proflavine in the Table). It would clearly be of interest to pursue the suggestion that the fraction of bound molecules intercalated may vary from drug to drug, preferably using homologous series of compounds where only a single substituent grouping is modified at a time. Studies with ethidium analogues of this type are in progress, using both kinetic analysis and the assay of changes in supercoiling of closed circular DNA reported here.

Acknowledgements

This work was aided by a Government grant from the Royal Society. Funds for travel to the Symposium were generously provided by the Wellcome Trust. I am indebted to many colleagues for helpful advice and criticism, and to the scientists and drug companies who willingly supplied samples of drugs.

References

Bauer, W., Vinograd, J.: The interaction of closed circular DNA with intercalative dyes. I. The superhelix density of SV40 DNA in the presence and absence of dye. J. molec. Biol. **33**, 141 (1968).

Behr, W., Honikel, K., Hartmann, G.: Interaction of the RNA polymerase inhibitor chromomycin with DNA. Europ. J. Biochem. **9**, 82 (1969).

Brenner, S., Barnett, L., Crick, F. H. C., Orgel, A.: The theory of mutagenesis. J. molec. Biol. **3**, 121 (1961).

Brock, T. D.: Streptomycin. Symp. Soc. Gen. Microbiol. **16**, 131 (1966).

Crawford, L. V., Waring, M. J.: (1) Supercoiling of polyoma virus DNA measured by its interaction with ethidium bromide. J. molec. Biol. **25**, 23 (1967).

— — (2) The supercoiling of papilloma virus DNA. J. gen. Virol. **1**, 387 (1967).

Fuller, W.: The molecular structures of the nucleic acids and their biological significance. Chapter in genetic elements (Symp. Federation of European Biochemical Societies), p. 17 (Shugar, D., Ed.). London: Academic Press 1966.

— Waring, M. J.: A molecular model for the interaction of ethidium bromide with DNA. Ber. Bunsenges. Physik. Chem. **68**, 805 (1964).

Hamilton, L. D., Fuller, W., Reich, E.: *X*-ray diffraction and molecular model building studies of the interaction of actinomycin with nucleic acids. Nature (Lond.) **198**, 538 (1963).

Hirschberg, E., Weinstein, I. B., Gersten, N., Marner, E., Finkelstein, T., Carchman, R.: Structure-activity studies on the mechanism of action of Miracil D. Cancer Res. **28**, 601 (1968).

Hirschman, S. Z., Leng, M., Felsenfeld, G.: Interaction of spermine and DNA. Biopolymers **5**, 227 (1967).

Hudson, B., Upholt, W. B., Devinny, J., Vinograd, J.: The use of an ethidium analogue in the dye-buoyant density procedure for the isolation of closed circular DNA: the variation of the superhelix density of mitochondrial DNA. Proc. nat. Acad. Sci. (Wash.) **62**, 813 (1969).

Kersten, W., Kersten, H., Szybalski, W.: Physicochemical properties of complexes between DNA and antibiotics which affect RNA synthesis (actinomycin, daunomycin, cinerubin, nogalamycin, chromomycin, mithramycin, and olivomycin). Biochemistry **5**, 236 (1966).

Lerman, L. S.: Structural considerations in the interaction of DNA and acridines. J. molec. Biol. **3**, 18 (1961).

— The structure of the DNA-acridine complex. Proc. nat. Acad. Sci. (Wash.) **49**, 94 (1963).

— Acridine mutagens and DNA structure. J. cell. comp. Physiol. **64**, Suppl. 1, 1 (1964).

Li, H. J., Crothers, D. M.: Relaxation studies of the proflavine-DNA complex: the kinetics of an intercalation reaction. J. molec. Biol. **39**, 461 (1969).

Mahler, H. R., Green, G., Goutarel, R., Khuong-Huu, Q.: Nucleic acid-small molecule interactions. VII. Further characterisation of DNA-diamino steroid complexes. Biochemistry **7**, 1568 (1968).

Müller, W., Crothers, D. M.: Studies of the binding of actinomycin and related compounds to DNA. J. molec. Biol. **35**, 251 (1968).

Neville, D. M., Jr., Davies, D. R.: The interaction of acridine dyes with DNA: an *X*-ray diffraction and optical investigation. J. molec. Biol. **17**, 57 (1966).

Newton, B. A.: Interaction of berenil with DNA and some characteristics of the berenil-DNA complex. Biochem. J. **105**, 50 P (1967).

O'Brien, R. L., Allison, J. L., Hahn, F. E.: Evidence for intercalation of chloroquine into DNA. Biochim. biophys. Acta (Amst.) **129**, 622 (1966).

Ohnishi, S., McConnell, H. M.: Interaction of the radical ion of chlorpromazine with DNA. J. Amer. chem. Soc. **87**, 2293 (1965).

Rosi, D., Peruzzotti, G., Dennis, E. W., Berberian, D. A., Freele, H., Tullar, B. F., Archer, S.: Hycanthone, a new active metabolite of lucanthone. J. med. Chem. **10**, 867 (1967).

SARTORELLI, A. C., IANNOTTI, A. T., BOOTH, B. A., SCHNEIDER, F. H., BERTINO, J. R., JOHNS, D. G.: Complex formation with DNA and inhibition of nucleic acid synthesis by methylglyoxal-*bis*-(guanylhydrazone). Biochim. biophys. Acta (Amst.) **103**, 174 (1965).
STREISINGER, G., OKADA, Y., EMRICH, J., NEWTON, J., TSUGITA, A., TERZAGHI, E., INOUYE, M.: Frameshift mutations and the genetic code. Cold Spr. Harb. Symp. quant. Biol. **31**, 77 (1966).
SUWALSKY, M., TRAUB, W., SHMUELI, U., SUBIRANA, J. A.: An *X*-ray study of the interaction of DNA with spermine. J. molec. Biol. **42**, 363 (1969).
WAGNER, T. E.: *In vitro* interaction of LSD with purified calf thymus DNA. Nature (Lond.) **222**, 1170 (1969).
WANG, J. C.: Variation of the average rotation angle of the DNA helix and the superhelical turns of covalently closed cyclic λ DNA. J. molec. Biol. **43**, 25 (1969).
WARD, D. C., REICH, E., GOLDBERG, I. H.: Base specificity in the interaction of polynucleotides with antibiotic drugs. Science **149**, 1259 (1965).
WARING, M. J.: Complex formation between ethidium bromide and nucleic acids. J. molec. Biol. **13**, 269 (1965).
— Structural requirements for the binding of ethidium to nucleic acids. Biochim. biophys. Acta (Amst.) **114**, 234 (1966).
— Nucleic acids. Ann. Rep. Chem. Soc. for 1968, **65 B,** 551 (1969).
— Variation of the supercoils in closed circular DNA by binding of antibiotics and drugs: evidence for molecular models involving intercalation. J. molec. Biol. **54,** 247 (1970).

Mode of Action of Miracil D

I. BERNARD WEINSTEIN and ERICH HIRSCHBERG

I. Introduction

PAUL EHRLICH, the father of "Molecular Pharmacology", formulated a strategy of chemotherapy based on designing toxic agents containing binding sites that specifically complex with receptor sites present in the parasite or cell type which is to be attacked (EHRLICH, 1890). Only within the past few years, however, has it become apparent that for certain drugs, including agents closely related structurally to some of the dyestuffs tested by EHRLICH, the cellular receptor site is DNA. One of these compounds is the subject of this report.

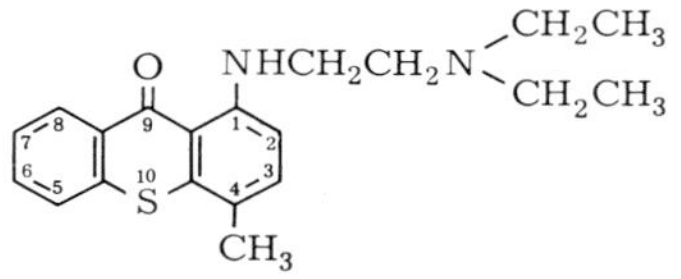

Fig. 1. Structural formula of Miracil D (1-diethylaminoethylamino-4-methyl-10-thiaxanthenone)

Miracil D (Lucanthone, Nilodin) is a 10-thiaxanthenone (Fig. 1). The three-ring heterocyclic system has a dialkylaminoalkylamino side chain which is similar to that of the antimalarials chloroquine and quinacrine. It was synthesized chemically in 1939 (MAUSS, 1948) and has been used for several decades as an effective drug in the treatment of schistosomiasis (NEWSOME, 1951; BLAIR, 1958; GOENNERT and KOELLING, 1962; STANDEN, 1963). In addition, Miracil D has antitumor activity for a variety of transplantable neoplasms in rodents (HACKMANN, GOENNERT and MAUSS, 1949; HIRSCHBERG, GELLHORN, MURRAY and ELSLAGER, 1959; BLANZ and FRENCH, 1963). Because of the biologic activity and clinical interest in this class of compounds, a few years ago we initiated studies to determine its mode of action. The fact that Miracil D shares certain structural features (namely a tricyclic-heterocyclic ring system) with proflavine, actinomycin D, and quinacrine, suggested to us that it might complex with nucleic acids and inhibit macromolecular synthesis. In the present paper we summarize evidence from our own laboratory, as well as that obtained by other investigators, indicating that this is the case. Certain structural and physiologic aspects of the interaction between Miracil D and nucleic acids will also be described.

II. Effects on Growth, Macromolecular Synthesis, and Enzyme Induction in Bacteria

Because of the complexity of schistosomicide and mammalian cell test systems, we first studied the drug in bacteria. We found that in the range of 10 to 20 μg/ml (3 to 6 × 10^{-5} M) it completely inhibited the growth of *B. subtilis* and *E. coli* (WEINSTEIN, CHERNOFF, FINKELSTEIN and HIRSCHBERG, 1965). When added at 6 × 10^{-5} M to an exponentially growing culture of *B. subtilis* it caused immediate and virtually complete inhibition of RNA synthesis, and a delayed and partial inhibition of protein synthesis (Fig. 2). With *E. coli* it also completely blocked RNA synthesis but in

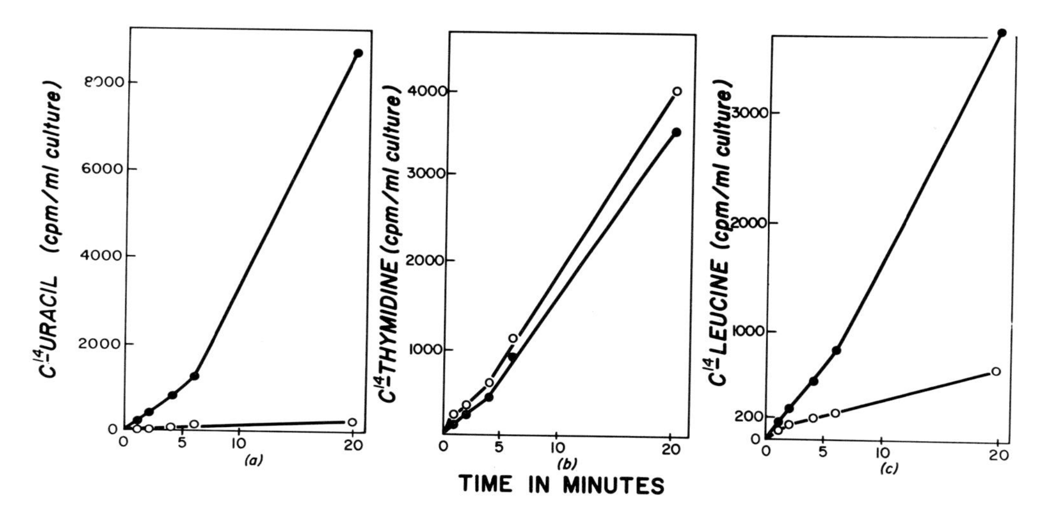

Fig. 2. Effect of Miracil D on RNA (a), DNA (b), and protein (c) synthesis; ●——●, control; ○——○, 6 × 10^{-5} M drug added at time zero to an exponentially growing culture of *B. subtilis*. For additional details see WEINSTEIN et al. (1965)

addition there was partial inhibition of DNA synthesis (WEINSTEIN, CARCHMAN, MARNER and HIRSCHBERG, 1967).

Since Miracil D inhibits RNA synthesis in intact *E. coli*, whereas Actinomycin D does not, it was of interest to examine its effect on the induction of β-galactosidase by methyl-β-D-thiogalactoside (TMG) in this bacterial system (WEINSTEIN et al., 1967). We found that when Miracil D 3 × 10^{-5} M was added together with the inducer there was only a slight inhibition of RNA synthesis and enzyme induction was not impaired. On the other hand, 6 × 10^{-5} M Miracil D caused greater than 85% inhibition of RNA synthesis and complete inhibition of β-galactosidase induction. Under these conditions, there was also partial inhibition of overall protein synthesis, though this in itself could not explain the complete inhibition of enzyme induction. The results provided evidence that β-galactosidase induction requires *de novo* RNA synthesis.

The fact that the dialkylaminoalkylamino side chain of Miracil D resembles aliphatic polyamines, as well as evidence that polyamines can antagonize the action of proflavine (KAY, 1959) and quinacrine (SILVERMAN and EVANS, 1944; MILLER and PETERS, 1945), led us to test the possibility that spermine might prevent the toxic effects of Miracil D in *E. coli*. We found that 8×10^{-3} M spermine completely protected a culture of *E. coli* against growth inhibition by 6×10^{-5} M Miracil D (WEINSTEIN et al., 1967). When instead of adding the two agents simultaneously, spermine was added either 1, 2, 5 or 10 min after Miracil D there was a progressive decrease in the ability of the polyamine to protect cells from growth inhibition by Miracil. Spermine also protected *E. coli* from the inhibitory effects of Miracil D on RNA synthesis and β-galactosidase induction. Studies described below indicate that these effects of spermine might be attributed to its ability to prevent the interaction between Miracil D and DNA. In addition, it appears that spermine also inhibits the cellular uptake of the drug (WEINSTEIN et al., 1967).

More recently, HAIDLE, BRINKLEY and MANDEL (1970) have confirmed the fact that when added to a mid-log phase culture of *Bacillus subtilis* Miracil D (6×10^{-5} M) caused immediate inhibition of growth. The turbidity and viability count began to decrease after 15 to 20 min, and the incorporation of ^{3}H-thymidine into DNA proceeded normally until that time. Electron micrographs of sectioned cells which had been grown for 5 to 15 min in the presence of Miracil D revealed striking condensation of the nucleoid and poorly defined or entirely absent mesosomes (Fig. 3). These results, as well as transformation experiments using donor DNA from cells grown in the presence of Miracil, led HAIDLE et al. to suggest that in *B. subtilis* the drug permits those DNA molecules in the process of replication to complete the process, but prevents the reinitiation of chromosome replication. The prevention of initiation of chromosome replication is ascribed by these authors to the aforementioned rapid inhibition by Miracil D of RNA and protein synthesis. Other interpretations have not been excluded.

III. Physical Interaction with DNA

The above *in vivo* results suggested that Miracil might act by complexing with DNA, thereby inhibiting RNA synthesis and enzyme induction. We therefore sought *in vitro* evidence for a physical interaction between the drug and DNA.

Effects of Nucleic Acids on the Absorption Spectrum of Miracil D. At neutral pH Miracil D has absorption maxima at 330 and 442 nm (WEINSTEIN et al., 1965). Both native and denatured DNA caused a shift in the previous absorption maxima to 337 and 447 nm. Several RNAs, both natural and synthetic, produced qualitatively similar changes in the absorption spectrum of Miracil D. The relative order of activity (in decreasing order) was DNA $>$ sRNA $>$ poly $A >$ poly $U \gg$ poly C (Fig. 4). The strong activity of poly A, particularly at low concentrations, suggests that the drug has a high affinity for adenylic acid residues or that it complexes particularly well with the ordered structure of poly A. HAIDLE and MANDEL (personal communication) have tested the effect of several bacterial DNAs, ranging in base composition from 27 to 69% $G+C$, on their ability to induce spectral shifts in Miracil D and found no correlation between activity and $G+C$ content. These results are reminiscent of those obtained with chloroquine (COHEN and YIELDING, 1965) and are in contrast to the dependence on $G+C$ content of DNA in the interaction of natural DNAs with Actinomycin D (REICH, 1964).

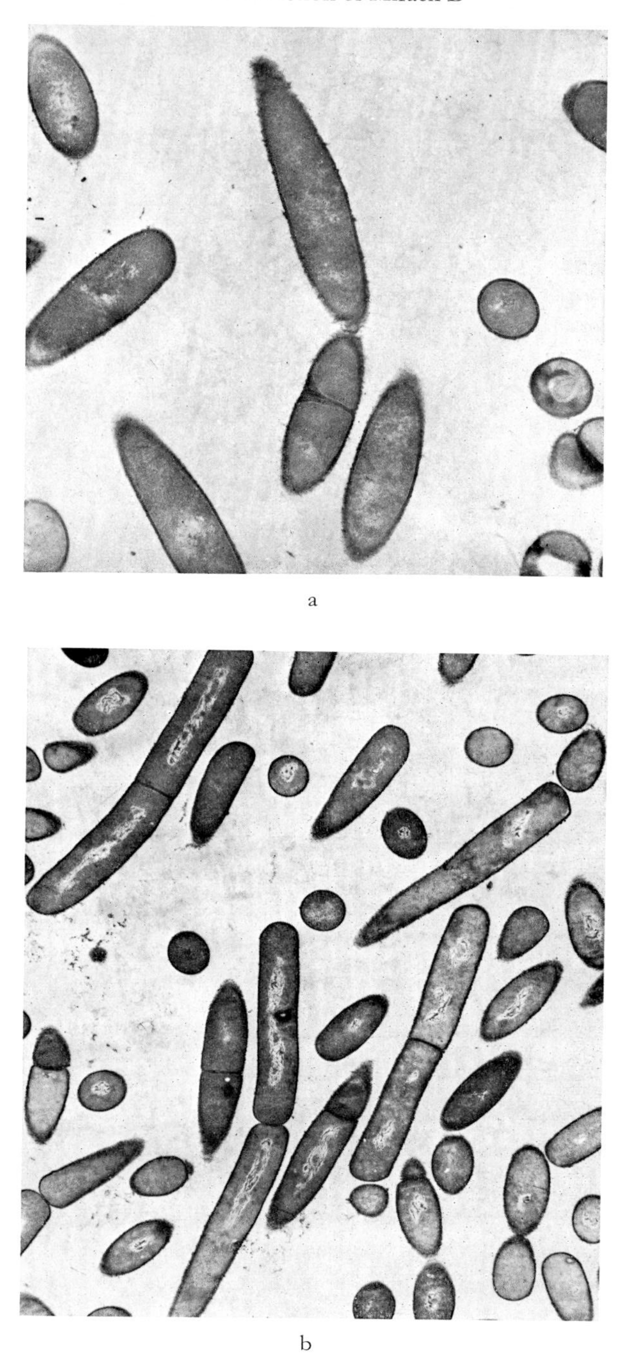

Fig. 3. a Longitudinal section of a normal *B. subtilis* cell Mag. × 62,500. b Longitudinal section of *B. subtilis* treated for 10 min with 6 × 10^{-5} M Miracil D. Mag. × 25,000. (Reprinted from HAIDLE et al., 1970)

When the 2′,3′-mononucleotides of adenine, guanine, cytosine, and uridine were tested in amounts equivalent (with respect to moles of nucleotide residues) to 2, 10, 25 and 50 μg of nucleic acid, they did not significantly affect the absorption of Miracil D in the range of 330 to 450 mμ. These results indicate the importance of the polynucleotide structure for effective interaction with Miracil D.

Effects on Heat Denaturation of DNA. Since the spectral studies provided evidence for a physical interaction between Miracil D and nucleic acids it was of interest to determine the effect of this interaction on the secondary structure of the nucleic acid. When assayed in low ionic strength buffer (3×10^{-3} M NaCl – 1×10^{-4} M Na_2EDTA

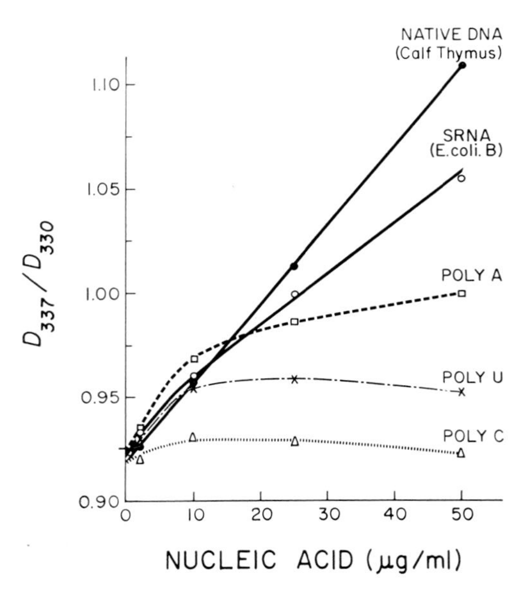

Fig. 4. Shift in absorption maximum of Miracil D 3×10^{-5} M induced by various nucleic acids. (For details see WEINSTEIN et al., 1965)

– 5×10^{-4} M sodium phosphate, pH 6.8), the heat denaturation profile of calf thymus DNA (25 μg/ml) revealed a T_m (50% of total hyperchromicity) of 56°. The presence of only 6×10^{-6} M Miracil caused a marked increase in heat stability with a shift in T_m to approximately 71° (Fig. 5). Miracil D at 1.2×10^{-5} M increased the T_m to 79°. The slope of the melting curve was considerably less steep in the presence of the drug. In the presence of Miracil D, the total hyperchromicity was also somewhat greater than in its absence. In part, this may simply reflect the contribution by the drug to the total absorption at 260 nm. When studied in standard saline citrate (0.15 M sodium chloride – 0.015 M sodium citrate, pH 7.0), Miracil D did not appreciably affect the T_m of calf thymus DNA. Additional evidence that the interaction of Miracil D and DNA is inhibited in media of high ionic strength is presented below.

Viscosity Studies. The interaction between Miracil D and DNA caused an appreciable increase in the relative viscosity of the DNA when measured in a buffer of low

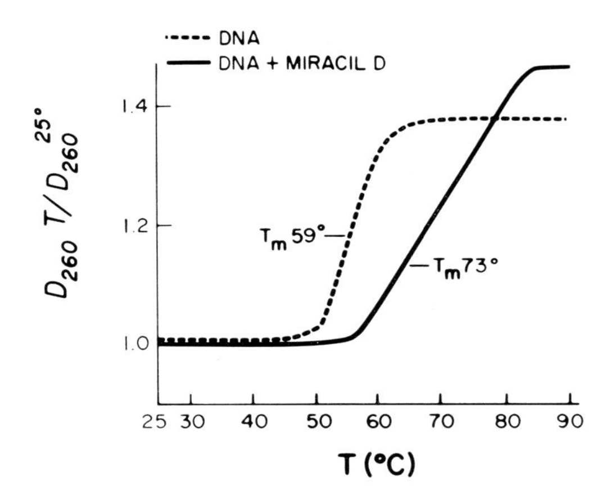

Fig. 5. Effect of Miracil D on heat denaturation of native calf thymus DNA. - - - - - - , DNA (25 μg/ml); ——— DNA + Miracil D (6×10^{-6} M). For additional details see WEINSTEIN et al. (1965)

Table. 1. *Effect of Miracil D and various cations on the relative viscosity of native DNA*

Cation Concentration (M)	Na^+	Mg^{2+}	Spermine
5×10^{-2}	1.00	—	—
1×10^{-2}	1.06	—	—
5×10^{-3}	1.19	1.00	—
1×10^{-3}	—	1.13	—
5×10^{-4}	1.43	1.20	—
1×10^{-4}	—	1.30	—
5×10^{-5}	1.41	1.50	1.00
2.4×10^{-5}	—	—	1.12
9×10^{-6}	—	—	1.25
1.7×10^{-6}	—	—	1.35
5×10^{-7}	—	—	1.43
None	1.49	1.49	1.43

Relative viscosity of calf thymus DNA in NaCl-EDTA phosphate buffer in the presence of 5×10^{-5} M Miracil D and the stated concentration of cation. For additional details see CARCHMAN et al. (1969).

ionic strength in a Zimm-Crothers low shear viscometer (CARCHMAN, HIRSCHBERG and WEINSTEIN, 1969). The drug also produced a 2 to 3 fold increase in intrinsic viscosity, but most of our data has been expressed as relative viscosity due to problems inherent in obtaining the intrinsic viscosity of DNA in low ionic strength buffers (CARCHMAN et al., 1969; LERMAN, 1961). The drug did not increase the viscosity of DNA when measured in standard saline citrate buffer, a finding consistent with our other evidence that interaction between Miracil D and DNA is abolished in media of high ionic strength. In view of these results, as well as of the ability of spermine to counteract the growth inhibitory effect of Miracil D for bacteria (WEINSTEIN et al., 1967), we also examined the effects of various cations on our viscosity measurements. Table 1 indicates that spermine (5×10^{-5} M), Mg^{2+} (5×10^{-3} M) or Na^{+} (5×10^{-2} M) completely prevented the increase in relative viscosity of DNA exerted by Miracil D (5×10^{-5} M). At these levels, the cations themselves have no effect on the viscosity of DNA. Comparable reversal was obtained when the cations were added to the DNA 10 min after the drug. There is an inverse relationship between the valence of the cation and the concentration required for reversal. The fact that spermine is 100 × more effective than Na^{+}, however, indicates that other factors, probably steric, are important.

IV. Effects on in Vitro RNA and Protein Synthesis

DNA-dependent RNA Polymerase. The inhibition of RNA synthesis in intact *E. coli*, coupled with the evidence of binding to DNA, suggested that the drug should

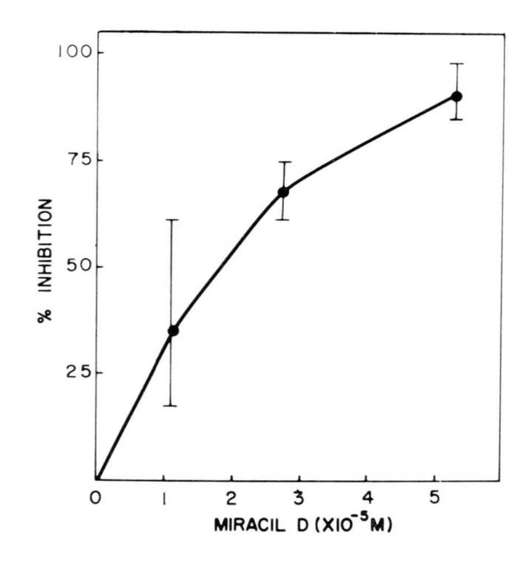

Fig. 6. Miracil D inhibition of the RNA polymerase reaction. Reaction systems contained per ml: 70 µg of salmon sperm DNA, 75 units of *E. coli* RNA polymerase, and Miracil D at the final concentrations indicated. The control system, i.e., in the absence of drug, incorporated 12.5 mµmol of (^{14}C) ATP. The data indicate the mean of three experiments. For additional details see WEINSTEIN et al. (1967)

effectively inhibit DNA transcription *in vitro*. Fig. 6 indicates that 3×10^{-5} M Miracil D produced greater than 50% inhibition of ^{14}C-ATP incorporation into RNA in a system containing purified *E. coli* B RNA polymerase and a limiting amount of DNA as template. Supplementation of the system with excess DNA reversed Miracil inhibition, whereas supplementation with excess enzyme did not, indicating that the target of drug action is DNA not enzyme. The presence of spermine (10^{-5} M) partially protected the system against inhibition by Miracil (WEINSTEIN et al., 1967). In recent experiments (WILSON, R. G. and CHURCH, J., personal communication), in a model RNA polymerase system from *Micrococcus lysodeikticus*, natural DNA was replaced as the template by poly *dAdT* or one of several ribo- and deoxyribohomopolymers. Miracil D inhibited the template activity of all polymers but in varying degrees. Those containing adenine residues were most readily affected, whereas the nature of the sugar moiety and the degree of strandedness made little difference.

Subcellular Protein Synthesis. In contrast to the pronounced effects on RNA synthesis, Miracil D (10^{-4} to 10^{-5} M) did not inhibit either endogenous or poly *U* directed phenylalanine incorporation in subcellular systems derived from *E. coli* or mammalian cells (WEINSTEIN et al., 1967). We have, as a matter of fact, seen a small but reproducible stimulation of both endogenous and poly *U* directed amino acid incorporation by Miracil D, as well as some of its analogs, in extracts of rat liver and rabbit reticulocytes (GERSTEN and WEINSTEIN, unpublished studies).

V. Effects in Other Biological Systems

The ability of Miracil D to control schistosomiasis in man and to inhibit transplanted rodent tumors has been mentioned. In a series of experiments with one of the latter, mouse leukemia L1210, *in vitro* incubation of tumor cells with Miracil D showed that the drug was an effective inhibitor of RNA synthesis. In contrast to our results with *B. subtilis*, however, concentrations of the drug required to inhibit RNA synthesis also caused an appreciable inhibition of ^{3}H-thymidine incorporation into DNA (HIRSCHBERG, WEINSTEIN and CARCHMAN, 1966). In further contrast to the findings in bacterial systems, the inhibition of RNA synthesis by the drug was not counteracted by spermine.

More recently, BASES and MENDEZ (1969) have analyzed in detail the effects of Miracil D on macromolecular synthesis in HeLa cells. They found that 1×10^{-5} M Miracil D inhibited RNA and DNA synthesis to 21 and 27% of control values respectively. During short exposures to the drug there was no apparent inhibition of protein synthesis or loss of cell viability, and the inhibitory effects were readily reversible when the cells were transferred to fresh media not containing the drug. BASES and MENDEZ have presented evidence that in HeLa cells the drug primarily inhibits the *de novo* synthesis of ribosomal RNA and tends to spare the synthesis of DNA-like or "messenger" RNA. The selective inhibition of ribosomal RNA resembles that seen with Actinomycin D; both drugs also appear to inhibit the maturation of RNA tumor viruses (BASES and MENDEZ, 1969). Bases has also found that Miracil D potentiates the lethal effects of radiation on HeLa cells, apparently by inhibiting post-radiation repair mechanisms (BASES, 1970). The latter effect may be related to the potentiation of radiation or alkylating agent toxicity described with chloroquine by YIELDING (this Symposium). OBE (1969) found that Miracil D induces achromatic lesions and chromatic breaks in human leukocyte chromosomes *in vitro* and LUERS (1955) described a mutagenic action of the drug in Drosophila.

Further comparative studies of Miracil D and Actinomycin D have been carried out in the cellular slime mold, *Dictyostelium discoideum* (HIRSCHBERG, CECCARINI, OSNOS and CARCHMAN, 1968). The concentrations of either drug required to block proliferation of the unicellular amoebae were significantly lower than those inhibiting aggregation to a multicellular organism. At these biologically effective levels only a small portion of total RNA synthesis in cells harvested at either stage was inhibited. In all respects, Miracil D was qualitatively similar but considerably less active than Actinomycin D.

VI. Structure Activity Studies

An understanding, at the molecular level, of the nature of the complex formed between DNA and the drug requires information on the structural features of Miracil D which are necessary for this interaction. We examined, therefore, a series of approximately 40 different 10-thiaxanthenones, as well as several more distantly related compounds, for their effects on the growth of intact *B. subtilis* and the heat-denaturation profile of native DNA. A smaller number of these same compounds were studied for their effects on the subcellular DNA-directed RNA polymerase from *E. coli*, and on the viscosity of native DNA. In general, there was a good correlation of the degree of activity of a given compound in the various *in vitro* test systems. The correlation between *in vitro* activity and growth inhibition of *B. subtilis* was less good but still apparent in the sense that compounds which had low activity in the *in vitro* systems were also poor growth inhibitors. For a complete description of these results the reader is referred to previously published studies (Hirschberg, Weinstein, Gersten, Marner, Finkelstein and Carchman, 1968; Carchman et al., 1969) but the salient features are summarized below.

Among the derivatives differing only in the dialkylaminoalkylamino side chain (Fig. 1) we found that: (1) replacement of the hydrogen on the proximal nitrogen (i.e. that attached to the ring system) by an alkyl group led to a marked decrease in activity; (2) absence of the terminal nitrogen also led to a significant decrease in activity; (3) the length of the carbon chain between the two nitrogens could be increased from 2 to 4 carbons with little effect on activity; and (4) the two ethyl residues on the terminal nitrogen could be replaced by H, -OH, or other alkyl residues without a consistent effect on activity.

With respect to modifications of the 10-thiaxenthenone ring system, we found that total absence of the ring system led to a loss of activity; the side chain alone, diethylaminoethylamine, did increase the T_m of DNA, but this is a well known effect for aliphatic polyamines (Mandel, 1962). Replacement of the three-ring system with a single benzene ring also led to a loss of activity in our test systems. Replacement of the 4-methyl group in the parent compound by -H, $-OCH_2CH_3$ or $-CH_2OH$ indicated that this residue enhances but is not absolutely essential for activity. Replacement of the ring sulfur with $C=O$ decreased but did not abolish activity; on the other hand sulfoxides, which are known *in vivo* metabolites (see below), were totally inactive.

A number of 6-chloro-Miracil derivatives were also studied. The findings were generally similar to those described above for the non-chlorinated derivatives, with the puzzling exception that in the 6-chloro-Miracils alkyl substitution on the proximal nitrogen of the side chain did not lead to a marked decrease in activity.

An additional series of Miracil D analogs have been studied with regard to inhibition of RNA polymerase *in vitro* and tumor growth *in vivo* (Hebborn, Bardos, Chmielewicz and Triggle, 1968). Activity was most pronounced when one of the ethyl groups on the terminal nitrogen was replaced by $-CH_2CH_2OH$ or $-CH_2CH_2Cl$.

Structure-activity relationships for inhibition of transplanted mouse tumors have also been examined for a large number of Miracil D derivatives (Blanz and French, 1963; Hirschberg et al., 1959). These results obtained in the intact animal have supported the conclusion (Blanz and French, 1963) that carcinostatic activity requires the presence of the intact 10-thiaxanthenone ring, bearing a compact substituent in the 4-position and a diamino side chain with an unsubstituted proximal nitrogen. Our results indicate that these features favor complexing with DNA and antibacterial activity as well.

VII. Search for the Schistosomicidal and Carcinostatic Metabolites of Miracil D

Independent lines of evidence indicated that conversion of Miracil D to one or more biologically active metabolites was required for both schistosomicidal activity

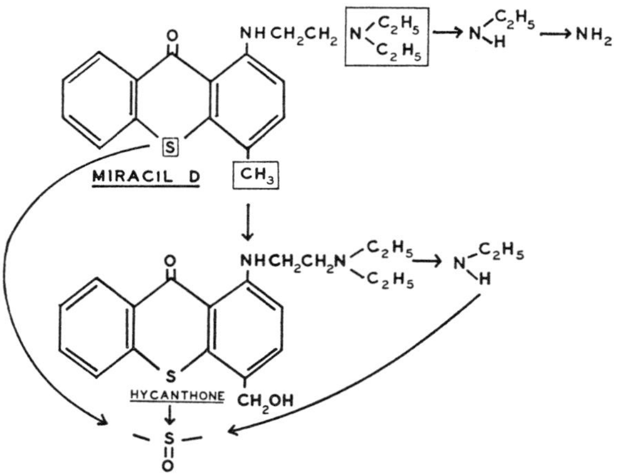

Fig. 7. Demonstrated metabolic pathways for Miracil D in mammalian systems (see text)

(Berberian, Freele, Rosi, Dennis, Archer, 1967; Goennert, 1961; Goennert and Koelling, 1962; Strufe, 1963) and for inhibitory activity against mouse leukemia L1210 (Hirschberg, Brindle and Semente, 1964). In mammalian organisms, there are three major metabolic pathways for Miracil D (Fig. 7): hydroxylation of the 4-methyl group, desethylation of the terminal side chain nitrogen, and sulfoxidation. The first of these appears crucial for the *in vivo* activation of Miracil D as a schistosomicide (Rosi, Peruzzotti, Dennis, Berberian, Freele, Tullar and Archer, 1967) since the 4-hydroxymethyl derivative, Hycanthone, is significantly more active than the parent compound.

The activities of a number of demonstrated and putative metabolites in the various test systems available to us were compared (Hirschberg, Weinstein, Carchman and Archer, 1968). The products of the first two pathways, i.e. Hycanthone (Rosi et al., 1967) and the monodesethyl and bisdesethyl derivatives of Miracil D, exhibit activity equivalent to, or greater than, those of the parent compound against growth of leukemia L1210 *in vivo*, RNA synthesis by L1210 cells *in vitro*, and growth of *B. subtilis*; they also complex with DNA *in vitro* as shown by increases in the T_m and relative viscosity of calf thymus DNA. Sulfoxide derivatives did not inhibit the growth of L1210 and were also inactive in our other test systems, thus excluding sulfoxidation as a pathway leading to the carcinostatic derivative of Miracil D. At the present time the actual structure of the proximate carcinostatic metabolite is not known. It is evident, however, that the metabolic conversion(s) favoring antitumor activity and schistosomicidal activity are not the same, since, among other things, the great increase in effectiveness of Hycanthone over Miracil D in the latter system is not found in the former.

It remains to be determined to what extent the structural requirements for antitumor activity of the various *in vivo* derivatives relate to their physiologic disposition and selective toxicity rather than to the primary mechanism of action. The same reservation applies to an even greater extent to an evaluation of the structure-activity relationships in tests for schistosomicidal potency, since these have revealed pronounced variations in biological activity and metabolic transformations by different assay procedures, in different animal hosts, and against the several species of flukes.

VIII. Discussion

Although our data do not define the precise structure of the Miracil D-DNA complex, taken together our results suggest the following hypothetical structure. It seems likely that the three-ring system of the drug intercalates between adjacent base pairs in DNA, in a manner analogous to the intercalation of the aminoacridines with DNA, originally described by Lerman [1961, 1963, 1964 (1)]. The following evidence is consistent with intercalation: (1) Both drugs produce a similar increase in the viscosity of DNA; (2) flow dichroism studies (kindly performed by Dr. Fred Hahn) suggest that the Miracil D chromophore is oriented perpendicular to the long axis of the DNA, i.e. is coplanar to the bases, as is the case with aminoacridines [Lerman, 1964 (1)]; (3) Hycanthone, a Miracil D derivative, causes uncoiling of supercoiled circular DNA, as do other intercalating drugs (see paper by M. Waring, this Symposium); (4) the interactions of Miracil D or aminoacridines with DNA are prevented in media of high ionic strength and are particularly inhibited by polycations, probably because under these conditions the DNA helix is held in a more rigid configuration [Liquori, Costantino, Crescenzi, Elia, Giglio, Puliti, De Santis Savino, Vitagliano, 1967; Lerman, 1964 (2)] which impairs the relaxation of structure required for intercalation (ionic forces may also play a role in the interaction of the Miracil D side chain with residues in DNA as discussed below). Other drugs which can apparently intercalate with DNA include quinacrine (Lerman, 1963), chloroquine (O'Brien, Allison and Hahn, 1966), 4-nitroquinoline (Nagata, Kodama, Tagashira and Imamura, 1966), certain polycyclic hydrocarbons [Lerman, 1964 (2)], the alkaloid berberine (Krey and Hahn, 1969), and ethidium bromide (see papers by M. Waring and W. Bauer, this Symposium).

In addition to the 10-thiaxanthenone ring system, it is apparent that the diamino side chain of Miracil D plays an important role in its interaction with DNA. We propose that the terminal nitrogen extends outward to the periphery of the DNA helix to interact with the DNA backbone (probably ionically with phosphate residues), thereby stabilizing the intercalated drug — DNA complex. This interpretation is consistent with the fact that the terminal nitrogen is essential for activity yet the nature of the substituents on it can vary considerably without abolishing activity. We think it is unlikely that the proximal nitrogen of the side chain interacts with phosphate residues, since if the ring system is completely intercalated this nitrogen would not be in the vicinity of the phosphate backbone. Nor is it likely that the diamino side chain forms interstrand bridges between two phosphate groups on complementary strands, as has been suggested by O'Brien, Olenick and Hahn (1966) for the quinacrine-DNA complex, since in Miracil D the two amino groups of the side chain are separated by only 4 to 5 Å and therefore could not bridge this interstrand distance. A striking feature of our analog studies is the fact that substitution of the hydrogen on the proximal nitrogen of the side chain with a CH_3 group markedly inhibits activity. Zilversmit (1970) has extensively studied this aspect and has suggested that this methyl substitution results in steric hindrance with the oxygen atom at C_9, necessitating a rotation of the side chain about the C_1 to proximal nitrogen bond, associated with reduced conjugation across the latter bond and a change in electron density on this nitrogen. These changes in conformation of the side chain with respect to the ring system (loss of coplanarity) and/or the altered reactivity of the proximal nitrogen, might hinder optimal "fit" of the drug between base pairs during intercalation. We have also considered the possibility that a hydrogen on the proximal nitrogen is essential for activity because it (and perhaps the quinoidal oxygen) might normally be involved in hydrogen bonding with base residues in the DNA (Hirschberg et al., 1968). This seems less likely since with complete intercalation of the drug it is not apparent to which residues in DNA this H-bonding might occur. Alternative mechanisms of interaction have not been excluded and further studies are required to define more precisely the molecular details of the Miracil D-DNA complex. At the present time we are at a loss to explain why 6-chloro Miracils are active even when they bear an alkyl residue on the proximal nitrogen, unless we assume that the 6-chloro Miracils form a much different type of complex with DNA.

Finally, it is of interest to compare the activities and specificities of Miracil D to that of other drugs known to complex with nucleic acids. Miracil D resembles Actinomycin D (Reich, 1964) in the sense that both are three-ring heterocyclic compounds which complex with DNA and thereby inhibit its template function in RNA synthesis. Miracil D, however, is only about one-tenth as potent as Actinomycin D, it produces variable inhibition of DNA replication, and also binds appreciably to RNA. In addition, the binding of Actinomycin D to nucleic acids, in contrast to Miracil D, has an absolute requirement for *G* residues. At the same time, Miracil D is more specific in its functional effects on nucleic acids than proflavine. The binding of proflavine to DNA results in inhibition of both DNA replication and transcription. In addition, the binding of proflavine to RNA inhibits its functions during protein synthesis (Weinstein and Finkelstein, 1967) whereas this is not the case with Miracil (Weinstein et al., 1967). Despite the structural similarities between Miracil D, chloroquine and quinacrine, it is of interest that the binding of chloroquine or quinacrine to DNA exerts a stronger inhibition on DNA replication rather than transcription (Cohen and Yielding, 1965; O'Brien et al., 1966), whereas Miracil has a stronger effect on transcription. These effects must reflect subtle and as yet inapparent differences in the structures of the DNA-drug complexes, as well as differences in the functional requirements of the DNA template in terms of transcription and replication, and the RNAs during translation.

The remarkable specificity of the above agents should encourage the search for a drug which would bind specifically to double stranded RNA, or to a replicative intermediate of viral RNA, and not to DNA. Such an agent would in a sense be the converse of Actinomycin D and it might find clinical application in the inhibition of replication of RNA viruses.

Acknowledgements

The authors wish to acknowledge the valuable contributions to these studies made by Miss R. CHERNOFF, Mr. R. CARCHMAN, Mr. T. FINKELSTEIN and Miss N. GERSTEN. We also wish to thank Drs. S. ARCHER, R. BASES, R. ZILVERSMIT, M. MANDEL, F. HAHN and R. WILSON for sharing with us results of their studies prior to publication as well as providing stimulating discussions on several aspects of the present topic.

This research was supported by U.S. Public Health Service Research Grant No. CA-02332 from the National Cancer Institute and the Alma Toorock Memorial for Cancer Research.

References

BASES, R.: Enhancement of x-ray damage in Hela cells by exposure to Lucanthone (Miracil D) following radiation. Cancer Res. **30**, 2007 (1970).

BASES, R., MENDEZ, F.: Reversible inhibition of ribosomal RNA synthesis in Hela by Lucanthone (Miracil D) with continued synthesis of DNA-like RNA. J. cell. Physiol. **74**, 283 (1969).

BERBERIAN, D. A., FREELE, H., ROSI, D., DENNIS, E. W., ARCHER, S.: Schistosomicidal activity of lucanthone hydrochloride, hycanthone and their metabolites in mice and hamsters. J. Parasit. **53**, 306 (1967).

BLAIR, D. M.: Lucanthone hydrochloride. A review. Bull. Wld Hlth Org. **18**, 989 (1958).

BLANZ, E. J., JR., FRENCH, F. A.: A systematic investigation of thiaxanthen-9-ones and analogs as potential antitumor agents. J. med. Chem. **6**, 185 (1963).

CARCHMAN, R. A., HIRSCHBERG, E., WEINSTEIN, I. B.: Miracil D: Effect on the viscosity of DNA. Biochim. biophys. Acta (Amst.) **179**, 158 (1969).

COHEN, S. N., YIELDING, K. L.: Inhibition of DNA and RNA polymerase reactions by chloroquine. Proc. nat. Acad. Sci. (Wash.) **54**, 521 (1965).

EHRLICH, P.: See: The collected papers: Including a complete bibliography, 4 vol. (HIMMELWEIT, F., Ed.). New York: Pergamon 1956—1957.

GOENNERT, R.: The structure-activity relationship in several schistosomicidal compounds. Bull. Wld Hlth Org. **25**, 702 (1961).

— KOELLING, H.: Thioxanthones and related compounds in experimental schistosomiasis. In: Drugs, parasites, and hosts, pp. 29—42 (GOODWIN, L. G., NIMMO-SMITH, R. H., Eds.). London: J. and A. Churchill Ltd. 1962.

HACKMANN, C., GOENNERT, R., MAUSS, H.: Untersuchungen über die tumorhemmende Wirkung des Miracils. Naturwissenschaften **36**, 29 (1949).

HAIDLE, C. W., BRINKLEY, B. R., MANDEL, M.: Effect of Miracil D on marker frequency ratio and cytotoxicity in *Bacillus subtilis*. J. Bacteriol. **102**, 835 (1970).

HEBBORN, P., BARDOS, T. J., CHIMIELEWICZ, Z. F., TRIGGLE, D. J.: Miracil D analogs: DNA-dependent RNA polymerase inhibitory activity and *in vivo* biological activity. Proc. Amer. Ass. Cancer Res. **9**, 29 (1968).

HIRSCHBERG, E., BRINDLE, S. D., SEMENTE, G.: Development and properties of mouse leukemia L1210 resistant to Miracil D. Cancer Res. **24**, 1733 (1964).

— Gellhorn, A., Murray, M. R., Elslager, E. F.: Effects of Miracil D, amodiaquin, and a series of other 10-thiaxanthenones and 4-aminoquinolines against a variety of experimental tumors *in vitro* and *in vivo*. J. nat. Cancer Inst. **22**, 567 (1959).
— Weinstein, I. B., Carchman, R.: Miracil D. An inhibitor of deoxyribonucleic and ribonucleic acid synthesis in mouse leukemia L1210 ascites. Proc. Ninth Intern. Cancer Cong. Tokyo, p. 347 (1966).
— — — Archer, S.: Search for the carcinostatic metabolite of Miracil D. Proc. Amer. Ass. Cancer Res. **9**, 30 (1968).
— — Gersten, N., Marner, E., Finkelstein, T., Carchman, R.: Structure-activity studies on the mechanism of action of Miracil D. Cancer Res. **28**, 601 (1968).
— Ceccarini, C., Osnos, M., Carchman, R.: Effects of inhibitors of nucleic acid and protein synthesis on growth and aggregation of the cellular slime mold *Dictyostelium discoideum*. Proc. nat. Acad. Sci. (Wash.) **61**, 316 (1968).
Kay, D.: The inhibition of bacteriophage multiplication by proflavin and its reversal by certain polyamines. Biochem. J. **73**, 149 (1959).
Krey, A., Hahn, F. E.: Berberine: Complex with DNA. Science **166**, 755 (1969).
Lerman, L. S.: Structural considerations in the interaction of DNA and acridines. J. molec. Biol. **3**, 18 (1961).
— The structure of the DNA-acridine complex. Proc. nat. Acad. Sci. (Wash.) **49**, 94 (1963).
— (1) Acridine mutagens and DNA structure. J. cell. comp. Physiol. (Suppl. 1) **64**, 1 (1964).
— (2) The combination of DNA with polycyclic aromatic hydrocarbons. In: Proceedings Fifth National Cancer Conference, pp. 39—48. Philadelphia: J. B. Lippincott Co. 1964.
Liquori, A. M., Costantino, L., Crescenzi, V., Elia, V., Giglio, E., Puliti, R., de Santis Savino, M., Vitagliano, V.: Complexes between DNA and polyamines: a molecular model. J. molec. Biol. **24**, 113 (1967).
Luers, H.: Biologisch-genetische Untersuchungen über die Wirkung eines Thioxanthonderivates an *Drosophila melanogaster*. Z. Vererbungsl. **87**, 93 (1955).
Mandel, M.: The interaction of spermine and native deoxyribonucleic acid. J. molec. Biol. **5**, 435 (1962).
Mauss, H.: Über basisch substituierte Xanthon- und Thioxanthon-Abkömmlinge; Miracil, ein neues Chemotherapeuticum. Chem. Ber. **81**, 19 (1948).
Miller, A. K., Peters, L.: Antagonism by spermine and spermidine of antibacterial action of quinacrine and other drugs. Arch. Biochem. **6**, 281 (1945).
Nagata, C., Kodama, M., Tagashira, Y., Imamura, A.: Interaction of polynuclear aromatic hydrocarbons, 4-nitroquinoline 1-oxides, and various dyes with DNA. Biopolymers **4**, 409 (1966).
Newsome, J.: Recent investigations into treatment of schistosomiasis by Miracil D in Egypt. Trans. roy. Soc. trop. Med. Hyg. **44**, 611 (1951).
Obe, G.: Zur Wirkung von Miracil auf menschliche Leukozytenchromosomen *in vitro*. Molec. Gen. Genetics **103**, 326 (1969).
O'Brien, R. L., Olenick, J. G., Hahn, F. E.: Reactions of quinine, chloroquine, and quinacrine with DNA and their effects on the DNA and RNA polymerase reactions. Proc. nat. Acad. Sci. (Wash.) **55**, 1511 (1966).
— Allison, J. L., Hahn, F. E.: Evidence for intercalation of chloroquine into DNA. Biochim. biophys. Acta (Amst.) **129**, 622 (1966).
Reich, E.: Actinomycin: Correlation of structure and function of its complexes with purines and DNA. Science **143**, 684 (1964).
Rosi, D., Peruzzotti, G., Dennis, E. W., Berberian, D. A., Freele, H., Tullar, B. F., Archer, S.: Hycanthone, a new active metabolite of lucanthone. J. med. Chem. **10**, 857 (1967).
Silverman, M., Evans, E. A., Jr.: Effect of spermine, spermidine, and other polyamines on growth inhibition of *Escherichia coli* by atabrine. J. biol. Chem. **154**, 521 (1944).
Standen, O. D.: Chemotherapy of helminthic infections. In: Experimental chemotherapy, vol. 1, pp. 701—892 (Schnitzer, R. J., Hawking, F., Eds.). New York: Academic Press 1963.
Strufe, R.: Stoffwechsel-Untersuchungen mit Miracil D. Med. Chem. Abhandl. Med.-Chem. Forschungsstätten Farbwerke Hoechst A.G. **7**, 337 (1963).

WEINSTEIN, I. B., CHERNOFF, R., FINKELSTEIN, I., HIRSCHBERG, E.: Miracil D: An inhibitor of ribonucleic acid synthesis in *Bacillus subtilis*. Molec. Pharmacol. **1**, 297 (1965).
— CARCHMAN, R., MARNER, E., HIRSCHBERG, E.: Miracil D. Effects on nucleic acid synthesis, protein synthesis and enzyme induction in *Escherichia coli*. Biochim. biophys. Acta (Amst.) **142**, 440 (1967).
— FINKELSTEIN, I.: Proflavine inhibition of protein synthesis. J. biol. Chem. **242**, 3757 (1967).
ZILVERSMIT, R.: Thioxanthones: Structural differences between Lucanthone and its N-methyl derivative. Molec. Pharmacol. **6**, 172 (1970).

Molecular Biological Effects of Nitroakridin 3582 and Related Compounds

Alan David Wolfe

I. Introduction

Synthesis of 87 6-(3)-nitro-9-aminoacridines was originally carried out (Schnitzer and Silberstein, 1929) in the hope of obtaining effective trypanocides. These efforts were only moderately successful, but such acridines were later found (Albert, Rubbo, Goldacre, Davey and Stone, 1945; Albert, 1966) "to outstrip all other acridines in antibacterial activity". The molecular processes responsible for these intense antibacterial effects remained unexplored, however, until the present studies were undertaken to elucidate the mode of action of Nitroakridin 3582 (NA, see Fig. 1), and thereby to suggest a mechanism of action potentially common to all aminonitroacridines.

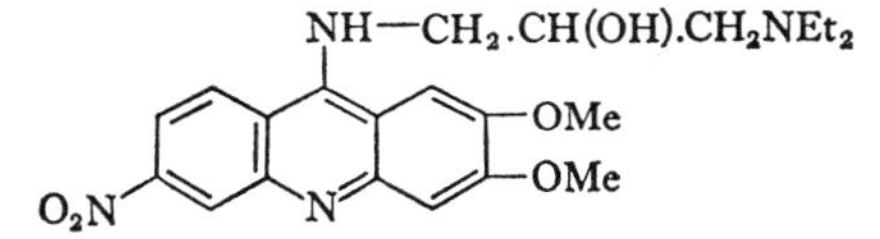

Fig. 1. Structure (Albert, 1966) of Nitroakridin 3582 (NA) (1-Diethylamino-3-[(2,3-dimethoxy-6-nitro-9-acridinyl)amino]propanol)

NA has been reported to be effective against *Rickettsiae* (Fussganger, 1945; Smadel, Snyder, Jackson, Fox and Hamilton, 1947) and *Streptococci* (Schnitzer, 1936), and to inhibit synthesis of the inducible enzyme β-galactosidase in *E. coli* (Hahn and Wisseman, 1951). Low concentrations of a related compound, Nitroakridin 3663, have been found (Denes and Polgar, 1960) to inhibit the assemblage of bacteriophage T2. The present studies have shown NA to complex strongly with deoxyribonucleic acid (DNA) and transfer ribonucleic acid (tRNA), and to inhibit cellular and cell free DNA, RNA and protein synthesis. NA and related compounds also labilized ribosomes to heat. Since DNA synthesis was more sensitive to NA than other processes, nitroacridines may be antibacterial primarily by virtue of their strong interaction with DNA.

II. Binding of Nitroakridin to DNA

Fig. 2 presents an adsorption isotherm (Scatchard, 1949) derived from spectrophotometric titrations of NA with DNA. The use of unusually low concentrations of polymer and ligand was necessitated by the property of NA to precipitate DNA. Results indicate that NA was bound to DNA by two processes, one of stronger and one of lesser intensity. The stronger process had an apparent association constant of 7×10^7 M^{-1}, and involved binding of 0.24 mol of NA per mole of nucleotide, or one molecule of NA for approximately four nucleotides. This result is analogous to

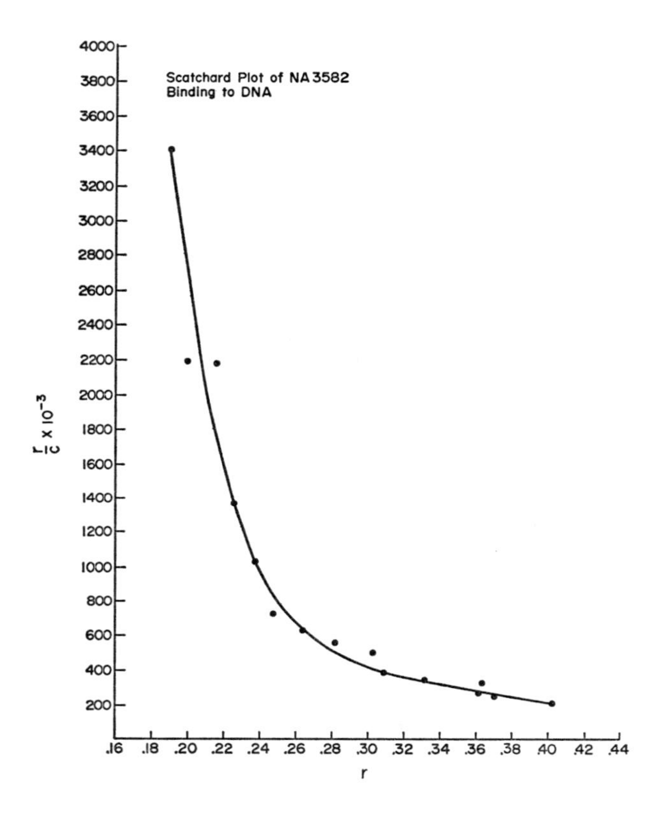

Fig. 2. Adsorption isotherm (Scatchard, 1949) derived from spectrophotometric titration of NA with *E. coli* DNA. A Cary Model 14 recording spectrophotometer and 10 cm cuvettes were used to measure absorbancy changes at 325 nm occurring as a result of repetitive addition of 10 μl aliquots of 2×10^{-3} M DNA (constitutive nucleotides) to 2.1×10^{-6} M NA in 27.8 ml of 5×10^{-2} M Tris-HCl, pH 7.2

the strong binding of proflavine (Peacocke and Skerret, 1956) and quinacrine (Kurnick and Radcliffe, 1962) to DNA and suggests that the biological effects of nitroacridines is caused by their interactions with DNA. NA also appeared to bind to DNA by a weaker process which had an apparent association constant of 2×10^6 M^{-1}, and entailed binding of one molecule of NA for approximately three nucleotides. Biophysical experiments are in progress to gain a better understanding of these drug-DNA interactions.

III. Effects of Nitroakridin on Biosyntheses in Cell-free Systems

Fig. 3 illustrates the effect of NA on the cell-free DNA and RNA polymerase reactions carried out by enzymes from *Micrococcus lysodeikticus* and using DNA from *E. coli* as template and primer. NA inhibited both reactions, although much higher concentrations of NA were required to achieve inhibition of RNA polymerization compared to inhibition of DNA polymerization. These effects were similar to those

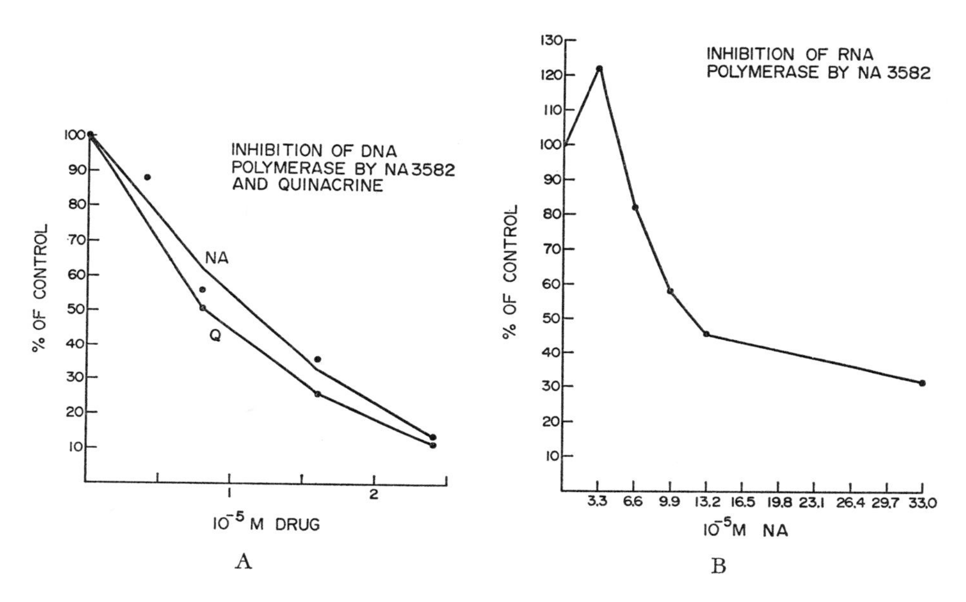

Fig. 3A and B. NA inhibition of cell-free DNA and RNA polymerization. The controls contained 1205 cpm ^{3}H-ATP polymerized, and 1391 cpm ^{14}C-ATP polymerized. Reaction systems (0.25 ml) for synthesis of DNA (Fig. 3A) contained per ml reaction mixture: potassium phosphate buffer, 70 μmoles at pH 7.0; $MgCl_2$, 7.0 μmoles; B-mercaptoethanol, 10 μmoles; deoxyribonucleoside triphosphates (dGTP, dCTP, TTP, and ^{3}H-dATP) 100 μmoles (sp. act. 10 μc/μmole); *E. coli* DNA, 100 nmoles (constituent nucleotides); and 4 units *M. lysodeikticus* DNA polymerase (sp. act. 3000 units/mg). Mixtures were incubated for 15 min at 37 °C, stopped by addition of cold 5 % TCA containing 0.5 % ATP, filtered, the filters washed with additional aliquots of TCA, and counted. Systems for synthesis of RNA (Fig. 3B) (0.30 ml) were comprised of 4×10^{-2} M Tris-HCl, pH 7.9; 4×10^{-3} M $MgCl_2$, and 1×10^{-3} M $MnCl_2$; ribonucleoside triphosphates (GTP, CTP, UTP and ^{14}C-ATP) 0.4 μmoles (sp. act. 5 μc/μmole); *E. coli* DNA 0.6 nmoles (constituent nucleotides), and a total of 18 units *M. lysodeikticus* RNA polymerase (sp. act. 2000 units/mg). Procedures were identical to those used with DNA polymerase. Enzymes were purchased from Biopolymers Laboratory, Dover, N.J., ^{3}H-dATP and ^{14}C-ATP from Schwarz Bioresearch, and unlabelled triphosphates from Calbiochem

produced by the acridines, proflavine (Hurwitz, Furth, Malamy and Alexander, 1962) and quinacrine, and by the quinoline, chloroquine (O'Brien, Olenick and Hahn, 1966). The 50% effective dose (ED_{50}) for inhibition of *in vitro* DNA synthesis was 1×10^{-5} M NA, while the ED_{50} for inhibition of RNA synthesis was

1.2×10^{-4} M NA. In bacterial cells DNA and RNA biosynthesis occur in an identical environment and, therefore, the effect of NA on the DNA polymerase reaction was also determined under conditions resembling those utilized for investigating the RNA polymerase reaction, Fig. 3 B. Table 1 shows that in a system optimal for RNA synthesis, the ED_{50} for NA inhibition of DNA synthesis was increased to 2.5×10^{-5} M NA. This concentration was still well below the ED_{50} for NA inhibition of the RNA polymerase reaction and, therefore, NA preferentially inhibited DNA polymerization. Similar results have been obtained with quinacrine and chloroquine (O'BRIEN, OLENICK and HAHN, 1966). Fig. 3 A also compares the inhibition of the DNA polymerase reaction by NA with that by quinacrine, and shows that quinacrine was slightly more potent than NA.

Table 1. *The effect of NA 3582 on DNA synthesis under conditions optimal for RNA polymerase*

System	CPM ^{3}H-dATP polymerized	Minus background	%
Control	3180	3016	100
Background (blank)	164	—	—
1.6×10^{-5} M NA	2164	2000	66
3.3×10^{-5} M NA	1365	1201	40

System contained: d-GTP, d-CTP, TTP, and ^{3}H-dATP (sp. act. 10 μc/μmole) 0.4 μmoles, 0.6 μmoles DNA/ml, 12 units DNA polymerase (sp. act. 3000 units/mgm) and Tris-HCl 4×10^{-2} M, pH 7.9; $MgCl_2$, 4×10^{-3} M; and $MnCl_2$, 1×10^{-3} M. Volume 0.30 ml.

Since NA had been reported to inhibit synthesis of the inducible enzyme β-galactosidase in *E. coli* (HAHN and WISSEMAN, 1951) i.e. of a protein, the effect of NA on cell-free synthesis of polyphenylalanine, and selected component reactions in this pathway, were investigated (WOLFE, COOK and HAHN, 1969). The effects of several other N-heterocyclic amines were studied for comparison. Fig. 4 depicts the effects of NA, ethidium bromide (EB) and chloroquine (CQ) on synthesis of polyphenylalanine in an amino acid polymerization system of *E. coli* origin (NIRENBERG and MATTHEI, 1961; WOLFE and HAHN, 1965). EB inhibited synthesis of polyphenylalanine severely, NA exerted moderate inhibition, and CQ actually stimulated synthesis. The ED_{50} for NA was 7×10^{-4} M, i.e. considerably greater than the concentration required for 50% inhibition of DNA and RNA synthesis *in vitro*. The relative insensitivity of polyphenylalanine synthesis to NA must be viewed against the high concentration of magnesium (1.4×10^{-2} M) required in this artificial system. *In vitro* protein synthesis utilizing natural initiation mechanisms (LEDER and NAU, 1967) requires approximately 8×10^{-3} M Mg^{+2}, and NA might be more effective in such a system. Recently EB was reported (GRINSTED, 1969) to inhibit protein synthesis in a thermophilic gram-positive bacterium; this inhibition was ascribed to interaction between EB and RNA.

Fig. 5 depicts the effects of NA, chloroquine and quinacrine on formation of ribosome-poly *U*-^{14}C-phenylalanine tRNA complexes, measured by membrane filter entrapment (NIRENBERG and LEDER, 1964) of these ternary complexes. NA strongly inhibited the codon-directed binding of tRNA^{Phe}, although this observation does not

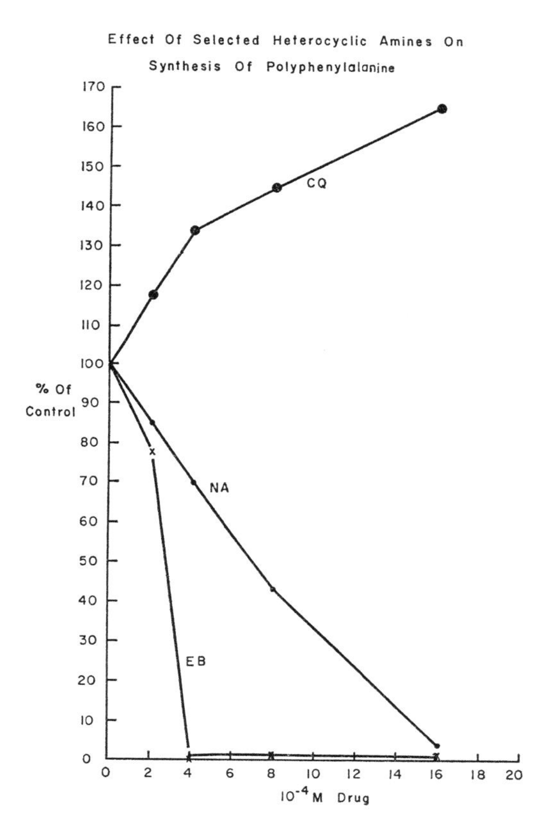

Fig. 4. Inhibition of polyphenylalanine synthesis by NA and ethidium bromide (EB), and stimulation of polyphenylalanine synthesis by chloroquine (CQ). The control contained approximately 57,000 cpm above background. System and methodology have been described elsewhere (WOLFE and HAHN, 1965)

permit the unambiguous interpretation that NA inhibited codon-anticodon recognition directly. Suppression of tRNA^{Phe} binding might also have been caused indirectly by inhibition of binding of poly *U* to ribosomes or alteration in the conformation of either tRNA or ribosomes. Nonetheless, NA strongly interfered with the formation of these ternary complexes.

Fig. 5 A shows that omission of poly *U* caused membrane filters to retain tRNA^{Phe} as a function of the concentration of NA, suggesting that NA could bind to tRNA.

This was verified by spectrophotometric determination of NA-tRNA adsorption isotherms (WOLFE, 1970), which indicated that NA binds to tRNA by two or more processes, one of which is very strong. In separate experiments (WOLFE, 1970) utilizing 1.4×10^{-2} M Mg^{+2}, NA marginally suppressed the charging of tRNA with

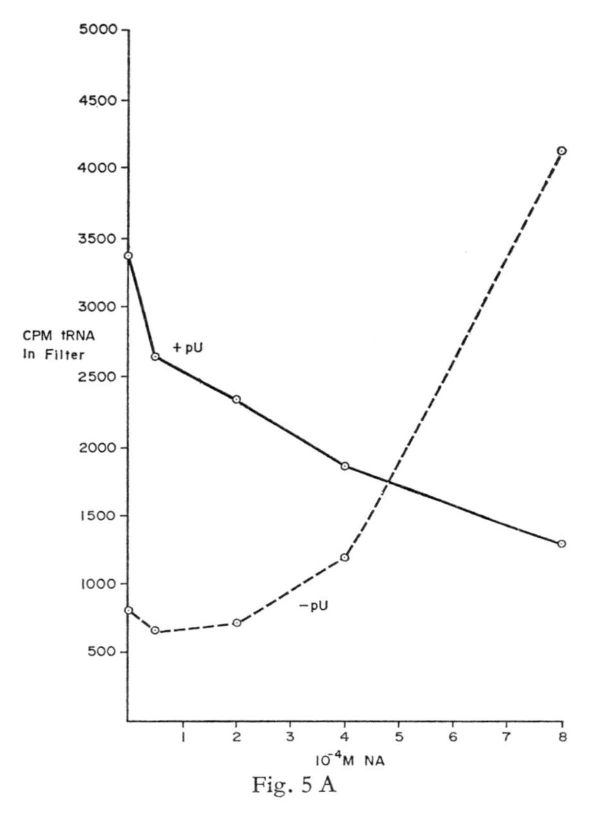

Fig. 5 A

Fig. 5. The effect of Nitroakridin 3582, chloroquine, and quinacrine on formation of ribosome-poly U-^{14}C-phenylalanine tRNA complexes and on filter retention of ^{14}C-phenylalanine tRNA in the absence of poly *U*. Complexes were isolated by filter retention (NIRENBERG and LEDER, 1964). Reaction systems (0.50 ml) were comprised of: 2.0×10^{-2} M Tris-HCl, pH 7.2; 1.4×10^{-2} M MgAc; 6.0×10^{-2} M KCl; 15.6 A_{260} units *E. coli* C-2 ribosomes; 60 µg poly *U*, and approximately 43 µg of ^{14}C-phenylalanine tRNA containing 1 µc/5.8 mg tRNA (about 16,200 cpm). Mixtures were incubated at 24 °C for 20 min, diluted with fresh aliquots of identical cold buffer containing unlabelled *E. coli* tRNA, filtered, and the filters washed and counted

^{14}C-phenylalanine. Thus the binding of NA to tRNA may well contribute on multiple levels to inhibition of polyphenylalanine synthesis. Proflavine (WEINSTEIN and FINKELSTEIN, 1967) has been shown to inhibit polyphenylalanine synthesis through interaction with tRNA, and EB also is known to complex strongly with tRNA (BITTMAN, 1969), to inhibit formation of ternary complexes (WOLFE, 1970), and to inhibit charging of tRNA with phenylalanine (LANDEZ, ROSKOSKI and COPPOC, 1969; WOLFE, 1970).

Fig. 5 also illustrates the effect of chloroquine and quinacrine on formation of ribosome-poly U-^{14}C-phe tRNA complexes. CQ stimulated formation of these com-

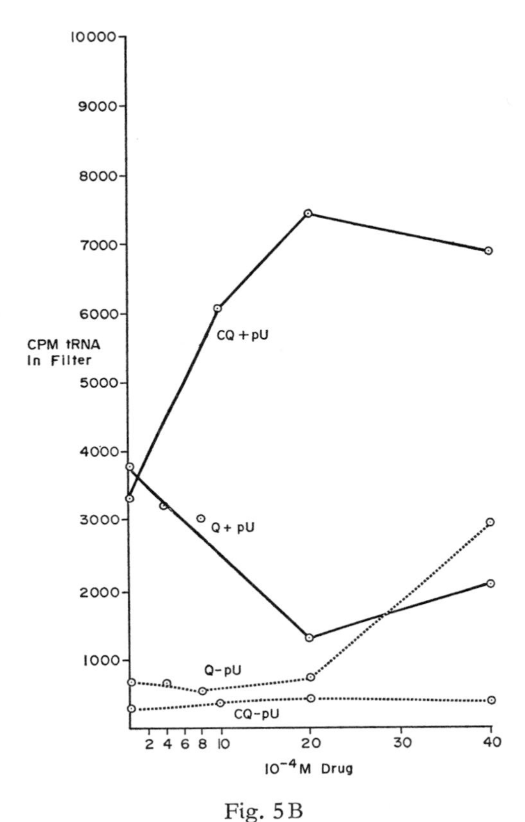

Fig. 5B

plexes, while quinacrine (Q) at similar high concentration inhibited their formation. CQ stimulation of this triple complex formation may well account for the enhancement of polyphenylalanine synthesis, but the ultimate cause of these effects remains to be determined. CQ activates tRNA^{tryp} (MUENCH, this volume) but chloroquine had little effect at these concentrations on charging of tRNA with phenylalanine (WOLFE, 1970). Indeed, high concentrations of chloroquine have been reported to suppress aminoacylation of tRNA (LANDEZ et al., 1969). Comparison of the effects of CQ with those of Q are of interest, for these molecules vary principally in the size of their planar area. The stimulatory effect of CQ is perhaps a function of the aliphatic side chain of the drug molecule, while the inhibitory effect of Q most probably is a function of the acridine ring. Another interesting facet of these results lies in the relatively greater affinity of NA than Q for tRNA (WOLFE, 1970). Examination of

Fig. 5 suggests that appreciably lower concentrations of NA than of Q were required for equal filter retention of ^{14}C-phe tRNA; the affinity of the respective drugs for sites in the filters was not determined.

IV. Nitroakridin and Ribosomes

Acridines also bind to ribosomes. Yeast ribosomes (Morgan and Rhoads, 1965) have been shown to induce shifts in the absorption spectrum of acridine orange, and corresponding shifts have been used (Furano, Bradley and Childers, 1966) in an attempt to test for helicity of RNA in ribosomes from *E. coli*. Both groups of workers considered their results compatible with a binding of acridine orange to 80% of the phosphate groups of the RNA in ribosomes. The possibility exists that acridines alter the structure or stability of ribosomes. Acridines, for example, increase the stability of double-helical DNA; evidence for this has been obtained through observation of the thermal denaturation profiles of DNA in the absence and presence of acridines. Therefore, the effect of acridines on ribosomes (Wolfe, Cook and Hahn, 1969); was investigated spectrophotometrically. Ribosomes maintained at 52 °C in a suitable ionic medium exhibited hyperchromicity at 260 nm which could be correlated with the loss of primary and secondary structure of their RNA (Wolfe, 1968; unpublished observations). When approximately 50% hyperchromicity had occurred, ribosomal protein precipitated; such precipitation was observed as light scattering causing increased absorbancy. In the present experiments, ribosomes were micropipetted into preheated buffer, or buffered drug solutions, and hyperchromicity recorded as a function of time at 52 °C. Such ribosomes were thus removed from a stabilizing (1×10^{-2} M Mg^{+2} and 2×10^{-2} M Tris, pH 7.2) ionic environment, and suddenly diluted into a destabilizing environment in which heat, low divalent cation concentration and high counterion concentration all contributed toward ribosome disassemblage. This test for drug interaction with ribosomes had been previously used to show that ribosomes from streptomycin killed *E. coli* were more heat stable than ribosomes from control cells (Wolfe, 1968; Wolfe and Hahn, 1968).

In contrast to streptomycin, acridines have been found to decrease markedly the thermal stability of ribosomes (Wolfe, Cook and Hahn, 1969; Wolfe, 1970). These results appear to be diametrically opposite to the stabilization of DNA by acridines. Fig. 6 presents the effects of three concentrations of NA upon ribosomes maintained at 52 °C. In this diagram, the actual absorbancy is represented on the ordinate, while the time of heating at 52° is shown on the abscissa. Ribosomes were obtained from *E. coli* A 19, RNAase I^- (Gesteland, 1966). Such ribosomes are labilized more slowly and to a lesser extent than ribosomes possessing associated RNAase I, but hyperchromicity patterns and cation influences are similar (Wolfe, 1968; unpublished observations). The thermal denaturation profiles, Fig. 6, reveal that NA at low concentrations reduced the heating interval which precedes the onset massive hyperchromicity and increased the hyperchromic rate, i.e. destruction of ribosomes occurred more rapidly in the presence than in the absence of NA. Higher concentrations of NA further decreased the onset and increased the rate of hyperchromicity but not in direct proportion to the NA concentration. When the 30S and 50S ribosomes from *E. coli* A 19 were tested individually (Wolfe, 1970), each exhibited more rapid breakdown in the presence of low concentrations of NA than in the absence of the drug. 70S ribosomes and their subunits from wild strains behaved similarly (WOLFE, COOK and HAHN, 1969; WOLFE, 1970, unpublished observations).

The effect of related compounds were then investigated using *E. coli* RNAase I+ ribosomes. Each compound tested, quinacrine, chloroquine, ethidium bromide, proflavine, and the 9-methylamino analogue of quinacrine, caused increases in the rate of hyperchromicity indicative of ribosome labilization. The ability of the two latter compounds to decrease the thermal stability of ribosomes suggested that such effects were primarily a function of the heterocyclic rings. This idea was reinforced when the aliphatic polyamine, spermine, was tested. Spermine was found to increase the

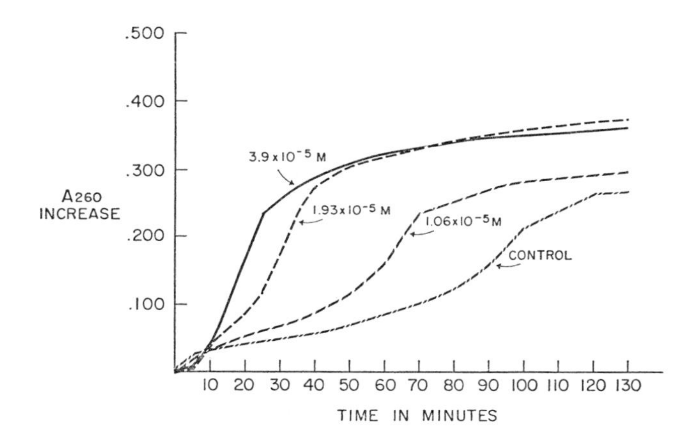

Fig. 6. NA enhancement of thermal denaturation of *E. coli* ribosomes. The increase in absorbancy at 260 nm is plotted as a function of time. Reaction systems contained approximately 1.0 A_{260} units of ribosome per ml, while the buffer was comprised of 3×10^{-2} M Tris-HCl, pH 7.2; 1×10^{-3} M MgAc; and 6×10^{-2} M KCl. Ribosomes were micropipetted into aliquots of preheated buffer. Experiments were carried out in a Gilford Model 2000 recording spectrophotometer, equipped with a Haake circulating flow heater

thermal *stability* of ribosomes, and to *retard* destruction of ribosomes by EB and NA. The effects of the planar aromatic molecules and the linear aliphatic polyamine were opposite. Further, in separate spectrophotometric experiments at room temperature (WOLFE, unpublished observations) spermine partially restored the absorption spectrum of EB and NA which had been decreased in intensity by ribosomes. Polylysine has also been shown to displace acridine orange from *E. coli* ribosomes (FURANO et al., 1966). The results suggest that N-heterocycles and aliphatic polyamines are bound at the same ribosomal sites, but thermal labilization experiments show their effects to be opposite.

The extent of ribosome destabilization by N-heterocyclic amines could be represented as a function of the planar areas of these compounds. Equimolar concentrations of EB, Q, and CQ were compared for their ability to accelerate thermal labilization of ribosomes (WOLFE, 1970). Results are shown in Fig. 7. The compounds caused ribosomes to be dissimilated more rapidly and each compound initiated hyperchromic increases at distinctly different times. EB had the earliest effect, Q had an intermediate effect, and CQ had the least effect. At similar concentrations, actions of proflavine and NA resembled that of Q. When the results shown in Fig. 7 were plotted for endpoint disassemblage times vs. planar area of labilizing drugs, the diagram Fig. 8 was obtained (WOLFE, 1970). Enhancement of thermal labilization of ribosomes by N-heterocyclic amines was a linear function of the total planar area of these compounds.

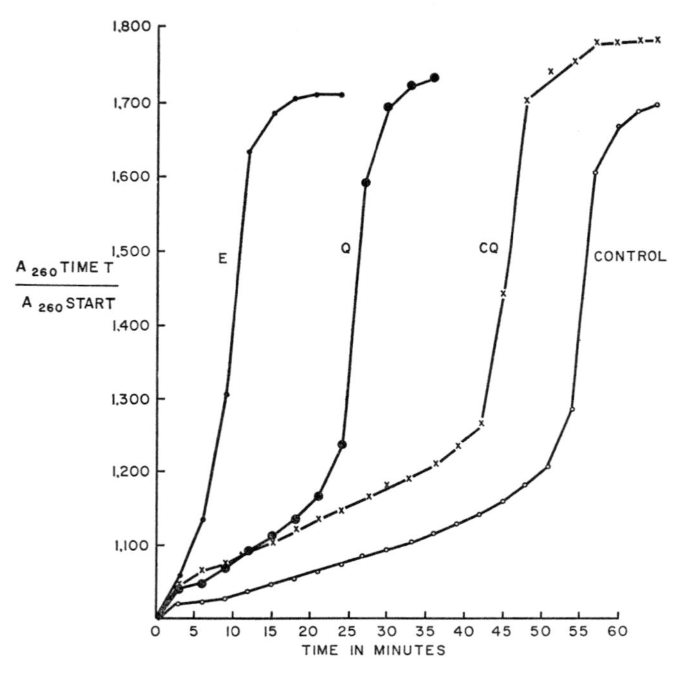

Fig. 7. Comparison of the effects of ethidium bromide, quinacrine and chloroquine on thermal labilization of ribosomes. Methodology as in Fig. 6, while drug concentrations were 6.6×10^{-6} M

Thus N-heterocycles bind to, and weaken, ribosomes. One possible mechanism of this destabilization may be a displacement of, or interference with, ribosomal aliphatic polyamines (COHEN and LICHTENSTEIN, 1960). The function and location of these compounds in the ribosome is unknown but they may serve to strengthen association between RNA and protein, while N-heterocycles may reverse such effects. Recently, proflavine was reported (HULTIN, 1969) to alter mammalian ribosome structure sufficiently to allow the dye procion blue to interact with a ribosomal protein which is otherwise non-reactive. A second, not necessarily exclusive, mechanism is suggested by the ability of heavy metal cations (EICHHORN, 1965) to labilize RNA through interaction with phosphoric acid groups. N-heterocyclic amines are organic cations which are thought to bind principally if not entirely to the phosphoric acid groups of RNA in ribosomes. The consequences of cation binding may be a labilization of the RNA in ribosomes; this was the earliest detectable event in the thermal disassemblage of ribosomes in drug-free control experiments (WOLFE, 1968, unpublished observations). Breakdown of ribosomal RNA appears to be a necessary prerequisite to precipitation of ribosomal protein (WOLFE, 1968). N-heterocycles might also cause release from ribosomes of associated (TAL and ELSON, 1963; SINGER, 1967) hydrolytic enzymes, including RNAases I and II, and polynucleotide phosphorylase, thereby enhancing thermal labilization. The possibility also exists that the reduction

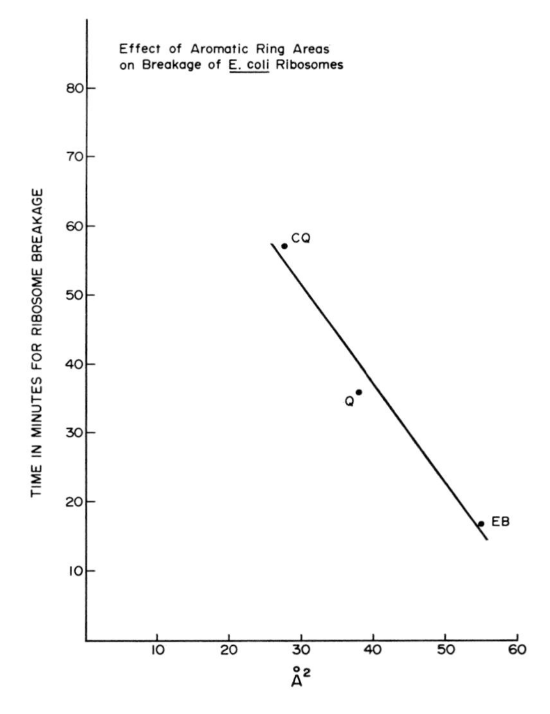

Fig. 8. Correlation between ribosome breakage time in the presence of drugs to ring area in $Å^2$ of the N-heterocycles used. Data is taken from Fig. 7. Ordinate: time taken between introduction of ribosomes into heated buffer and endpoint of hyperchromicity; abscissa: total ring area of compounds

of the thermal stability of ribosomes by N-heterocyclic compounds is one manifestation of a more general effect on nucleic acids in view of the recent report (GOLDRING, GROSSMAN, KRUPNICK, CRYER and MARMUR, 1970) that ethidium bromide caused decreases in the molecular weight of yeast mitochondrial DNA during induction of petite mutants.

V. Effects of Nitroakridin on Bacterial Growth and Biosyntheses

Finally, the effects of NA on cellular multiplication, viability, and macromolecular synthesis in *E. coli* were studied, and inhibition of macromolecular synthesis *in vivo* by NA was compared with corresponding inhibitions *in vitro*. NA suppressed cell multiplication, caused exponential death, and inhibited macromolecular synthesis in bacteria, with dosage responses similar to those found in cell-free model systems.

Fig. 9 illustrates the effects of graded concentrations of NA on multiplication and/or viability of *E. coli* C-2. Reductions in the rate of multiplication occurred at

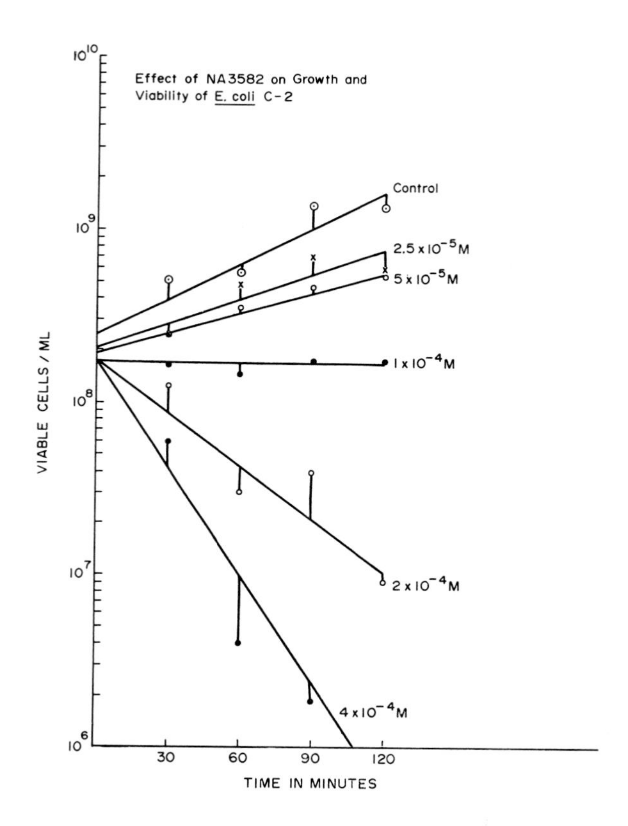

Fig. 9. The effect of graded concentrations of NA on multiplication and viability of *E. coli* C-2. Aliquots of bacteria growing exponentially in a glucose-mineral-salts medium (FISHER and ARMSTRONG, 1947) were exposed to different concentrations of NA, samples withdrawn at time intervals, serially diluted, plated on nutrient agar, incubated, and colonies counted. Data for this diagram and Fig. 10 were obtained in a single typical experiment (WOLFE, 1970)

low concentrations of NA, bacteriostasis was observed at 1×10^{-4} M NA, and higher NA concentrations produced loss of viability exponentially with time. Bacterial filaments, characteristic of unbalanced growth, were formed at bacteriostatic concentrations of NA, while exponential death occurred at NA concentrations which strongly suppressed synthesis of RNA and protein as well as of DNA.

Radioactive tracer studies (WOLFE, 1970; WOLFE, COOK and HAHN, 1970) to determine the effect of NA on bacterial biosyntheses of RNA, protein, and DNA were carried out by measurements of the incorporation into trichloroacetic acid insoluble fractions of ^{14}C-uracil, ^{14}C-isoleucine, and ^{14}C-thymidine, respectively. Thymidine incorporation into DNA was preferentially suppressed, this is consistent with results in cell-free DNA polymerase systems and with observations of the formation of filaments.

Fig. 10 compares the responses of macromolecular synthesis and cell multiplication to the dosage of NA using a graphic probit transformation. This method of analysis

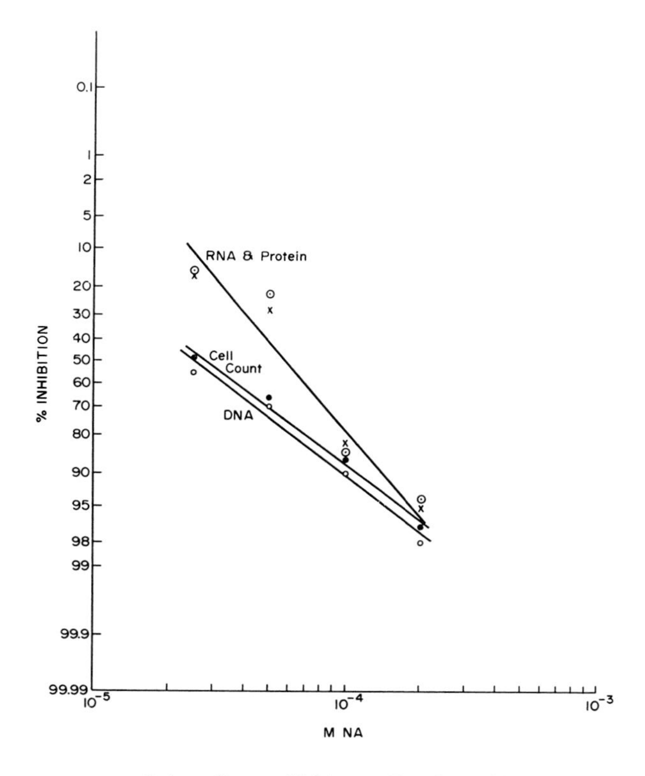

Fig. 10. Dosage-response of the effects of NA on *E. coli* multiplication, RNA, DNA and protein biosynthesis. Per cent inhibition of these processes is represented on the ordinate as a function of a normal distribution, while the logarithm of the NA concentration is represented on the abscissa

was originally introduced (TREFFERS, 1956) to render antibiotic dose-response correlations linear. Percent inhibitions are plotted on the ordinate as a function of a normal distribution, while the log of the NA concentration is represented on the abscissa. Data so handled revealed that inhibition of cell multiplication closely followed inhibition of DNA synthesis, while inhibitions of RNA and protein synthesis were weaker and, in turn, similar to each other. These results are consistent with the current idea that cell division is related to DNA synthesis, while messenger RNA formation is a controlling influence on protein synthesis.

VI. Concluding Remarks

NA has been shown to bind strongly to DNA in a manner similar to that of other acridines. The consequences of this binding include inhibition of DNA and RNA polymerization *in vitro* and of DNA and RNA biosynthesis *in vivo*. The antibacterial effects of acridines can be ascribed to their binding to DNA although it is not entirely clear whether they induce bacteriostasis and bacterial death by the same mechanism.

Suppression of DNA biosynthesis by low concentrations of NA resulted in bacteriostasis and filamentous growth, while higher concentrations caused loss of viability as an exponential function of time. This effect was produced by NA and also is produced by quinacrine (Ciak and Hahn, 1967) at concentrations which inhibit RNA and protein biosynthesis to a considerable extent.

The qualitative distinction between bacteriostatic and bactericidal effects may point to the existence of two different modes of action of acridines, conceivably related to different effects upon DNA. Furthermore, acridines may not only bind to DNA but also to other cell components such as ribosomes or cell membranes. The present results suggest that NA produced bacteriostasis by binding to DNA but the mechanism of NA's bactericidal action may be more complex.

References

Albert, A.: The acridines, 2nd edition New York: St. Martin's Press 1966.

— Rubbo, S. D., Goldacre, J. J., Davey, M. E., Stone, J. D.: The influence of chemical constitution on antibacterial activity. Part II: A general survey of the acridine series. Brit. J. exp. Path. **26**, 160 (1945).

Barner, H. D., Cohen, S. S.: The induction of thymine synthesis by T2 infection of a thymine requiring mutant of *Escherichia coli*. J. Bact. **68**, 80 (1954).

Bittman, R.: Studies on the binding of ethidium bromide to transfer ribonucleic acid: absorption, fluorescence, ultracentrifugation and kinetic investigations. J. molec. Biol. **46**, 251 (1969).

Caldwell, P. C., Hinshelwood, C.: The nucleic acid content of *Bact. lactis aerogenes*. J. chem. Soc. **1950**, 1415.

Ciak, J., Hahn, F. E.: Quinacrine (Atebrin): Mode of action. Science **156**, 655 (1967).

Cohen, S. S., Lichtenstein, J.: Polyamines and ribosome structure. J. biol. Chem. **235**, 2112 (1960).

Denes, G., Polgar, L.: Effect of Nitroakridin 3663 on synthesis of deoxyribonucleic acid and multiplication of T2 phage. Nature (Lond.) **185**, 386 (1960).

Eichhorn, G. L., Butzow, J. J.: Degradation of polyribonucleotides by lanthanum ions. Biopolymers **3** (1), 79 (1965).

Furano, A. V., Bradley, D. F., Childers, L. G.: The conformation of the ribonucleic acid in ribosomes. Biochemistry **5**, 3044 (1966).

Fussganger, R.: Pharmaceuticals and insecticides: I.G. Farbenindustrie A.G., Hoechst (Main). Combined Intelligence Objectives Sub-Committee Report **24/4** (1945).

Gesteland, R. F.: Isolation and charcterization of ribonuclease I mutants of *Escherichia coli*. J. molec. Biol. **16**, 67 (1966).

Goldring, E. S., Grossman, L. I., Krupnick, D., Cryer, D. R., Marmur, J.: Ethidium bromide-induced breakdown of yeast mitochondrial DNA during induction of petites. Fed. Proc. **1970**, 2710.

Grinsted, J.: Antimicrobial drugs and RNA. Biochim. biophys. Acta (Amst.) **179**, 268 (1969).

Hahn, F. E., Wisseman, C. L.: Inhibition of adaptive enzyme formation by antimicrobial agents. Proc. Soc. exp. Biol. (N.Y.) **76**, 533 (1951).

HULTIN, T.: The use of procion blue as a molecular probe in the study of ribosome structure. Europ. J. Biochem. **9**, 579 (1969).
HURWITZ, J., FURTH, J., MALAMY, M., ALEXANDER, M.: The role of deoxyribonucleic acid in ribonucleic acid synthesis. III. The inhibition of the enzymatic synthesis of ribonucleic acid and deoxyribonucleic acid by actinomycin D and proflavine. Proc. nat. Acad. Sci. (Wash.) **48**, 1222 (1962).
KURNICK, N. B., RADCLIFFE, I. E.: Reactions between DNA and quinacrine and other antimalarials. J. Lab. clin. Med. **60**, 669 (1962).
LANDEZ, J. H., ROSKOSKI, R., COPPOC, G. L.: Ethidium bromide and chloroquine inhibition of rat liver cell-free aminoacylation. Biochim. biophys. Acta (Amst.) **195**, 276 (1969).
LEDER, P., NAU, M. M.: Initiation of protein synthesis, III. Proc. nat. Acad. Sci. (Wash.) **58**, 774 (1967).
MORGAN, R. A., RHOADS, D. J.: Binding of acridine orange to yeast ribosomes. Biochim. biophys. Acta (Amst.) **102**, 311 (1965).
MUENCH, K.: This volume.
NIRENBERG, M., LEDER, P.: RNA codewords and protein synthesis. Science **145**, 1399 (1964).
— MATTHEI, J. H.: The dependence of cell-free protein synthesis in *Escherichia coli* upon naturally occurring or synthetic polyribonucleotides. Proc. nat. Acad. Sci. (Wash.) **47**, 1588 (1961).
O'BRIEN, R. L., OLENICK, J. G., HAHN, F. E.: Reactions of quinine, chloroquine, and quinacrine with DNA and their effects on the DNA and RNA polymerase reactions. Proc. nat. Acad. Sci. (Wash.) **55**, 1511 (1966).
PEACOCKE, A. R., SKERRETT, J. N. H.: The interaction of aminoacridines with nucleic acids. Trans. Farad. Soc. **52**, 261 (1956).
SCATCHARD, G.: The attractions of proteins for small molecules and ions. Ann. N.Y. Acad. Sci. **51**, 660 (1949).
SCHNITZER, R.: Zum Wirkungsmechanismus bakterizider Chemotherapeutica. Med. u. Chem. **1936**, 34.
— SILBERSTEIN, W.: Über neue trypanocide Acridinfarbstoffe. Untersuchungen an homologen Reihen von 6 Nitro-9-Aminoacridinen. Z. Hyg. Infekt-Kr. **109**, 519 (1929).
SINGER, M. F., TOLBERT, G.: Isolation and properties of a potassium-activated phosphodiesterase (RNAase II) from *Escherichia coli*. Biochemistry **4**, 1319 (1965).
SMADEL, J. E., SNYDER, J. C., JACKSON, E. B., FOX, J. P., HAMILTON, H. L.: Chemotherapeutic effects of acridine compounds in experimental rickettsial infections in embryonated eggs. J. Immunol. **57**, 155 (1947).
TAL, M., ELSON, D.: The location of ribonuclease in *E. coli*. Biochim. biophys. Acta (Amst.) **36**, 40 (1963).
TREFFERS, H. P.: The linear representation of dosage-response curves in microbial-antibiotic assays. J. Bacteriol. **72**, 108 (1956).
WEINSTEIN, I. B., FINKELSTEIN, I. H.: Proflavine inhibition of protein synthesis. J. biol. Chem. **242**, 3757 (1967).
WOESE, C., NAONO, S., SOFFER, R., GROS, F.: Studies on the breakdown of messenger RNA. Biochem. biophys. Res. Commun. **11**, 435 (1963).
WOLFE, A. D.: Degradation of bacterial ribosomes at constant elevated temperature. Fed. Proc. **27**, 806 (1968).
— Thesis, University of Maryland 1970.
— COOK, T. M., HAHN, F. E.: Bact. Proc. **1969**, 146.
— — — Bact. Proc. **1970**, 70.
— HAHN, F. E.: Mode of action of chloramphenicol. IX. Effects of chloramphenicol upon a ribosomal amino acid polymerization system and its binding to bacterial ribosomes. Biochim. biophys. Acta (Amst.) **95**, 146 (1965).
— — Stability of ribosomes from streptomycin-exposed *Escherichia coli*. Biochem. biophys. Res. Commun. **31**, 945 (1968).

Binding of Naphthylvinylpyridines to DNA

Helen L. White, James R. White and Chester J. Cavallito

I. Introduction

The suggestion that planar, heterocyclic bases might be inserted between the bases of DNA was first put forth by Michaelis (1947), several years before the double helical structure of DNA had been proposed. In 1961 the phenomenon of intercalation

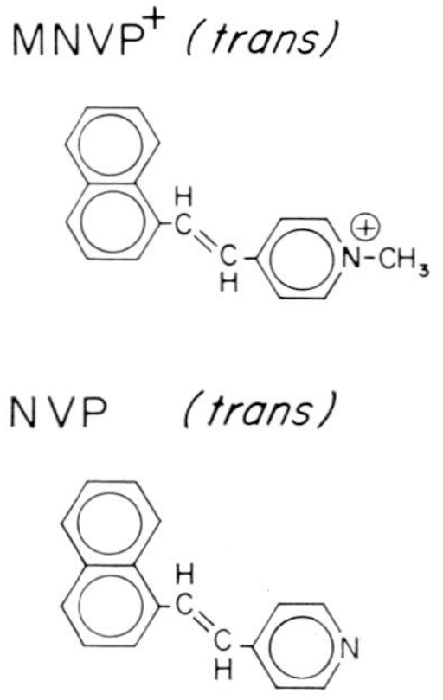

MNVP$^+$ *(trans)*

NVP *(trans)*

Fig. 1. Structures of trans-N-methyl-4-(1-naphthylvinyl) pyridinium cation (MNVP$^+$) and 4-(1-naphthylvinyl) pyridine (NVP)

was clearly described and characterized in the classical work of Lerman (1961) with proflavine. Since then many papers have appeared in which intercalation has been invoked as a mode of DNA-ligand interaction for aromatic compounds and heterocyclic bases. However, even with the simple structure of proflavine there is still disagreement about the nature of the interaction — in particular, whether the intercalated molecule is partially or completely intercalated, and the relative importance of the electrostatic, hydrophobic, and hydrogen bonding forces which may stabilize the complex (Blake and Peacocke, 1968; Gilbert and Claverie, 1968).

For mechanistic studies of intercalation, a ligand that is water-soluble, planar, aromatic or heteroaromatic, and having a structure resembling a purine-pyrimidine base pair in overall size might be ideal. A series of naphthylvinylpyridine analogues has recently been synthesized as part of a search for potent inhibitors of the enzyme choline acetyltransferase (CAVALLITO, YUN, SMITH and FOLDES, 1969). Because these compounds fulfill all of the above requirements, they appear to be useful model compounds for intercalation studies. A combination of data, particularly from absorption spectrometry, viscometry, and velocity sedimentation supports this premise.

The present report describes experiments with the two compounds shown in Fig. 1. Both of these compounds were prepared as *trans* isomers. In solution they are extremely sensitive to photoisomerization (WHITE and CAVALLITO, 1970). If exposed to wavelengths shorter than 400 nm, a photo-stable mixture results, which contains approximately 20% of the *trans* isomer in equilibrium with the *cis* form. In the *cis* isomer the two ring systems are not coplanar. To prevent such isomerization, all work with the *trans* isomers was done under illumination by 75 watt pink light bulbs.

The naphthylvinylpyridines were crystallized as the iodides or chlorides. The nature of the anion had no effect on DNA binding properties of the ligand. The amine, 4-(1-naphthylvinyl)pyridine (NVP), had a pK_a of 5.5 in aqueous solution, and therefore its spectrum was strongly dependent on pH. In the protonated form its spectrum resembled that of the N-methyl quaternary analogue, $MNVP^+$.

It was thought that binding studies with NVP and $MNVP^+$ under conditions of varied pH might determine the influence of the ligand's positive charge on binding parameters. Moreover NVP in the unprotonated form would not interact electrostatically with the negatively charged phosphates of the DNA double helix, and therefore, with this compound one might observe binding by intercalation only, uncomplicated by electrostatic contributions.

II. Evidence for Intercalation

1. Sedimentation and Viscosity

LERMAN (1961) showed that an intercalation complex between ligand and DNA should have an intrinsic viscosity that is greater than that of DNA alone, and a sedimentation rate that is smaller. Recent work (MÜLLER and CROTHERS, 1968) has indicated that DNA of low molecular weight is preferable in experiments of this kind. Therefore a solution of calf thymus DNA was fragmented by sonic oscillation to a molecular weight of about 5×10^5 by the method of DOTY, MCGILL and RICE (1958). Intrinsic viscosity determinations were performed with a Cannon 4-bulb Ubbelohde viscometer, and sedimentation was measured in a Beckman Model E analytical ultracentrifuge equipped with UV optics. At low ionic strength (0.005 M) and at a phosphate to ligand molar ratio of 10, the intrinsic viscosity of "sonicated" DNA was increased by a factor of 1.54 with $MNVP^+$ and 1.45 with NVP, whereas with both ligands the sedimentation rate was decreased to 92% of the value in absence of ligand. At higher ionic strength (0.2 M), intrinsic viscosity was increased by a factor of about 1.3 with both ligands and sedimentation rate was either decreased or unchanged. Complicating effects of cooperative binding external to the double helix at the higher ionic strength probably are responsible for the apparently unchanged sedimentation.

2. Melting Transitions

Further support for an intercalation mechanism was derived from studies of DNA melting transitions, using a Gilford Model 2000 multiple sample recording spectrophotometer equipped with a linear thermosensor control. With $MNVP^+$ at concentrations of 5 to 10 μM and a DNA phosphate to ligand ratio of 10, an increase in the mid-point of the melting transition (T_m) by 10 to 14 °C was found with DNA from various sources having guanine + cytosine ($G+C$) contents ranging from 30 to 72%. This stabilization of the double helix did not appear to be dependent on $G+C$ content. NVP at pH 4.5 produced the same stabilizing effect as the N-methyl quaternary compound, but at pH 7.4 NVP caused a much smaller increase in T_m, apparently because heating resulted in a dissociation of the DNA-NVP complex.

An intercalative mechanism for binding of a ligand to DNA is consistent with a stabilization of the double helix and consequently an increase in T_m, provided the ligand binds more strongly to native than to denatured DNA at the temperature of the transition. Such an increase, however, does not constitute proof of intercalation. A ligand bound strongly to phosphates (Mg^{++}) or across either groove of the double helix will also increase the temperature of the transition. But, when considered with the evidence of the previous section, showing increased viscosities and decreased sedimentation rates, and evidence presented later that these compounds bind less strongly to denatured than to native DNA, one may conclude that the large increase in T_m points to intercalation binding.

3. *Trans-Cis* Isomerization

The ionized naphthylvinylpyridines photoisomerize easily from *trans* to noncoplanar *cis* forms. It is unlikely that the *cis* structure can intercalate because in this isomer the aromatic and heterocyclic moieties are at an angle less than 90° to one another. Furthermore one would predict that with an intercalated *trans* molecule, photoisomerization would be essentially prevented because of the steric constraints imposed upon it by the planar DNA bases lying above and below it. This prediction was tested experimentally by exposing solutions of $MNVP^+$ in a quartz cuvette to a Westinghouse 275 watt sunlamp for successive periods of 10 sec. Since the absorbance of the *cis* isomer is much lower at 377 nm than that of the *trans* compound, it was possible to follow at this wavelength the rate of photoisomerization. The unbound *trans* isomer exhibited a half-life of 10 sec under conditions of the experiment. The presence of DNA greatly reduced the apparent rate of isomerization, and it was possible to show that the remaining light sensitivity of a DNA-$MNVP^+$ solution could be accounted for by isomerization of the free or weakly bound forms of the ligand. At low ionic strength, with high phosphate to ligand ratios, no significant isomerization was detected after 60 sec exposure. It should be pointed out that UV light absorption by DNA was not a factor in preventing isomerization, because the light source employed did not emit an appreciable amount of light below 300 nm.

The protective effect of DNA was demonstrated in another experiment by starting with a solution of $MNVP^+$ which had come to *trans-cis* equilibrium in room light. One would expect that if DNA were added to such an equilibrium mixture the small amount of $MNVP^+$ that was initially present as the *trans* form would bind rapidly to the DNA, and that thereafter *cis-trans* photoisomerization would continuously replenish the unbound *trans* form until a new equilibrium was reached. These events were in fact observed visually. Upon addition of excess DNA, the *trans-cis* mixture gradually turned yellow in room light as *trans* form was generated and bound to DNA.

III. Spectrophotometric Studies

1. Binding to Native DNA

Fig. 2 shows a family of spectra obtained for $MNVP^+$ in the presence of increasing concentrations of calf thymus DNA, and is representative of many experiments which were performed under various conditions of pH and ionic strength. The spectrum of the free compound was first determined, and this solution was then titrated with 10 μl aliquots of a DNA-ligand solution having the same initial ligand concentration. At the same time 10 μl aliquots of DNA were added to the reference buffer. In Fig. 2

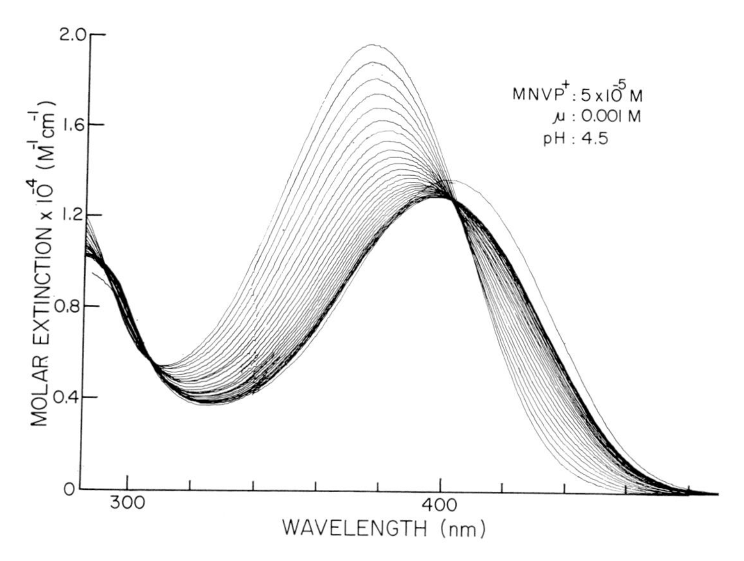

Fig. 2. Titration of $MNVP^+$ with calf thymus DNA. Buffer was 0.1 mM EDTA and sodium phosphate at ionic strength of 1 mM. The free ligand exhibited a maximum at 377 nm, while each spectrum below this was obtained after successive 10 μl additions of a DNA-ligand solution having a concentration of 5.14 mM DNA_p and 50 μM $MNVP^+$. DNA at the same concentration was added in 10 μl aliquots to the reference cuvette. The spectrum of fully bound ligand is the curve whose maximum is shifted furthest toward the red. These spectra were obtained using a Cary Model 15 spectrophotometer

the wavelength maximum of the free ligand at 377 nm was shifted to 402 nm for the fully bound species, a difference of 25 nm. This strong bathochromic shift constitutes further evidence of intercalation binding (LESKO, SMITH, Ts'o and UMANS, 1968). There were several near isosbestic points, at 403, 307, 291, and 240 nm, but these were not distinct, implying the presence of more than one bound species. That there may be interaction between bound ligand molecules was indicated by the fact that, as DNA was added, small changes in absorbance continued to occur at 400 nm after no further change occurred at 377 nm. This is similar to observations made with ethidium bromide (WARING, 1965). When the ratio of bound ligand to nucleotide phosphate was calculated at 377 nm from the spectra of Fig. 2, the Scatchard plot shown in Fig. 3 resulted. There appeared to be a strong binding component with an association constant (K) of 7×10^6 M^{-1} and maximum number of binding sites (n) representing about one in ten nucleotides. There is some question about the validity of the Scatchard analysis, in particular about how curvature in these plots should be interpreted (CROTHERS, 1968). However, when the data fit this type of plot, it is still the best method available for separating a very strong binding type from weaker modes and for giving an indication of relative K and n values.

When a titration experiment like that of Fig. 2 was performed at pH 7.4, spectra of free and bound forms were very similar to those in Fig. 2, but the maximum

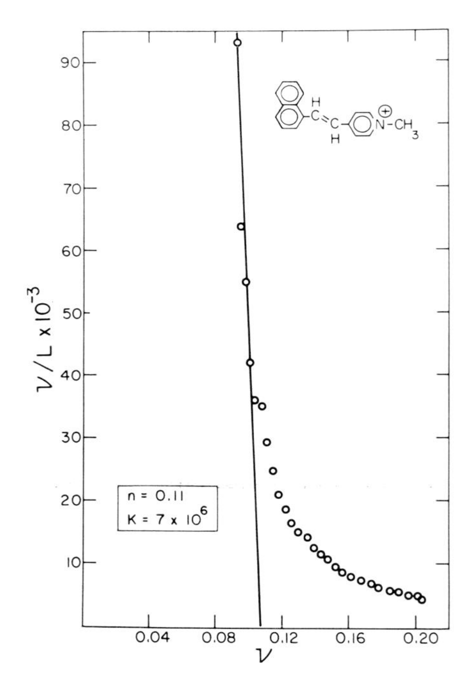

Fig. 3. Scatchard plot obtained from experiment of Fig. 2, with calculations at 377 nm

number of strong binding sites, n, increased to 0.2, that is one bound ligand per five nucleotides of calf thymus DNA. The weaker binding components at both pH values gave an estimated association constant at least 100 times smaller than the strong binding and a total maximum number of binding sites at pH 7.4 of about one per nucleotide pair.

When similar titrations were performed with NVP at low ionic strength and pH 7.4, conditions under which the compound is essentially uncharged, the family of spectra shown in Fig. 4 was obtained. The free ligand absorbed at 330 nm at this pH, and the bound species, at 400 nm, a shift of 70 nm. The isobestic point at 363 nm was quite distinct, while at 293 nm it was not, indicating that, while more than one bound form may have been present, these did not differ in extinction in the longer wavelength region. One would not expect that external binding by electrostatic interactions contributed significantly to spectra in this instance. The strongest binding again involved a maximum of one ligand per five DNA nucleotides, but the association constant was only 2.3×10^5, i.e. much less than for the quaternary analogue.

Since the fully bound species of $MNVP^+$ and NVP both exhibited absorption maxima near 400 nm, it was thought that NVP at pH 7.4 might be binding as the protonated species, even though it exists essentially as the unprotonated species in solution. This possibility was further supported by data obtained at pH 4.5 for NVP, below its pK_a of 5.5. The spectra of free and DNA-bound ligand and calculated binding constants were very similar at pH 4.5 for $MNVP^+$ and protonated NVP.

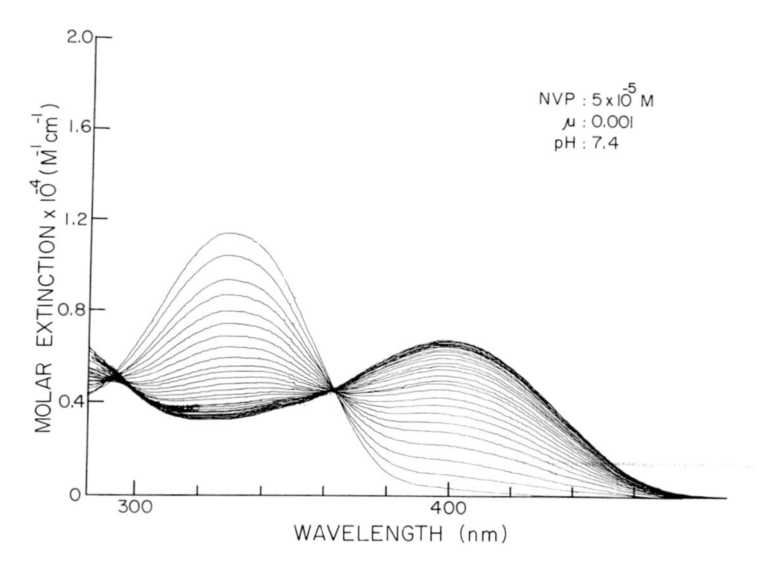

Fig. 4. Titration of NVP with calf thymus DNA. Experimental procedure was the same as for Fig. 2, except that the concentrations in the DNA-ligand stock solution were 4.87 mM DNA_P and 50 μM NVP. The free ligand exhibits a maximum at 330 nm and fully-bound species, at 400 nm

As ionic strength was raised to 0.2, the number of sites for strong binding of NVP and $MNVP^+$ decreased 50 to 100 fold. Isosbestic points were not distinct, and weaker binding was still evident. In addition, the data were complicated at higher ν values by a cooperative or aggregate type of binding, resulting in sigmoidal plots for ν *vs* free ligand concentration. This cooperative binding was much more apparent with unprotonated NVP at high ionic strength and tended to diminish as pH was lowered. The enhancement of cooperative or aggregate binding with increased ionic strength indicated that ligand-ligand stacking interactions were promoted under these conditions.

The spectrum of fully bound NVP at pH 7.4 depended strongly on ionic strength, in contrast to that of $MNVP^+$, which was only slightly blue-shifted by comparable increases in salt concentration. As ionic strength was varied from 0.001 to 0.2 with DNA-NVP, there was a distinct isosbestic point at 367 nm, as bound species at low ionic strength (with a maximum at 400 nm) were converted to a form or forms which predominate at high ionic strength, with a maximum at 336 nm. The spectrum of the unbound species did not pass through this isosbestic point. The maximum at 336 nm may represent a bound uncharged species, but spectra at ionic strength of 0.2 and higher were complicated by the cooperative effect, and no reliable binding parameters could be determined. Increasing ionic strength has the effect of decreasing the ratio of protonated to unprotonated ligand, and is equivalent to raising pH above 7.4. A similar effect on spectra of DNA-bound NVP at pH 7.4 was produced by adding magnesium ions. Approximately 1 mM Mg^{++} produced the same effect as 0.1 M Na^{+}. This would imply that the bound species at 400 nm depends on the presence of anionic DNA phosphate moieties, in agreement with the earlier suggestion that NVP at this wavelength is bound as the protonated species.

2. Denatured DNA

Spectrophotometric titration with heat-denatured calf thymus DNA produced results which indicate that the strong binding component was approximately ten-fold weaker than with native DNA, but that the maximum number of binding sites increased at least two-fold. In addition, the cooperative binding at higher ionic strengths was diminished in comparison with native DNA, a result which is reasonable if one considers the cooperative binding to be reinforced by the regular, relatively rigid local structure of a section of native double helix. The denatured polymer, in a more coil-like configuration, would not present so favorable a surface for the cooperative external stacking of ligand molecules.

IV. Specificity of Binding

1. Base Specificity

The binding of $MNVP^{+}$ and NVP to several ribonucleotide homopolymers was studied to ascertain whether there was any specificity for a particular nucleotide base. Fig. 5 shows spectra for $MNVP^{+}$ at ionic strength of 0.2 and an initial phosphate to ligand ratio of 300. The polymer was present at the same concentration in both sample and reference cuvettes. Therefore the spectra reflect only the ligand in free and bound form. The pyrimidine homopolymers, polyribocytidylic acid and polyribouridylic acid, caused only a small depression and no shift of the absorption maximum at 377 nm. Polyphosphate under the same conditions gave a similar effect. Polyriboadenylic acid showed a relatively small bathochromic shift; polyriboinosinic acid, a rather large shift; and polyriboguanylic acid an even larger shift to 420 nm. This bathochromic shift was markedly greater than that found with native or denatured calf thymus DNA.

With NVP, analogous spectral changes were found with the same homopolymers, and a maximum in the presence of polyriboguanylic acid occurred at 420 nm. However, with the monomeric compounds, guanosine and guanylic acid, only a scarcely detectable spectral change was obtained under these conditions, a shift of not more than one or 2 nm. Therefore the polymeric structure, combining a purine, particularly guanine, with an anionic phosphate backbone appears to favor the binding of naphthylvinylpyridines. It is well-known, however (MICHELSON, MASSOULIÉ and GUSCHLBAUER, 1967), that the homopolymers assume rather specific configurational forms, and the possible participation of secondary structures in the binding of ligands to the homopolymers must be kept in mind. For example, the small bathochromic shift with polyriboadenylic acid was not seen at pH 4.5, where the polymer is thought to exist in a more rigid and perhaps double helical configuration.

Binding studies with polyriboguanylic acid and $MNVP^+$ indicated that, in comparison with denatured DNA, binding was somewhat weaker, but involved more

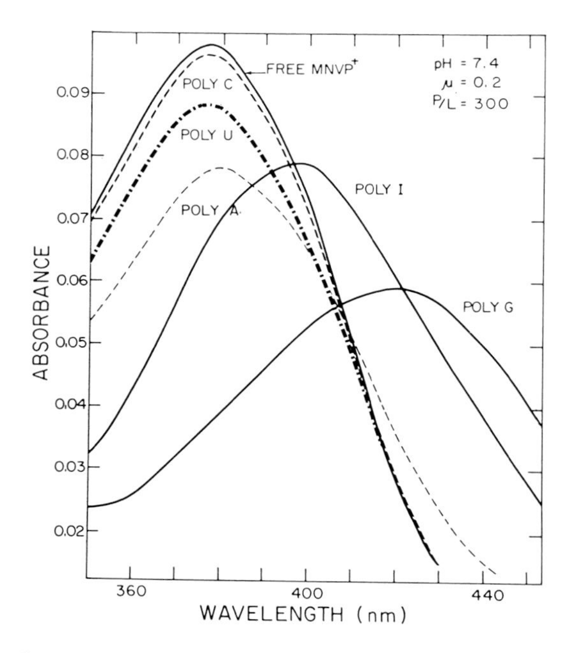

Fig. 5. Interaction of $MNVP^+$ with ribonucleotide homopolymers. Concentration of $MNVP^+$ was 50 μM in phosphate-EDTA buffer, ionic strength = 0.2, pH 7.4. Polymer was present at a phosphate to ligand ratio of 300. Reference cuvettes contained the same amount of polymer. Spectra were determined using a Cary Model 15 recording spectrophotometer

sites. Ionic strength strongly influenced the parameter *n*. At low ionic strength (0.001) a binding of one ligand per nucleotide was detected, with an association constant of about 10^5 M^{-1}.

In summary, the number of available sites for binding of the naphthylvinylpyridinium compounds to nucleic acids was found to increase in the sequence native DNA < denatured DNA < polyriboguanylic acid. The number of available binding sites appeared similar with native DNA samples from *Micrococcus lysodeikticus* (72% $G+C$) and calf thymus (42%), but was decreased with DNA of *Clostridium perfringens* (30%). While an interaction with the guanine moiety is definitely indicated by the observations with homopolymers, other bases in DNA must also participate. An electrostatic interaction between phosphates of DNA and pyridinium moieties of the ligand may be reinforced by interactions with one or two purine moieties on the opposite strand of the double helix.

The different interactions of $MNVP^+$ with the homopolymers may be correlated with ionization potentials or electron donor properties of purine and pyrimidine bases. Guanine is considered the strongest electron donor, followed by hypoxanthine, and then adenine, while pyrimidines are poor electron donors (PULLMAN and PULLMAN, 1963; KORNICKER and VALLEE, 1969).

2. Ligand Modifications

The importance of a positively charged pyridinium moiety for strong binding of this class of compounds to DNA has been discussed above. The presence of an N-methyl group instead of a proton apparently has little effect on binding parameters, since similar results were obtained with NVP and $MNVP^+$ at pH 4.5.

Analogues of $MNVP^+$ were prepared having phenyl or phenanthryl moieties in place of naphthyl. Association constants for strong binding to DNA increased as follows: phenyl $\ll$ naphthyl $<$ phenanthryl. The maximum number of strong binding sites was about one per five bases for all three compounds at pH 7.4 and low ionic strength.

V. Discussion

Evidence from binding studies of a number of intercalating ligands has inspired the description of several possible models of DNA intercalation. The intercalating ligand in all models is a planar, usually three-ringed, heteroaromatic structure. This, in the model proposed by LERMAN (1961) for proflavine, is inserted across the double helix, perpendicular to the helix axis, so that it may interact with base pairs of both DNA strands. Modifications of this model allow for the possibility that an intercalating chromophore may not be exactly perpendicular to the helix axis (MÜLLER and CROTHERS, 1968). In another model, which has been described by PRITCHARD, BLAKE and PEACOCKE (1966), the proflavine molecule is inserted between bases on one side of the double helix so that the proflavine ring nitrogen may interact with an anionic phosphate of DNA. This latter model is thought to be more consistent with the finding that proflavine binding to both native and denatured DNA exhibits nearly identical binding parameters. However, it must be remembered that heat-denatured DNA is typified by a considerable amount of double-strandedness (FELSENFELD and SANDEEN, 1962).

The binding of the naphthylvinylpyridines differs from that of proflavine in that with the former compounds, binding to native DNA is much stronger and involves fewer sites than binding to heat-denatured DNA. It is reasonable to conclude, therefore, that the mode of binding of naphthylvinylpyridines may be changed upon heat denaturation.

In addition, the effect of increasing ionic strength was to decrease the number of available sites for the strongest binding of the naphthylvinylpyridines without significantly decreasing the weaker binding. In fact, the tendency for cooperative or aggregate binding was increased. These observations are in contrast to those with proflavine (BLAKE and PEACOCKE, 1968) for which increased ionic strength was found to reduce more profoundly the weaker electrostatic binding external to the double helix than the stronger intercalative type. Therefore, in the case of the naphthylvinylpyridines, the effect of ionic strength indicates that the strongest mode of binding

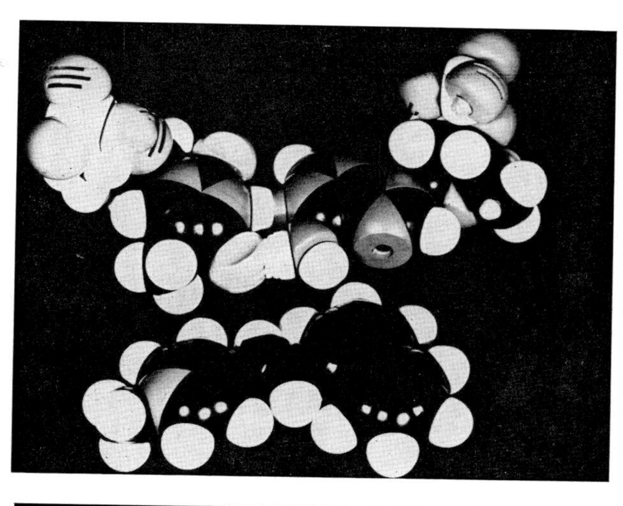

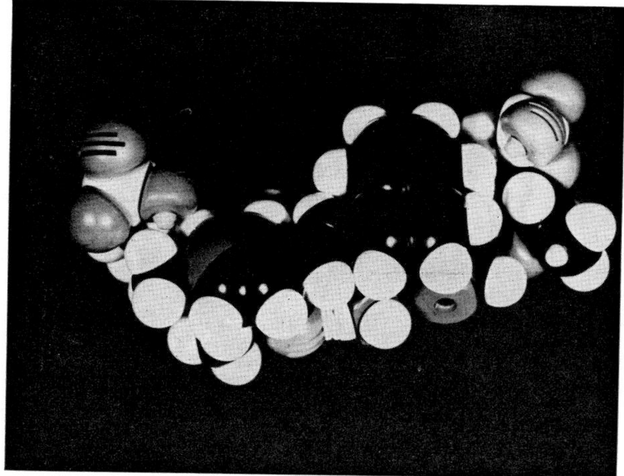

Fig. 6. A possible mode of intercalation by $MNVP^+$. Above are shown KOLTUN, PAULING, COREY space-filling molecular models of a deoxyadenylate-thymidylate base pair (top) and $MNVP^+$ (bottom). Below, the $MNVP^+$ model has been placed on top of the base pair in order to illustrate an appropriate fit

may be largely electrostatic in nature. The anionic character of DNA phosphates may be required for interaction with pyridinium moieties. Following this interaction, the whole NVP molecule may become intercalated, its size being such that it optimally fits the space between two consecutive base pairs (Fig. 6). The pyridinium moiety may at the same time remain in a position of interaction with a DNA phosphate, consistent with the finding that NVP at pH 7.4 is apparently bound as the protonated molecule. However, once a ligand has been intercalated in this way, the electrostatic interaction would be reinforced by base-ligand hydrophobic interactions, and, in analogy with parameters for acridine intercalation, one would not expect the observed large dependence on ionic strength. Accordingly it may also be of importance to consider the effect of ionic strength on DNA configuration. At very low ionic strength the repulsion of neighboring DNA phosphate groups tends to allow an extension of the double helix. In this extended, more rigid configuration the insertion of intercalating ligand molecules would be facilitated.

The preceding model would be characterized by a relatively high binding energy and might be quite sensitive to denaturating influences, which tend to disturb the precise character of the native double helix. It might also be restricted according to the nearest neighbor exclusion hypothesis (CROTHERS, 1968). For the naphthylvinylpyridinium compounds, free energy changes of 8 to 10 kcal per mole for binding to native calf thymus DNA were calculated from association constants, w th a maximum binding of one ligand per five bases of DNA. Lower binding energies were found with denatured DNA and polyriboguanylic acid.

If an $MNVP^+$ molecule interacting electrostatically with a phosphate moiety on DNA is prevented from intercalating, perhaps by nearest neighbor exclusion or by contraction of the double helix due to increased ionic strength, alternative modes of binding might occur. With native DNA at pH 7.4, the maximum number of sites found by extrapolation of either Scatchard or reciprocal plots, for the total of strong and weak binding, was estimated at one ligand per base pair. About 40% of the binding was by the strongest mode at low ionic strength, while more than 99% was by weaker mechanisms at high ionic strength. Since raising the ionic strength did not appear to decrease the weaker binding of naphthylvinylpyridines, which was characterized by free energy changes of from 5 to 7.5 kcal per mole, there is no justification for assuming that this binding represents simply electrostatic interactions between anionic DNA phosphates and cationic ligands. Although such electrostatic interactions may participate, one might suppose that these are somehow reinforced by specific interactions in one groove of the double helix and/or perhaps by partial intercalation. The estimated association constants for the weaker binding of both NVP and $MNVP^+$ at pH 7.4 were very similar, indicating that a discrete cationic charge on the ligand is not an absolute requirement. A mechanism providing for stacking of ligand molecules in one groove of the double helix would allow a prediction of at least one bound ligand per base pair, in agreement with experiment. Partial intercalation might not require the expanded DNA configuration achieved at low salt concentrations and hence might be less sensitive to effects of ionic strength. At the same time increased ionic strength might promote stacking by encouraging ligand-ligand interactions.

The strong intercalative binding observed with proflavine differs considerably from that found with the naphthylvinylpyridines. It may be of significance that the compounds, NVP and $MNVP^+$, although energetically favoring a coplanar configuration, possess a degree of flexibility around the exocyclic double bond, a feature which is lacking in proflavine. The ability of the ligand to undergo some deviation from coplanarity under external influences may permit it to adapt to a more energetically favored bonding relationship. Additional work with modified naphthylvinylpyridine analogues may help to characterize further the mechanism of intercalation by these compounds.

Acknowledgement

This work was assisted by U.S.P.H.S. grants NBO 7932, AIO 6211, and by N.S.F. equipment grant GB-4577.

References

Blake, A., Peacocke, A. R.: The interaction of aminoacridines with nucleic acids. Biopolymers **6**, 1225 (1968).

Cavallito, C. J., Yun, H. S., Smith, J. C., Foldes, F. F.: Choline acetyltransferase inhibitors. Configurational and electronic features of styrylpyridine analogs. J. med. Chem. **12**, 134 (1969).

Crothers, D. M.: Calculation of binding isotherms for heterogeneous polymers. Biopolymers **6**, 575 (1968).

Doty, P., McGill, B. B., Rice, S. A.: The properties of sonic fragments of deoxyribose nucleic acid. Proc. nat. Acad. Sci. (Wash.) **44**, 432 (1958).

Felsenfeld, G., Sandeen, G.: The dispersion of the hyperchromic effect in thermally induced transitions of nucleic acids. J. molec. Biol. **5**, 587 (1962).

Gilbert, M., Claverie, P.: Recherches théoriques sur l'intercalement des aminoacridines dans l'ADN. In: Molecular associations in biology, p. 245 (Pullman, B., Ed.). New York: Academic Press 1968.

Kornicker, W., Vallee, B. L.: Metallocinium cations, nucleic acids and proteins. Ann. N.Y. Acad. Sci. **153**, 689 (1969).

Lerman, L. S.: Structural considerations in the interaction of DNA and acridines. J. molec. Biol. **3**, 18 (1961).

Lesko, S. A., Smith, A., Ts'o, P. O. P., Umans, R. S.: Interaction of nucleic acids, IV. Biochemistry **6**, 434 (1968).

Michaelis, L.: The nature of the interaction of nucleic acids and nuclei with basic dyestuffs. Cold Sprg Harb. Symp. quant. Biol. **12**, 131 (1947).

Michelson, A. M., Massoulié, J., Guschlbauer, W.: Synthetic polynucleotides. In: Progress in nucleic acid research and molecular biology, p. 83 (Davidson, J. N., Cohn, W. E., Eds.). New York: Academic Press 1967.

Müller, W., Crothers, D. M.: Studies of the binding of actinomycin and related compounds to DNA. J. molec. Biol. **35**, 251 (1968).

Pritchard, N. J., Blake. A., Peacocke, A. R.: The interaction of aminoacridines with nucleic acids. Nature **212**, 1360 (1966).

Pullman, B., Pullman, A.: Quantum biochemistry, p. 217. New York: Interscience Publishers 1963.

Waring, M. J.: Complex formation between ethidium bromide and nucleic acids. J. molec. Biol. **13**, 269 (1965).

White, H. L., Cavallito, C. J.: Photoisomerization of styrylpyridine analogues in relation to choline acetyltransferase and cholinesterase inhibition. Biochim. biophys. Acta **206**, 242 (1970).

Formation of Yeast Mitochondria. V. Ethidium Bromide as a Probe for the Functions of Mitochondrial DNA

H. R. Mahler, B. D. Mehrotra and P. S. Perlman

I. Introduction

The last decade has seen a considerable amount of research effort directed towards the indentification and characterization of the components required for a separate genetic apparatus in large cellular organelles (see Roodyn and Wilkie, 1967). As a consequence of these studies we now know that mitochondria and chloroplasts possess their own DNA as well as an entire and separate transcriptional and translational machinery (S. Nass, 1969). The information content of mitochondrial DNA is, however, not sufficient to code for all of the enzymes, RNA species and regulatory factors that are required for the biosynthesis of all the macromolecules of the organelle, much less to code for all factors necessary for mitochondrial function. It has, therefore, been the aim of our own research program to define the extent and nature of the genetic message contained in and expressed by the mitochondria of baker's yeast (*Saccharomyces cerevisiae*).

This system offers a number of practical and technical advantages due to its being derived from a unicellular eucaryotic organism capable of rapid growth, readily responsive to changes in physiological conditions and capable of producing mutant strains deficient in the most important mitochondrial function, that of electron transport and respiration. Specifically our studies have taken advantage of the following characteristics of *S. cerevisiae* for the design of various kinds of experiments:

a) Repressibility of mitochondrial development by glucose (catabolites) (Ephrussi, Slonimski, Yotsuyanagi and Tavlitzki, 1956).

b) Existence of extrachromosomal, cytoplasmic mutants, usually referred to as ϱ^-. These mutants have provided biochemists with an important conceptual and experimental tool since, in certain instances at least, their mitochondrial DNA is of such a nature as to preclude its serving as the genetic blueprint of *any* mitochondrial function.

c) Use of specific inhibitors — Inhibitors are now known which specifically inhibit either the mitochondrial *or* the nucleocytoplasmic protein synthesizing machinery (Huang, Biggs, Clark-Walker and Linnane, 1966; Clark-Walker and Linnane, 1967).

In the studies to be presented here we briefly touch on the exploitation of characteristic b) and more completely on c) particularly as it pertains to the mode of action of the quaternary diamino phenanthridinium dye, ethidium bromide (EB), an agent known for its trypanocidal action.

It has received a great deal of attention in recent years since certain of its properties render it particularly useful as a rather specific probe in molecular biology. At low concentrations, its primary mode of action appears to be a strong, relatively base composition-independent, probably intercalative binding to double helical DNA (WARING, 1965; LEPECQ and PAOLETTI, 1967); this interaction, particularly if the DNA molecule is covalently circular, results in profound alterations in its tertiary structure (RADLOFF, BAUER and VINOGRAD, 1967; BAUER and VINOGRAD, 1968) and interferes with its action as a template in the reactions catalyzed by both the DNA and RNA polymerase (ELLIOT, 1963).

Our active interest in EB was aroused — within the context of our continuing studies on the biogenesis of mitochondria in yeast — by two findings: (1) the discovery by SLONIMSKI, PERRODIN and CROFT (1968) that the dye was a particularly effective and selective mutagen for bringing about the $\varrho^+ \rightarrow \varrho^-$ conversion in this organism, a mutation that has been localized in the mitochondria for both phenotype and genotype; and (2) our own observation that EB at low concentrations was a particularly potent inhibitor of the RNA polymerase resident in and utilizing the DNA of yeast mitochondria (SOUTH and MAHLER, 1968). It should also be mentioned that certain new, and probably highly relevent results have been obtained in animal systems: PENMAN and his collaborators have found that EB specifically inhibits mitochondrial (and not nuclear) RNA synthesis (ZYLBER, VESCO and PENMAN, 1969) and MEYER and SIMPSON have reported that the purified mitochondrial DNA polymerase is more sensitive to the drug than is the nuclear enzyme (MEYER and SIMPSON, 1969).

We have now extended these studies with yeast with regard to the following questions: What is the nature of the primary event (or events) when yeast cells and their mitochondria are exposed to EB; and what are its consequences both on the level of mitochondrial DNA and its expression which in their aggregate lead to the conversion of a wild-type (ϱ^+) into a petite (ϱ^-) cell.

II. The ϱ^- Genotype

Before dealing with the questions just raised it is appropriate to deal briefly with what we have discovered on the molecular level, i.e. at that of mitochondrial DNA with one particular extreme sub-class of the extensive class of cytoplasmic, respiration deficient ϱ^- mutants. We need this information in defining one of the two stable end points in our investigations, the other being of course the point of departure, the respiration sufficient wild-type.

The considerable efforts of many researchers using a wide variety of biological systems during the last ten years support the generalization that mitochondria contain DNA which is distinct from nuclear DNA in terms of information content and (usually) physical properties (RABINOWITZ, 1968; M. NASS, 1969). The results of certain researchers have indicated that yeast mit-DNA is unique among all other mit-DNAs studied, being linear and heterogeneous in length (BILLHEIMER and AVERS, 1969). Recent studies, however, by GUÉRINEAU and colleagues and particularly in Borst's lab (GUERINÉAU, GRANDCHAMP, YOTSUGANAGI and SLONIMSKI, 1968; HOLLENBERG, BORST, THURING and VAN BRUGGEN, 1969; HOLLENBERG, BORST and VAN BRUGGEN, 1970, in press; and BORST, 1969) indicate that yeast mit-DNA can exist as a covalently closed circular molecule with a molecular weight of about 50 million daltons (25 μ in length). It is likely, then, that the information content and/or level of redundancy of mit-DNA is greater for yeast than for any animal cell tested to date.

Since our studies of mit-DNA have already been published (MEHROTRA and MAHLER, 1968) we shall only summarize them here (Table 1).

We have concluded from these studies that mitochondrial DNA in such strains consists predominantly of alternating *dA* and *dT* with interspersed short stretches containing *dG* and *dC*. It is evident that the information content of such a molecule must be greatly reduced, compared to its wild-type counterpart or may even be absent entirely. BORST (1969) has recently reported that the mit-DNA of such an extreme example of petite character contains less than 5% of the information resident in the

Table 1. *Properties of mitochondrial DNA. Experimental details are presented in* MEHROTRA *and* MAHLER, *1968*

Strain	Constitution	Amount[a] (μg/mg)	Density[b] (g/ml)	T_m (°C, in SSC)	% $(A+T)$ analytical
4D	haploid, $P\varrho^+$	7.8	1.675	73.0	83[c]
2A-184	haploid, $P\varrho^-_{50}$	3.6	1.668	66.5	96[c]
4D-21	haploid, $P\varrho^-_{10}$	4.3	1.666	66.5	
FLEISCHMANN	diploid, $P\varrho^+$	0.84	1.678	75.0	79[d]
M-2[e]	diploid, $P\varrho^-$		1.670	68	
Crab $d(A\text{-}T)$			1.671	65.6	96

[a] All per mg mitochondrial protein under conditions of maximal de-repression.
[b] In neutral CsCl, at 20 °C vs *E. coli* DNA at 1.702.
[c] By paper chromatography of labeled hydrolysate.
[d] By paper electrophoresis.
[e] Ethidium bromide mutagenesis.

parental mit-DNA. Nevertheless so far as concerns a) several important morphological criteria, b) the majority protein accounting for mitochondrial mass and c) a number of characteristic mitochondrial enzymes, these strains appear identical to others in the general class of ϱ^- mutants including ones not possessing as severe a lesion in their mit-DNA and closely resemble their parent wild-type [PERLMAN and MAHLER, 1970 (2)]. We shall return to this mutant phenotype in due course.

III. Effects of EB on Genotype and DNA

1. Kinetics of Mutagenesis $\varrho^+ \rightarrow \varrho^-$

We have confirmed the observations of SLONIMSKI et al. (1968) in all important respects with our diploid strain of yeast, in particular with respect to two significant details: That mutagenesis can be obtained by exposing to the drug starved, non-growing cells which can then proliferate into mutant colonies in the absence of extracellular drug and that the conversion of $\varrho^+ \rightarrow \varrho^-$ cells is characterized by kinetics which operationally consist of a lag, followed by a pseudo-first order phase with respect to time (Fig. 1). By using mid-exponential cells grown on lactate and starved for 3 h in phosphate buffer (0.1 M, pH 6.5) we have been able to obtain virtually quantitative (> 99.5% without lethality) conversion. The kinetics can be interpreted using a model postulating single mutagenic events (hits) taking place independently on a number of sites (targets) (ATWOOD and NORMAN, 1949). A computer program (least Squares fit) was used to analyze our data in terms of Eq. 1 derived from this model, which related the time of exposure "t" to the logarithm of the number of wild-type survivors in order to derive "k'" the pseudo-first order constant and "n",

$$[\varrho^+/(\varrho^+ + \varrho^-)] = 1 - (1 - e^{-k't})^n \quad (1)$$

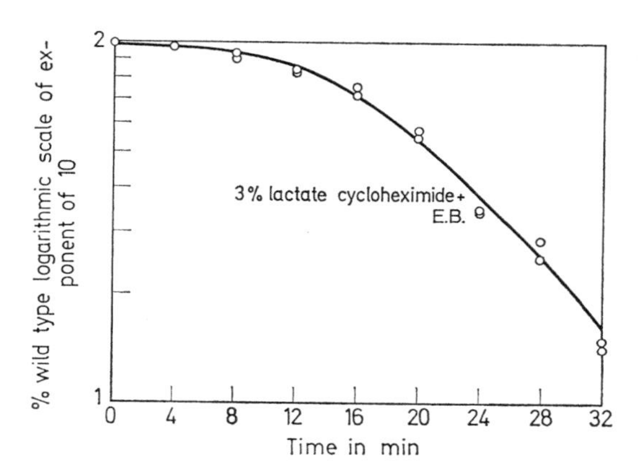

Fig. 1. Time course of mutagenesis by ethidium bromide. Cells were grown and treated as described in the legend to Table 2. The EB concentration was 22 μM. The plot is computer generated to fit Eq. 1

the number of targets. In order to obtain a meaningful estimate of "n" we must assume that the drug reaches an effective intracellular concentration immediately upon addition of EB to the culture. Preliminary results (Table 2 A) show that "n" is independent of the state of repression — and hence the level of functional mitochondria and respiratory capacity — of the cells, provided that all cellular proliferation during the exposure is inhibited by the addition of cycloheximide; "k'" and not "n" is affected by varying the concentration of EB (Table 2), a reasonable outcome if we assume that "k'" is proportional to the number of EB molecules which have formed (or are capable of forming) complexes susceptible to mutagenic change. At cell concentrations between 10^6 and 10^7/ml "k'" and "n" are not affected by the exact concentration used. This observation suggests that the kinetics of mutagenesis in growing cells ought to be subject to a similar analysis provided that at no time the cell concentration is greater than 10^7/ml.

For some experiments, particularly on phenotypic effects it is useful to have available observations on the action of the mutagen added directly to exponentially growing cells. Preliminary evidence shows that such treatment results in mutagenesis analogous to that shown in Fig. 1.

Current experiments show that under, at least, certain conditions (growing cells) the assumption required for the use of Eq. (1) may not be met; therefore additional careful work must be done before all of the data of Table 2 may be evaluated completely.

As mentioned earlier nuclear DNA from yeast consists of about 63% $A+T$ while mit-DNA is 75 to 85% $A+T$. Cytoplasmic petite strains of *S. cerevisiae* have been found to contain mit-DNA with a nearly normal base composition, with a grossly abnormal one (96% $A+T$), or with intermediate ones. No correlation between final base composition and the mode of mutagenesis of such strains has yet been reported.

Table 2. *Kinetic studies of EB mutagenesis*

Standard conditions: Cells were grown to 10^7/ml on 3% lactate or 4% glucose medium, starved for 3 h in 0.1 M phosphate buffer (pH = 6.5) at 10^7/ml. Starved cells were pelleted and resuspended in fresh phosphate buffer at 10^7/ml and were treated with EB as noted in the table at 30 °C. The fraction of ϱ- cells in the population was determined as described by SLONIMSKI et al. (1968) and checked with tetrazolium agar (OGUR, ST. JOHN and NAGAI, 1957). The selective agar used (0,1% glucose + 2% glycerol or 3% lactate) made the tetrazolium overlay superfluous. Data were analyzed according to Eq. 1 by sub-routine "STEP IT", copyright by J. P. CHANDLER (1965), distributed by Quantum Chemistry Program Exchange, Indiana University.

EB (μM)	Carbon source	Inhibitor present during mutagenesis	k (min^{-1})	n
A 21.8	3% lactate	cyclo (2 μg/ml)	0.117	7.9
21.8	4% glucose	cyclo (2 μg/ml)	0.121	5.1
21.8	4% glucose	none	0.0922	16.8
B 10.9	3% lactate	cyclo (2 μg/ml)	0.0485	8.8
21.8	3% lactate	cyclo (2 μg/ml)	0.109	9.8
43.6	3% lactate	cyclo (2 μg/ml)	0.14	6.4

2. Effects of EB on Mitochondrial DNA

The respiratory deficient mutants obtained by treatment with EB, as are those isolated after exposure to other intercalating dyes, appear to be exclusively of the cytoplasmic ϱ^- rather than the chromosomal p variety (SLONIMSKI, P., personal communication). Furthermore the phenotypic consequences of treatment with EB, to be described in the next section appear to rule out effects on chromosomal (nuclear) DNA. However, quite direct evidence is also available by two separate means. Two independently isolated mutant clones were sub-cultured and a sufficient quantity of cells obtained to permit isolation and characterization of their mitochondrial DNA by means of its buoyant density in CsCl. The isolated DNAs exhibited a density of 1.668 g/ml characteristic of the completely aberrant type previously found in certain mutants (MEHROTRA and MAHLER, 1968).

In growing yeast cells the amount of mit-DNA as a fraction of the total is relatively small and determined principally by genetic (FUKUHARA, 1969) rather than by physiological parameters (MOUSTACCHI and WILLIAMSON, 1966). Therefore a test of the effect of EB on mit-DNA synthesis in exponentially growing cells would necessarily be rather insensitive.

Mitochondrial DNA can be labeled preferentially over nuclear DNA in either of two ways. As found by GROSSMAN, GOLDRING and MARMUR (1969) interference with cytoplasmic protein synthesis, in particular by the inhibitor cycloheximide, leads to this kind of preferential incorporation. However, the resultant DNA has only been

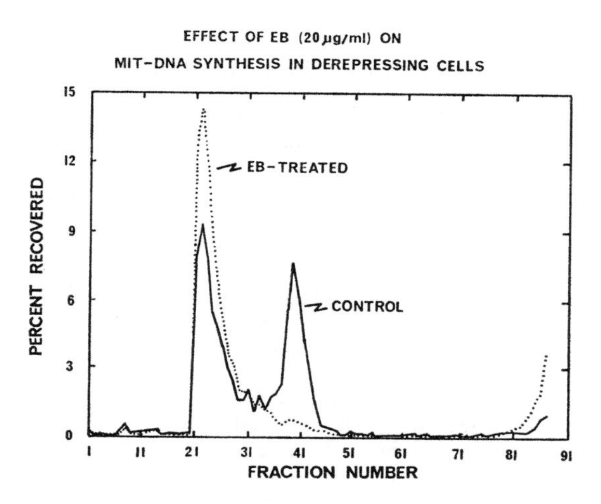

Fig. 2. Effect of EB (20 µg/ml) on mit-DNA synthesis in derepressing cells. Wild-type cells were grown on 1% glucose. Thirty-five min before isotope was added, EB (20 µg/ml) was added to an aliquot of cells. ^{3}H-adenine was added to the EB treated cells and ^{14}C-adenine to the control (untreated cells). The cells were allowed to derepress for 2 h after which time they were harvested and washed. The cells were mixed and converted to spheroplasts which were then lysed with sodium lauryl sarcosinate. A concentrated solution of CsCl in .02 M Tris-Cl (pH 8.5) was added until the refractive index (30 °C) of the solution was 1.3980. The solution was centrifuged at 75,000 xg at 20 °C for 60 h in a 40 rotor using an L-2 ultracentrifuge. The gradient was fractionated into 86 fractions of 0.05 ml each; fractions were prepared for counting as described in MEHROTRA and MAHLER (1968). Settings on the Packard Tri-Carb scintillation counter for counting mixtures of ^{14}C and ^{3}H so that spillover and quench may be estimated were established by the method of HENDLER (1964). All calculations were done using the Indiana University CDC 3600 computer. ——— ^{14}C control cells, ... ^{3}H EB-treated cells. The small peak at the top of the gradient (fractions 81 to 86) is of unknown origin (it is inhibited by cyclo)

characterized by its buoyant density and may not be representative of the mitochondrial DNA present under more normal physiological conditions. We have therefore explored alternative paradigms. One such consists of conducting labeling experiments while mitochondria are actively proliferating subsequent to release from glucose repression. An analogous preferential synthesis of mit-DNA occurs upon addition of molecular oxygen to an anaerobic culture of yeast (RABINOWITZ, GETZ, CASEY and SWIFT, 1969). A representative experiment indicating both the extensive incorporation (35%) into the low density mitochondrial DNA and its preferential inhibition by EB, is shown in Fig. 2. No more than 7% of the total DNA is mitochondrial in this strain grown on 3% lactate (fully derepressed) or 1% glucose (fully repressed).

3. Pedigree of Conversions Resulting in Mutant DNA

We know that transient exposure of a cell to EB results in an event with the following properties: the probability of its occurrence increases with time; more than

Table 3. *Fate of parental DNA subsequent to EB mutagenesis (% DPM retained)*

Two aliquots of actively derepressing cells were labelled for 2 h, one with ^{3}H-adenine and the other with ^{14}C-adenine. The cells were washed and suspended in 0.1 M phosphate buffer, pH 6.5 ($t = 0$). Both samples were starved for 1 h after which EB (20 μg/ml) was added to one sample; the other was treated identically except that EB was omitted. After $1^1/_4$ h of EB treatment the population was 98% ϱ^- and after $2^1/_2$ h it was >99.9%. After $2^1/_2$ h of treatment with or without EB, the cells were washed and suspended in fresh 1% glucose medium containing no drug. After 1, 3 and 4 generations of growth aliquots of EB-treated and control cells were harvested and washed. The EB-treated population remained > 97% ϱ^- throughout the experiment.

The corresponding cell samples were mixed (e.g., control, 1 generation sample [^{14}C] + EB-treated, 1 generation sample [^{3}H]) and converted to spheroplasts using the glusulase procedure as described by LEON and MAHLER (1968).

Angle-rotor preparative CsCl gradients were prepared, centrifuged, fractionated counted and the counts corrected for spillover and quench as described in the legend to Fig. 2. In each gradient the mit-DNA peak was well separated from the nuclear peak so that the total counts in each region could be estimated. Counts in each column are expressed as the percentage of the initial counts retained.

Radioactivity in the mit-DNA peak of the control cells was constant during four generations of growth while that in the nuc-DNA peak in both control and EBIC increased to the same extent; this latter observation can be interpreted in terms of separate adenine pools for nuclear and mit-DNA under these particular conditions. This observation has been made on four separate occassions; it is *not* severely influenced by the presence of a 1 mM adenine chase.

Treatment	Mit-DNA Control	Mit-DNA EB	Nuclear DNA Control	Nuclear DNA EB
$t = 0$	100	100	100	100
1 h starved	98.5	106	107	117
$1^1/_4$ h EB treated	104	75.5	102	116
$2^1/_2$ h EB treated	107	70	103	120
1 generation	101	28.1	121	141
3 generation	99	<6.32[a]	191	204
4 Generation	106.8	<1[a]	204	219

[a] No peak was present. Counts above the baseline in the mit-DNA region were totaled.

one but fewer than ten such occurrences must take place before its consequences become manifest; all mitochondria in the line of descent eventually become mutant and carry the aberrant DNA.

Can we subject the crucial sequence of events to analysis on the molecular level? We have attempted to do this by pre-labeling the parental DNA during active derepression with ^{3}H-adenine prior to exposure to EB, removing the mutagen and growing the cells on 1% glucose medium in the absence of added EB for three generations. The fate of the parental DNA molecules was traced during mutagenesis and subsequent growth.

As an internal control an aliquot of the derepressing cells was labeled with ^{14}C-adenine and treated identically except that the EB treatment was omitted. EB-treated and control cells which were similarly treated were mixed and the radioactivity in nuclear and mit-DNA was analyzed by counting the fractions of a preparative CsCl gradient (see MEHROTRA and MAHLER, 1968). Table 3 summarizes the results of this experiment; we conclude that EB mutagenesis results in the degradation and eventually complete destruction of mit-DNA. No such effects were observed for nuclear DNA in EB-treated cells or for nuclear and mit DNA in control cells.

In conjunction with the demonstration of a preferential inhibition of mit-DNA synthesis by EB these observations suggest the following model: EB treatment inhibits irreversibly all further synthesis of parental wild type mit-DNA. Destruction of this species begins during the mutagenic treatment and becomes quantitative during the first few generations of growth. It is worth noting, at this point, that we were completely unable to detect any shift of the label in this parental DNA towards a "lighter" position in the gradient, i.e. in the direction expected for a petite DNA enriched in $A+T$. Thus the latter probably does not arise in the course of a semiconservative mode of replication. Experiments designed to determine just when and how EBIC begin making this, or any other, aberrant mit-DNA are in progress.

4. Genotypic Effects on Other Yeast Species

All of the results described so far have been obtained with *Saccharomyces cerevisiae*, a yeast that is classified as facultatively anaerobic and petite positive. These two properties are not independent parameters. Only because such a yeast is able to satisfy all its energy requirements by glycolysis is it able to grow anaerobically and therefore the (intrinsically anaerobic) ϱ^- genotype is viable as long as it is provided with a readily fermentable energy source. It would constitute a lethal mutation if it were restricted to the use of non-fermentable carbon sources.

What then is the effect of exposing to EB, under comparable conditions, petite-negative obligately aerobic species? Provided that the respiration of such species is *entirely* linked to mitochondrial cytochromes interference of the continued production of functional mitochondria as a consequence of the homologue of the ϱ^- mutation should be lethal, even on glucose. This appears to be case for yeasts *Hansenula wingei* and *Torulopsis utilis*. When a culture of either strain (grown on 5% glucose — these strains are not subject to catabolite repression by glucose) is treated with 50 μM EB under growing *or* non-growing conditions, the viable count decreases as a function of time until as few as 5% of the cells remain capable of forming a colony on nutrient agar. In each case cells "killed" by EB treatment are capable of limited growth so that interference with mitochondrial and not nuclear gene function is likely regardless of the kind of organism used. KELLERMAN, BIGGS and LINNANE have reported (1969) that an obligately aerobic yeast, *Candida parapsilosis*, is capable of growth in the presence of EB and that growth results in the formation of grossly abnormal mitochondrial morphology. They did not report any lethal effects of the agent.

IV. The ϱ^- Phenotype

Before discussing our analysis of the phenotypic effects of EB treatment a brief description of important biochemical features of the end product, ϱ^- cells, is in order.

Cytoplasmic petite mutants of *Saccharomyces cerevisiae* have been available for genetic and biochemical analysis for a number of years (EPHRUSSI et al., 1949; SLONIMSKI, 1953 and reviewed by NAGAI, YANAGISHIMA and NAGAI, 1961). Treatment of wild-type cells with agents such as fluorouracil (MOUSTACCHI and MARCOVICH, 1963), particular acridine dyes (EPHRUSSI, HOTTINGUER and CHIMENES, 1949), ultraviolet light (ALLEN and MACQUILLEN, 1969) and many other chemical agents (SARACHEK, 1959; NAGAI, 1962) results in very high levels of petite mutants in the treated population. The mutation also occurs spontaneously with certain common strains characterized by mutational frequencies as high as 10^{-2} and hypermutable ones reaching levels an order of magnitude higher.

Very little is known about the primary molecular lesion that arises as a consequence of mutagenesis nor is it known whether this lesion varies with the mutagenic agent; only the end product has been studied in the form of the eventual phenotype and even that in a fashion that still leaves much detail unexplored (see, for example, SLONIMSKI and EPHRUSSI, 1949; REILLY and SHERMAN, 1965; SHERMAN and SLONIMSKI, 1964; and MACKLER, DOUGLAS, WILL, HAWTHORNE and MAHLER, 1965).

In general, cytoplasmic petites contain mitochondria which appear to lack inner membrane — or at least which have no cristae (YOTSUYANAGI, 1962). Although membrane fractions from ϱ^- cells have been reported to contain cross-reacting material (CRM) to cyt ox (KRAML and MAHLER, 1967) and NADH: cyt *c* reductase (MAHLER, MACKLER, SLONIMSKI and GRANDCHAMP, 1964) antisera they lack the absorbance peaks and enzymatic activity normally associated with the CRM. The DNA contained in petite mitochondrial fractions, while different in buoyant density from nuclear DNA is also less dense (higher $A+T$ content) than normal mit-DNA (MOUNOLOU, JAKOB and SLONIMSKI, 1966; MEHROTRA and MAHLER, 1968; CARNEVALI, MORPUGO and TECCE, 1969); such a shift in buoyant density rules out point mutation and is highly suggestive of a massive deletion mutation and a great diminution in information content. The absence of mit-RNA in petites has been reported (WINTERSBERGER and VIEHSHAUSER, 1968) although another laboratory has reported the presence in petite cells of RNA species which hybridize with petite mit-DNA (FUKUHARA, FAURES and GENIN, 1969). These conflicting reports serve to accentuate the likelihood that the information content of petite (ϱ^-) mit-DNA may vary all the way from close to normal to virtually zero in spite of the close similarity or identity of their customary respiration deficient phenotype.

Thus to a first approximation we may connect the failure to synthesize both respiratory enzymes and components necessary for a structured (cristate) inner membrane with the absence of mitochondrial genes, or more generally, with the absence of a large portion of normal mit-DNA.

A recent study from our laboratory [PERLMAN and MAHLER, 1970 (2)] describes characteristics of highly purified mitochondrial fractions isolated from an acridine-induced petite derivative of a haploid strain of *S. cerevisiae*, the mit-DNA of which shows the most severe lesion reported to date, 96% $A+T$.

Such respiratory deficient particles resemble their functional counterparts in such physical parameters as buoyant density, sedimentation coefficient, and observable size. They have a well-defined but non-cristate inner membrane. Studies *in vivo* (SCHATZ and SALTZGABER, 1969) and *in vitro* (DAWIDOWICZ, unpublished observation) indicate that the chloramphenicol-sensitive amino-acid incorporating system is absent from mitochondria of such ϱ^- cells.

Except for the complete absence of the enzymes of the respiratory chain (except perhaps SDH) these particles contain normal, or nearly so, levels of MDH, IDH, citrate synthase and F_1 ATPase. All those enzymes that are *present* may, then, be excluded from further consideration as having been encoded in mitochondrial genes while all those enzymes and functions that are *absent* are related in some manner to mitochondrial gene products.

V. Effects of EB on Phenotype

1. Properties of EB-induced Petites

Four independent isolates of EB-induced petites were grown in quantity on 1% glucose and their content of respiratory enzymes measured. Table 4 compares them to the wild-type, to an authentic cytoplasmic petite (haploid-strain 4D-21) and to a haploid wild-type strain (4D) isogenic to the petite. No significant differences in enzyme content are apparent among the four EB petites. They are, like an authentic ϱ^- strain, characterized by the complete absence of cytochrome-linked respiratory enzymes but contain measurable and often high levels of a cold stable ATPase (F_1) MDH, GDH, and IDH.

2. Effects on Established Phenotypes

The rapid and quantitative mutagenesis of wild-type, respiration sufficient cells under non-growing conditions may be a potent probe for studying mitochondrial gene function provided that the drug treatment does not reduce the viable count or in any way interfere with the proper functioning of the respiratory apparatus present in the cells before drug treatment.

The exposure of cells to EB even after extensive starvation and for prolonged periods of time has no effect on the activity of the existing complement of a wide variety of soluble and membrane bound enzymes (Table 5). Effects on expression of nuclear genes can be dissociated from those that may be a consequence wholly or in part for the mutagenic alteration of mitochondrial genotype by studying the consequence of exposing to EB cells of a genotype that has already been altered: since the growth and enzyme complement of established ϱ^- petites is completely unaffected by the presence of the drug we conclude that it does not inhibit the expression of nuclear genes. Similar results are also obtained when an alteration in the mitochondrial genotype, although present, proves to be irrelevant, because the mitochondrial functions are dispensable under these particular conditions. This is the case, for instance when *S. cerevisiae*, mutagenized by EB, is placed on 1% galactose, a fermentable, weakly repressing sugar: again there is no effect on the growth of such cells. This kind of analysis can be refined and extended to the level of synthesis of individual enzymes

Table 4. *Enzyme levels*[a] *in homogenates*

All cells were grown for 20 to 24 h on 1% glucose to stationary phase (~ 5 × 10^7/ml for wild-type and ~ 2 × 10^7/ml for petites). The EB petites were single colony isolates from a single batch of mutagenized cells which had been subcultured twice before they were used in this experiment. Activities were assayed in a mechanically prepared cell-free homogenate [PERLMAN and MAHLER, 1970 (1)]. The petites all contain a cold-stable, oligomycin-insensitive, F_1-inhibitor-sensitive, ATPase in their mitochondria.

	Succ:cyt c red	SDH[b]	Cyt ox	Anti a sens NADH:cyt c red	MDH	NAD GDH	NADP IDH	Catalase
Diploid wild-type (Grande)	45	61	230	57	3120	140	100	2620
Stable EB 1	0	0	0	0	830	21.5	22	150
Petites 2	0	0	0	0	890	21.8	22.5	120
3	0	0	0	0	1000	19.0	24.1	136
4	0	0	0	0	823	19.9	19.2	101
Haploid wild-type (4D)	72	76	150	81	3460	50.9	86	5750
Petite (4D-21)	0	0	0	0	469	30	35.6	1052

[a] nmoles · min^{-1} · mg^{-1}.

[b] SDH activity in petites could not be measured in homogenates. Preliminary evidence using highly purified mitochondrial fractions from petite cells reveals a very low, but measurable level of SDH activity in 4D-21.

by studying examples of classes that previous experience suggests are specified by nuclear genes and synthesized on extramitochondrial ribosomes: these include soluble enzymes such as L-glutamate dehydrogenase (GDH) and those which are localized both in the cytosol and the easily solubilized (matrix) compartment of the mitochondria such as L-malate dehydrogenase (MDH) and the NADP-linked isocitrate dehydrogenase (IDH). As can be seen from Fig. 3 EB is completely without effect on the biosynthesis of this class of enzymes whether we use *S. cerevisiae* exposed continuously to EB while growing on lactate (Fig. 3 A), the same yeast previously mutagenized by this agent and subsequently grown on galactose (Fig. 3 B), or the aerobe *T. utilis* treated continuously with EB while growing on glucose (Fig. 3 C).

Table 5. *Effects of starvation and EB on enzyme activities*

A sample of yeast was grown in 1.8% glucose for 17 h. All manipulations were done using sterile equipment. The cells were starved of a nitrogen source in 0.1 M PO_4, pH 7.4 + 1.8% glucose for 5, 10 or 20 h with shaking at 30° C. The cells were centrifuged and resuspended in 0.1 M PO_4, pH 7.4 for treatment with EB at a concentration of 21 μM. A cytoplasmic extract was prepared from each sample by mechanical cell breakage using glass beads. It was centrifuged once at 2000 xg in order to remove cell debris and unbroken cells. Enzyme assays were as described in PERLMAN and MAHLER [1970 (1)].

		Cyt ox	SDH	Succ: cyt c red	MDH	Anti A sens NADH: cyt c red
Grown	17 h	97.3	61	34	2280	47
Starved	5 h	226	49	40	2818	41
	10 h	222	38	34	2130	62
	20 h	155	61	39	2980	63
EB	0	230	53	44	3340	76
	15 min	247	63	45	3300	96
	60 min	210	52	44	2780	103
	5 h	302	60	55	3340	123
	10 h	213	43	38	2230	58
	σ	57.5	8.8	6.53	488	27.65
	σ (%)	27	16.6	16	17.5	37
Avg		211	53	41	2800	74

3. Effects on Emergent Phenotypes

With the experiments just described as controls we can now address ourselves to the question: what are the phenotypic consequences of the interaction of EB specifically with *mitochondrial* DNA? In particular we want to know whether (1) the rate and manner of appearance of the emergent (mitochondrial) phenotype can be accounted for strictly as a consequence of the prior alteration in genotype and (2) whether these phenotypic alterations emerge in a generalized coordinate fashion or whether we can discern indications of selective in the sense of interference with the products of certain activities to the exclusion of others. To anticipate and summarize the results; exposure to EB either continuously, in the case of growing cells —

either of *S. cerevisiae* or of *T. utilis*, or only for a period sufficient for mutagenesis in the absence of growth leads to rapid block in the continued synthesis of a single detectable respiratory activity, namely, cytochrome *c* oxidase (cyt ox). That of all others continues, under appropriate conditions, for several generations.

The answer to question 1 is clearly in the negative in the sense that transcription even from the parental strands *must* have been interfered with irreversibly as a result

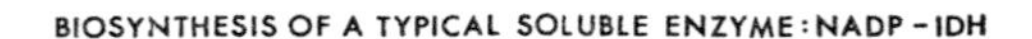

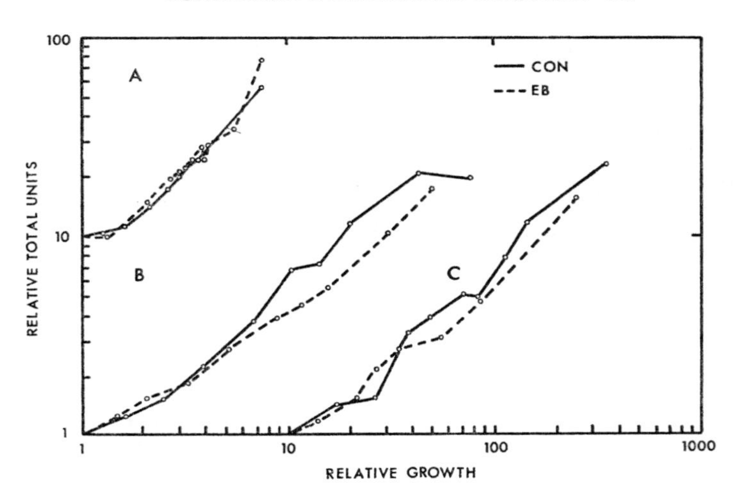

Fig. 3

	Initial specific activity nmoles · min^{-1} · mg^{-1}	Initial cell concentrations (turbidity)	Cells
A	105	0.5	*S. cerevisiae* growing on 3% lactate
B	75	0.092	*S. cerevisiae* EBIC growing on 1% galactose
C	250	0.17	*T. utilis* growing on 5% glucose

of the initial exposure to EB. As a corollary DNA replication subsequent to this exposure, in the absence of the drug, cannot have been semiconservative, since the relevant information of the daughter produced under these conditions has been lost. The verification of this point is in progress. The answer to question 2 is in favor of the second alternative. Hence, as a corollary the *initial* consequence of interfering with the mitochondrial genome and its expression with respect to the biosynthesis of the respiratory chain is restricted to just one segment: the elaboration of an active cytochrome oxidase. These conclusions are based on a number of different experiments which will now be described in some detail.

4. Length of Phenotypic Lag

One way of performing such experiments, which at the same time allows one to draw conclusions concerning the above questions, is to place mutagenized cells on a selective medium which does not support the growth of the established ϱ^- phenotype, e.g. on lactate. Again, all inferences drawn will be strengthened if we compare results

Table 6. *Comparison of enzyme levels in cells growing on 3% lactate; relative total units*[a]

A. Cells grown on 3% lactate for 30 h to a final cell concentration of 5×10^7/ml were starved for 3 h in 0.1 M phosphate buffer (pH 6.5) and treated with 78.5 μM EB for 2 h at a cell concentration of 2.6×10^6/ml. The washed mutagenized (and control) cells were then diluted in fresh 3% lactate medium at 3.4×10^6/ml for growth. Samples were harvested periodically and enzyme synthesis was measured. Specific activities at t_0 were similar for EB treated and control cells and are reported in parentheses in the table.

B. Cells growing on 3% lactate were split into two aliquots at a cell concentration of 4×10^6/ml EB at 50 μM was added to one aliquot and samples of cells were collected periodically and enzyme activities were measured. During the first hour of EB treatment the ϱ^+— ϱ^- conversion took place quantitatively.

A. Cells previously mutagenized

	t (h)	Growth	Cyt ox	SDH	MDH	% ϱ^-
EB	0	1	1(356)[b]	1(43.6)	1(2620)	99.5
	4	2	.57	.68	1.12	99
	8	4	1.34	5.3	3.5	93
Control	0	1	1(238)	1(36)	1(2410)	<1
	3.5	2	1.1	.76	.44	<1
	8	4	2.8	3.2	3.0	<1

B. Cells growing in the presence of EB

	t (h)	Growth	Cyt ox	SDH	IDH	+ ϱ^-
EB	0	1	1(165)	1(21)	1(105)	<1
	3	2	1.4	4.4	1.5	99.5
	10	4	2.4	10	2.8	99.5
Control	0	1	1(165)	1(21)	1(105)	<1
	4.7	2	3.0	4.4	1.3	<1
	10	4	5.3	12	2.8	<1

[a] Relative total units are total units (per ml of culture) normalized to $t = 0$.
[b] Numbers in parentheses are specific activity (nmol · min^{-1} · mg^{-1}) at $t = 0$.

with non-growing, previously mutagenized cells growing in the absence of drug with continuously exposed cells: growth data for the two sets of conditions, using a variety of parameters (turbidity, and proportion of ϱ^- cells in the population) and some relevant enzyme data are summarized in Table 6. Two sets of conclusions appear justified. (1) The petite phenotype emerges rapidly and discontinuously; mutagenized cells grow reasonably rapidly but for only a limited number of generations (or mass doublings): this number is between two and three. (2) The levels of all enzymes studied increase isometrically except for cytochrome oxidase which is produced at a reduced rate in both cases. These findings by themselves rule out any models in which a genetically and phenotypically competent parent throws off daughters deficient in one or both these regards.

Limited growth of cells which are genetically petite but phenotypically wild-type under non-repressing conditions coupled with the selective inability to synthesize cyt ox activity suggest that growth ought to proceed for something more than two

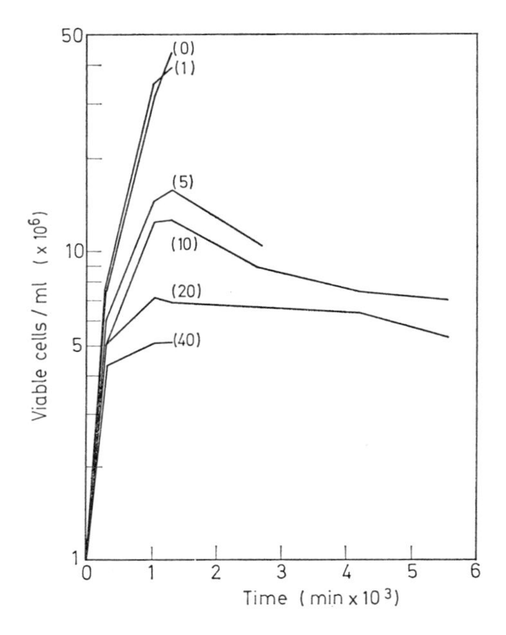

Fig. 4. Cells of *S. cerevisiae* growing on 3% lactate were treated with various EB concentrations (µg/ml shown in brackets) beginning at 10^6 cells/ml. Viable counts and % ϱ^- were measured periodically. The maximum length of time required to reach > 90% ϱ^- was 900, 300, 90, 50, and 30 min corresponding to 1, 5, 10, 20 and 40 µg EB/ml respectively. Differences in total growth may be correlated with the length of time required for quantitative mutagenesis so that about 2 to 3 generations of growth occurred after mutagenesis was complete in each case. The aliquot treated with 1 µg EB/ml showed normal growth and was not completely mutagenized until the carbon source was nearly exhausted

generations but with an ever decreasing rate. In the experiments reported in Table 6 growth appeared to cease precipitously after the second generation of growth. Therefore we set out to examine in detail the amount of growth possible under these conditions and the cause of the unexpected cessation of growth.

In a separate experiment aliquots of cells growing on 3% lactate were treated with several different EB concentrations (0, 1, 5, 10, 20 and 40 µg EB/ml), each of which requires a different length of time to reach, say, $\geq 90\%$ ϱ^- (indicated in legend). Growth (viable count) continued to increase for 2 to 3 generations *after* the population consisted entirely of petite-colony forming cells and then stopped. Viable count was measured for up to 3 days after the cessation of growth and no further increases were detected (Fig. 4).

Cells which had grown for 22 h in the presence of EB were then analyzed for their content of respiratory and soluble enzymes and compared to the activity present at the time of drug addition (Fig. 5).

Cells which had been treated with mutagenically effective EB concentrations showed a marked ability to synthesize the respiratory enzymes Succ:cyt *c* reductase,

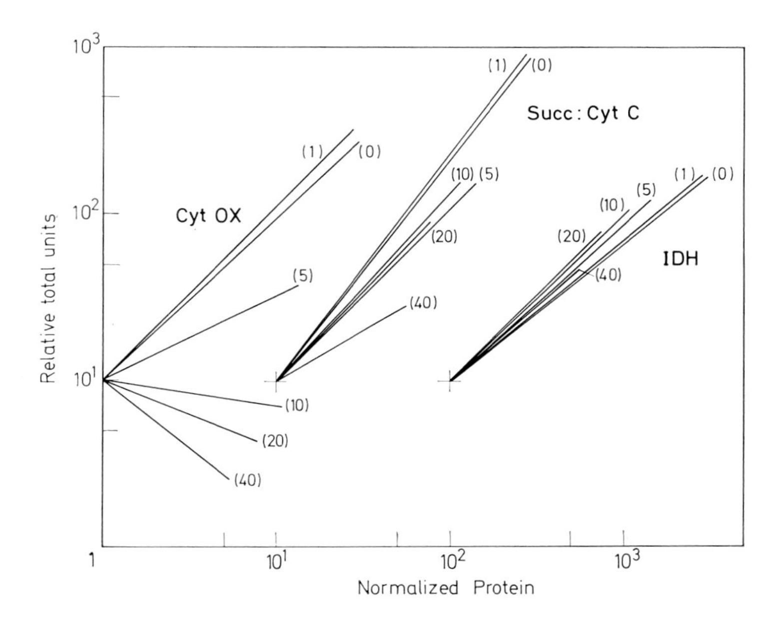

Fig. 5. Duplicate aliquots of the cultures whose growth is shown in Fig. 4 were harvested after 1300 min of growth after the addition of EB. Enzyme activities in cell free homogenates were assayed. The point at which the population consisted of $> 90\%$ ϱ^- cells is indicated on the cyt ox panel by arrows. Specific activity at $t = 0$ was 187, 29.1 and 115 nmoles · min^{-1} · mg^{-1} for cyt ox, succ:cyt c reductase and IDH respectively. Cell protein concentration at $t = 0$ was 0.02 mg/ml

SDH and NADH:cyt *c* reductase and the soluble enzyme IDH. The inability to synthesize cyt ox activity, however, was total for the three samples which were mutagenized very rapidly. In the aggregate, data presented in Table 6 and in Figs. 4 and 5 indicate that cyt ox synthesis is severely inhibited once mutagenesis is completed and that the activity of that enzyme is stable during growth. As the rate of growth declines cyt ox activity becomes labile leading to the rapid cessation of growth.

The alteration in mitochondrial function indicated by the changed enzymatic composition may be correlated with a profound morphological change (Fig. 6). Mitochondria of cells grown for 22 h on lactate in the presence of 25 μM EB possess a greatly diminished cristae content and contain some electron-dense granules of unknown origin and function (see also Fig. 9). Since mitochondria of nuclear mutants which lack cyt ox activity have a cristate inner membrane, we may conclude that a functioning mitochondrial genome is essential for cyt ox synthesis and cristae formation and that these processes are separate.

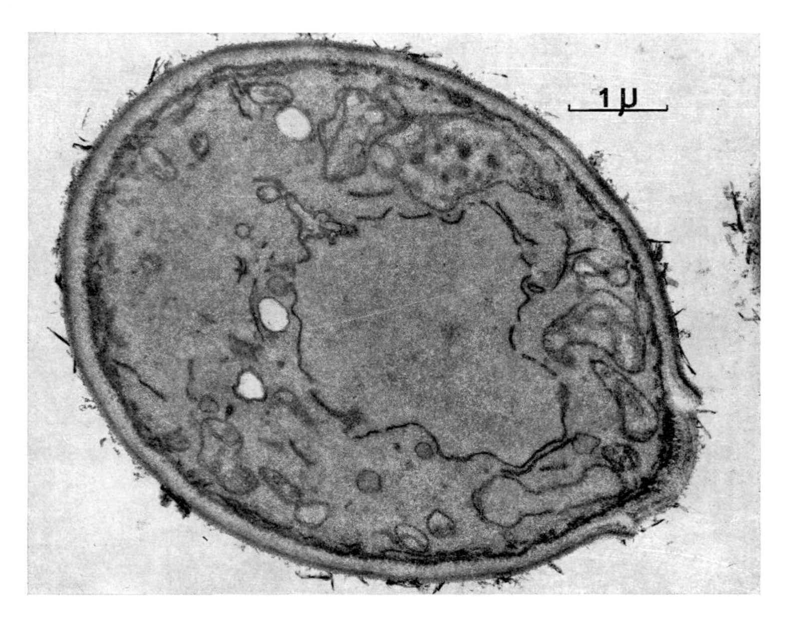

Fig. 6. Electron micrograph of a typical cell of *S. cerevisiae* grown in the presence of 10 μg EB/ml for 22 h on 3% lactate. Mitochondrial shape is more irregular than in the control samples and cristae content is evident but is much reduced. Cells were prepared for microscopy ($KMnO_4$ fixation, uranyl acetate and lead citrate stain) as described in PERLMAN and MAHLER [1970 (1)]

5. Comparison of Transcriptional with Translational Blocks

As indicated earlier placing previously mutagenized cells on galactose allows continued observation of the emergent phenotype since the (changing) mitochondrial contributions are gratuitous. A representative experiment is shown in Fig. 7. The sequence of events is that formation of cytochrome oxidase is blocked virtually completely from the start while that of all other respiratory enzymes continues for a limited period of time, with the time of cessation and the total number of units formed by that moment differing for different activities in the order SDH > succinate:cyt *c* reductase > NADH:cytochrome *c* reductase. Synthesis of soluble enzymes (GDH, IDH and MDH were assayed) was unaffected (Fig. 3). In other words, in these experiments as in the previous ones, treatment with EB must have led to interference with the expression of the mitochondrial genome of the parent prior to the first cellular (and mitochondrial division), presumably at the transcriptional level (perhaps by the same event that also results in the altered genotype). Subsequent to it (no EB present) the alteration in the genotype is itself necessary and sufficient to maintain this block which is at first manifested solely by a lesion in the formation of cytochrome oxidase. With time, this lesion (and perhaps others in parallel not evident at the level of resolution employed)* leads to secondary consequences which are reflected by an

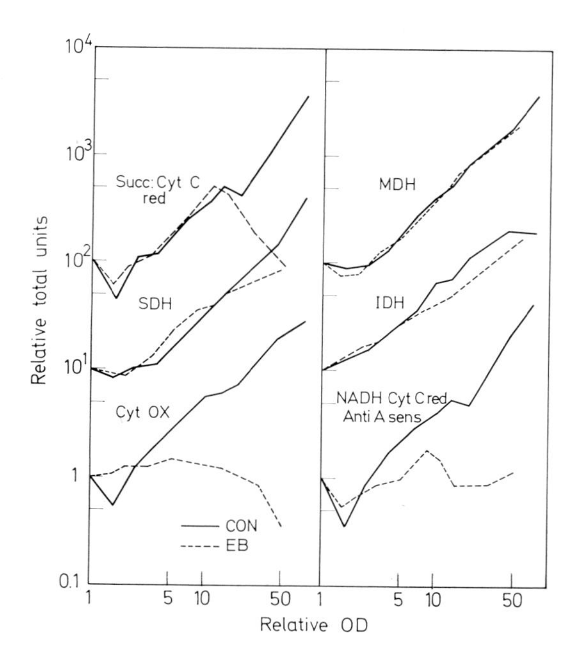

Fig. 7. Cells of *S. cerevisiae* grown to 2 × 10^7/ml on 3% lactate were starved and mutagenized for 2 h in 22 μM EB. A control sample was treated the same except for deletion of the EB treatment. Cells were then inoculated into fresh 1% galactose medium at an initial A_{600} of 0.09. Cell samples were collected periodically (the experiment lasted for 12 h) until 5 to 6 generations of growth had occurred. In this experiment A_{600} was a valid measure of growth. During the entire experiment the percentage of petite colony forming cells was < 1 for the control and > 99 for EBIC. Homogenates were prepared by grinding with glass beads and enzyme activity for the various marker enzymes was measured. Initial specific activity (nmoles · min^{-1} · mg^{-1}) of succ:cyt c reductase, SDH, cyt ox, MDH, IDH and NADH cyt c red was (for control cells) 84, 92, 286, 2780, 76 and 134 and (for EBIC) 88, 93, 281, 2620, 79 and 140 respectively. The decrease in total activity for several enzymes observed during the first few hours has been examined in some detail and appears to correspond to destruction of activity due to the metabolic step-up. It is not influenced by EB treatment or by the carbon source into which the starved cells are placed

inability of the resultant aberrant mitochondria to accomodate and integrate other respiratory enzymes synthesized elsewhere. Finally, cessation of synthesis or integration reflects itself in an apparent destruction of various activities. This sequence of events is also manifest if we examine mitochondrial competence (or rather emergent

* A block in their synthesis may not lead to demonstrable consequences if their apoproteins are present in excess initially.

*in*competence) not at the level of function, but at that of structure: the loss of fine structure in the inner membrane and cristae — which we might regard as an inverted morphogenetic sequence — is apparent from electron micrographs taken at the different times subsequent to resumption of growth after EB treatment (similar to Figs. 6 and 9).

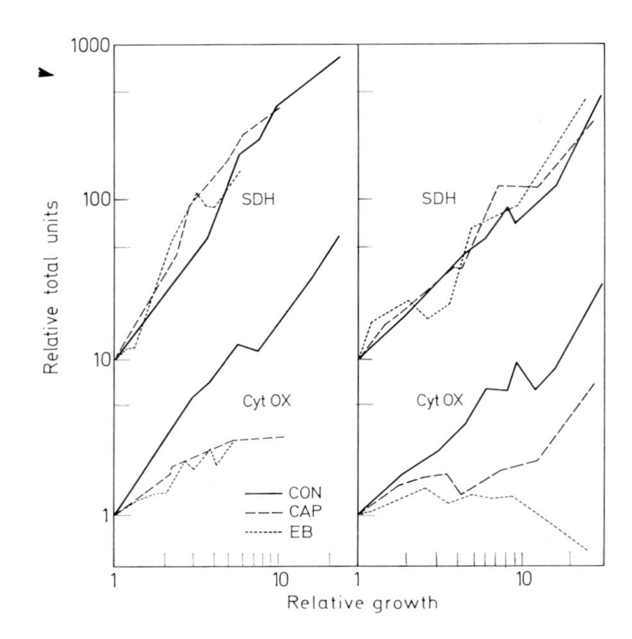

Fig. 8. The effects of CAP (4 mg/ml) and EB (20 μg/ml) were compared in two strains of yeast, one a facultative anaerobe *(S. cerevisiae)* and the other an obligate aerobe *(T. utilis)* both growing in the complete absence of catabolite repression (3% lactate and 5% glucose, respectively). Initial specific activity (nmoles · min^{-1} · mg^{-1}) of SDH and cyt ox were (for *S. cerevisiae*) 21 and 165 and (for *T. utilis*) 35 and 147. Initial A_{600} was 0.09 and 0.17 respectively. Samples were collected periodically and enzyme activity measured in a cell free homogenate. Each experiment lasted about 10 h; however the last *T. utilis* samples was incubated for 12 additional h. The increase in cyt ox activity in the last point of the *T. utilis* (CAP treated) curve is due to a decrease in effectiveness of CAP as the pH of the medium dropped below pH 6.0 (to 4.0). By comparing Figs. 5 and 8 we observe that the initial effect of EB at the level of enzyme synthesis is the inhibition of cyt ox synthesis. A period of marked lability occurs only upon prolonged treatment with EB and is specific for cyt ox

The model postulated for the primary phenotypic effect of EB (prior, during, or just subsequent to the first division cycle when applied to cells mutagenized under non-growing conditions, or when added to growing cells) is an interference with mitochondrial gene expression at the transcriptional level. Provided that no message crucial for mitochondrial function leaves the particle and all such messages are translated there, we would then expect, to a first approximation, that interference with this process, i.e. of gene expression at the translational level, should lead to very similar consequences. That this is, indeed, the case both for *S. cerevisiae* growing on lactate, and for *T. utilis* growing on glucose is shown in Fig. 8 where we compare the effect of the addition of EB and the translational inhibitor chloramphenicol (CAP) on the biosynthesis of the mitochondrial activities, succinate dehydrogenase and

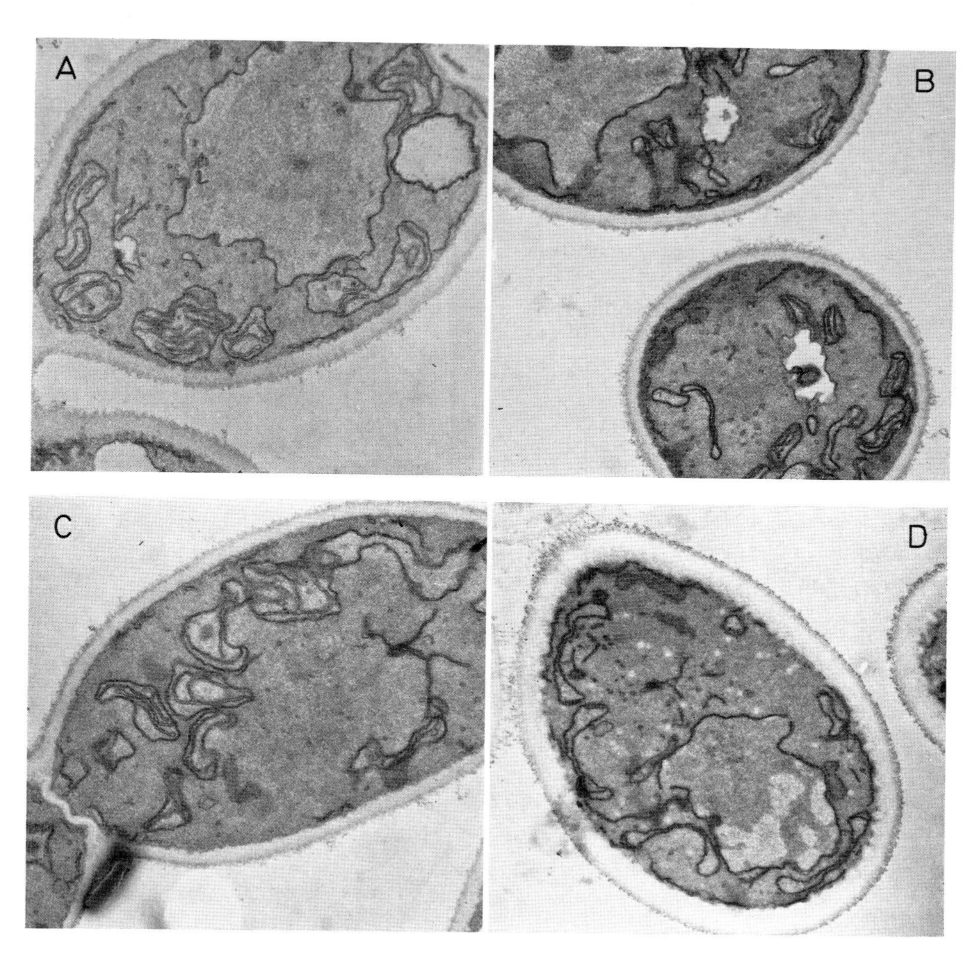

Fig. 9. Cells of *T. utilis* grown in the presence of 20 μg EB/ml were prepared for electron microscopy as in Fig. 6. Length of time growing in EB (h) and normalized growth are A) 2,2; B) 4, 3.6; C) 8,9; D) 19.5, 25, respectively. Control cells are indistinguishable from frame A. Cristae are present in all samples although there is a progressive loss of them and an increasing irregularity of size and shape of mitochondria. The cells of frames C and D were unable to form colonies on nutrient agar. Enzyme data for this series of samples are presented in Figs. 3 and 8

cytochrome oxidase. The first is representative of soluble matrix, and most respiratory chain enzymes and remains unaffected: the second is again unique. Finally, the morphological equivalent is shown in Fig. 9 for the case of *T. utilis* exposed continuously to EB for 2, 4, 8 and $19^1/_2$ h (9 A, B, C, and D rsepectively).

The primary effect of chloramphenicol is inhibition of protein synthesis on mitoribosomes. It does not inhibit mit-DNA synthesis (GROSSMAN et al., 1969 and PERLMAN and MAHLER, in preparation) and its effects are fully reversible. Inhibition of mit-RNA synthesis by CAP has not been reported.

VI. Conclusions

From what we have been able to convey in this presentation and from a number of related experiments we think that we are justified in concluding the following.

1. Using mitochondriogenesis in living yeast cells of several unrelated species as the paradigm EB has been shown to be a highly selective agent capable of blocking mitochondrial function exclusively.

2. Its primary mode of action is a very rapid block of both the continued synthesis, and the transcription of mitochondrial DNA.

3. The initial consequence becomes manifest as an inability to continue production of functional cytochrome oxidase: All other mitochondrial functions including ones tightly linked to the inner membrane and the cristae of the organelle continue to be made at normal rates and in normal amounts.

4. The eventual outcome is the accumulation of aberrant mitochondrial DNA and of mitochondria exhibiting the pleiotropic respiratory lesion characteristic of ϱ^- strains. Such mutants are viable in a facultative anaerobe such as *S. cerevisiae*; they may be lethal in an obligate aerobe such as *T. utilis*.

Acknowledgements

We acknowledge the excellent and essential technical assistance of Mrs. KAREN WALKER and Mrs. CAROL WILLIAMS as well as their assistance in the preparation of this manuscript. We also thank Mr. MICHAEL ONDRIK for consultation on computer applications essential for these studies; and Mr. HARRY COHEN, for the electron microscope studies and photographic work.

Supported by Research Grant GM 12228 from the National Institute of General Medical Sciences, National Institutes of Health, U.S. Public Health Service.

References

ALLEN, N. E., MACQUILLAN, A. M.: Target analysis of mitochondrial genetic units in yeast J. Bact. **97**, 1142 (1969).

ATWOOD, K. C., NORMAN, A.: On the interpretation of multi-hit survival curves. Proc. nat. Acad. Sci. ((Wash.) **35**, 696 (1949).

BAUER, W., VINOGRAD, J.: The interaction of closed circular DNA with intercalative dyes. I. The superhelix density of SV40 DNA in the presence and absence of dye. J. molec. Biol. **33**, 141 (1968).

BILLHEIMER, F. E., AVERS, C. J.: Nuclear and mitochondrial DNA from wild-type and petite yeast: circularity, length, and buoyant density. Proc. nat. Acad. Sci. (Wash.) **64**, 739 (1969).

BORST, P.: Mitochondrial DNA: structure, information content, replication and transcription. Soc. exp. Biol., Symp. **24**, (1969) (in press).

CLARK-WALKER, G. D., LINNANE, A. W.: The biogenesis of mitochondria in *Saccharomyces cerevisiae*. A comparison between cytoplasmic respiratory-deficient mutant yeast and chloramphenicol-inhibited wild type cells. J. Cell Biol. **34**, 1 (1967).
ELLIOTT, W. H.: The effects of antimicrobial agents on deoxyribonucleic acid polymerase. Biochem. J. **86**, 562 (1963).
EPHRUSSI, B., HOTTINGUER, H., CHIMENES, A.: Action de l'acriflavine sur les levures. I. La mutation «petite colonie». Annales de l'Institut Pasteur **76**, 351 (1949).
— SLONIMSKI, P. P., YOTSUYANAGI, Y., TAVLITZKI, J.: Variations physiologiques et cytologiques de la levure au cours de cycle de la croissance aerobie. C. R. Lab.Carlsberg **26**, 87 (1956).
FUKUHARA, H.: Relative proportions of mitochondrial and nuclear DNA in yeast under various conditions of growth. Europ. J. Biochem. **11**, 135 (1969).
— FAURES, M., GENIN, C.: Comparison of RNA's transcribed *in vivo* from mitochondrial DNA of cytoplasmic and chromosomal respiratory deficient mutants. Molec. gen. Genet. **104**, 264 (1969).
GROSSMAN, L. S., GOLDRING, E. S., MARMUR, J.: Preferential synthesis of yeast mitochondrial DNA in the absence of protein synthesis. J. molec. Biol. **46**, 367 (1969).
HENDLER, R. W.: Procedure for simultaneous assay of two β-emitting isotopes with the liquid scintillation counting technique. Anal. Biochem. **7**, 110 (1964).
HOLLENBERG, C. P., BORST, P., THURING, R. W. J., VAN BRUGGEN, E. F. J.: Size, structure, and genetic complexity of yeast mitochondrial DNA. Biochim. biophys. Acta (Amst.) **186**, 417 (1969).
HUANG, M., BIGGS, D. R., CLARK-WALKER, G. D., LINNANE, A. W.: Chloramphenicol inhibition of the formation of particulate mitochondrial enzymes of *Saccharomyces cerevisiae*. Biochim. biophys. Acta (Amst.) **114**, 434 (1966).
KELLERMAN, G. M., BIGGS, D. R., LINNANE, A. W.: Biogenesis of mitochondria. XI. A comparison of the effects of growth-limiting oxygen tension, intercalating agents, and antibiotics on the obligate aerobe, *Candida parapsilosis*. J. Cell Biol. **42**, 378 (1969).
KRAML, J. R., MAHLER, H. R.: Biochemical correlates of respiratory deficiency — VIII. A precipitating antiserum against cytochrome oxidase of yeast and its use in the study of respiratory deficiency. Immunochemistry **4**, 213 (1967).
LEON, S. A., MAHLER, H. R.: Isolation and properties of mitochondrial RNA from yeast. Arch. Biochem. **126**, 305 (1968).
LEPECQ, J. B., PAOLETTI, C.: A fluorescent complex between ethidium bromide and nucleic acids. J. molec. Biol. **20**, 87 (1967).
MACKLER, B., DOUGLAS, H. C., WILL, S., HAWTHORNE, D. C., MAHLER, H. R.: Biochemical correlates of respiratory deficiency. IV. Composition and properties of respiratory particles from mutant yeasts. Biochemistry **4**, 2016 (1965).
MAHLER, H. R., MACKLER, B., SLONIMSKI, P. P., GRANDCHAMP, S.: Biochemical correlates of respiratory deficiency II. Antigenic properties of respiratory particles. Biochemistry **3**, 677 (1964).
MEHROTRA, B. D., MAHLER, H. R.: Characterization of some unusual DNAs from the mitochondria from certain "petite" strains of *Saccharomyces cerevisiae*. Arch. Biochem. **128**, 685 (1968).
MEYER, R. R., SIMPSON, M. V.: DNA biosynthesis in mitochondria: differential inhibition of mitochondrial and nuclear DNA polymerase by the mutagenic dyes ethidium bromide and acriflavin. Biochem. biophys. Res. Commun. **34**, 238 (1969).
MOUSTACCHI, E., WILLIAMSON, D. H.: Physiological variations in satellite components of yeast DNA detected by density gradient centrifugation. Biochem. biophys. Res. Commun. **23**, 56 (1966).
NAGAI, S.: Production of respiration-deficient mutants of yeast by some quinone-imine dyes. Exp. Cell Res. **27**, 14 (1962).
— YANAGISHIMA, N., NAGAI, H.: Advances in the study of respiration-deficient (RD) mutation in yeast and other microorganisms. Bact. Rev. **25**, 404 (1961).
NASS, M. M. K.: Mitochondrial DNA: advances, problems, and goals, studies of size and structure of mitochondrial DNA related to biogenesis and function of the organelle. Science **165**, 25 (1969).

NASS, S.: The significance of the structural and functional similarities of bacteria and mitochondria. Int. Rev. Cytol. **55**, 55 (1969).
OGUR, M., ST. JOHN, R., NAGAI, S.: Tetrazolium overlay technique for population studies of respiration deficiency in yeast. Science **125**, 98 (1957).
PERLMAN, P. S., MAHLER, H. R.: (1) Intracellular localization of enzymes in yeast. Arch. Biochem. **136**, 245 (1970).
— — (2) Formation of yeast mitochondria III: biochemical properties of mitochondria isolated from a cytoplasmic petite mutant of yeast. J. Bioenerget. **1**, 113 (1970).
RABINOWITZ, M.: Extranuclear DNA. Bull. Soc. Chim. biol. (Paris) **50**, 311 (1968).
— GETZ, B. S., CASEY, J., SWIFT, H.: Synthesis of mitochondrial and nuclear DNA in anaerobically grown yeast during the development of mitochondrial function in response to oxygen. J. molec. Biol. **41**, 381 (1969).
RADLOFF, R., BAUER, W., VINOGRAD, J.: A dye-buoyant-density method for the detection and isolation of closed circular duplex DNA: the closed circular DNA in HeLa cells. Proc. nat. Acad. Sci. (Wash.) **57**, 1514 (1967).
REILLY, C., SHERMAN, F.: Glucose repression of cytochrome a synthesis in cytochrome-deficient mutants of yeast. Biochim. biophys. Acta (Amst.) **95**, 640 (1965).
ROODYN, D. B., WILKIE, D.: The biogenesis of mitochondria. London: Methuen 1968.
SARACHEK, A.: The induction by Mn^{++} of heritable respiratory deficiency in non-dividing populations of *Saccharomyces*. Biochim. biophys. Acta (Amst.) **33**, 227 (1959).
SCHATZ, G., SALTZGABER, J.: Protein synthesis by yeast promitochondria *in vivo*. Biochem. biophys. Res. Commun. **67**, 996 (1969).
SHERMAN, F., SLONIMSKI, P. P.: Respiration-deficient mutants of yeast II. Biochemistry. Biochim. biophys. Acta (Amst.) **90**, 1 (1964).
SLONIMSKI, P. P.: Recherches sur le formation des enzymes respiratoires chez le levure. Paris: Masson 1953.
— EPHRUSSI, B.: Action de l'acriflavine sur les levures V. Le systeme des cytochromes des mutants «petite colonie». Ann. Inst. Pasteur **77**, 47 (1949).
— PERRODIN, G., CROFT, J. H.: Ethidium bromide induced mutation of yeast mitochondria: complete transformation of cells into respiratory deficient non-chromosomal "petites". Biochem. biophys. Res. Commun. **30**, 232 (1968).
SOUTH, D. J., MAHLER, H. R.: RNA synthesis in yeast mitochondria: a derepressible activity. Nature (Lond.) **218**, 1226 (1968).
WARING, M. J.: Complex formation between ethidium bromide and nucleic acids. J. molec. Biol. **13**, 269 (1965).
YOTSUYANAGI, Y.: Etudes sur le chondriome de la levure II. Chondriomes des mutants a deficience respiratoire. J. Ultrastruct. Res. **7**, 141 (1962).
ZYLBER, E., VESCO, C., PENMAN, S.: Selective inhibition of the synthesis of mitochondria-associated RNA by ethidium bromide. J. molec. Biol. **44**, 195 (1969).

Studies on tRNA Structure Using Covalently and Noncovalently Bound Fluorescent Dyes

Charles R. Cantor, Kenneth Beardsley, James Nelson, Terence Tao and Kenneth W. Chin

I. Introduction

For the past few years, we have been interested in developing techniques which can provide information about the tertiary structure of nucleic acids in solution. The use of fluorescent labels would appear at this time to be an advantageous approach because of the large variety of detailed information which could be obtained from a single set of measurements (Stryer, 1968). Fluorescent dyes can be attached to a nucleic acid either covalently or noncovalently. If a single dye is attached to the nucleic acid, fluorescence polarization (Weber, 1966) or decay of fluorescence anisotropy measurements (Tao, 1969) can provide information about the size and shape of the macromolecule and degree of flexibility near the point of attachment of the label (Wahl, Paoletti and LePecq, 1970). If a dye which is responsive to its environment, such as 1-Anilino-8-naphthalenesulfonate (Stryer, 1965; Weber and Laurence, 1954) or ethidium bromide (LePecq and Paoletti, 1967), is used, considerable information may be gained about the local structure near the point of attachment of the dye. When two or more dyes are present on a nucleic acid molecule, it may be possible to determine the distance between them by energy transfer measurements (Stryer and Haugland, 1967). If the dyes are attached at known points in the primary structure of the nucleic acid, the resulting distance measurements will provide a severe constraint on possible tertiary structures (Beardsley and Cantor, 1970).

All of the techniques outlined above have been used in our laboratory to provide information about the tertiary structure of transfer RNA. The principal purpose of this paper is to describe progress to date and to indicate the expectations for future advances using these techniques. Some of the problems involved in this work are, relevant to the main theme of this symposium on complexes of biologically active molecules with nucleic acids. First of all, many of the molecules we have used as fluorescent probes are also biologically active. For example, ethidium bromide is a trypanocidal drug (Newton, 1964) and many of the dyes of the acridine series are potent mutagens (Lerman, 1964).

But there is a more important relationship between studies on the use of fluorescent probes to provide structural information and studies on the nature of complexes between biologically active substances and nucleic acids. In the former type of work, one hopes to minimize the possible perturbation of the structure by the fluorescent probe. In the latter work, it is precisely the perturbation which becomes the major aspect of interest. In all of our conformational studies with fluorescent dye probes, we have been concerned about the possibility of the dye introducing a large structural perturbation and thus yielding results of limited interesting. In this paper, we are concerned with experiments which yield information on whether the presence of a dye is perturbing the very structure it is being used to probe.

For those whose principal interest is drug action, let us briefly outline the types of information which are potentially available from fluorescence studies. It would be unlikely that all of these could be obtained simultaneously on a single drug system; and most of these techniques will not be useful unless the drug is fluorescent. The principal advantage of fluorescence spectroscopy is sensitivity in detecting very small amounts of material. With a moderately fluorescent substance, measurements can be made at less than one hundredth the concentration required for conventional absorption spectrographic measurements. The techniques described here can be carried out in dilute aqueous solutions at controlled temperature. This should permit, in principle, a study of the interaction of fluorescent drugs with complicated biological systems. Almost any information that can be obtained from absorption measurements can also be obtained from fluorescence measurements. Thus, by standard techniques, it is possible to obtain information concerning the number and types of binding sites, the strength of binding (ELLERTON and ISENBERG, 1969), and sometimes whether any covalent change has occurred to the drug on binding. Some of the special characteristics of fluorescence, however, lend themselves to such studies more readily than absorption measurements. For example, if fluorescence lifetimes are studied, it becomes possible to detect the presence of two or more different kinds of binding sites even though the spectrum of the bound dye in each might have an identical shape. From fluorescence polarization or nanosecond depolarization studies, the degree to which the dye is held rigidly attached to a macromolecule can be determined. If enough is known about the size and shape of the macromolecules, it may be possible to say something definite about the orientation of the dye with respect to the principal axes of the macromolecule. From quenching experiments, one can obtain information about the environment of the dye (VAUGHAN and WEBER, 1970). Finally, if the macromolecule contains chromophores which absorb in the region of the spectrum near where the dye fluoresces, it may be possible to learn something about the location of the dye relative to these chromophores by energy transfer measurements. Similarly, if one is so fortunate as to have a naturally fluorescent biological macromolecule, all of these techniques can be used in reverse with a nonfluorescent drug molecule to determine some of the aspects of its binding.

II. Preparation of Fluorescently Labeled tRNA

The nucleotide sequence of a typical transfer RNA molecule, $tRNA^{phe}$ from yeast (RAJBHANDARY, CHANG, STUART, FAULKNER, HOSKINSON and KHORANA, 1967), is shown in Fig. 1 in the cloverleaf pattern of secondary structure. Although such a tRNA must have binding sites which permit it to be located specifically on the ribosome and also permit it to be recognized by the correct amino-acyl synthetase and various modifying enzymes, little or nothing is known yet about the precise location of these sites (LENGYEL and SÖLL, 1969). Hence, it would be rather difficult at this stage to design a fluorescent molecule which would bind noncovalently to a tRNA and label a specific region of the molecule. Such approaches have been quite successful with proteins (STRYER, 1965), but with nucleic acids the present lack of knowledge only permits two more general ways of noncovalent labeling. The first of these is to use a dye such as ethidium bromide which is known to intercalate into double stranded helical regions preferentially (WARING, 1966). In a long, uninter-

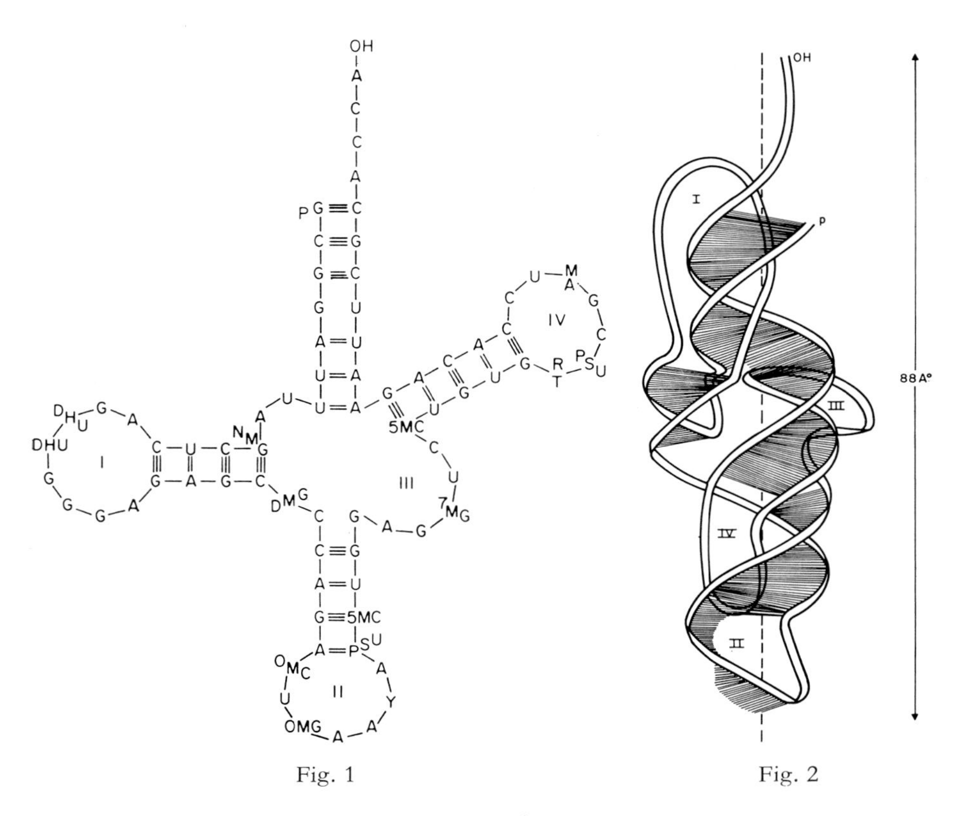

Fig. 1 Fig. 2

Fig. 1. The nucleotide sequence of yeast tRNA^{phe} as determined by RAJBHANDRAY et al. (1967). The roman numerals I, II and IV, respectively, identify the regions usually called dihydrouridine loop, anticodon loop, and pseudouridine loop

Fig. 2. A possible tertiary structure for tRNA. This drawing was adapted from the model proposed by CONNORS et al. (1969), based on the results from small angle X-ray scattering studies. The roman numerals correspond to the same regions shown in Fig. 1

rupted, double helix, like a DNA molecule, the dye can probably intercalate anywhere. If the tRNA structure were quite open, like that shown in Fig. 1, one might expect a relatively large number of possible sites for ethidium bromide intercalation. However, if the transfer RNA were folded into a tertiary structure, such as that shown in Fig. 2, most of the potential sites for ethidium bromide intercalation may be inaccessible (CONNORS, LABANAUSKAS and BEEMAN, 1969). This could be due to two reasons: The double helices themselves might become buried, or their lengths might become fixed by constraints at either end and binding would become unlikely because of the impossibility of an overall increase in helix length which must occur when a dye enters a double helix. Since a variety of types of evidence indicates a compact tertiary structure for transfer RNA (FRESCO, ADAMS, ASCIONE, HENLEY and LINDAHL, 1966; CANTOR, JASKUNAS and TINOCO, 1966; LAKE and BEEMAN, 1968; CRAMER, DOEPNER, V. D. HAAR, SCHLIMME and SEIDEL, 1968; YANIV, FAVRE and BARRELL, 1969), one might expect limitations in the number of possible bindings sites for ethidium bromide.

A second technique for specifically noncovalently labeling a nucleic acid with a dye exists potentially, although, to our knowledge, has never been applied. It would consist of attaching a fluorescent dye covalently to one specific oligonucleotide sequence which could then be complexed with a large macromolecule such as a single strand DNA or RNA specifically to position the label. By choosing a sufficiently long oligonucleotide sequence, it might be possible to design a label which would bind, at most, to one or two sites on the macromolecule.

At present, the most attractive possibilities for specifically introducing fluorescent dyes into nucleic acids is covalent labeling. Specific chemical techniques exist for attaching a covalent label either to the 5′ end of a nucleic acid or to the 3′ end. The most successful approach used thus far has been to oxidize the 3′ end of tRNA with periodate (ZAMECNIK, STEPHENSON and SCOTT, 1960) and to react the resulting bis aldehydes with either an amine to form a Schiff base, a hydrazine to form a hydrazone, or a hydrazide to form an acyl hydrazone. These reactions are shown schematically below.

RNA–C(=O)–H + H_2N … N^+–CH_3 … NH_2 → RNA(H)C=N … N^+–CH_3 … NH_2

RNA–C(=O)–H + $NH{-}NH_2$ … N → RNA(H)C=N–N(H)– … N

RNA–C(=O)–H + H_2N … N^+–CH_2–C(=O)–$NH{-}NH_2$ … NH_2 → RNA(H)C=N–N(H)–C(=O)–CH_2–N^+ … (NH_2, NH_2)

In this manner, ethidium bromide, acriflavine, proflavinyl acetic acid hydrazide, and 9-hydrazinoacridine have all been specifically attached to the 3′ end of tRNA. Depending on the dye used, a mono or bis adduct is obtained (BEARDSLEY and CANTOR, 1970).

A second possible approach involves reacting the free α-amino group of an aminoacyl tRNA with an N-hydroxysuccinimide ester. This technique, which has been used successfully in purification of specific tRNA species (GILLAM, BLEW, WARRINGTON, VON TIGERSTROM and TENER, 1968), is illustrated below.

$$C_6H_5NH_2 + HO{-}\overset{O}{\overset{\|}{P}}{-}O{-}RNA \;(O^-) \xrightarrow{DCC} C_6H_5{-}NH{-}\overset{O}{\overset{\|}{P}}{-}O{-}RNA \;(O^-)$$

Preparation of suitable fluorescent N-hydroxysuccinimide esters is currently in progress in our laboratory.

Techniques for specifically labeling the 5′ phosphate of nucleic acids have not yet been so well developed as those used with the 3′ end. RALPH, YOUNG and KHORANA (1963) have reported a technique which uses an amine to label the 5′ phosphate of a tRNA as shown below.

$$RNA{-}\overset{O}{\overset{\|}{C}}{-}CHR{-}NH_2 + C_{10}H_7{-}O{-}CH_2{-}\overset{O}{\overset{\|}{C}}{-}O{-}N(\text{succinimide}) \rightarrow RNA{-}\overset{O}{\overset{\|}{C}}{-}CHR{-}NH{-}\overset{O}{\overset{\|}{C}}{-}CH_2{-}O{-}C_{10}H_7$$

We have had limited success in attaching proflavine to the 5′ phosphate of tRNA using this technique but the yields thus far have been discouraging.

There are modification reactions known for some of the unusual nucleosides of tRNA. Unfortunately, in most cases the specificity is not sufficient for these reactions to be useful for attachment of fluorescent labels. Eventual improvements in the chemistry of labeling nucleic acids will hopefully permit fluorescent dyes or other informative labels to be introduced into positions occupied by bases such as pseudouridine, dihydrouridine, or 4-thiouridine.

It is indeed fortunate that there exists one naturally occurring tRNA with a fluorescent base on the anticodon loop. The base, whose structure is still unknown, is called *Y* and exists in the phenylalanine tRNAs of yeast (RAJBHANDARY et al., 1967), rat liver (FINK, GOTO, FRANKEL and WEINSTEIN, 1968) and wheat germ (DUDOCK, KATZ, TAYLOR and HOLLEY, 1969), but not of *E. coli*. The location of this base in the sequence is shown in Fig. 1. This base is necessary for amino acid transfer in protein synthesis on the ribosome. However, ZACHAU and his coworkers have shown that the base can be selectively removed without impairing the ability of activating enzymes to place phenylalanine on the tRNA (THIEBE and ZACHAU, 1968). The *Y* base has an absorption maximum at 320 nm which is sufficiently far to the red of the main absorption band of tRNA to permit selective excitation. Therefore it is an ideal naturally occurring fluorescent label and has been used in many of the experiments described below.

When covalently attached dyes like acriflavine are used as fluorescent probes, it is essential to prove that the dye is in fact covalently linked to the tRNA and not just intercalated. To do this, we must rely on purification procedures which can eliminate the chance of contamination by any noncovalently bound dye. We have used five to ten successive alcohol precipitations as well as sephadex chromatography to remove any intercalated dye. Controls which contained only noncovalently bound dye show clearly that this is far more than sufficient to remove any observable traces of intercalated dye (BEARDSLEY and CANTOR, 1970).

III. Singlet-Singlet Energy Transfer Experiments

The process of singlet-singlet energy transfer between two fluorescent chromophores is shown schematically in Fig. 3. A necessary constraint for singlet-singlet

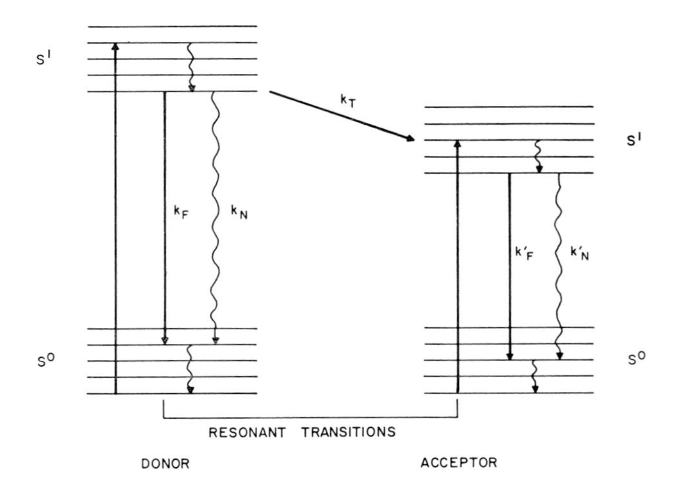

Fig. 3. Schematic drawing of singlet-singlet energy transfer. The vertical axis represents increasing energy. Several vibronic levels of the ground and first excited singlet states of the donor and acceptor are shown. The unlabeled solid arrows depict allowed absorptions of the donor and acceptor; K_F and K_F' are the rates of fluorescence of the donor and the acceptor; K_N and K_N' are the rates of all other processes except fluorescence and singlet-singlet transfer which can depopulate the first excited singlet states; K_T is the rate of transfer between the resonantly coupled levels of the donor and acceptor. The short unlabeled wavey arrows show vibrational relaxation to the ground vibronic state of each singlet manifold

energy transfer to occur is that the fluorescence spectrum of the donor molecule overlap the absorption spectrum of the acceptor molecule. The theory of singlet-singlet energy transfer was developed in detail by FORSTER (1966) and recently a number of the details of the theory have been explicitly shown to be correct by the elegant work of STRYER and his collaborators (STRYER and HAUGLAND, 1967; HAUGLAND, YGUERABIDE and STRYER, 1969). Forster's theory predicts that the efficiency of singlet-singlet energy transfer defined below in terms of rate constants shown in Fig. 3, can be described by the following equation:

$$E = \frac{K_T}{K_T + K_F + K_N} = \frac{R_0{}^6}{R_0{}^6 + R^6} . \tag{1}$$

In this equation, R is the distance between the two chromophores and R_0 is a constant which depends on the spectral overlap of the normalized donor emission and the acceptor absorbance, J, the quantum yield of the donor, ϕ, the index of refraction of the medium, n, and the average angular orientation of the electronic transition moments of the donor and the acceptor, K^2.

$$R_0 = 9.79 \times 10^3 \, (K^2 J \phi n^{-4})^{1/6} . \tag{2}$$

The sixth power dependence shown in Eq. 1 means that measurements of efficiency can yield very accurate determinations of the distance. Similarly, the sixth power dependence in Eq. 2 means that the quantities which make up R_0 need not be known with precision to generate a reliable value for the constant R_0.

The presence of singlet-singlet energy transfer can be detected either by measurements of static fluorescence or by following the time course of photon emission (HAUGLAND et al., 1969). In this paper, we shall discuss only applications of the former technique. Since transfer competes with fluorescence of the donor in depopulating the excited singlet state, the donor fluorescence will be quenched. This provides one measurement of the efficiency of energy transfer, E.

$$f_D = f_D^0 \, (1 - E) . \tag{3}$$

Here f_D^0 is the fluorescence of the donor in the absence of quenching, f_D is the observed donor fluorescence in the presence of the acceptor. A second manifestation of singlet-singlet energy transfer is sensitized fluorescence of the acceptor. The acceptor should show an enhanced fluorescence excitation spectrum centered in the absorption band of the donor. As shown in the equation below, if the acceptor absorbs very little at wavelengths where the donor absorption is large, a very accurate measurement of the efficiency of energy transfer should be possible.

$$f_A = f_A^0 \, (1 + E \, \varepsilon_D / \varepsilon_A) . \tag{4}$$

In Eq. 4, f_A^0 is the fluorescence of the acceptor in the absence of energy transfer f_A is the observed fluorescence in the presence of the donor, $\varepsilon_D/\varepsilon_A$ is the ratio of extinction coefficients of the donor and acceptor at the wavelength at which the donor was excited. One important aspect of energy transfer processes should be clear. Since the phenomenon can be observed either according to Eq. 3 or to Eq. 4, it is not necessary to have a fluorescent acceptor. Any absorbing molecule can quench the fluorescence of a donor. However, data obtained through the use of Eq. 4 inevitably become more accurate as $\varepsilon_D/\varepsilon_A$ becomes large. This emphasizes the desirability of using a fluorescent energy acceptor.

To measure singlet-singlet energy transfer by Eqs. 3 and 4, it is apparent that three different experimental systems are needed: A macromolecule containing only the donor, both donor and acceptor, and only the acceptor. In this way, environmental perturbations of the fluorescence of the donor and acceptor can be minimized. Only fluorescence changes due to actual energy transfer will be observed if the two chromophores are not capable of causing a structural perturbation large enough so that one can sense directly the presence of the other. For studies on transfer RNA, the *Y* base of purified phenylalanine tRNA from yeast was used as the energy donor. Three different fluorescent dyes, acriflavine, proflavinyl acetic acid hydrazide, and 9-hydrazino acridine were covalently attached to the periodate-oxidized 3′ end of tRNA to serve as acceptors (BEARDSLEY and CANTOR, 1970). The absorption and emission spectra of one of these dyes, proflavinyl acetic acid hydrazide, and the excitation and emission spectra of the *Y* base are shown in Fig. 4. It is apparent that the overlap

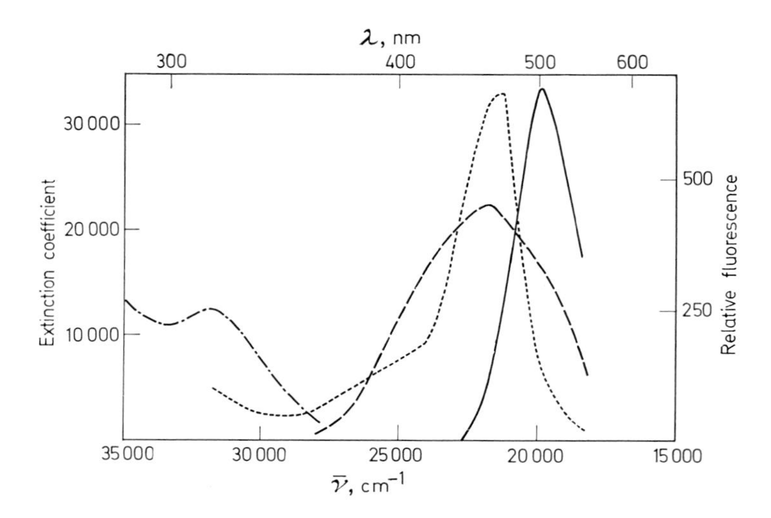

Fig. 4. Spectral properties of two of the chromophores used in singlet-singlet energy transfer studies on yeast tRNA^{phe}; —·—· uncorrected excitation spectrum of the *Y* base in tRNA^{phe} (× 2); — — — corrected emission spectrum of *Y* (× 3); ...... absorption spectrum of proflavinyl-acetic acid hydrazide (PAH) covalently attached to tRNA; ——— corrected emission spectrum of PAH attached to tRNA. Similar results are abtained with the other energy acceptors used (BEARDSLEY and CANTOR, 1970; BEARDSLEY, 1969)

between the emission of *Y* and the absorption of acriflavine is nearly perfect. The overlap between the emission of *Y* and the emission of the acriflavine is a complicating feature and this has been corrected for by procedures described elsewhere (BEARDSLEY and CANTOR, 1970).

Both the quenching of *Y* by the dye and sensitized fluorescence of the dyes were measured at a temperature of 25 °C in pH 6.2 to 6.6 buffered solutions containing 0.01 *M* Mg^{++} and the experimental results are summarized in Table 1. Both the quenching and the sensitized fluorescence were very small. Compare the results shown in the fourth and fifth columns of the table with what would be expected as shown in the fifth column if the distance between the chromophores were equal to R_0. It is clear that the apparent distance between *Y* and the dye attached to the 3′ terminus of tRNA is far greater than R_0. To interpret these results meaningfully, we must

Table 1[a]. *Summary of energy transfer experiments on yeast tRNA*^{phe}

Dye, loading	R_0, Å	Exciting wavelength	% Change of dye due to apparent transfer	% Change of Y due to apparent transfer	Calculated % change[b] for $R = R_0$	Distance, Å
Acriflavine — sample 1	30	315	+30		+213	46
Loaded 0.15:1		330	+31		+255	47
Acriflavine — sample 2	30	310	+19		+113	45
Loaded 0.2:1		320	+24		+167	46
		330	+25		+200	47
9-Hydrazino acridine	24[c]	315		−2	− 50	45
Loaded 2:1						
PAH — sample 1	30	320		−7	− 50	46
Loaded 1:1						
PAH — sample 2	30	315	+ 8		+140	54
Loaded 0.5:1		325	+ 6		+183	59
PAH — sample 3	30	315	+10		+140	52

[a] This table was adapted from Beardsley and Cantor (1970). [b] Corrected for purity of tRNA^{phe} samples.
[c] Corrected for 2:1 loading.

make an estimate of the parameter R_0 from Eq. 2. The quantum yield of the Y base has been shown to be 0.07 (BEARDSLEY, 1969). The refractive index of the medium between the two dyes can be estimated to be around 1.4. The spectral overlap is easily calculated by numerical integration. This leaves K^2 as the only known parameter. If the angle between the donor and acceptor transition moments is randomized by rapid motion, the value of K^2 will tend towards two thirds. This is the value we have used to calculate R_0. It can be justified on several grounds. Firstly, fluorescence polarization measurements of the Y base have shown that this base is capable of moving with respect to the bulk of the tRNA (BEARDSLEY, TAO and CANTOR, 1970). Secondly, a comparison of the rotational correlation time of noncovalently intercalated ethidium bromide, discussed below, with the fluorescence depolarization results for covalently bound acriflavine (MILLAR and STEINER, 1966) suggests that a dye attached to the 3′ terminus of tRNA is not rigidly oriented relative to the tRNA structure. Finally, quite consistent results are obtained with all three dyes as shown in Table 1 when the value of K^2 equals two-thirds is used. This would be unlikely if the dyes were held rigid relative to the Y base.

Using calculated values of the R_0 for energy transfer between Y and the three dyes, which range between 24 and 30 A°, it is possible to make estimates of the distance between the Y base and the 3′ end of the molecule. These estimates, shown in the last column of Table 1, range from 45 to 59 A°. If we conservatively estimate that the true range of possible distances is between 40 and 60 A°, a large number of possible tertiary structures for tRNA can be eliminated. This includes the hairpin structure (BROWN and ZUBAY, 1960), and a number of ways of folding cloverleaves or variants (CANTOR et al., 1966; MELCHER, 1969; ARMSTRONG, HAGOPIAN, INGRAM and WAGNER, 1966). The range of distances derived probably represents a lower estimate for several reasons. Experiments with rigid model systems have shown that calculated R_0's tend to be smaller than ones actually derivable from these experiments (LATT, CHEUNG and BLOUT, 1965; STRYER and HAUGLAND, 1967). Secondly, if there is any conformational flexibility, the distance dependence shown in Fig. 1 will tend to sample very strongly short distances and sample rather long distances not at all. Thus, the value of R derived from the experimental results is likely to be considerably shorter than be mean value. If a typical tertiary structure, such as that shown in Fig. 2, is compared with our results, the distance derived from fluorescence measurements is a little shorter than that proposed model building based on a variety of physical techniques. In view of the uncertainties involved in all methods, this discrepancy is not serious, but we will return later to a discussion of a possible reason for it.

IV. Experiments with Intercalated Ethidium Bromide

The presence of the Y base in $tRNA^{phe}$ allows for a second type of singlet-singlet energy transfer experiment to be performed. It is well known that ethidium bromide binds noncovalently to nucleic acid double helices (WARING, 1966). We were interested to see if we could determine anything about the location of this binding in the tRNA. This was possible because ethidium bromide can serve as an energy acceptor for the Y base. The R_0 for this chromophore couple is calculated to be 23 A° using the assumptions mentioned before. Ethidium bromide fluoresces very weakly in aqueous solution, but the fluorescence shows a marked enhancement when the dye binds to double helices (LEPECQ and PAOLETTI, 1967). Almost all of the fluorescence observed in a solution containing bound and free ethidium bromide arises from intercalated molecules. Thus, fluorescence can be used simultaneously to follow the binding of the dye and to look for possible binding sites near the *Y* base by quenching of the *Y* fluorescence. When these experiments were performed at relatively low salt concentration, a surprising result was obtained. This is shown in Fig. 5. Binding of ethidium bromide causes a drastic quenching of the *Y* fluorescence which parallels in

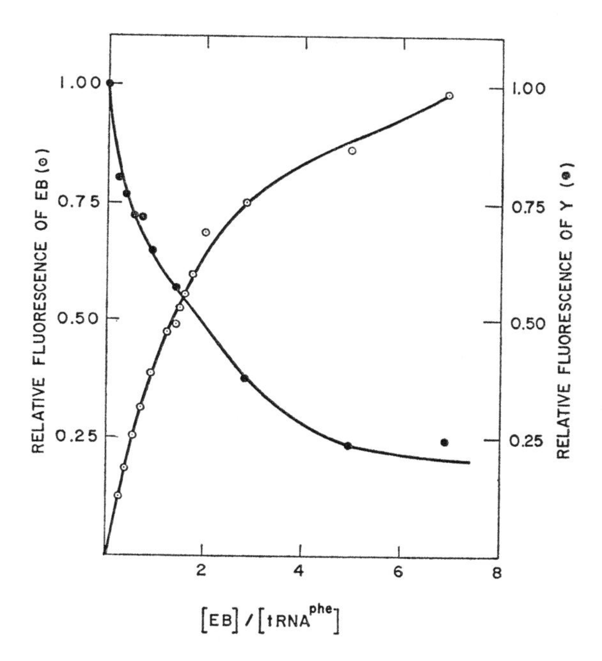

Fig. 5. Relative fluorescence of ethidium bromide (solid circles) and the *Y* base of yeast tRNA^{pke} (hollow circles) as a function of mole ratio of dye added to a 1.25×10^{-5} MtRNA^{phe} solution. These solutions did not contain magnesium ion. The ethidium bromide fluorescence was monitored at 581 nm by excitation at 520 nm. The *Y* fluorescence was measured at 430 nm from a 320 nm excitation

detail the shape of the binding curve. A Scatchard plot of the fluorescence data shows that, for ethidium bromide binding to yeast tRNA^{phe}, there is only one binding site and the strong quenching shown in Fig. 5 argues that this site must be quite near the *Y* base (TAO, NELSON and CANTOR, 1970). To study this in more detail, the technique of fluorescence lifetime measurement was used. Under usual circumstances, the fluorescence lifetime, τ, is related to the quantum yield, ϕ, as shown by Eq. 5.

$$\phi = \tau/\tau_R \tag{5}$$

In Eq. 5, τ_R is the radiative lifetime of the emitting state. It can be estimated from the integrated absorption intensity of the fluorescent group (PARKER, 1968) and is fairly constant under most circumstances. Thus, quenching of Y should be accompanyied by a decrease in the fluorescence lifetime. The fluorescence lifetime of Y in tRNA^{Phe} in low salt as a function of the amount of dye added is shown in Fig. 6. This result was obtained by the technique of single photon counting described elsewhere (TAO et al., 1970). Contrary to what one would have predicted, there was no change in the fluorescence lifetime of Y as dye was added. There is only one way in which the two different types of results can be reconciled, i.e. to assume that quenching is an all-or-none process: the lifetime does not change because all of the emission

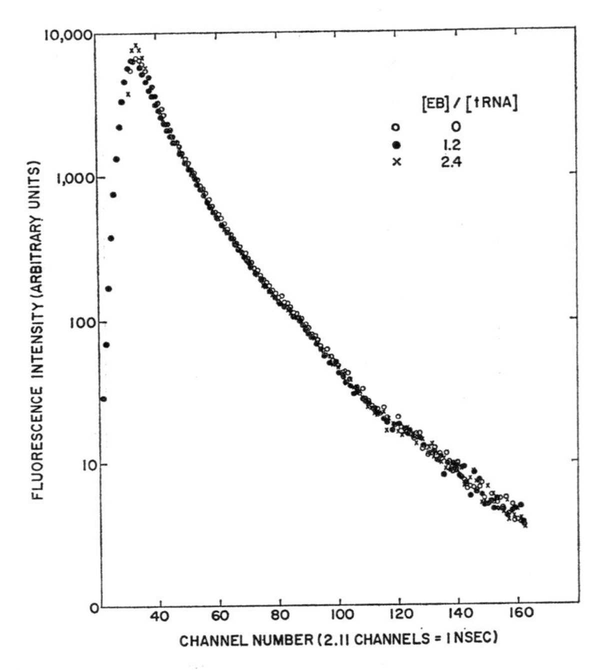

Fig. 6. Fluorescence decay of the Y base of tRNA^{Phe} as a function of the mole ratio of ethidium bromide (EB) added to a dilute solution. These results, which were obtained by the single photon counting technique, are described in more detail in TAO et al. (1970). These solutions contained no magnesium ion

comes from Y residues located on tRNA without any bound ethidium bromide. Any tRNA which contains bound ethidium bromide shows complete quenching of the Y fluorescence. For this to happen, the ethidium bromide binding site must be very near to the Y base, probably within 10 A°. Quite different results, however, were obtained in the presence of high salt as discussed elsewhere (TAO et al., 1970).

The surprising finding that conditions existed under which ethidium bromide bound selectively to a site very near to the Y base caused us to reexamine the nature of the ethidium bromide binding site to ascertain that the binding was, indeed, occurring by intercalation between base pairs. We used two techniques. The first was to compare the circular dichroism, CD, induced in ethidium bromide bound to unfractionated yeast tRNA with the CD found in the dye absorption bands when the dye was bound to DNA where intercalation is unquestioned. These results, shown in Fig. 7, argue that the overwhelming majority of the ethidium bromide bound to this heterogeneous sample of tRNA is intercalated in a double helical site. This general conclusion about the nature of the binding sites is in good agreement with the recent report of BITTMAN (1969) who studied the binding of ethidium bromide by absorption spectroscopy, ultracentrifugation, static fluorescence, and temperature jump kinetics. He used unfractionated *E. coli* tRNA and reports more binding sites than the one we detected by our static fluorescence studies with purified yeast $tRNA^{phe}$. Thus, it may be misleading to generalize from experiments on tRNA mix-

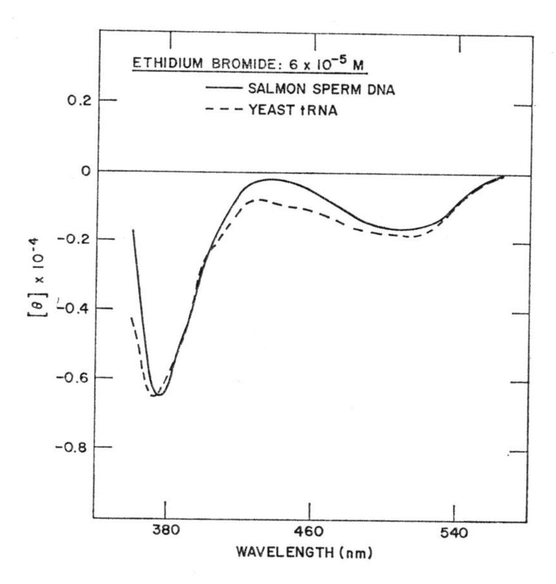

Fig. 7. Circular dichroism in the visible ethidium bromide absorption band of ethidium bromide noncovalently bound to nucleic acids. This figure was taken from TAO et al. (1970)

tures to the results found for $tRNA^{phe}$. It was not possible to measure the CD induced in ethidium bromide bound to yeast $tRNA^{phe}$ because of the large sample that would have been required. Instead, a second way of assessing the nature of the ethidium bromide binding site was used. This involved measuring the fluorescence lifetime of the bound dye. We found that ethidium bromide bound to phenylalanine tRNA in the presence of low salt shows a single exponential fluorescence decay corresponding to a lifetime of 26.2 nanoseconds. This is very close to the value of 23.3 nanoseconds found for the dye bound to DNA (TAO et al., 1970). The similarity of the two values suggests that binding indeed was occurring by intercalation between base pairs of the tRNA. The finding that only a single exponential lifetime exists, suggests that there is either a single binding site or else one class of very similar binding sites.

The energy transfer results so obtained have given interesting insight into the structure of tRNA. To really put faith in them, however, one would like to be sure that the presence of the dyes used in the experiments has not so disrupted the structure as to make all of the results meaningless. Several ways of assessing the extent of dye perturbation are discussed in the next section.

V. Does the Presence of Bound Dyes Perturb the tRNA Structure?

The dyes we have used as fluorescent probes are all planar aromatic hydrocarbons containing a positively charged group. There are a number of ways in which these dyes might be expected to perturb a nucleic acid structure once they are covalently attached to the 3′ terminus of the macromolecule. If the 3′ end is in a very tightly packed region of the structure, then it would be conceivable that insertion of a dye could cause massive unfolding of the tRNA near the 3′ end. If, on the other hand, the dye is bound in a relatively open region of the structure, the potential exists for it to attract a double helix into its proximity so that it can intercalate. This might occur either by movement of one of the loops or arms of the tRNA or by movement of the 3′ stem. Considering the strong binding of these dyes to double stranded nucleic acids, the possibility of perturbation by such a mechanism seems real. Finally, if the dye cannot find a site in which to intercalate on the parent tRNA, it may be able to cause an aggregation of two molecules so that a dye on one can intercalate to a helix on the other. In addition to all these possible sources of structural perturbation, it is not impossible that just the oxidation of the 3′ terminus to form a pair of reactive aldehydes could seriously perturb the structure of tRNA. Fortunately, the work of YARUS and BERG (1969) has shown that periodate oxidized tRNA is a very good competitive inhibitor for amino acyl tRNA in the synthetase reaction. This suggests that at least this last alternative is not likely to be a serious problem.

How can one tell whether the dye has perturbed the structure? The idea that a gross unfolding might be caused by the presence of a single dye molecule seems unlikely on thermodynamic grounds. While there is always the example of allosteric proteins to consider, there exists a number of simple ways in which large structural changes can be detected unequivocally. The data of MILLAR and STEINER (1966) on acriflavine conjugates of unfractionated tRNA suggest that the presence of the dye attached to the 3′ terminus does not perturb the total amount of secondary structure as measured by optical methods. We have made measurements of the rotational correlation time of tRNA containing a noncovalently bound ethidium bromide using the technique of nanosecond fluorescence depolarization (TAO et al., 1970). From this we can estimate the size and shape of the tRNA molecule; these estimates are in good agreement with others found from standard hydrodynamic techniques (ADAMS, LINDAHL and FRESCO, 1967) and from studies of small angle *X*-ray scattering (LAKE and BEEMAN, 1968; KINGBAUM and GODWIN, 1968; CONNORS et al., 1969). Both of these approaches rule out the possibility of drastic structural perturbation of tRNA by the presence of a dye.

In the case of $tRNA^{phe}$ from yeast, the presence of the *Y* base permits an especially sensitive way of detecting possible changes in conformation brought about by the presence of a dye covalently attached to the 3′ end of the molecule. Previous work in our laboratory has shown that the fluorescence of the *Y* base is altered by even very small changes in molecular conformation (BEARDSLEY et al., 1970). For example, as the magnesium concentration in a $tRNA^{phe}$ sample is dropped from 10^{-2} molar to 10^{-4} molar, in the presence of 0.1 molar sodium chloride, the fluorescence of the *Y* base drops by more than a factor of two. In the same range of experimental conditions the extinction coefficient and the ultraviolet circular dichroism changed by only a few percent and the size and shape of the tRNA molecule, as judged by fluorescence depolarization, using bound ethidium bromide, changed not at all. This indicates that the fluorescence of *Y* is able to reflect conformational changes which are too small to be detected by other techniques.

This observation permits one to return to the energy transfer results discussed darlier to look for possible evidence of structural perturbation by the presence of the eye. Any major structural perturbation which was propagated through the structure should have caused a substantial increase or decrease in the *Y* fluorescence. However, the changes in the *Y* fluorescence described in the table are small indeed. It is not likely that these small changes are the result of two cancelling effects, a decrease in fluorescence due to energy transfer and an increase in fluorescence due to an environmental change. If this were the case, one would expect the calculated distances from *Y* quenching data to be quite different from those calculated from sensitized dye emission. This was not observed. We think it likely that the attachment of the dye to the 3′ terminus does not cause a major conformational refolding of a tRNA. Small changes at the 3′ terminus cannot, however, be ruled out by this experiment.

The fluorescence depolarization experiments of MILLAR and STEINER (1968), and CHURCHICH (1963) showed clearly that a dye covalently attached to the 3′ end of tRNA is held somewhat rigidly with respect to the bulk of the polynucleotide structure. Full rigidity is ruled out by the fact that the rotational correlation times they measured by static polarization of the fluorescence of covalently attached acriflavine are about 10 to 15 nanoseconds in our units depending on the amount of magnesium ion in the sample. This is considerably shorter than the rotational correlation time of 25 nanoseconds we measured from the decay of fluorescence anisotropy of intercalated ethidium bromide. It is comparable to the value of 9 to 10 nanoseconds found for the rotational correlation time of the *Y* base (BEARDSLEY et al., 1970). These results suggest a model for tRNA in which the ends of loops or the 3′ stem are somewhat flexibly attached to the bulk of the structure. A small amount of vibrational or torsional motion is permitted which causes some depolarization of the fluorescence from both *Y* on the anticodon loop and acriflavine on the stem. Note, however, that both structures are still quite stiff. If there was any real extent of rapid rotational motion or extensive torsional motion about the numerous covalent bonds in either structure, one would expect the correlation times to be much shorter than observed (WALLACH, 1967; GOTTLIEB and WAHL, 1963). The partial rigidity of the bound acriflavine chromophore could arise in two different ways. The 3′ end of the tRNA may naturally be coordinated fairly tightly to the rest of the structure or the presence of a dye on the 3′ end may cause a structural perturbation which results in intercalation of the dye and loss of much of the normal conformational flexibility at the 3′ end. The second alternative, while representing an unfortunate complication from the viewpoint of fluorescence probes of structure, may be the kind of effect to be expected when biologically active drugs interact with nucleic acids. To explore this point, we have recently performed a number of experiments with ethidium bromide covalently attached to the aldehydes resulting from periodate oxidation of the 3′ end of tRNA.

This dye was chosen because of the large increase in fluorescence lifetime and quantum yield exhibited when intercalation occurs, or when the dye is otherwise transported to a nonpolar medium (LEPECQ and PAOLETTI, 1967). When ethidium bromide is allowed to react with periodate oxidized tRNA for 5 h at 37°, approximately 20% of the tRNA molecules contain a covalently attached ethidium bromide. The visible spectrum of the dye is red-shifted compared to that of dye and closely resembles the spectrum of noncovalently bound ethidium bromide intercalated into the tRNA structure. To explore further the environment of the ethidium bromide molecule, we measured the fluorescence lifetime of the covalently attached dye. The result was 25.4 nsec, a value which is remarkably similar to the fluorescence lifetime of non-covalently bound ethidium bromide, 26.5 nsec. These results suggest that a covalently attached ethidium bromide molecule can find an intercalation site.

In their studies of the static polarization and fluorescence intensity of acriflavine conjugates of tRNA, MILLAR and STEINER (1966) observed that upon heating, the fluorescence intensity of the dye changed radically before any noticeable changes in fluorescence polarization had occurred. Taking their results together with those reported here for ethidium bromide conjugates, a general model can be developed. At low temperatures, a dye covalently attached to the 3′ terminus intercalates into a nearby double helical region. As the temperature is raised, the intercalative binding is disrupted but the dye is still held relatively stiffly due to coordination of the *CpCpA* end with the rest of the tRNA. At still higher temperatures, the tRNA structure itself is disturbed and the polarization of dye fluorescence diminishes.

The major question which remains unanswered is the nature of the intercalation site. There are several possibilities: (1) The normal tertiary structure of tRNA may contain a piece of double helix right near the CCA terminus into which a covalently attached dye will naturally fit. Such a structure has been suggested by CRAMER et al. (1968). (2) The presence of a dye covalently attached to the 3′ terminus may perturb the tertiary structure of tRNA either by causing the CCA tail to fold up close to the body of the tRNA or by causing a loop to move closer to the 3′ terminus to enable intercalation of the dye. In many of the possible tertiary structures that have been suggested for tRNA, the CCA end is left dangling (LEVITT, 1969; CONNORS et al., 1969). Notice that if the structure shown in Fig. 2 were correct, a dye attached to the *3′* terminus could easily diminish the distance between this point and the *Y* base by up to 20 A°. This could explain the relatively small value obtained for this distance from energy transfer measurements. (3) It is remotely possible that the presence of a dye on the 3′ end causes two or more tRNA molecules to aggregate and permits the dye of one tRNA to intercalate into a double helix of the other. In fact, it is even possible that the two dyes themselves may tend to interact since dyes like ethidium bromide are known to have a strong tendency to aggregate. To distinguish possibility (3) from the first two, the relative quantum yield of dye fluorescence was measured as a function of the concentration of tRNA over a range of several orders of magnitude of concentration (R. LANGLOIS, unpublished results). These results show quite clearly that there is no concentration dependence of fluorescence. All of these measurements were made at concentrations sufficiently low to exclude any artifacts due to self-absorption or concentration quenching. With possibility (3) eliminated, the problem which remains is to decide whether any structural perturbation at the 3′ end has occurred. Experiments are in progress with a view to resolving this problem.

VI. Summary

A variety of fluorescence techniques have been used to provide information about the tertiary structure of tRNA. Singlet-singlet energy transfer between the *Y* base of yeast $tRNA^{phe}$ and several dyes covalently attached to the 3′ terminus has been observed. This allows the distance between these two fixed points to be measured. In addition, energy transfer measurements between the *Y* base and a noncovalently intercalated ethidium bromide molecule permits the location of the dye to be estimated. The fluorescence of the *Y* base itself is a sensitive indicator of structure near the anticodon loop. From studies of the fluorescence of dyes covalently attached to the 3′ terminus, it is possible to conclude that a dye in this position can probably intercalate into a double helical region. Whether this occurs through a minor perturbation of the native tRNA tertiary structure is not yet known, but major structural reorganization following attachment of the dye can be ruled out.

Acknowledgements

This work was supported by a grant, GM 14825, from the U.S. Public Health Service. The authors are indebted to Professor Richard Bersohn for many helpful discussions and for making much fluorescence equipment available to us. We thank Mr. Richard Langlois and Mr. Frederick Schachat for permitting us to cite some of their unpublished data.

References

Adams, A., Lindahl, T., Fresco, J. R.: Conformational differences between the biologically active and inactive forms of a transfer ribonucleic acid. Proc. nat. Acad. Sci. (Wash.) **57**, 1684 (1967).

Armsrtong, A., Hagopian, H., Ingram, V. M., Wagner, E. K.: Chemical studies on amino acid acceptor ribonucleic acids. VII. Partial ribonuclease digestion of alanine and valine soluble ribonucleic acid from yeast. Biochemistry **5**, 3027 (1966).

Beardsley, K. P.: Fluorescence studies on the structure of phenylalanine tRNA. Ph. D. Thesis, Columbia University 1969.

Beardsley, K., Cantor, C. R.: Studies of transfer RNA tertiary structure by singlet-singlet energy transfer. Proc. nat. Acad. Sci. (Wash.) **65**, 39 (1970).

— Tao, T., Cantor, C. R.: Studies on the conformation of the anti-codon loop of $tRNA^{phe}$. Effect of environment on the fluorescence of the *Y* base. Biochemistry **9**, 3524 (1970).

Bittman, R.: Studies of the binding of ethidium bromide to transfer ribonucleic acid: Absorption, fluorescence, ultracentrifugation and kinetics. J. molec. Biol. **46**, 251 (1969).

Brown, G. L., Zubay, G.: Physical properties of soluble RNA of *Escherichia coli*. J. molec. Biol. **2**, 287 (1960).

Cantor, C. R., Jaskunas, S. R., Tinoco, I., Jr.: Optical properties of ribonucleic acids predicted from oligomers. J. molec. Biol. **20**, 39 (1966).

Churchich, J. E.: Fluorescence studies on soluble ribonucleic acid labeled with acriflavine. Biochim. biophys. Acta (Amst.) **75**, 274 (1963).

Connors, P. G., Labanauskas, M., Beeman, W. W.: Structural studies on transfer RNA: The molecular conformation in solution. Science **166**, 1528 (1969).

Cramer, F., Doepner, H., v. d. Haar, F., Schlimme, E., Seidel, H.: On the conformation of transfer RNA. Proc. nat. Acad. Sci. (Wash.) **61**, 1384 (1968).

Dudock, B. S., Katz, G., Taylor, E. K., Holley, R. W.: Primary structure of wheat germ phenylalanine transfer RNA.Proc. nat. Acad. Sci. (Wash.) **62**, 941 (1969).

Ellerton, N. F., Isenberg, I.: Fluorescence polarization studies of DNA-proflavine complexes. Biopolymers **8**, 767 (1969).

Fink, L. M., Goto, T., Frankel, F., Weinstein, I. B.: Rat liver phenylalanine tRNA: column purification and fluorescence studies. Biochem. biophys. Res. Commun. **32**, 963 (1968).
Förster, T.: Delocalized excitation and excitation transfer. In: Modern quantum chemistry, Istanbul lectures, p. 93 (Sinanoglu, O., Ed.). New York: Academic Press 1966.
Fresco, J. R., Adams, A., Ascione, R., Henley, D., Lindahl, T.: Tertiary structure in transfer ribonucleic acids. Cold Spr. Harb. Symp. quant. Biol. **31**, 527 (1966).
Gillam, I., Blew, D., Warrington, R. C., von Tigertsrom, M., Tener, G. M.: A general procedure for the isolation of specific transfer ribonucleic acids. Biochemistry **7**, 3459 (1968).
Gottlieb, Y. Y., Wahl, P.: Etude théorique de la polarization de fluorescence des macromolécules portant un groupe emetteur mobile autour d'un axe de rotation. J. Chim. Phys. (Paris) **60**, 850 (1963).
Haugland, R. P., Yguerabide, J., Stryer, L.: Dependence of the kinetic of singlet-singlet energy transfer on spectral overlap. Proc. nat.Acad. Sci. (Wash.) **63**, 23 (1969).
Krigbaum, W. R., Godwin, R. W.: Small angle X-ray study of alanine transfer ribonucleic acid and bulk yeast transfer ribonucleic acid. Macromolecules **1**, 375 (1968).
Lake, J. A., Beeman, W. W.: On the conformation of yeast transfer RNA. J. molec. Biol. **31**, 115 (1968).
Latt, S. A., Cheung, H. T., Blout, E. R.: Energy transfer. A system with relatively fixed donor-acceptor separation. J. Amer. chem. Soc. **87**, 995 (1965).
Lengyel, P., Söll, D.: Mechanism of protein synthesis. Bact. Rev. **33**, 265 (1969).
Le Pecq, J. B., Paoletti, C.: A fluorescent complex between ethidium bromide and nucleic acids. J. molec. Biol. **27**, 87 (1967).
Lerman, L. L.: The structure of the DNA-acridine complex. Proc. nat. Acad. Sci. (Wash.) **49**, 94 (1963).
Levitt, M.: Detailed molecular model for transfer ribonucleic acid. Nature (Lond.) **224**, 759 (1969).
Melcher, G.: On the tertiary structure of transfer ribonucleic acid. FEBS Letters **3**, 185 (1969).
Millar, D. B., Steiner, R. F.: The effect of environment on the structure and helix-coil transition of soluble ribonucleic acid. Biochemistry **5**, 2289 (1966).
Newton, B. A.: Mechanisms of action of phenanthridine and aminoquinaldine trypanocides. In: Advances in Chemotherapy, p. 35 (Goldin, A., Hawkins, F., Eds.). New York: Academic Press 1964.
Parker, C. A.: Photoluminescence of solutions, p. 22—28. Amsterdam: Elsevier Publishing Co. 1968.
RajBhandary, U. L., Chang, S. H., Stuart, A., Faulkner, R. D., Hoskinson, R. M., Khorana, H. G.: Studies on polynucleotides, LXVIII. The primary structure of yeast phenylalanine transfer RNA. Proc. nat. Acad. Sci. (Wash.) **57**, 751 (1967).
Ralph, R. K., Young, R. J., Khorana, H. G.: Studies on polynucleotides, XXI. Amino acid acceptor ribonucleic acids (2). The labeling of terminal 5'-phosphomonoester groups and a preliminary investigation of adjoining nucleotide sequences. J. Amer. chem. Soc. **85**, 2002 (1963).
Stryer, L.: The interaction of a naphthalene dye with apomyoglobin and apohemoglobin. A fluorescent probe of nonpolar binding sites. J. molec. Biol. **13**, 482 (1965).
— Fluorescence spectroscopy of proteins. Science **162**, 526 (1968).
— Haugland, R. P.: Energy transfer: A spectroscopic ruler. Proc. nat. Acad. Sci. (Wash.) **58**, 719 (1967).
Tao, T.: Time-dependent fluorescence depolarization and brownian rotational diffusion coefficients of macromolecules. Biopolymers **8**, 609 (1969).
— Nelson, J. H., Cantor, C. R.: Conformational studies on tRNA. Fluorescence lifetime and nanosecond depolarization measurements on bound ethidium bromide. Biochemistry **9**, 3514 (1970).
Thiebe, R., Zachau, H. G.: A specific modification next to the anticodon of phenylalanine transfer ribonucleic acid. Europ. J. Biochem. **5**, 546 (1968).
Vaughan, W. M., Weber, G.: Oxygen quenching of pyrenebutyric acid fluorescence in water. A dynamic probe of the microenvironment. Biochemistry **9**, 464 (1970).

WAHL, P., PAOLETTI, J., LE PECQ, J. B.: Decay of fluorescence emission anisotropy of the ethidium bromide-DNA complex. Evidence for an internal motion in DNA. Proc. nat. Acad. Sci. (Wash.) **65**, 417 (1970).
WALLACH, D.: Effect of internal rotation on angular correlation functions. J. chem. Phys. **47**, 5258 (1967).
WARING, M. J.: Structural requirements for the binding of ethidium to nucleic acids. Biochim. biophys. Acta (Amst.) **114**, 234 (1966).
WEBER, G.: Polarization of the fluorescence of solutions. In: Fluorescence and phosphorescence analysis, p. 217 (HERCULES, D., Ed.). New York: Interscience Publishers 1966.
— LAURENCE, D. J. R.: Fluorescent indicators of adsorption in aqueous solution and in the solid phase. Biochem. J. **56**, 31P (1954).
YANIV, M., FAVRE, A., BARRELL, B. G.: Structure of transfer RNA: Evidence for interaction between two non-adjacent nucleotide residues in $tRNA^{val}$ from *Escherichia coli*. Nature (Lond.) **223**, 1331 (1969).
YARUS, M., BERG, P.: Recognition of tRNA by isoleucyl-tRNA synthetase. Effect of substrates on the dynamics of tRNA-enzyme interaction. J. molec. Biol. **42**, 171 (1969).
ZAMECNIK, P. C., STEPHENSON, M. L., SCOTT, J. F.: Practical purification of soluble RNA. Proc. nat. Acad. Sci. (Wash.) **46**, 811 (1960).

Antimutagens and Antimicrobial Drug Resistance

S. J. De Courcy, Jr.

I. Introduction

Paul Ehrlich (1913) observed, first, the occurrence of "relapsing crops" of parasites in animals under experimental chemotherapy, i.e. of strains which had developed resistance to chemotherapeutic drugs. This discovery was subsequently forgotten until the advent of antibacterial chemotherapy and the emergence of bacterial strains which were resistant either to sulfonamides or to antibiotics.

Two hypotheses have been forwarded as explanations of acquired drug resistance. The first is known as the phenotypic adaptation hypothesis (Davies, Hinshelwood and Pryce, 1944; Dean and Hinshelwood, 1966) which assumed that bacteria develop resistance by adaptive processes *in the presence* of the respective drug. The second is known as the mutation-selection hypothesis which holds that bacteria mutate spontaneously, i.e. *in the absence* of drugs, to drug-resistance, and that such resistant mutants are subsequently perpetuated selectively under drug pressure, i.e. under conditions in which the wild-type parent organisms are eliminated. Experimental evidence for phenotypic adaptation of entire bacterial cultures to drugs *upon contact* has been reviewed by Moyed (1964); additional examples have been discovered for tetracycline (Connamacher, Mandel and Hahn, 1967) and for erythromycin and lincomycin (B. Weisblum, personal communication, 1970). In the absence of the drugs, the entire cultures soon revert to drug sensitivity. It is evident that one is dealing here with unusual strain-specific induction phenomena which can not be invoked as general explanations of the acquisition of drug resistance in chemotherapy. The evidence for the validity of the mutation-selection hypothesis has been reviewed by Bryson and Szybalski (1955).

II. The Antimutagen Hypothesis

Sevag and De Courcy [reviewed 1970 (1, 2)] have suggested that antimicrobial drugs are capable of inducing drug resistance in bacteria which then becomes a hereditary property of such organisms. They have considered that in drug-sensitive bacterial populations there exist certain cells which possess the capability of mutating to drug resistance under the mutagenic influence of drugs, i.e. that mutation to drug resistance is not necessarily a spontaneous and random event. It was reasoned that substances which form stable complexes with chromosomal DNA might prevent the occurrence of such mutations and, therefore, act as *antimutagens*. Johnson and Bach (1966) have shown that spermine and quinacrine suppress spontaneous mutations to streptomycin resistance, and Novick and Szilard (1952), Novick (1956) and Webb and KUBITSCHEK (1963) have anticipated the antimutagen hypothesis for instances in which chemical substances act as mutagens to produce bacteriophage-resistant bacterial mutants. It appears, therefore, that the antimutagen hypothesis is not absolutely predicated upon the acceptance of the suggestion from Sevag's group that wild-type bacterial cultures contain certain mutation-prone cells which come under the mutagenic action of chemotherapeutic drugs upon contact.

III. Spermine and Spermidine as Antimutagens

Since spermine and spermidine, both naturally occcurring polyamines, are known to complex with DNA (LEPECQ, TALAER, FESTY and TRUHAUT, 1962), these were tested for antimutagenic activity (SEVAG and DRABBLE, 1962; DRABBLE and SEVAG, 1962).

Table 1. *Selective elimination of streptomycin-sensitive Aerobacter aerogenes and the ability of 2 streptomycin-resistant cells to survive and multiply in the presence of streptomycin plus spermine* (DRABBLE and SEVAG, 1962)

Growth media drug supplements (μg/ml)		Normal inoculum (X 10^8)	Resistant cells added	Turbidity readings at hours of: (KLETT-SUMMERSON, Filter No. 56)		
				0	23	47
Drug-free control		1.3	0	6	115	—
Streptomycin (SM)	100	1.3	0	27	25	24
SM	100	1.3	2	27	125	—
SM 100 + spermine 100		1.3	0	27	25	21
Sm 100 + spermine 100		1.3	2	23	130	—

When liquid media containing either streptomycin, penicillin or erythromycin were inoculated with drug-sensitive cells of *Staphylococcus aureus*, *Aerobacter aerogenes* or *Escherichia coli*, all cultures eventually yielded populations of bacteria resistant to the respective antibiotic used. In contrast, when similar test cultures containing antibiotic were supplemented with either spermine or spermidine, killing of the bacteria resulted without the emergence of resistance. When a test system containing both an antibiotic and spermine was inoculated with a large inoculum (10^8 cells) of sensitive cells *plus two cells of proven resistance to that antibiotic* (Table 1), a resistant population emerged in every instance. It is concluded that, initially, the inocula of normal drug-sensitive cells did not contain resistant cells. Evidently, had as few as two preexisting resistant cells been present in the inoculum of 10^8 wild-type drug-sensitive cells, the test used would have permitted their selective multiplication.

IV. Atebrin as an Antimutagen

A search for compounds which could substitute for spermine was undertaken since spermine, although manifesting antimutagenic activity *in vitro*, is not useful as a practical chemotherapeutic agent. Atebrin, the principal antimalarial drug of World War II, had been shown to form stable complexes with DNA (Kurnick and Radcliffe, 1962) and to prevent the enzymatic degradation of DNA by deoxyribonuclease (LePecq *et al.*, 1962). Also, the pharmacology of this compound was known (Osol, Pratt and Altschule, 1967), and its wide use had produced few undesirable side effects. Accordingly, atebrin was employed in the experimental procedures used in the spermine studies outlined above as well as in additional experiments. When cultures of normal drug-sensitive *E. coli* or *S. aureus* were exposed to combinations of atebrin plus a variety of antibiotics or sulfathiazole (Table 2), the entire inocula were killed without producing resistant mutants (Sevag, 1964; Sevag and Ashton, 1965). In the absence of atebrin, drug resistant-populations emerged in every instance. Further, a streptomycin-resistant strain of *E. coli*, although able to grow in the combination of high levels of streptomycin plus atebrin, was sensitive to as little as 1 μg of sulfathiazole plus 100 μg of atebrin per ml, thereby demonstrating the specificity of acquired resistance and the general capacity of atebrin to prevent the development of resistant strains (Sevag and Ashton, 1964).

V. Synergism Versus Antimutagenesis

The capacity of combinations of antimicrobial drugs with spermine or atebrin to prevent the development of resistant organisms in drug-sensitive bacterial populations can not be ascribed to combined antibacterial actions since atebrin was equally active in combination with each of a variety of antimicrobial agents of widely different structures and mechanisms of action in preventing the emergence of resistant organisms. The drug would have to be considered as non-specifically synergistic, an interpretation which would be contradictory to the known specificity of synergistic drug combinations.

Concurrently with the above findings of Sevag and co-workers, it was demonstrated in other laboratories that both spermine (Johnson and Bach, 1965; Johnson and Bach, 1967) and quinacrine (Atebrin) (Johnson and Bach, 1966; Johnson and Bach, 1967), acting as antimutagens, suppressed random mutations in bacteria, and decreased the mutagenic action of caffeine. Both spermine and quinacrine suppressed in *E. coli* spontaneous and 2-aminopurine-induced mutations to streptomycin resistance. More recently quinacrine has been shown to reduce the frequency of resistance to cytarabine in L-1210 murine leukemic cells (Bach, 1969).

VI. Phenothiazines and Dibenzocycloheptenes as Antimutagens

Since *S. aureus* resistant to one antibiotic could be prevented from becoming resistant to a second antibiotic when the latter was used in combination with atebrin (Sevag and Ashton, 1964), the obvious medical implications of these observations prompted a search for additional compounds which act in a similar manner. It was also of theoretical as well as of practical interest to evaluate the behavior of various compounds possessing structural moieties comparable to those of atebrin in order to bring out the prototypical molecular features responsible for antimutagenic action. Restricting the search to drugs in current medical use, members of four groups of

Table 2. *Prevention of the emergence of antibiotic-resistant strains from the sensitive population of Staphylococcus aureus 3A* (SEVAG and ASHTON, 1964)

Growth Media: Additions (μg/ml)	Turbidity readings at hours of							
	0	14—18	43—46	65—90	114—118	138—148	168—186	209—312
1 None	3	276	248					
2 Atabrine 80	8	230	214					
3 Atabrine 160	5	194	204					
4 Chloromycetin 7	8	26	38	37	—	232		
5 Chloromycetin 8	3	17	22	19	—	20	150	
6 Chloromycetin 7 + atabrine 160	0	16	23	23	—	24	31	
7 Chloromycetin 8 + atabrine 160	6	21	23	25	—	27	33	
8 Tetracycline 1.0	5	18	16	16	—	235	232	
9 Tetracycline 2.5	3	16	18	18	—	21	22	250
10 Tetracycline 1.0 + atabrine 160	2	16	20	17	—	20	24	20
11 Tetracycline 2.5 + atabrine 160	3	15	21	20	—	26	28	30
12 Erythromycin 2.5	2	15	13	98	—	226		
13 Erythromycin 2.5 + atabrine 80	0	13	14	14	—	20	—	25
14 Erythromycin 2.5 + atabrine 160	8	26	26	26	—	26	26	26
15 Novobiocin 5	7	14	20	214				
16 Atabrine 100	8	216	220					
17 Novobiocin 5 + atabrine 100	9	18	28	27	—	25	24	
18 Penicillin 0.2 units (inoculum 0.5 × 10^8/5 ml)	7	153	132					
19 Penicillin 0.5 units	9	62	152					
20 Atabrine 100	7	218	216					
21 Penicillin 0.2 units + atabrine 100	3	—	—	5	5	5	—	11
22 Penicillin 0.5 units + atabrine 100	0	—	—	2	2	4	—	8
Sulfathiazole (inoculum 25 cells/5 ml)								
23 PABA 5—40	0	32	256					
24 Atabrine 100	0	0	0	5	—	27	180	
25 Atabrine 100 + PABA 5—40	0	0	0	3	—	157	158	
26 Sulfathiazole 100	0	0	37	157	215			
27 Sulfathiazole 100 + PABA 5	0	47	236					
28 Sulfathiazole 100 + atabrine 100	0	0	1	8	—	15	22	
29 Sulfathiazole 100 + atabrine 100 + PABA 5	0	3	8	72	—	185	193	

Growth medium: Salts-glucose + 1% casein hydrolysate + tryptophan 20 μg + 1 μg thiamine + 1 μg nicotinamide/ml.
Inoculum: 1 × 10^8 cells/5 ml medium collected from a culture grown in 0.5% glucose broth.

compounds structurally or functionally related to atebrin were examined (Heller and Sevag, 1966). These groups were (a) *antimalarials with a quinoline nucleus:* chloroquine and hydroxychloroquine; (b) *the acridine derivatives:* acridine orange, acriflavin and a number of experimental compounds kindly provided by Smith, Kline and French Laboratories; (c) *the phenothiazine tranquilizers:* promazine, chlorpromazine, promethazine, levomepromazine and stelazine; (d) *the antidepressants* (dibenz-azepine and dibenzocycloheptene derivatives): tofranil, cyclobenzaprine, elavil, protriptyline and 3-chlorodibenzocycloheptene.

Using *S. aureus* (ATCC 3528) and *E. coli* B as representative gram-positive and gram-negative bacteria, and streptomycin, sulfathiazole, tetracycline and chloramphenicol as antimicrobial agents, each of the compounds selected for potential antimutagenic properties was tested along with atebrin. With the exception of the quinolines, chloroquine and hydroxychloroquine, which were antimutagenically inactive, all other compounds, when combined with each of the antimicrobial agents used, prevented the emergence of drug resistant organisms from wild-type drug-sensitive bacterial populations. It is significant that each compound, manifesting antimutagenic activity, possessed a flat heterocyclic system of three fused rings. Based on these findings, it was proposed that such a structural moiety was required for antimutagenic activity, although side-chain substitution was of modifying influence.

VII. Coumadin as an Antimutagen

Since coumadin was remindful of structures of compounds already shown to possess antimutagenic activity, it was also studied.

This study (De Courcy, Barr and Blakemore, 1969; De Courcy, Barr, Blakemore and Mudd, 1971) showed that coumadin, although not manifesting any antibacterial activity *in vitro* at the concentrations employed, was effective in precluding the development of resistance to polymyxin B in *P. aeruginosa* and to streptomycin in *E. coli* and *S. aureus* when applied concomitantly with the antibiotics. A single pre-exposure of *S. aureus* to a sub-inhibitory concentration of streptomycin abolished the ability of the coumadin-streptomycin combination to prevent the development of streptomycin resistant bacteria. It is concluded that coumadin too is antimutagenic.

VIII. Specificity of Drug Induced Resistance and Absence of Cross Resistance

A penicillin-resistant strain of *S. aureus* 3 A was capable of surviving and multiplying in the combined presence of penicillin and spermine but was unable to survive in the combined presence of streptomycin and spermine; the original penicillin- and streptomycin-sensitive parent strain, on the other hand, was unable to survive in the presence of either the penicillin-spermine combination or the streptomycin-spermine combination (Sevag and Drabble, 1962). In a similar experiment (Sevag and Ashton, 1964) a drug-sensitive parent strain of *E. coli* was killed by the combination of streptomycin plus atebrin or sulfathiazole plus atebrin. Conversely, a derived streptomycin-resistant strain, although able to survive and multiply in the streptomycin-atebrin combination, was killed by the sulfathiazole-atebrin combination. Since these results suggested drug-specific resistance in which cross resistance did not occur, a study in detail (DE COURCY and SEVAG, 1966) was undertaken, to test if, in general, a single pre-exposure to a sub-inhibitory level of one antibiotic could evoke cross-resistance to other antibiotics of different chemical structures and mechanisms of action.

Sets of growth media, each containing one of five or six different antibiotics, with or without atebrin, were prepared in duplicate. One set was inoculated with aliquots of a suspension of cells harvested from a replica colony grown on drug-free agar medium; the second set was inoculated with suspensions of cells harvested from the corresponding replica colony grown on agar supplemented with sub-inhibitory concentrations of streptomycin (Table 3), erythromycin (Table 4) or novobiocin (Table 5). Without exception, resistant organisms developed from *both* inocula in all test media containing antibiotic alone. Inocula without prior drug exposure failed to survive in *all* test media containing antibiotics plus atebrin. However, while a single sub-inhibitory pre-exposure to either streptomycin, erythromycin or novobiocin rendered bacterial cells capable of surviving and multiplying in the specific test media, containing high levels of the respective homologous antibiotic plus atebrin, such pre-exposure failed to produce survival in test media containing other antibiotics plus atebrin. In no experiment in this study was there any evidence of cross resistance.

It was conjectured from these results that the apparent selection of pre-existing drug resistant individuals in normal bacterial populations under drug pressure consists, in reality, of *inducing* with the "selective" drug certain bacteria which possess a disposition to mutate to drug resistance. Accordingly, an antibiotic may induce the emergence of specific drug resistance to the inducing compound.

IX. Effect of the Nutritional Environment on the Antimutagenic Activity of Atebrin

Development of resistance of *Pseudomonas aeruginosa* to polymyxin B depends upon the composition of the culture medium in which the inoculum is grown as well as on that in which exposure to the drug takes place (HAAS and SEVAG, 1953). Therefore, it was investigated what influence different culture media might have on the efficacy of atebrin, as an antimutagen, to prevent the development of resistance to polymyxin B. It was found (DE COURCY and MUDD, 1969) that the nutritional environment not only influenced the level of polymyxin resistance which developed, but also the length of incubation time required for resistant organisms to emerge. However, the capacity of atebrin to prevent the development of polymyxin resistance was unaffected by differences in the culture media. The use of combinations of different concentrations of atebrin with different concentrations of polymyxin distiuguished between combinations which permitted the emergence of resistance as opposed to those which were lethal to the bacterial population. In order for atebrin to prevent the emergence of polymyxin resistance, its concentration must be adjusted to the concentration of polymyxin present. A polymyxin-atebrin combination capable of eliminating a normal drug sensitive population of *Ps. aeruginosa* was ineffective in preventing the emergence of resistance in a culture which had received a single sub-inhibitory pre-exposure to polymyxin.

Table 3. *Specificity of the resistance reaction to the combined action of streptomycin and atabrine of clones (that is, respective inocula) from replicated sister colonies grown without and with prior exposure only to streptomycin, and absence of cross resistance to other antibiotics in combination with atabrine*

(DE COURCY and SEVAG, 1966)

Growth systems	Exp.	*A* Inoculum from a sister colony grown on drug-free agar							*B* Inoculum from a sister colony grown on agar with 10 μg of streptomycin/ml						
		Growth turbidity readings at hours of:							Growth turbidity readings at hours of:						
Additions (μg/ml)		0	24	48	72	96	132—168	204—240	0	24	48	72	96	132—168	204—240
1 None	IX *A*	3	320						5	320					
2 Atabrine 150	IX *A*	3	93	250					0	120	244				
3 Streptomycin 150	IX *A*	0	—	92	272				0	0	71	250			
4 Streptomycin 150 + atabrine 150	IX *A*	0	0	0	0	—	—	0	0	0	5	50	80	222	
5 Tetracycline 1	IX *A*	8	13		25	220			6	10	14	41	226		
6 Tetracycline 1 + atabrine 150	IX *A*	2	0	0	14	14	7	14	1	6	9	6	8	11	2
7 Penicillin 0.5 units	IX *A*	3	0	15	141	—	—	—	7	9	14	135	—	—	—
8 Penicillin 0.5 units + atabrine 150	IX *A*	1	2	5	3	—	—	11	0	2	7	6	—	—	0
9 Erythromycin 0.5	IX *Ar*	10	10	14	—	15	300	—	9	14	10	96	179	280	—
10 Erythromycin 0.5 + atabrine 150	IX *Ar*	12	13	14	—	—	15	21	13	13	12	—	—	—	11
11 Chloromycetin 10	IX *Ar*	3	39	76	123	153	—	—	3	25	47	71	155	—	—
12 Chloromycetin 10 + atabrine 150	IX *Ar*	5	7	9	—	10	—	15	7	5	9	6	—	—	7

Inoculum size for experiment IX *A* = 1.8×10^7 cells and for experiment IX *Ar* = 3.1×10^7 cells/5 ml medium.

Inoculum size for experiment IX *A* = 2.4×10^7 cells and for experiment IX *A* = 2.6×10^7 cells/5 ml medium.

Medium: 8 g of nutrient broth (Difco) per litre were supplemented by: sodium chloride 8.5 g, sodium monohydrogen phosphate 7.58 g,potassium dihydrogen phosphate 1.82 g and, following sterilization, sterile glucose solution added to render 0.5%. Final pH was 7.34.

Table 4. *Specificity of the resistance reaction to the combined action of erythromycin and atabrine (that is, respective inocula) from replicated sister colonies grown with and without prior exposure only to erythromycin, and absence of cross resistance to other antibiotics in combination with atabrine* (DE COURCY and SEVAG, 1966)

Growth systems	Exp.	*A* Inoculum from a sister colony grown on drug-free agar							*B* Inoculum from a sister colony grown on agar with 0.1 μg erythromycin/ml							
		Growth turbidity readings at hours of:							Growth turbidity readings at hours of:							
Additions (μg/ml)		0	24	48	72	96	168—204	204—240	0	24	48	72	96	132—168	168—204	204—240
1 None	IX *Er*	2	305						2	225	—	270				
2 Atabrine	IX *Er*	12	12	46	265				17	15	16	17	32	252	—	288
3 Erythromycin 0.5	IX *Er*	14	17	18	40	128	300	—	17	155	310	—	—	—	—	—
4 Erythromycin 0.5 + atabrine 150	IX *Er*	19	25	22	20	—	—	18	22	20	20	17	15	15	232	305
5 Tetracycline 1.0	IX *Er*	14	40	85	147	290			15	17	29	286	305	—	—	—
6 Tetracycline 1.0 + atabrine 150	IX *Er*	15	—	—	—	—	15	15	15	—	—	—	—	—	—	10
7 Chloromycetin 10	IX *Er*	12	32	100	—	219	—	—	13	25	115	230	—	—	—	—
8 Chloromycetin 10 + atabrine 150	IX *Er*	9	9	—	—	—	—	11	11	11	—	—	—	—	—	5
9 Penicillin 0.5 units	IX *Er*	13	11	15	125	—	—	—	15	19	110	125	—	—	—	—
10 Penicillin 0.5 units + atabrine 150	IX *Er*	12	12	—	—	—	—	8	13	12	—	—	—	—	—	7
11 Streptomycin 150	IX *E*	4	3	0	125	260	—	—	9	8	55	180	232	—	—	—
12 Streptomycin 150 + atabrine 150	IX *E*	4	2	—	—	—	—	10	5	4	—	—	—	—	—	10
13 "Novobiocin" 5	IX *E*	11	160	286	300	—	—	—	5	125	226	260	—	—	—	—
14 "Novobiocin" 5 + atabrine 150	IX *E*	11	11	—	—	—	—	16	9	9	—	—	—	—	—	15

A: Inoculum size for experiment IX $Er = 3.2 \times 10^7$ cells and for experiment IX $E = 3.3 \times 10^7$ cells/5 ml medium.

B: Inoculum size for experiment IX $Er = 6 \times 10^6$ cells and for experiment IX $E = 2.6 \times 10^7$ cells/5 ml medium.

Table 5. *Specificity of the resistance reaction to the combined action of novobiocin and atabrine (that is, respective inocula) from replicated sister colonies grown with and without prior exposure only to novobiocin, and the absence of cross resistance to other antibiotics in combination with atabrine*

Growth systems	Exp.	*A* Inoculum from a sister colony grown on drug-free agar							*B* Inoculum from a sister colony grown on agar with 1.5 μg novobiocin/ml						
		Growth turbidity readings at hours of:							Growth turbidity readings at hours of:						
Additions (μg/ml)		0	22—26	43—46	67—69	98—142	167—170	218—224	0	22—26	43—46	67—69	98—142	167—170	218—224
1 None	IX *B*	10	305	—	—	—	—	—	2	220	275	—	—	—	—
2 Atabrine 150	IX *B*	19	19	246	—	—	—	—	9	45	252	—	—	—	—
3 Novobiocin 5.0	IX *Br*	19	79	300	—	—	—	—	15	270	—	—	—	—	—
4 Novobiocin 5.0 + atabrine 150	IX *Br*	14	15	16	16	15	23	16	9	19	96	280	—	—	—
5 Streptomycin 150	IX *Br*	16	13	50	180	236	—	—	15	13	84	127	230	269	—
6 Streptomycin 150 + atabrine 150	IX *Br*	17	17	20	19	18	21	18	14	11	13	12	12	17	10
7 Chloromycetin 10	IX *Br*	10	29	60	74	171	—	—	6	40	56	102	170	—	—
8 Chloromycetin 10 + atabrine 150	IX *Br*	10	9	13	11	11	15	9	10	11	11	11	11	15	7
9 Tetracycline 1.0	IX *Br*	15	32	34	230	300	—	—	13	25	23	180	290	—	—
10 Tetracycline 1.0 +atabrine 150	IX *Br*	13	16	17	17	15	18	18	21	20	20	18	19	22	19
11 Erythromycin 0.5	IX *Br*	15	18	18	13	350	—	—	13	15	16	15	94	365	—
12 Erythromycin 0.5 + atabrine 150	IX *Br*	17	18	23	19	17	20	18	20	18	17	15	15	17	16
13 Penicillin 0.5 unit	IX *B*	6	10	22	134	—	—	—	5	5	75	120	—	—	—
14 Penicillin 0.5 unit +atabrine 150	IX *B*	10	13	12	12	10	10	8	4	5	5	6	8	5	3

Similarly, a study (De Courcy, 1969) on gentamicin with atebrin as the antimutagen, demonstrated that the nutritional environment in which cells make initial contact with gentamicin influenced quantitatively the organism's susceptibility to this drug. For example, in salts-glucose-nutrient broth medium, the minimal inhibitory concentration (MIC) of gentamicin *per se* was 6.3 μg per ml. Conversely, *in salts-glucose medium* the combination of 0.5 MIC of gentamicin (25 μg per ml) plus 100 μg of atebrin per ml, was lethal to the entire population with no resistance developing while in *salts-glucose-nutrient broth medium* the combination of 0.5 MIC of gentamicin (3.2 μg per ml) plus 100 μg of atebrin per ml permitted the emergence of gentamicin resistant bacteria, albeit after a prolonged incubation period.

X. The Exclusion of Selection in the Fluctuation Test by Atebrin

The fluctuation test, originally designed by Luria and Delbrück (1943) to study mutation from phage sensitivity to phage resistance in *E. coli*, was subsequently adapted by Demerec (1945) to investigate in *S. aureus* mutation from penicillin sensitivity to penicillin resistance. Observations in a number of such studies (Bryson and Szybalski, 1955) were interpreted as evidence for the spontaneous mutation-selection hypothesis. It is principally this experimental test that gave rise to the general acceptance of the view that specific agents such as viruses and toxic agents do not induce specific resistance which is expressed in micro-organisms as the result of a mutation. However, the design of these experiments renders it impossible to identify a colony as resistant to an agent without exposing the original cell to that agent.

The fluctuation test is carried out by inoculating aliquots of about 200 bacterial cells from a liquid culture into each of a series of tubes containing 1 ml of medium. Following overnight incubation, the cultures from each tube, now containing about 10^8 cells which grew out in the absence of drug, are introduced into separate agar plates supplemented with antibiotic. Some plates yield hundreds of resistant colonies whereas others yield only very few, or even none. It is reasoned that spontaneous mutations responsible for the observed deviation from Poisson statistics in the number of colonies on the test plates will occur spontaneously either early in some tubes or late in other tubes and are subsequently detected by exposure to drug after these mutants have multiplied.

We have re-evaluated the fluctuation test (De Courcy and Sevag, 1967). The question under consideration was whether resistant colonies appearing on drug-supplemented agar in the final step of the test are descended from spontaneous mutants which preceded, and were independent of, drug action, or, conversely, whether they are descendants of progenitors possessing the *potentiality* for mutation and are induced to actual resistance by direct contact with drug during the final step of the test. Accordingly, fluctuation tests were performed in which one-half of the contents of each fluctuation tube was introduced into a culture plate containing only streptomycin, and the other half into a corresponding plate containing streptomycin plus atebrin.

Results of tests performed without incorporating atebrin showed, typically, that individual cultures varied widely with respect to the number of cells which possessed the ability to form colonies in the presence of streptomycin. However, parallel plates containing streptomycin plus atebrin were almost free of colonies. That rare occasional colonies appeared on plates containing the drug combination was interpreted by assuming that their progenitor cells possessed the potentiality of being induced by streptomycin to levels of resistance higher than could be overcome by 100 μg of streptomycin per ml.

In order to support, further, our interpretation that colonies which arise on plates containing streptomycin only, but which fail to appear on plates containing streptomycin plus atebrin, originate from *potentially* resistant progenitor cells rather than from streptomycin resistant mutants of spontaneous origin, another series of fluctuation tests was performed. In this experimental design, the contents of each fluctuation tube were apportioned equally to three culture plates, one containing streptomycin only, and the remaining two containing streptomycin plus atebrin. To the third plate containing the drug combination, known numbers of streptomycin resistant cells were introduced additionally.

Aliquots of fluctuation tube cultures failed to produce colonies in the presence of the drug combination, but the parallel plates which had received exogenous cells of proved resistance in addition to aliquots of fluctuation tube cultures showed colonies *corresponding in number to the number of resistant cells added.* It is inferred from these results that, had there been any spontaneous streptomycin resistant mutants present in the fluctuation tube cultures of normal cells, they would have yielded colonies. Because such colonies were absent we conclude that colonies grown in the presence of streptomycin originate from progenitor cells of a potentially resistant character rather than from *de facto* resistant cells of spontaneous mutational origin. This supports the hypothesis that the development of resistance can be causally related to exposure to the respective antibacterial drugs.

XI. Application of the Principle of Antimutagenesis to Cases of Chronic and Acute Infections

Once specific resistance has become established in a bacterial strain as the result of exposure to sub-lethal concentrations of antimicrobial drug, the mutagen-antimutagen principle no longer applies to the *same* antimicrobial drug. However, a *different* antimicrobial drug to which the organism has received no prior exposure, and to which it is susceptible, *plus an antimutagen*, eliminated such populations *in vitro*. This suggested the potential usefulness of drug-antimutagen combinations in the clinical treatment of chronic bacterial infections.

Applying the principle of antimutagenesis in a limited clinical study, Sharda, Cornfeld and Michie (1966), using atebrin plus appropriate antimicrobial drugs, reported chemotherapeutic successes in 8 of 10 patients suffering from persistent chronic urinary infections. It is particularly noteworthy that, prior to undergoing this combined therapy for their chronic infections, each of these patients had received, singly or in combination, five or more antibacterial drugs, all of which had chemotherapeutically failed. Failures in every case had resulted from the development of resistant strains.

More recently, preliminary results of a similar study of larger scope and employing a more rigid protocol, were reported (Horwitz, Eshelman, Sevag, De Courcy, Mudd and BLAKEMORE, 1968). It was concluded that antimicrobial drug therapy of urinary infections with atebrin as an adjunct is more effective than that with the same antimicrobial drug given alone. In a report of an extension of these studies (ESHELMAN, HORWITZ, DE COURCY, MUDD and BLAKEMORE, 1970), it is shown (Table 6) that in 34 patients treated with selected antimicrobial drugs alone, 41% of the cases remained bacteriologically negative, whereas, 59% of the infections recurred within six months; in 37 patients treated with selected antimicrobial drugs plus atebrin, 85% of the cases remained negative with only 15% of the infections recurring within six months. It was concluded from this study that atebrin, in adjunctive antimicrobial drug therapy of urinary tract infections, affords a significant therapeutic advantage over that of antimicrobial drugs used alone in the prevention of emergence of drug resistant bacteria.

The demonstrated effectiveness of chemotherapeutically applicable antimutagenic compounds in preventing the emergence of resistant bacteria and the encouraging results of preliminary clinical trials utilizing atebrin as an antimutagen in adjunctive chemotherapy of urinary tract infections, renders the mutagen-antimutagen principle attractive as a possible basis of circumventing the problem of acquired microbial drug resistance.

References

BACH, M. K.: Reduction in the frequency of mutation to resistance to cytarabine in L 1210 murine leukemic cells by treatment with quinacrine hydrochloride. Cancer Res. **29**, 1881 (1969).

BRYSON, V., SZYBALSKI, W.: Microbial drug resistance. Advanc. Genet. **7**, 1 (1955).

CONNAMACHER, R. H., MANDEL, H. G., HAHN, F. E.: Adaptation of populations of *Bacillus cereus* to tetracycline. Molec. Pharmacol. **3**, 586 (1967).

DAVIES, D. S., HINSHELWOOD, C. N., PRYCE, J. M. G.: Studies in the mechanism of bacterial adaptation. Trans. Faraday Soc. **40**, 397 (1944).

DEAN, A. C. R., HINSHELWOOD, C.: Growth function and regulation in bacterial cells. Oxford: Clarendon Press 1966.

DE COURCY, S. J., JR.: Discussion to session 1. International Symposium on Gentamicin. J. Infect. dis. **119**, 385 (1969).

— MUDD, S.: Effect of the nutritional environment on the development of resistance to polymyxin B in *Pseudomonas aeruginosa* and its prevention by atebrin. Antimicrobial Agents and Chemotherapy **1968**, 72 (1969).

— SEVAG, M. G.: Specificity and prevention of antibiotic resistance in *Staphylococcus aureus*. Nature (Lond.) **209**, 373 (1966).

— — Population dynamics and results of fluctuation tests in a study of the role of atebrin as an antimutagen in preventing streptomycin resistance in *Staphylococcus aureus*. Antimicrobial Agents and Chemotherapy **1966**, 235 (1967).

— BARR, M. M., BLAKEMORE, W. S.: Prevention of emergence of drug resistance by coumadin. Bact. Proc. M-79 (1969).

— — — MUDD, S.: Prevention of antibiotic resistance *in vitro* in *Staphylococcus aureus*, *Escherichia coli* and *Pseudomonas aeruginosa* by coumadin. J. Infect. dis., **123**, 11 (1971).

DEMEREC, M.: Production of *Staphylococcus* strinas resistant to various concentrations of penicillin. Proc. nat. Acad. Sci. (Wash.) **31**, 16 (1945).

DRABBLE, W. T., SEVAG, M. G.: Prevention of the development of microbial resistance to drugs. Antimicrobial Agents and Chemotherapy **1962**, 649 (1962).

EHRLICH, P.: Chemotherapeutics: Scientific principles, methods and results. Lancet **1913**, 445.

ESHELMAN, J. L., HORWITZ, M. R., DE COURCY, S. J., JR., MUDD, S., BLAKEMORE, W. S.: Atebrin as an adjuvant in chemotherapy of urinary tract infections. J. Urol. (Baltimore) **104**, 902 (1970).

HAAS, G. J., SEVAG, M. G.: Critical role of amino acids on the sensitivity and development of resistance to polymyxin B. Arch. Biochem. **43**, 11 (1953).

HELLER, C. S., SEVAG, M. G.: Prevention of the emergence of drug resistance in bacteria by acridines, phenothiazines and dibenzocycloheptenes. Appl. Microbiol. **14**, 879 (1966).

HORWITZ, M. R., ESHELMAN, J. L., SEVAG, M. G., DE COURCY, S. J., JR., MUDD, S., BLAKEMORE, W. S.: Effect of combined atebrin-antimicrobial drug therapy on urinary tract infections. *In vitro* and *in vivo* studies. Surg. Forum **19**, 532 (1968).

JOHNSON, H. G., BACH, M. K.: The antimutagenic action of polyamines: Suppression of the mutagenic action of an *E. coli* mutator gene and of 2-aminopurine. Proc. nat. Acad. Sci. (Wash.) **55**, 1453 (1966).

— — Apparent suppression of mutation rates in bacteria by spermine. Nature (Lond.) **208**, 408 (1965).

— — On the mechanism of antimutagen action. Proc. 5th Internat. Cong. Chemother. **4**, 427 (SPITZY, K. H., HASCHEK, H., Eds.) Vienna: Verlag der Wiener Medizinischen Akademie. 1967.

KURNICK, N. B., RADCLIFFE, I. E.: Relation between DNA and quinacrine and other antimalarials. J. Lab. clin. Med. **60**, 669 (1962).

LE PECQ, J. B., LE TALAER, J. Y., FESTY, B., TRUHAUT, R. C.: Inhibition of deoxyribonuclease (DNAase) by complexing deoxyribonucleic acid with dyes. Compt. Rend. **254**, 3918 (1962).

LURIA, W. E., DELBRÜCK, M.: Mutations of bacteria from virus sensitivity to virus resistance. Genetics **28**, 491 (1943).

MOYED, H. S.: Biochemical mechanisms of drug resistance. Ann. Rev. Microbiol. **18**, 347 (1964).

NOVICK, A.: Mutagens and antimutagens. Brookhaven Symp. Biol. **8**, 201 (1956).

— SZILARD, L.: Antimutagens. Nature (Lond.) **170**, 926 (1952).

OSOL, A., PRATT, R., ALTSCHULE, M. D.: The United States Dispensatory and Physicians Pharmacology, 26th Ed., p. 1075. Philadelphia: J. B. Lippincott Co. 1967.

SEVAG, M. G.: Prevention of the emergence of antibiotic resistant strains of bacteria by atebrin. Arch. Biochem. **108**, 85 (1964).

— ASHTON, B.: Evolution and prevention of drug resistance. Nature (Lond.) **203**, 1323 (1964).

— — Prevention of antibiotic and sulfonamide resistance by atebrin. Antimicrobial Agents and Chemotherapy **1964**, 410 (1965).

— DE COURCY, S. J., JR.: (1) Origin and prevention of drug resistance in microorganisms. In: Topics in medicinal chemistry, Vol. 3 (RABINOWITZ, J. L., MYERSON, R. M., Eds.). New York: John Wiley and Sons, Inc. p. 107 (1970).

— — (2) Biochemical events underlying the evolution and prevention of drug resistance in microorganisms. In: Functional biochemistry of cell structures (The N. M. Sissakian Memorial Volume) (OPARIN, A. I., Ed.). Moscow: Academy of Sciences of the U.S.S.R. Publishing House "Nauka", p. 369 (1970) (in Russian).

— DRABBLE, W. T.: Prevention of the emergence of drug resistant bacteria by polyamines. Biochem. biophys. Res. Commun. **8**, 446 (1962).

SHARDA, D. C., CORNFELD, D., MICHIE, A. J.: Effect of mepacrine (Atebrin) on the success of antibacterial treatment of urinary infections. Arch. Dis. Childh. **41**, 400 (1966).

WEBB, R. B., KUBITSCHEK, H. E.: Mutagenic and antimutagenic effects of acridine orange in *Escherichia coli*. Biochem. biophys. Res. Commun. **13**, 90 (1963).

Some Studies on the Antimutagenic Action of Polyamines

MICHAEL K. BACH and HERBERT G. JOHNSON

I. Introduction

Cells of all organisms contain one or more basic substances. In the bacteria and the viruses these are polyamines while in animal cells the polyamines are usually supplemented by basic proteins such as the histones. These basic molecules have been implicated in a large number of biochemical processes. They are known to affect RNA synthesis and protein synthesis (MUENCH, 1969, 1967; NICHOLSON and PEACOCKE, 1966; SCHOLTISSEK and BRECHT, 1966; O'BRIEN, OLENICK and HAHN, 1966, and many others). In the viruses they are presumed to play a key role in the folding of the viral nucleic acids to permit their encapsulation in the limited space afforded inside the viral capsule. Similarly, the histones appear to be synthesized in mammalian cells in perfect balance with the synthesis of DNA (BAZILL and PHILPOT, 1963; BUTLER and COHN, 1963; LITTLEFIELD and JACOBS, 1965) and they occur as complexes with DNA. They are likely to play an important role in the regulation of the availability of the DNA genome for translation into RNA and protein (BONNER and HUANG, 1966).

Recognizing this general property of the polyamines, we felt many years ago that a study of the binding of polyamines to DNA might lead to meaningful new insights into the control of the genetic machinery of the cell and thus to possible new kinds of therapeutic manipulations. Not long after the initiation of our studies there appeared the first of a series of papers by the late M. G. SEVAG and his collaborators (SEVAG and DRABBLE, 1962) reporting the delay in the tendency of bacterial cultures to become resistant to antibiotics when the polyamine spermine was included in the culture medium. Over the years, this early report was followed by others from the same laboratory (SEVAG, 1964; SEVAG and ASHTON, 1964; HELLER and SEVAG, 1966; DE COURCY and SEVAG, 1966), as well as by reports from other laboratories using other test systems and other methods for the measurement of the effect of the polyamines (DUBININ, 1963; DUBININ, CHEREZHANOVA and BULCHNIKOVA, 1963; WEBB and KUBITSCHEK, 1963; MAGNI, VON BORSTEL and SORA, 1964; PUGLISI, 1967; and ZAMENHOF, 1969). It is our purpose here to review our own results with antimutagens [JOHNSON and BACH, 1964, 1965 (1, 2), 1966, 1967, 1969; BACH, 1969] in the context of studies from other laboratories and to discuss possible mechanisms of action of these compounds in the light of what is known of their subcellular distribution and biochemical properties.

II. Antimutagen Activity in Various Systems

While SEVAG and DRABBLE speculated on the antimutagenic nature of the activity of spermine in their bacterial cultures, and suggested that this might be due to the binding of the spermine to the bacterial DNA, neither of these points was proven in their studies. To demonstrate that the action of polyamines was, indeed, on the mutational event itself, we studied the effect of the growth of bacterial cultures in the presence of certain polyamines on the random fluctuation test of LURIA (1946). Using this test we found (JOHNSON and BACH, 1965) a 2 to 4 fold reduction in the mutation rate of *E. coli* and *S. aureus* when two completely unrelated mutational events were studied. Thus, both a forward mutation, the development of resistance to streptomycin, and a back mutation from trytophan-requirement to wild type, were suppressed by non toxic concentrations of spermine in *E. coli* and the suppression of the first of these mutations was also confirmed in *S. aureus*. The random fluctuation tests were shown to be valid because the inclusion of spermine in the culture medium did not select for either the wild type or the mutants. The facts that 'antimutagenic' activity was seen regardless of the mutational end point being measured, that the growth rate of the mutants was unaffected by the spermine when the mutants were isolated and their growth rate measured, and that the antimutagenic activity was only demonstrable when the cultures were grown in the presence of the polyamine (and not when the compound was added at the time of selective plating) all argue against selection and in favor of a true antimutagenic activity of the polyamines.

Our detailed understanding of the type of events which can lead to mutations has increased greatly in the last two decades. The random mutations we had studied initially are the most poorly defined. We felt that we might gain a better understanding of the mechanism of action of the antimutagens by attempting to define the organisms and the types of mutational events which are susceptible to inhibition by these materials (Table 1). Turning first to the bacterial systems, we found (JOHNSON and BACH, 1965, 1966) that cultivation in the presence of non-toxic concentrations of the polyamine spermine or of quinacrine hydrochloride markedly decreased mutation rates when mutations were caused by the purine analog, 2-aminopurine, or when an *E. coli* strain containing Treffer's mutator gene (TREFFERS, SPINELLI and BELSER, 1954) was used. Both these agents are known to cause point mutations by specific base exchanges in the DNA (YANOFSKY, COX and HORN, 1966) and thus the prevention of these mutations must reflect a stabilizing force which is active at the instant of DNA replication. It was unexpected and very interesting, therefore, when we found that the polyamines also inhibited mutations caused by ultraviolet irradiation and by the inclusion of caffeine in the cultures. Mutagenesis by ultraviolet light and, presumably by 'radiomimetic drugs' is believed to be caused by the formation of pyrimidine dimers or cross links in the DNA which, during subsequent replication of the DNA, give rise to errors due to misreading provided the repair process, which is known to exist in the cells, has not deleted the errors before replication. The mutagenic activity of caffeine is believed to be due to the inhibition of the repair process (SHIMADA and TAKAGI, 1967; SIDEROPOULOS and SHANKEL, 1968; LIEB, 1964) and thus reflects an indirect increase in the expression of the mutations caused by ultraviolet light. Thus, in order to counteract this mutational event, the polyamines must be presumed to promote repair or, alternatively, to minimize the likelihood of misreading a strand which has uncorrected errors in it. The latter seems more likely since the effect was seen in the presence of caffeine, where the repais process is presumably impaired.

In addition to the mutational events mentioned above, the acridines have been known for some time to 'cure' bacterial cultures from cytoplasmic genes, such as the *R* factor of *F* factor, which they may harbor (CUZIN and JACOB, 1966; ROWND, NAKAYA and NAKAMURA, 1966; HASHIMOTO, KONO and MITSUHASHI, 1964). The mechanism involved here may be different from the antimutagenic effect we have been studying since the 'cure' appears to consist of the preferential inhibition of the replication of the extrachromosomal genes without inhibiting the growth of the host

Table 1. *The antimutagenic action of polyamines*

Organism	Antimutagen (μg/ml)	Mutational[b] end-point	Mutagenic agent	[a]Mutation rate in controls	Ratio of mutation rate in (control) to mutation rate in (plus spermine)
E. coli	Spermine, 150	SM^r	Random	1.3×10^{-10}	3.7
E. coli	Spermine 150 at time of plating	SM^r	Random	1.2×10^{-10}	1.1 (not significant)
E. coli	Spermine, 150	Trypt → Trypt	Random	6.4×10^{-10}	2.9
S. aureus	Spermine, 150	SM^r	Random	1.1×10^{-10}	2.5
E. coli	Spermine, 150	SM^r	Caffeine, 150 μg/l	15×10^{-10}	11
E. coli	Spermine, 150	SM^r	UV (99.5 % kill)	22×10^{-10}	8.5
E. coli	Spermine, 150	SM^r	2-Amino-purine	36×10^{-10}	5.0
E. coli	Spermine, 150	SM^r	Mutator gene	12×10^{-10}	1.7
E. coli	Quinacrine, 10	SM^r	Mutator gene	12×10^{-10}	3.9
D-98	Quinacrine, 0.06	Azg^r	Random	0.01	2.9
D-98	Quinacrine, 0.06 added at time of plating	Azg^r	Random	0.01	1.0
L-1210	Quinacrine, 0.10	Ara-C^r	Random	0.01×10^{-4}	5—16

[a] Mutation rate, $m = -(2.303 \log_{10} P_0)/N$ where N = total number of cells per culture and P_0 = proportion of cultures which were free of mutants.
[b] SM^r = resistance to streptomycin, Azg^r = resistance to 8-azaguanine, ara-C^r = resistance to cytarabine.

cells. In our experiments, on the other hand, the antimutagenic effect was seen without any effect on growth. In a recent paper, ZAMENHOF (1969) confirmed the antimutagenic effect of non-toxic concentrations of quinacrine hydrochloride and spermine on random mutations in *E. coli.* By contrast, this study demonstrated that the mutations caused by the mutator gene Ast-1 in *E. coli,* which is distinguishable from the TREFFERS mutator gene because it is not base pair-specific and because it can reverse its own mutations, were not affected by these compounds. The antimutagenic effect is thus clearly not universal even in *E. coli*.

Finally it should be pointed out that the very same acridine derivatives (including quinacrine) which we have shown to be antimutagenic are the prototypes for "frame shift" mutagens (LERMAN, 1964; BROCKMAN and GOBEN, 1965; MALLING, 1967; CARLSON, SEDEROFF and COGAN, 1967; AMES and WHITFIELD, 1966). We do not understand the full reason for this discrepancy. It is clear, however, that most of the mutagenic effects have been obtained in bacteriophages and not in bacteria. Furthermore, the usual means for selecting mutants with the aid of these materials is by placing a crystal of the solid chemical on an agar plate and selecting the colonies which grow in the presence of the selecting agent (also in the agar) regardless of any toxic effect of the 'mutagen' on the culture as a whole. Thus, in most instances the concentrations of acridines which are used are, in fact, quite toxic to the cells. Finally, as demonstrated by WEBB and KUBITSCHEK (1963) and explained by DELMELLE and DUCHESNE (1967), acridine orange can be converted from an antimutagen to a mutagen by moving the cultures from total darkness to light, a variable which has not been rigorously controlled in most experiments.

A working hypothesis for explaining the antimutagenic activity of the polyamines was to propose that the added polyamines displaced the polyamines which are naturally present in the respective cells, and took their place by binding to the DNA in the cells. The added polyamines caused a greater stabilization of the DNA than the naturally present polyamines. Partial support for this hypothesis comes from the fact that, while spermine was active in bacterial cultures, the addition of neither spermidine nor putrescine, the polyamines which naturally occur in the cells, ever showed activity. It is also tempting to speculate that the extreme difficulty which has been encountered in efforts to isolate polyamine deficient mutants (T. T. TCHEN, personal communication) may be due to the fact that such mutants, if they ever occurred, would be highly unstable or lethal since there would be no polyamines present in them to 'stabilize' their DNA at all.

In contrast to the bacteria, mammalian cells are endowed not only with spermine in addition to spermidine and putrescine, but also with strongly basic proteins (histones) which are believed to occupy those sites on DNA which might be available to exogenously added polyamines in the bacterial cells. It can be postulated, then, that it would be necessary to use materials which compete much more effectively for the DNA binding sites than spermine if antimutagenesis is to be demonstrated in mammalian cells. Using an adaptation of the random fluctuation test and single cell cloning techniques, we succeeded in demonstrating that inclusion of quinacrine into the culture medium of two different mammalian cell lines in culture markedly reduced their spontaneous mutation rate (JOHNSON and BACH, 1969; BACH, 1969). One of these cell lines was a murine leukemic cell (L-1210) and the mutational event we measured was the development of a therapeutically detrimental degree of resistance to cytarabine, a drug which is highly effective in prolonging the life of mice having the strain of leukemia. It was of interest to see, therefore, if the suppression of the development of resistance with quinacrine could be demonstrated as an increased 'cure' rate following treatment with cytarabine *in vivo*. The design of an experiment to test this possibility is depicted in Fig. 1. The work of SKIPPER, SCHABEL and WILCOX (1964, 1967) has demonstrated that treatment with relatively high doses of cytarabine every 3 h on every fourth day affords the maximum rate of killing L-1210 cells in mice. Thus if mice are inoculated with a very small number of leukemic cells, and are treated daily with quinacrine, our *in vitro* studies suggest that there might be a time when the antimutagen-treated mice will, on the average, have no cytarabine-resistant leukemic cells in them while the untreated mice will have at least one resistant cell present. If, at that point, vigorous treatment with cytarabine is initiated we can expect to kill essentially all the sensitive L-1210 cells, leaving alone any resistant cells. Since a single resistant cell is sufficient to kill a mouse (SKIPPER et al., 1964), we tried to show that treatment with quinacrine affects the number of mice which are 'cured'

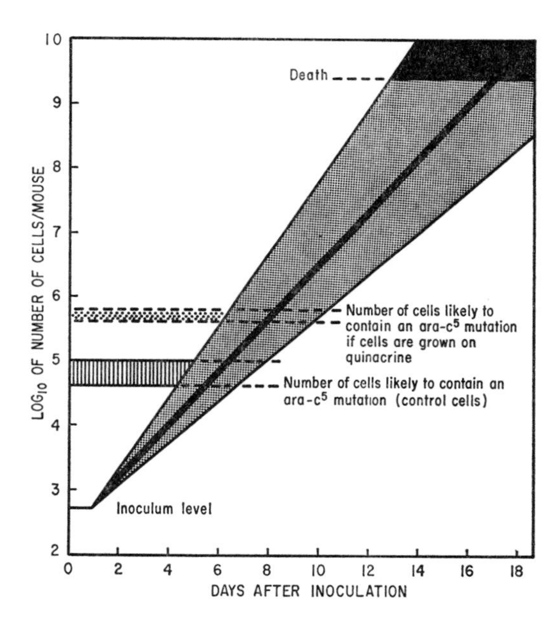

Fig. 1. Schematic to explain design for experiment attempting to demonstrate antimutagenesis *in vivo*

(i.e., those which had not even a single resistant cell in them) while not affecting the survival time of the mice which ultimately die since this is only a function of the number of leukemic cells which survive the cytarabine treatment in those animals where any cells survive at all. It should be clear that, since each mouse acts as a single culture in this 'random fluctuation test', a huge number of mice are required for any meaningful experiment. Two experiments were run, each involving multiple injections into a total of 500 mice. In both experiments, the mice which were treated with quinacrine and with cytarabine on the 'critical' day (day 5) had significantly more survivors than did the groups which were not treated with quinacrine, or those which were treated with cytarabine earlier or later. Because of an unexplained anomaly in one of the experiments, the controls which were not treated with cytarabine at all also showed a significant increase in number of 'cures' in the quinacrine-treated group.

This was not found in the second experiment, nor had we ever seen an effect of quinacrine on 'curing' in numerous other studies. The overall results of the two experiments show a statistically significant 'antimutagenic' effect at the 5% level. Thus, the data are suggestive of the fact that the antimutagenic effect of quinacrine can be demonstrated in an *in vivo* situation if the conditions are carefully controlled.

III. Mechanism of Action: Evidence for Binding to DNA

A minimal requirement for proof of the hypothetical explanation of the mode of action of the polyamines in these systems is that they must bind to the DNA inside the living cells. We attempted to show such a binding in the bacterial system using tritiated spermine rather than the heterocyclic compounds (such as quinacrine) as we felt this would be a more meaningful test. It has been known for some time that polyamines bind to nucleic acids *in vitro*. The stabilization of DNA in dilute salt solutions against thermal denaturation by these materials is well known (MAHLER, MEHROTRA and SHARP, 1961; MEHROTRA and MAHLER, 1964; LIQUORI, CONSTANTINO, CRESCENZI, ELIA, GIGLIO, PULITI, SAVINO and VITAGLIANO, 1967). Attempts to demonstrate similar stabilization *in vitro* under more physiological conditions (e.g. isotonic saline with or without the addition of mM $CaCl_2$ and $MgCl_2$) failed because spermine and the other aliphatic polyamines which were tested are virtually displaced by the sodium and divalent metal ions. No stabilization against thermal denaturation could be demonstrated above that caused by the increased salt concentration itself. This is not so for the heterocyclic compounds such as quinacrine which are capable of intercallating (LERMAN, 1964; O'BRIEN, ALLISON and HAHN, 1966) between paired bases in the double stranded helix but, in view of the fact that both the aliphathic and the aromatic amines are antimutagenic, any explanation of their mode of action must be applicable to both types of compounds.

Because of the failure to demonstrate binding to DNA in physiological solutions *in vitro*, we attempted to gather evidence for the binding of exogenously added polyamines to DNA inside the bacterial cells by the physical isolation of the cellular constituents following the uptake of tritiated spermine by the cells. Since the reaction of spermine with DNA is really the formation of a salt between a polycation and a polyanion, and since the cells contain many other polyanions in addition to DNA it was to be expected that a large portion of the spermine would be found attached to other cellular constituents such as membranes, ribosomes, RNA. Similarly, since we had already found that the attachment of spermine to DNA in isotonic saline is virtually abolished it was essential to limit the isolation steps being used to those involving very dilute salt solutions.

In confirmation of results of TABOR and TABOR (1966) we found that a portion of the spermine which was taken up by the *E. coli* in our studies was bound in such a way that it was not exchangeable with unlabelled spermine in the washing medium. Active respiration had to take place for this uptake. We further found that by far most of the spermine taken up was found in or on the spheroplasts which could be isolated from these cells by lysozyme and EDTA. Virtually all the spermine which remained attached to the bacterial cell walls during this fractionation was readily exchangeable with unlabeled spermine in the washing medium (JOHNSON and BACH, 1968). In attempting to follow the sub-cellular distribution of the tritiated spermine within the spheroplasts we encountered large losses in bound spermine which occurred during almost any manipulation we attempted. This made it impossible to apply the customary methods of separating ribosomes, or of separating RNA from DNA. Nevertheless, we succeeded (Table 2) in isolating a nucleic acid-rich fraction in which the loss of spermine appeared to parallel the loss of DNA following a variety of enzymatic treatments. This fraction was obtained by preparing a phenol extract of the bacterial cells in a low salt medium, deproteinizing the phenol extract by the method of Marmur (1961), and then isolating that portion of the total extract which was sedimented

Table 2. *Subcellular localization of absorbed spermine-^3H in E. coli spheroplasts: Solubilization by enzymatic digestion of a nucleic acid fraction and controls for artifactual binding and reequilibration*

Treatment	% Spermine retained in fraction	Specific activity		
		cpm/A_{260}	(cpm/µg)	
			RNA	DNA
A. Control	70	—	413	1020
Plus Trypsin 100 µg/ml × 30 min × 37°	45	—	310	915
Plus RNAse, 10 µg/ml × 30 min × 37°	58	—	687	965
Plus DNAse, 100 µg/ml × 30 min × 37°	30	—	181	925
Plus Trypsin and RNAse as above	49	—	660	875
Plus trypsin and DNAse as above	12[a]	—	80[a]	1410[a]
B. Control-spermine-labeled cells	100	530		
5 mg/ml RNA added during isolation	16	295		
5 mg/ml sonicated DNA added during isolation	18	396		
Control-unlabeled cells, spermine-^3H added during isolation	7	24		
Control-unlabeled cells + sonicated DNA added during isolation	4	11		

[a] These are based on virtually no radioactivity and a negligible amount of residual DNA.

upon centrifugation at 105,000 × *g* for 24 h in a 15 to 30% sucrose gradient in the presence of only 0.1 M Tris buffer. It should be noted that the composition of this phenol-isolated fraction was approximately 67% RNA and 33% DNA. Nevertheless, the solubilization of spermine from the fraction followed the solubilization of DNA and was virtually independent of the solubilization of RNA since the specific activity of the residual pellet (cmp/µg DNA) remained virtually constant while the specific activity based on RNA content varied over more than 8-fold. It is also worth pointing out that it was necessary to use a combination of trypsin and DNAse to obtain good solubilization of DNA and spermine. Neither of these enzymes, by themselves, was particularly effective. The need for using trypsin may explain why we succeeded in isolating a specifically spermine-rich nucleic acid fraction in the first place since, as we already indicated, the complex of spermine with DNA, when formed *in vitro* did not survive the isolation conditions which were used in isolating this fraction.

Possibly the spermine was bound in some sort of a ternary complex involving materials other than DNA perhaps cell membrane components) along with DNA.

A necessary control for understanding the results presented in Table 2 is the demonstration that spermine added during the isolation of the nucleic acid fraction will not bind to the same sites as the spermine which had been taken up by the living cells. Furthermore, it is essential to rule out any possible redistribution of the bound spermine during the isolation process. In an experiment which was designed to rule out these possibilities, we found a negligible amount of tritiated spermine in the nucleic acid fraction which was pelleted upon prolonged centrifugation when tritiated spermine was added to the cells together with phenol. Similarly, the addition of a 10-fold excess of exogenous RNA or sonicated DNA, neither of which was sufficiently large to sediment during the centrifugation step, resulted in only a small change in the specific activity of the pellet fraction. These results are presented in the lower half of Table 2.

In viewing the quantitative aspects of the binding of spermine to cellular DNA it appears that, assuming one spermine molecule can be bound for every 4 nucleotides (LIQUORI et al., 1967), the maximum amount of spermine which could conceivably have been bound to the cellular DNA was only 3% of the amount of spermine which was taken up by the cells. Viewed conversely, the fraction of the spermine which was susceptible to solubilization by the action of DNAse and trypsin was about the same. While this certainly does not account for a major portion of the spermine in the cells, and while we cannot ascribe to DNA a unique role as the site of binding of this substance, it is worth stressing that the externally supplied spermine effectively replaced all the endogenous cations of the DNA under our conditions. Thus the results satisfy the primary requirement for demonstrating the binding of *some* spermine to DNA under conditions where spermine acts as an antimutagen.

IV. Structure-Activity Studies

It was felt that further insight into the mechanism of action of the antimutagens might be obtained from a detailed study of the structural requirements for activity in these molecules. Using a culture of *E. coli* carrying the mutator gene of TREFFERS, we developed a quantitative assay for antimutagenicity. The dose-response curve for quinacrine hydrochloride, which was used as the reference compound in this study, is shown in Fig. 2. All together some 200 compounds were tested at several concentrations each for their effect in this system. Although the dose-response plots did not all run parallel to that for quinacrine, the slopes were sufficiently similar that activity could be expressed as the ratio of the molar concentration of quinacrine to that of the substance being studied, which resulted in the same inhibition of mutation rate. The results of this study are summarized in the next three tables (Tables 3, 4, and 5). Table 3 considers the effect of the nature of the alkyl side chain on the activity. While the side chains may reduce the toxicity of the compounds, as reflected by the relatively high maximum concentrations at which they could be tested, they did not necessarily enhance their activity. This makes it unlikely that quinacrine is active in this system entirely because of the spermine-like side chain (O'BRIEN, OLENICK and HAHN, 1965). Part II of the table shows that the introduction of hydrophylic groups in the vicinity

Table 3. *Effect of the presence and structure of an amino alkyl substituent on antimutagen activity in E. coli having a mutator gene*

Heterocyclic bases used:

R, OCH_3, Cl, N and R, Cl, N

A = 6-Chloro-2 methoxyacridine Q = 7-Chloroquinoline

Base	R	Relative[a] activity	Maximum conc.[b] tested (μg/ml)
I. Effect of presence vs absence of side chain			
A	$-NH_2$	10	1.0
A	$-NH-CH(CH_3)-[CH_2]_2-CH_2-N(CH_2-CH_3)_2$	1.0	25
A	$-NH-C_6H_4-CH(OH)-CH_2NH_2$	10	10
Q	$-NH_2$	1.1	10
Q	$-NH-CH(CH_3)-[CH_2]_2-CH_2-N(CH_2-CH_3)_2$	3.6	10
Q	$-NH-C_6H_4-CH(OH)-CH_2NH_2$	1.5	10
II. Effect of hydrophylic vs hydrophobic groups in side chain			
Q	$-NH-CH_2-(C_5H_9)N-CH_3$	0.04	10
Q	$-NH-CH_2-CO-(C_5H_9)N-CH_3$	0.49	10
Q	$-NH-CH(CH_3)-[CH_2]_2-CH_2-N[CH-CH_3]_2$	3.6	10
Q	$-NH-CH_2-CH(OH)-CH_2-N[CH_2-CH_3]_2$	13.3	50
Q	$-NH-C_6H_4-CH_2-N[CH_2-CH_2-Cl]_2$	0.15	50
Q	$-NH-C_6H_3(OCH_3)(CH_2-NH-CH_2-CH_2-Cl)$	1.6	10

[a] Relative activity is the ratio of the concentration of quinacrine to the concentration of the drug which was used, which would elicit the same effect. Concentrations are computed in units of $\mu M/10^9$ viable cells as shown in Fig. 2.

[b] Maximum concentration tested was 50 μg/ml unless this concentration was too toxic.

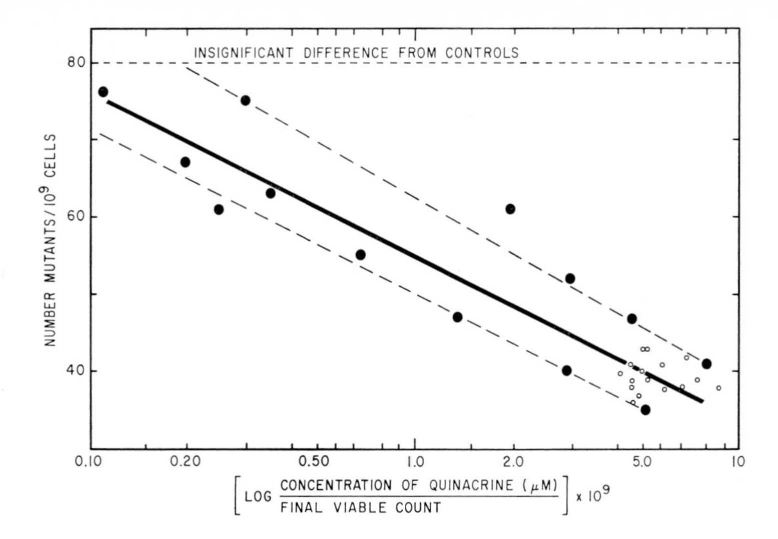

Fig. 2. Dose response relation in the prevention of mutations to streptomycin resistance by quinacrine. Plot is for log μM/N $\times 10^9$ vs m/N $\times 10^9$ where N = final ave viable count at drug conc. shown, m = ave. number mutants observed/plate. The heavy line is the line with computed slope taking as initial point m/N = 40 for μM/N = 5.0. Small points around 5.0 are control points obtained for *all* the runs made

Table 4. *Effect of the presence and location of various substituents on the antimutagen activity of heterocyclic compounds in a strain of E. coli having a mutator gene*

Remarks	Base used	Location and structure of *R* groups	Relative activity	Maximum conc. tested (μg/ml)
I. The location of the amino group	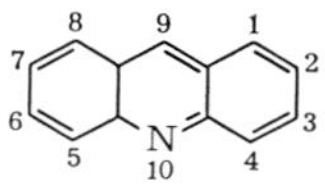	9-NH_2	Inactive	0.1
		2-NH_2, tetrahydro	0.29	10
		2-NH_2	0.028	10
		3-NH_2	0.37	1
		3,6-diamino	7.5	1
		6,9-diamino, 2-ethoxy	Inactive	1
II. Effect of presence of other ring substituents	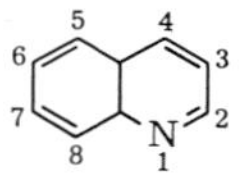	2-methyl-4-amino	Inactive	1
		7-Chloro-4-amino	1.1	10
		7-Chloro-4-mercapto	0.07	10
		-7-Chloro-4-[diethyl-amino-(1-methyl)-butyl-]amino]	3.6	10
		-7-Iodo-4-[diethyl-amino-(1-methyl)-butyl]amino]	0.35	50
		1-N-Methyl-4-[p-hydroxy]-anilino	1.0	10

Table 5. *The Antimutagen activity of compounds containing other flat heterocyclic bases in a strain of E. coli having a mutator gene*

Remarks	Base used	Location and structure of side chains	Relative activity	Maximum conc. tested (μg/ml)
I. Benzoquinolines- Phenanthridines	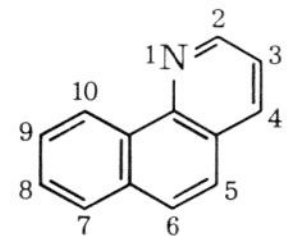	4-[3-(diethylaminopropyl)amino]	0.13	1
		2-hydroxy-4-[3-diethylaminopropyl)amino]	0.88	1
	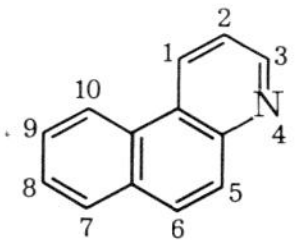	5-[2-(diethylaminoethyl)amino]	Inactive	1
	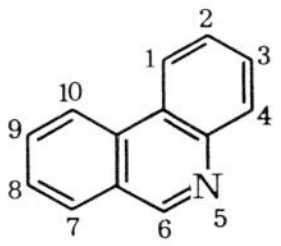	6-[2-(dimethylaminopropyl)amino]	0.70	1
II. Benz[c]acridines-varying substitution	Benz [c] acridine	7-[3-(diethylaminopropyl)-amino-]	0.43	1
		2-methoxy-7-[3-(diethylaminopropyl)-amino]	22	1
	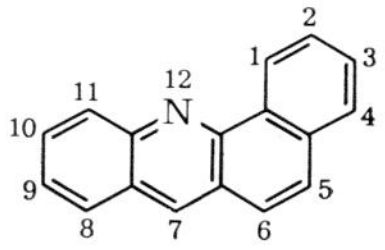	3-methoxy-7-[3-(diethylaminopropyl)-amino]	24	1
		4-methoxy-7-[3-(diethylaminopropyl)-amino]	Inactive	1
		4-methoxy, 10-chloro-7-[3-(diethylaminopropyl)-amino]	28	1
		10-Chloro-7-[3-(diethylaminopropyl)-amino]	Inactive	1
Benz[a] vs Benz[c]acridines	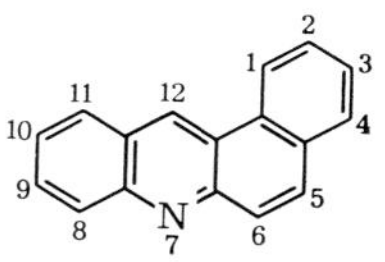 Benz [c] acridine	7-[3-(diethylaminopropyl)amino]-benz[c]acridine	0.43	1
		12-[3-(diethylaminopropyl)amino]-benz[a]acridine	0.14	1

of the side chain amino group in the compounds showed markedly potentiated activity. The same effect has also been observed in several other instances.

Table 4 summarizes the effect of changing the location of the amino group to several positions on the acridine nucleus and the introduction of 'activating' groups in the quinoline nucleus. It is seen that 9-amino acridine, the parent compound for quinacrine, was so toxic that no activity could be demonstrated. The reduction of one of the rings reduced the toxicity and permitted the demonstration of some activity. As could be predicted from the studies of ALBERT (1952) moving the amino group to different positions has markedly affected the activity and toxicity of the molecule. The discrepancy between the results with the two diamino compounds is difficult to understand in view of the known activity of 9-amino-2-methoxy acridine as well as the activity of 3-amino acridine. In part II of the Table, the 7-chloro substituent in the quinoline series is shown to have a pronounced activating effect while the 7-iodo compound is shown to have a considerably lower activity. It is also noteworthy that a quinoline compound, even without further activation, was an effective antimutagen. The same observation regarding the effect of the quarternization of the ring nitrogen (last line of Table) was also seen with one acridinium compound and especially with the two phenathridinium derivatives which were tested.

Results with some other molecules containing flat heterocyclic ring systems are summarized in Table 5. Activity in the unsubstituted amino alkyl benzoquinolines and phenanthridines was quite comparable to that in the unsubstituted acridines. The marked dependence of activity on the location of the substituents in the benzacridine series (part II of the table), and especially the interaction of two inactive substituents (10-chloro and 4-methoxy) to make an active compound are difficult to comprehend.

In contrast to the heterocyclic compounds described in these Tables, aliphatic polyamines were, as a group, less active. Suffice it to recall that spermine, one of the best of these compounds in our tests required a 15 times higher concentration to elicit an effect comparable to that of quinacrine hydrochloride.

V. The Lack of Structural Specificity in the Antimutagen Effect

We have already mentioned the fact that the antimutagens appeared to act equally well on mutations with different end points, and regardless of the mutation-inducing event. Nonetheless, it was of interest to establish whether two of these compounds might act synergistically. If these compounds have different preferred positions for attachment (i.e. act independently and therefore synergistically) then one can calculate the predicted effect of combining both drugs in the same cultures as the product of their individual effects if these are expressed as the fraction $\frac{\text{m/N treated}}{\text{m/N control}}$ where m is the number of mutants per plate and N the final viable count of cells plated. Two structually unrelated compounds were chosen. For each, the effect of doubling the dose could be calculated from the standard curves. Thus in Table 6, for an untreated control value of 90 mutants per 10^9 cells, and for a treated value of 40 the ratio for either compound by itself is 0.44 and the predicted ratio for the combination is 0.19. Multiplying this by the control value of 90 leads to a prediction of 17 mutants per 10^9 cells. On the other hand, if the compounds occupy the same positions, then the effect of using the combination of the lower doses of each compound should be the same as that of doubling that dose and using either compound alone since the second increment of binding (for either compound) would be expected to go to the next batch of unoccupied sites. As can be seen, the results agreed very well with the prediction based on independent binding sites. We conclude, therefore, that the activity of the two compounds does not depend on preferred attachment sites which are specific for each compound. Such preferred attachment sites could have been equated with specific base sequences in DNA or, more broadly, with regions of relative enrichment for one or the other base pair. The abundance of these sites, if present, is likely to differ from gene to gene. Thus the preferred attachment of these compounds to different sites would have implied a differential relative effect of the compounds on different mutational end-points. The demonstration that apparently there were no preferred attachment sites is therefore equivalent to another demonstration that the antimutagenic action is independent of the mutational end-point being determined, as long as the mutations are due to changes in the chromosomal DNA.

Table 6. *The effect of the use of combinations of antimutagens on the mutation frequency in E. coli having a mutator gene*

Compound and dose (μg/ml)	Mutation frequency × 10^9		
	Predicted		Observed
	additive	independent	
Ethidium bromide, 0.66	40	40	41 ± 6
Ethidium bromide, 1.42	30	30	34 ± 3
Quinacrine hydrochloride, 11	40	40	42 ± 6
Quinacrine hydrochloride, 22	30	30	30 ± 3
Quinacrine hydrochloride, 11 plus Ethidium bromide, 0.66	30	17	33 ± 3

VI. Lack of Correlation between Antimutagenic Activity and Ability to Protect DNA from Thermal Denaturation

As stated in the introduction, the antimutagenic action of polyamines has been alleged to reflect their ability to 'stabilize' DNA in solution. The stabilization of DNA is most commonly measured by means of a change in the melting temperature (T_m) of DNA when the compound in question is added to 'native' DNA in a dilute salt solution. We have attempted to quantitate the effectiveness of various compounds in bringing about this stabilization by measuring the lowest concentration of the compound which gave a measurable stabilization effect. The data in Table 7 show there was no correlation between the DNA stabilizing activity of these compounds and their activity as antimutagens. A possible reason for the lack of correlation is the salt dependence of the binding of the polyamines to DNA. Thus compounds which presumably act primarily by intercalation and have no secondary interaction of their side chains with the phosphate groups in the DNA chain are less susceptible to displacement by high salt concentrations, and are more likely to bind to DNA under physiological conditions.

Table 7. *Lack of correlation between the effect of various polyamines on the melting temperature of DNA (T_m) and their activity as antimutagens in a strain of E. coli containing a mutator gene*

Compound		Relative activity anti-mutagen	-log of min. conc. (M) affecting T_m [a]
I.	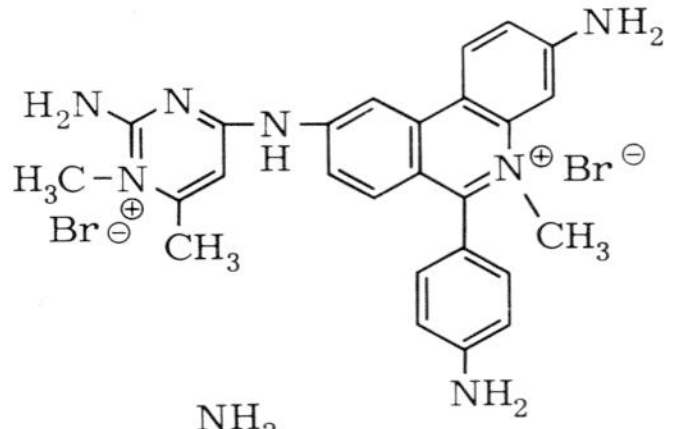 NH_2; H_2N, N, N–H; $H_3C-N^{\oplus}$ $Br^{\ominus}$; CH_3; $N^{\oplus}$ $Br^{\ominus}$; CH_3; NH_2	26	5.0
II.	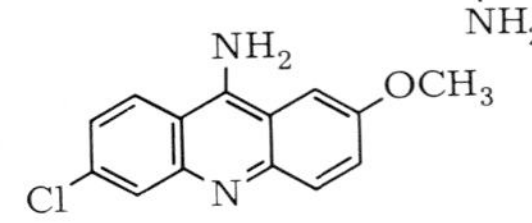 NH_2; OCH_3; Cl; N	10	6.0
III.	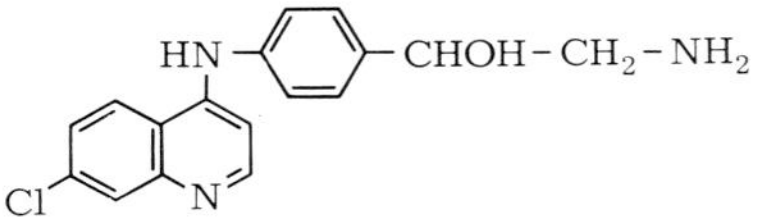 HN–C₆H₄–CHOH–CH_2–NH_2; Cl; N	1.5	6.0
IV.	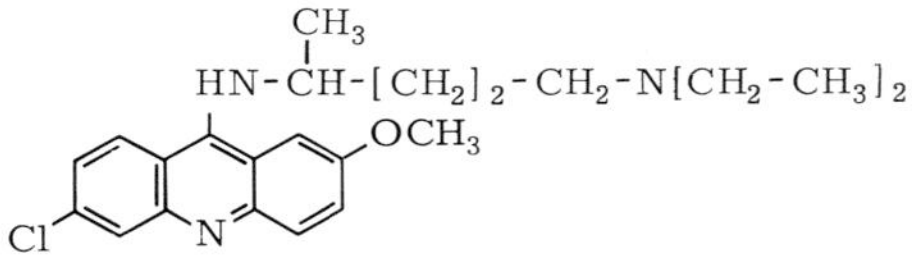 HN–CH(CH_3)–$[CH_2]_2$–CH_2–N$[CH_2-CH_3]_2$; OCH_3; Cl; N	1.0	5.5
V.	CH_3–NH–CH_2–CH_2–N(CH_3)–CH_2–CH_2–N(CH_3)–CH_2–CH_2–$NHCH_3$	0.21	5.0
VI.	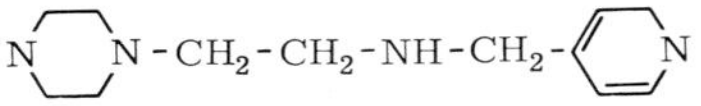 N N–CH_2–CH_2–NH–CH_2–(N)	0.15	5.0
VII.	$[CH_3]_2$–N–CH_2–CH_2–NH–CH_2–CH_2–N$[CH_3]_2$	0.56	5.2
VIII.	CH_3NH–CH_2–CH_2–CH_2–N(CH_3)–CH_2–CH_2–CH_2–NH–CH_3	0.038	6.0
IX.	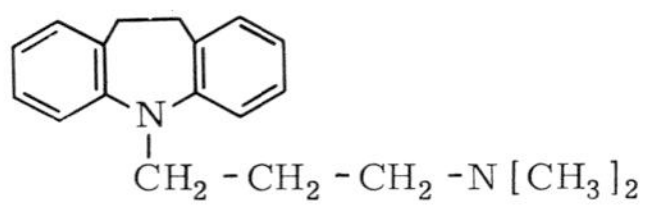 N–CH_2–CH_2–CH_2–N$[CH_3]_2$	0.017	<4.0
X.	NH_2–$[CH_2]_3$–NH–$[CH_2]_4$–NH$[CH_2]_3$–NH_2	0.010	6.0
XI.	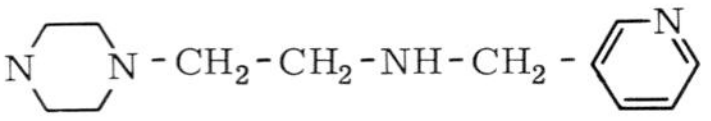 N N–CH_2–CH_2–NH–CH_2–(N)	0.001	5.0
XII.	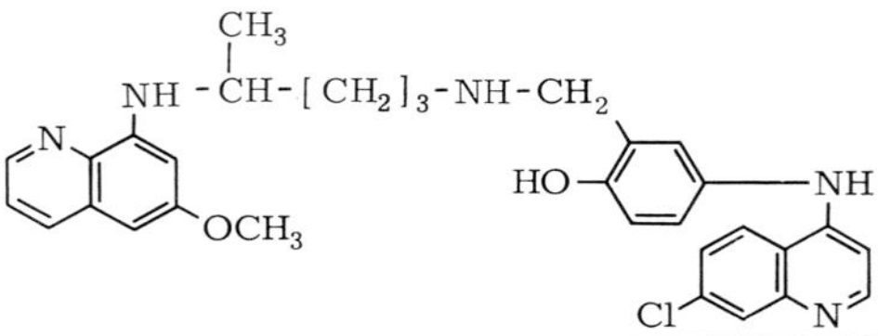 NH–CH(CH_3)–$[CH_2]_3$–NH–CH_2; N; OCH_3; HO; NH; Cl; N	Mutagen	6.0

[a] T_m of calf thymus DNA in 0.002 M salt. The conc. of compounds was decreased in 3-fold increments beginning at 10^{-4} M. The minimum effect (at min. conc.) was at least 2°.

VII. A Possible Alternative Explanation for the Antimutagenic Effect of Polyamines

Regardless of the lack of correlation between *in vitro* stabilization of DNA in dilute solution and antimutagenicity, the results presented thus far are consistent with the hypothesis that certain molecules which are capable of finding a firm binding site in the DNA chain in the form of intercallation, and which have additional amino

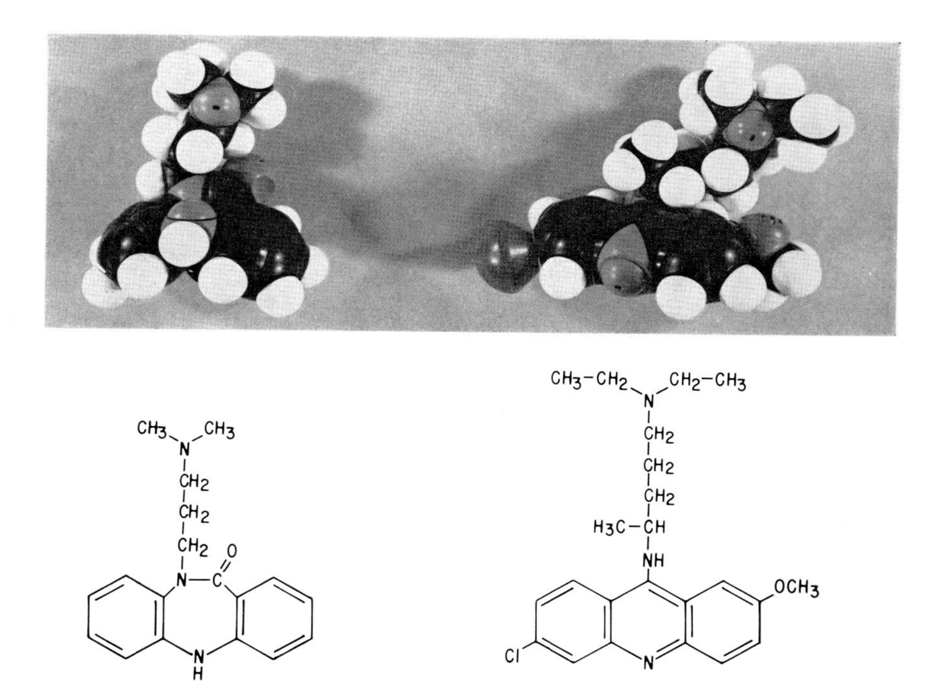

Fig. 3. CPK models of quinacrine and a diazepine to demonstrate the lack of planarity in the rings of the diazepine

groups which can interact with vicinal phosphate groups in the DNA backbone, are effective antimutagens. This is particularly so if conditions are favorable for a hydrophobic interaction as well (see Table 3 part II). This explanation is difficult to reconcile with the finding that a whole series of compounds, particularly certain derivatives of 10, 11 dihyro-5H-dibenzo (b, e) diazepine (Fig. 3) are clearly non-planar while being highly effective antimutagens. Also note in Table 4 that 9-amino-tetrahydroacridine, a non-flat molecule, was active. These molecules could not intercallate into DNA in the usual manner. The diazepines are not likely to act after the scission of the B ring since an analog of the most active compound tested, which had the amide bond between carbon 10 and nitrogen 11 cleaved to give a potentially planar two-ring system was slightly mutagenic. Furthermore, the magnitude of the activity of these compounds is several times larger than the best aliphatic polyamine we tested again making it unlikely that they act by virtue of their aliphatic side chain alone. Heller and Sevag (1966) also reported antimutagenic activity in non-planar molecules. An explanation of their bindings has been proposed by Pritchard, Blake and Peacocke (1966). The hydrophobic bonding which is possible with these molecules, as well as with the heterocyclic polyamines, may be the major reason for their antimutagenic effect and the intercalation potential may not be at issue. An alternative explanation comes from a completely different field of investigation, namely the effect of these substances on biological membranes. While it is, perhaps, inappropriate to bring this confused area into the present context, I believe it is essential to bear in mind that all the molecules which we have considered, planar and non-planar, are known to bind very strongly to living membranes (Grossowics, 1963). When living cells are allowed to take up chloroquine or acridine dyes the primary localization is in the membranous structures of the cells, particularly the lysosomes (Allison and Young, 1964; Wolf and Aronson, 1961). The antimalarials have been used for years in the control of collagen diseases (DuBois, 1954; Merwin and Winkelmann, 1962), and it has been suggested that they act by stabilizing the lysosomal membranes thus preventing the rupture of the lysosomes and the release of the hydrolytic enzymes which would cause an inflammatory response [Weissmann, 1964 (1, 2), 1965]. Finally, it must be recalled that, while we succeeded in obtaining evidence that spermine was found, in part, attached to DNA in bacterial cells, there was evidence that the DNA in the fraction in question was not free DNA since trypsin digestion markedly potentiated the solubilization of the DNA and of the spermine by DNAse. It seems tempting to suggest that here, too, we dealt with a DNA-membrane complex. Could it be, that biological membranes play a key role in the orderly replication of DNA, and that the polyamines assist in maintaining this orderliness by preserving the hydrophobic environment around the DNA strands?

References

Albert, A.: The acridines. London: Edward Arnold and Co. 1951.

Allison, A. C., Young, M. R.: Uptake of dyes and drugs by living cells in culture. Life Sci. **3**, 1407 (1964).

Ames, B. N., Whitfield, H. J., Jr.: Frameshift mutagenesis in salmonella. Cold Spr. Harb. Symp. quant. Biol. **31**, 221 (1966).

Bach, M. K.: Reduction in the frequency of mutation to resistance to cytarabine in L-1210 murine leukemic cells by treatment with quinacrine hydrochloride. Cancer Res. **29**, 1881 (1969).

Bazill, G. W., Philpot, J. St.-L.: Studies on the assay of primer DNA in the presence of histone and nucleoprotein and in isolated nuclei. Biochim. biophys. Acta (Amst.) **76**, 223 (1963).

Bonner, J., Huang, R.-C. C.: Methodology for the study of the template activity of chromosomal nucleohistone. Biochem. biophys. Res. Commun. **22**, 211 (1966).

Brockman, H. E., Goben, W.: Mutagenicity of a monofunctional alkylating agent derivative of acridine in neurospora. Science **147**, 750 (1965).

Butler, J. A. V., Cohn, P.: Studies on histone 6: Observations on the biosynthesis of histones and other proteins in regenerating rat liver. Biochem. J. **87**, 330 (1963).

Carlson, E. A., Sederoff, R., Cogan, M.: Evidence favoring a frameshift mechanism for ICR-170 induced mutations in *Drosophilla melanogaster*. Genetics **55**, 295 (1967).

CUZIN, F., JACOB, F.: Inhibition of genetic transfer by donor cells of *E. coli* K_{12} by acridines. Ann. Inst. Pasteur, V. 1, **111**, 427 (1966).
DE COURCY, S. J., JR., SEVAG, M. G.: Specificity and prevention of antibiotic resistance in *Staphylococcus aureus*. Nature (Lond.) **209**, 373 (1966).
DELMELLE, M., DUCHESNE, J. D.: On the mechanism of photomutagenic action. Compt. Rend. **264**, 138 (1967) (Series D).
DUBININ, N. P.: The problem of antimutagens in connection with mutagenesis and chemical protection. In: Proc. Internat. Congress Genetics, the Hague, p. 63, (V. I) (GEERTS, S. J., Ed.). New York: MacMillan 1963).
— CHEREZHANOVA, L. V., BULCHNIKOVA, E. K.: Control of mutation in cancer cells. Dokl. Akad. Nauk Sci. U.S.S.R. **146**, 1037 (1963).
DU BOIS, E. L.: Quinacrine (atabrine) in treatment of systemic and discoid lupus erythematosus. Arch. intern. Med. **94**, 131 (1954).
GROSSOWICZ, N.: Mechanism of protection of cells by spermine against lysozyme-induced lysis. J. Bact. **85**, 293 (1963).
HASHIMOTO, H., KONO, K., MITSUHASHI, S.: Elimination of penicillin resistance of *Staphilococcus aureus* by treatment with acriflavine. J. Bact. **88**, 261 (1964).
HELLER, C. S., SEVAG, M. S.: Prevention of the emergence of drug resistance in bacteria by acridines, phenothiazines and dibenzo cycloheptenes. Appl. Microbiol. **14**, 879 (1966).
JOHNSON, H. G., BACH, M. K.: Apparent suppression of random mutation rates by polyamines on bacteria. Bact. Proc. **1964**, G144.
— — Apparent suppression of mutation rates on bacteria by spermine. Nature (Lond.) **208**, 408 (1965).
— — Experiments on the site of binding of radioactive spermine in *Escherichia coli*. Bact. Proc. **1965**, G23.
— — The antimutagenic action of polyamines: Suppression of the mutagenic action of an *E. coli* mutator gene and of 2-aminopurine. Proc. nat. Acad. Sci. (Wash.) **55**, 1453 (1966).
— — On the mechanism of antimutagen action. Proc. 5th Internat. Congress Chemotherapy, **4**, 427. Vienna 1967.
— — Apparent antimutagenic activity of quinacrine hydrochloride in Detroit-98 human sternal marrow cells grown in culture. Cancer Res. **29**, 1367 (1969).
LERMAN, L. S.: Acridine mutagens and DNA structure. J. cell. comp. Physiol. **64** (Suppl. I), 1 (1964).
LIEB, M.: Dark repair of UV induction in K_{12} (λ). Virology **23**, 381 (1967).
LIQUORI, A. M., CONSTANTINO, L., CRESCENZI, V., ELIA, V., GIGLIO, E., PULITO, R., SAVINO, D. S., VITAGLIANO, V.: Complexes between DNA and polyamines: A molecular model. J. molec. Biol. **24**, 113 (1967).
LITTLEFIELD, J. W., JACOBS, P. S.: The relation between DNA and protein synthesis in mouse fibroplasts. Biochem. biophys. Acta (Amst.) **108**, 652 (1965).
LURIA, S. E.: Spontaneous bacterial mutations to resistance to antibacterial agents. Cold Spr. Harb. Symp. quant. Biol. **11**, 130 (1946).
MAGNI, G. E., VON BORSTEL, R. C., SORA, S.: Mutagenic action during meiosis and antimutagenic action during mitosis by 5-aminoacridine in yeast. Mutation Res. **1**, 227 (1964).
MAHLER, H. R., MEHROTRA, B. D., SHARP, C. W.: Effects of diamines on the thermal transition of DNA. Biochem. biophys. Res. Commun. **4**, 79 (1961).
MALLING, H. V.: The mutagenicity of acridine mustard C (ICR 170) and the structurally related compounds in *Neurospora*. Mutation Res. **4**, 265 (1967).
MEHROTRA, B. D., MAHLER, H. R.: Studies on polynucleotide-small molecule interactions: III. The effect of diamines on helical polynucleotides. Biochim. biophys. Acta (Amst.) **91**, 78 (1964).
MERWIN, C. F., WINKELMANN, R. K.: Antimalarial drugs in the therapy of lupus erythematosus. Proc. Mayo Clin. **37**, 253 (1962).
MUENCH, K. H.: Chloroquine-mediated conversion of transfer ribonucleic acid of *Escherichia coli* from an inactive to an active state. Cold Spr. Harb. Symp. quant. Biol. **31**, 539 (1967).
— Chloroquine and synthesis of aminoacyl transfer ribonucleic acids. Tryptophanyl transfer ribonucleic acid synthetase of *E. coli* and tryptophanyladenosine tryphosphate formation. Biochemistry **8**, 4872 (1969).

Nicholson, B. H., Peacocke, A. R.: Inhibition of ribonucleic acid polymerase by acridines. Biochem. J. **100**, 50 (1966).
O'Brien, R. L., Allison, J. L., Hahn, F. E.: Evidence for intercallation of chloroquine into DNA. Biochim. biophys. Acta (Amst.) **129**, 622 (1966).
— Olenick, J. G., Hahn, F. E.: Reactions of quinine, chloroquine and quinacrine with DNA and their effects on the DNA and RNA polymerase reactions. Proc. nat. Acad. Sci. (Wash.) **55**, 1511 (1966).
Pritchard, N. J., Blake, A., Peacocke, A. R.: Modified intercalation model for the interaction of aminoacridines and DNA. Nature (Lond.) **212**, 1360 (1966).
Puglisi, P. P.: Mutagenic and antimutagenic effects of acridinium salts in yeast. Mutation Res. **4**, 289 (1967).
Rownd, R., Nakaya, R., Nakamura, A.: Molecular nature of the drug resistance factors of the enterobacteriaceae. J. molec. Biol. **17**, 376 (1966).
Scholtissek, C., Brecht, H.: Action of acridines on RNA and protein synthesis and on active transport in chick embryo cells. Biochim. biophys. Acta (Amst.) **123**, 585 (1966).
Sevag, M. G.: Prevention of the emergence of antibiotic resistant strains of bacteria by atabrine. Arch. Biochem. **108**, 85 (1964).
— Ashton, B.: Evolution and prevention of drug resistance. Nature (Lond.) **203**, 1323 (1964).
— Drabble, W. T.: Prevention of the emergence of drug resistant bacteria by polyamines. Biochem. biophys. Res. Commun. **8**, 446 (1962).
Shimada, S., Takagi, Y.: The effect of caffeine on the repair of ultraviolet damaged DNA in bacteria. Biochim. biophys. Acta (Amst.) **145**, 763 (1967).
Sideropoulos, A. S., Shankel, D. M.: Mechanism of caffeine enhancement of mutations induced by sublethal ultraviolet dosages. J. Bact. **96**, 198 (1968).
Skipper, H. E., Schabel, F. M., Jr., Wilcox, W. S.: Experimental evaluation of potential anticancer agents: XIII. On the criteria and kinetics associated with "curability" of experimental leukemia. Cancer Chemother. Repts. **35**, 1 (1964).
— — — Experimental evaluation of potential anticancer agents. XXI. Scheduling of arabinosylcytosine to take advantage of its S-phase specificity against leukemia cells. Cancer Chemother. Repts. **51**, 125 (1967).
Tabor, C. W., Tabor, H.: Transport systems for 1, 4 diaminobutane, spermidine and spermine in *Escherichia coli*. J. biol. Chem. **241**, 3714 (1966).
Treffers, H. P., Spinelli, V., Belser, N. O.: A factor (or mutator gene) influencing mutation rates in *Escherichia coli*. Proc. nat. Acad. Sci. (Wash.) **40**, 1064 (1954).
Webb, R. B., Kubitschek, H. E.: Mutagenic and antimutagenic effects of acridine orange in *Escherichia coli*. Biochem. biophys. Res. Commun. **13**, 90 (1963).
Weissmann, G.: Lysosomes. New Engl. J. Med. **273**, 1143 (1965).
— Labilization and stabilization of lysosomes. Fed. Proc. **23**, 1038 (1964).
— Lysosomes. Blood **24**, 594 (1964).
Wolf, M. K., Aronson, S. B.: Growth, fluorescence and metachromasy of cells cultured in the presence of acridine orange. J. Histochem. Cytochem. **9**, 22 (1961).
Yanofsky, C., Cox, E. C., Horn, V.: The unusual mutagenic specificity of an *E. coli* mutator gene. Proc. nat. Acad. Sci. (Wash.) **55**, 274 (1966).
Zamenhof, P. J.: On the identity of two bacterial mutator genes: Effect of antimutagens. Mutation Res. **7**, 463 (1969).

Interaction and Linkage of Polycyclic Hydrocarbons to Nucleic Acids

STEPHEN A. LESKO, JR., HANS D. HOFFMANN, PAUL O. P. TS'O and VERONICA M. MAHER

I. Introduction

Ever since the discovery of the carcinogenicity of certain aromatic polycyclic hydrocarbons, workers in the field have been confronted with the challenging question about the origin of the biological activity and specificity of these compounds. How can these relatively chemically inert polycyclic hydrocarbons have such a profound biological effect? This question must be related to the basic mechanism of chemical carcinogenesis.

This laboratory has been actively engaged for some time in studies on the physico-chemical forces which determine the conformation and properties of nucleic acids. These studies have shown that hydrophobic forces and stacking interactions contribute significantly to the association of purine bases and nucleosides in solution, to the stability of nucleic acid formation, as well as to the binding of purine nucleosides and their analogues to nucleic acids [TS'O, HELMKAMP and SANDER, 1962 (1, 2); TS'O, HELMKAMP, SANDER and STUDIER, 1963 (1); TS'O, MELVIN and OLSON, 1963 (2); TS'O and CHAN, 1964; CHAN, SCHWEIZER, TS'O and HELMKAMP, 1964; SCHWEIZER, CHAN and TS'O, 1965; BROOM, SCHWEIZER and TS'O, 1967]. Following this line of research, we started to investigate the interaction of carcinogenic polycyclic hydrocarbons with DNA since these hydrocarbons are also hydrophobic compounds. It appeared to us that the rapid advances in nucleic acid chemistry and molecular biology might bring new insights into the problem of carcinogenesis.

The research from this laboratory in this direction was guided by the following general principles as a working hypothesis: (1) The interaction of polycyclic hydrocarbons with nucleic acids should have serious biological consequences although it cannot be ruled out that carcinogenesis may have multiple etiological origins and could be initiated by interaction of these hydrocarbons with other cellular components. (2) Both reversible and irreversible interactions of polycyclic hydrocarbons with nucleic acids may play a key role in determining their biological manifestations and therefore both types of interactions should be investigated.

Our findings on the physical binding (TS'O and LU, 1964; LESKO, SMITH, TS'O and UMANS, 1968) indicate that there is a strong affinity between polycyclic hydrocarbons and nucleic acids which is in accord with results from other laboratories (LIQUORI, DE LERMA, ASCOLI, BOTRE and TRASCIATTI, 1962; BOYLAND and GREEN, 1962; LERMAN, 1964; KODAMA, TAGASHIRA, IMAMURA and NAGATA, 1966; NAGATA,

Kodama, Tagashira and Imamura, 1966; Isenberg, Baird and Bersohn, 1967). The mechanism of this binding process can be understood from knowledge about the hydrophobic-stacking properties of nucleic acids gained from the aforementioned physicochemical studies. While the initial physical interaction of polycyclic hydrocarbons with the genetic material in a living cell may play a key role in the manifestation of their biological effect, this process is rather difficult to study *in vivo*. On the other hand, the chemical processes in the living system which lead to a covalent linkage of these hydrocarbons to nucleic acids may be more accessible for investigation. However, it is an enormous task to isolate and characterize the products from *in vivo* chemical reactions between carcinogens and DNA in treated cells or animals. Therefore, we concentrated our efforts first on studying these chemical reactions *in vitro*. The main object of our investigation was to find an informative model system with conditions sufficiently similar to those which exist in cells, and which exhibited a correlation between carcinogenicity and the extent of covalent attachment of hydrocarbons to DNA. This requires the search for suitable methods to activate the relatively inert hydrocarbons for the formation of a chemical linkage with nucleic acids under mild or essentially physiological conditions. As outlined below, activation can be achieved with physical agents (X-ray and uv irradiation) or with chemical reagents (iodine and hydrogen peroxide-ferrous ion). We are now able to introduce a covalent linkage between polycyclic hydrocarbons and DNA to an extent of about one hydrocarbon per 100 bases of DNA with various chemical activating systems in aqueous buffer or in a mixture of aqueous buffer and ethanol (2:1, *v*/*v*) at 5 to 40° and neutral pH.

The base specificity of these chemical reactions has been investigated by using homopolynucleotides as a model system. This is actually the first step necessary to elucidate the precise reaction mechanisms of our model systems. If these *in vitro* chemical systems sufficiently resemble biological systems, then these experiments will also provide the model compounds necessary for the identification and characterization of products resulting from reactions between hydrocarbons and DNA in biological systems.

What effect does the physical binding or chemical linkage of polycyclic hydrocarbons have on the biological function of nucleic acids? The effect most likely should involve the replication or transcription processes. The mutagenic effect of covalent attachment of polycyclic hydrocarbons to *B. subtilis* SB 19 transforming DNA in an *in vitro* reaction has been assayed in the *B. subtilis* system of Freese and Strack (1962). Such studies should provide information pertinent to the carcinogenic process at the molecular level.

II. Physical Binding

Our results and views on the physical binding of benzpyrene to nucleosides, nucleotides, native and denatured DNA, calf thymus nucleohistone, homopolyribonucleotides and yeast RNA have been previously published (Lesko et al., 1968). The important conclusions from that report will be described briefly below. The benzpyrene was assayed by a radio-chemical method after a careful examination of the radio-chemical purity of the tritium-labelled hydrocarbon.

1. Binding of [³H]3,4-Benzpyrene to Nucleosides, Nucleotides and Nucleic Acids

The binding of [³H]3,4-benzpyrene to native and denatured DNA is shown in Table 1. The result come from solubility measurements, and the binding constant, K, is defined as

$$K(M^{-1}) = \frac{([^3\mathrm{H}]\ 3,4\text{-benzpyrene solubilized})}{(\text{substrate})\ (\text{solubility of benzpyrene in } 10^{-2}\ \mathrm{M}\ \text{buffer})}$$

where the concentration of DNA (substrate) is expressed in terms of monomeric units. The binding constants for DNA are about 100-fold larger than those obtained

Table 1. *Physical binding of [³H]3,4-benzpyrene to native DNA, denatured DNA and thymus nucleohistone at 5°*

	Nucleotide conc. (mM)	Salt conc.	Molar ratio benzpyrene/base $\times 10^4$	Binding constant $K(\mathrm{M}^{-1}) \times 10^{-3}$
Native DNA	2.2	HMP[a]	4.16	25.0
	2.8	HMP[a]	8.33	50.0
	4.1	HMP[a]	5.88	35.0
	5.0	HMP[a]	10.00	55.0
	2.2	HMP + 0.5 M NaCl	1.49	9.0
	2.5	HMP + 1.5 M NaCl	1.42	8.8
	2.0	HMP + 1.5 M NaCl	1.04	6.5
Denatured DNA	2.2	HMP	4.16	25.0
	3.7	HMP	4.00	23.0
	6.0	HMP	4.76	27.0
	9.2	HMP	3.12	18.0
	2.2	HMP + 0.5 M NaCl	11.11	60.0
Nucleohistone	1.5	TMP[b]	3.33	22.6
	1.5	TMP[b]	3.12	18.0
	0.63	TMP + 1.5 M NaCl	0.90	5.5
	0.63	TMP + 1.5 M NaCl	1.03	6.1
	0.63	TMP + 1.5 M NaCl	1.14	6.8

[a] HMP = 1×10^{-2} M phosphate buffer, pH 6.8.
[b] TMP = 1×10^{-3} M phosphate buffer, pH 6.8.

for nucleosides and nucleotides. The monomers exhibited binding constants ranging from 100 to 600, clustering mainly around 400. It should be noted that the high binding constants (2 to $6 \times 10^4\ \mathrm{M}^{-1}$) reported in Table 1 for the binding of [³H]3,4-benzpyrene by DNA indicate an apparent standard free energy change of 5.5 to 6.0 Kcal in the binding process. This is indeed a very large free energy change in the binding of a small molecule by DNA in the absence of electrostatic interaction. The influence of salt on the binding of [³H]3,4-benzpyrene to DNA is strongly dependent on the secondary structure of the DNA. Addition of salt decreases the binding in the case of native DNA but increases the binding in the case of denatured DNA (Table 1).

Addition of excess nonradioactive 3,4-benzpyrene to DNA-[³H]3,4-benzpyrene solutions caused a rapid loss of radioactivity when assayed by the standard procedure.

After 4 h, 50 to 60% of the radioactivity was lost, while after 24 h, 75% was lost. These data indicate that the binding process is reversible. This conclusion was further substantiated when it was shown that 99.8% of the [^{3}H]3,4-benzpyrene could be removed by ethanol precipitation and washing of the physical complex with organic solvents (ethanol and ether).

In order to obtain conclusive evidence for the existence of a physical complex between DNA and [^{3}H]3,4-benzpyrene, the solutions, after filtration, were analyzed by sucrose density centrifugation and electrophoresis. Fig. 1 clearly illustrates that DNA, as measured by absorbance at 260 nm, and [^{3}H]3,4-benzpyrene migrate

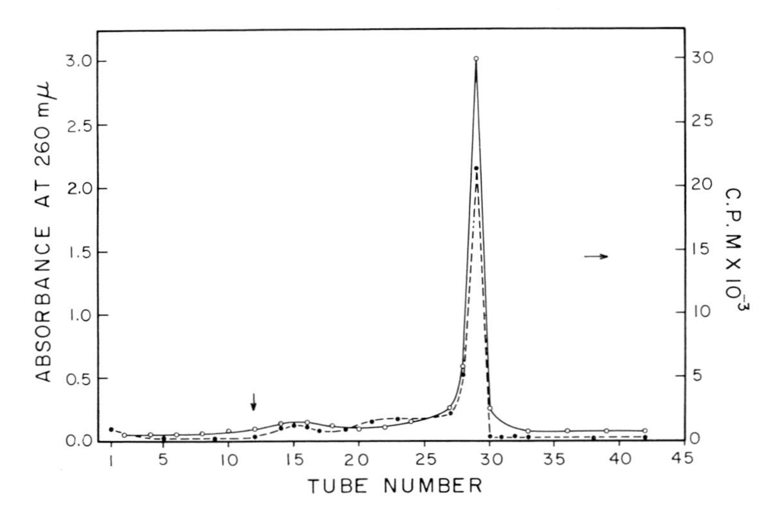

Fig. 1. Sucrose gradient electrophoresis pattern of a native DNA-[^{3}H]3,4-benzpyrene physical complex in 1×10^{-2} M phosphate buffer (pH 6.8), 1000 v, 4 ma, 2.5 h. One ml samples were collected and absorbance at 260 nm (●——●) and radioactivity (○ - - - - ○) were measured

together in an electrical field, thus demonstrating the existence of a physical complex. DNA and [^{3}H]3,4-benzpyrene also moved at the same rate during sucrose gradient centrifugation (LESKO et al., 1968), which again proved the existence of a physical complex.

2. Specificity of the Physical Interaction

The physical interaction of [^{3}H]3,4-benzpyrene and [^{3}H]1,2-benzpyrene with polynucleotides has been investigated using conditions identical to those adopted for DNA, and the results are shown in Table 2. It can be seen that poly *G*, helical poly *A* [formed at pH 5; Ts'o et al., 1962 (1); RICH, DAVIS, CRICK and WATSON, 1961] and poly *G* · poly *C* have a much higher affinity for [^{3}H]3,4-benzpyrene than neutral poly A, poly C and poly U. This observation indicates that purine polynucleotides and polynucleotides that have a considerable amount of secondary structure have the highest affinity for [^{3}H] 3,4-benzpyrene. The binding of [^{3}H] 1,2 benzpyrene, a nongenic isomer, to all the polynucleotides ranges from almost the same, within experimental error, to slightly less than the binding of [^{3}H]3,4-benzpyrene. Since the solubility of [^{3}H]1,2-benzpyrene and [^{3}H]3,4-benzpyrene in aqueous buffer is about the same (LESKO et al., 1968), these data suggest that the binding constants for the interaction of homopolynucleotides with [^{3}H]1,2-benzpyrene and [^{3}H]3,4-benzpyrene are very similar. The same results were obtained from a comparative study on the physical binding of [^{3}H]1,2-benzpyrene and [^{3}H]3,4-benzpyrene to DNA (LESKO et al., 1968) as well as on the physical binding of carcinogenic [^{3}H]1,2,5,6-dibenzanthracene and noncarcinogenic [^{3}H]1,2,3,4-dibenzanthracene to DNA. Therefore, the physical binding of these two pairs of isomeric polycyclic hydrocarbons to nucleic acids shows no specificity in relation to carcinogenic activity.

Table 2. *Physical binding of [^{3}H]3,4-benzpyrene and [^{3}H]1,2-benzpyrene to polyribonucleotides in 1×10^{-2} M phosphate buffer (pH 6.8) at 5°*

Polymer	Nucleotide conc. (mM)	Molar ratio 3,4-benzpyrene/ base × 10^4	Molar ratio 1,2-benzpyrene/ base × 10^4
Poly *A*	5.8	1.61	—
	2.6	2.50	—
	1.3	1.51	—
	3.0	3.40	2.50
Poly *C*	4.0	0.74	—
	3.0	0.59	—
	3.0	1.00	0.90
Poly *U*	4.1	0.33	—
	4.5	0.33	—
	3.0	0.50	0.40
Poly *G*	3.0	37.00	25.0
	2.7	20.00	14.92
Poly (*A*) · (*U*)	6.0	4.20	2.90
Poly (*G*) · (*C*)	6.0	12.00	12.40
Poly *A* (helical)[a]	2.8	33.00	—

[a] Helical poly *A* was formed in 5×10^{-2} M acetate buffer (pH 5.0).

3. Spectral Studies of 3,4-Benzpyrene Complexes with Purine, AMP and Nucleic Acids in Relation to the Mechanism of Physical Binding

It has been previously noted (LIQUORI et al., 1962; BOYLAND and GREEN, 1962; Ts'o and LU, 1964) that the absorption spectrum of 3,4-benzpyrene in DNA solutions exhibits a pronounced bathochromic shift of 10 nm. The spectral maxima of 3,4-benzpyrene in a variety of solvents and solutions are given in Table 3. These spectral data constitute a strong support, though indirect, for the model of intercalation as the mechanism of the physical binding of 3,4-benzpyrene to DNA (BOYLAND and GREEN,

1962). This model was originally proposed on the basis of stereochemical feasibility and its analogy to the acridine derivatives (LERMAN, 1964). Only recently, there were some preliminary data from flow dichroism (NAGATA, KODAMA, TAGASHIRA and IMAMURA, 1966) and from viscosity (LERMAN, 1964) in support of this point of view. The argument from the spectral data in support of this model of intercalation comes in three parts.

a) The spectral maxima of 3,4-benzpyrene from 350 to 410 nm are the same in a variety of solvents ranging from 50% aqueous ethanol to hexane, but are shifted 4 to 7 nm to longer wavelengths when dissolved in benzene or pyridine (Table 3). This suggests that the preferential stabilization of the excited state over the ground state of 3,4-benzpyrene occurs when the hydrocarbon interacts with a π-electron system, and is not easily induced through changes in the dielectric constant or hydrogen-bonding properties of the solvent. Further, the equivalence of the 3,4-benzpyrene spectrum in benzene and in pyridine indicates that the nitrogen lone-pair electrons play no significant role in the shift.

Table 3. *Spectral maxima of 3,4-benzpyrene in various solvents and solutions*

Solvent or solution	λ_{max} (nm)			
50% aqueous ethanol	403	384	363	347
Ethanol	403	384	363	347
Heptane	403	383	362	347
Cyclohexane	403	384	363	347
Benzene	405	389	369	351
Pyridine	405	390	370	352
0.2% denatured DNA	407	393	373	354
0.2% native DNA	408	394	374	355
0.1% poly *A*, 0.05 M acetate, pH 5.0	409	396	376	358
0.2 M 5′-AMP, pH 6.8 HMP	406	390	370	—
0.3 M purine, pH 6.8 HMP	405	389	370	—
0.3 M purine, ethanol	403	384	363	—

b) That this interaction arises from vertical stacking is strongly suggested by the spectral data for 3,4-benzpyrene in aqueous solutions of 5′-AMP and purine (Table 3), taken in conjunction with earlier spectroscopic, X-ray diffraction, vapor pressure osmometry, and proton magnetic resonance results. It is seen that the present 3,4-benzpyrene-purine and 3,4-benzpyrene-5′-AMP spectral data, as well as earlier spectra of 3,4-benzpyrene dissolved in 50% ethanol saturated with caffeine or with 1,3,7,9-tetramethyluric acid (BOOTH, BOYLAND and ORR, 1954), or of 3,4-benzpyrene dissolved in 0.2% aqueous caffeine solution (RESKE and STAUFF, 1965), exhibit comparable red shifts to those arising from 3,4-benzpyrene interaction with the benzene and pyridine π-electron systems. These data should be considered in conjunction with the following measurements. (i) The detailed analyses by X-ray diffraction of the molecular structure of cocrystals of 3,4-benzpyrene and tetramethyluric acid by LIQUORI's group (DAMIANI, DESANTIS, GIGLIO, LIQUORI, PULITI and RIPAMONTI, 1965; DAMIANI, GIGLIO, LIQUORI, PULITI and RIPAMONTI, 1966). Their results indicate a parallel stacking of one 3,4-benzpyrene plane and two tetramethyluric acid molecules at an average perpendicular separation of 3.45 Å. (ii) The extensive investigation by proton magnetic resonance and vapor pressure osmometry on the association of purine, 6-methylpurine [Ts'o et al., 1963 (2); Ts'o and CHAN, 1964; CHAN et al., 1964], nucleosides (SCHWEIZER et al., 1965; BROOM et al , 1967) and nucleotides, especially AMP (SCHWEIZER, BROOM, Ts'o and HOLLIS, 1968). These results showed conclusively that all these compounds associate extensively in solution by mode of vertical stacking of the bases. A similar conclusion was also reached independently by JARDETZKY (1965). (iii) Our proton magnetic resonance studies of purine in ethanol at varying purine concentrations reveal no characteristic downfield shift of the aromatic proton lines with decreasing concentration, thus indicating the absence of stacking interactions. Further, we find that the red shift in the visible spectrum of 3,4-benzpyrene is absent when it is dissolved in a 0.3 M purine solution in ethanol. It thus appears that the presence of the observed 3,4-benzpyrene red shift is fully dependent upon the presence of strong stacking interactions of the bases, which take place in aqueous solution but not in organic solvents (CHAN et al., 1964). This conclusion is in agreement with that of VAN DUUREN (1964). From a careful fluorescence study, he concluded that there is very little interaction of tetramethyluric acid with 3,4-benzpyrene when dissolved in benzene, even though such an interaction can be observed in aqueous alcohol solution (BOOTH et al., 1954). On the basis of all this information, it is safe to conclude that the bathochromic shift of the spectra of 3,4-benzpyrene with purine and with 5'-AMP is directly associated with the formation of face-to-face vertical stacks.

c) The bathochromic shift (7 to 11 nm) in the spectrum of 3,4-penzpyrene when complexed with DNA (Table 3) is well known, and there is also a large shift (11 to 13 nm) for the complex of 3,4-benzpyrene-helical poly *A*. This represents a considerable increase in the magnitude of the 3,4-benzpyrene red shift accompanying a change in environment from stacked monomeric π-electron systems to a relatively rigid polymer system. If the 3,4-benzpyrene in the polymer system were no longer involved in face-to-face interactions, but were now orientated, for example, coplanar with or perpendicular to the DNA bases, one would expect a reduction in the base-induced perturbation of the 3,4-benzpyrene π-electron system due to a reduction in the overlap of 3,4-benzpyrene molecular orbitals with base molecular orbitals. While one cannot accurately predict the accompanying spectral changes, it is highly unlikely that this alteration in 3,4-benzpyrene environment would lead to the large observed increase in the red shift. Rather, this spectral increase is considerably more compatible with a model whereby 3,4-benzpyrene in a polynucleotide environment associates by vertical stacking with the bases in a manner similar to that of the monomer system. As the polynucleotide helix provides a more rigid and regular structure than the stacks of monomers, the 3,4-benzpyrene may be considered to be more tightly locked in its face-to-face arrangement with the bases. This provides a more stable geometry for the action of the intermolecular π-electron dispersion forces, and could thereby lead to the observed increase in the magnitude of the associated red shift. It should be remembered that because of the large base to 3,4-benzpyrene ratio (from 1000 to 3000) there is virtually no direct interaction between bound 3,4-benzpyrene molecules in the 3,4-benzpyrene-DNA complex, a situation very different from the acridine dye-DNA complex which can exhibit a low base to dye ratio of from 1 to 20. The fluorescence studies of ISENBERG et al. (1967) give support to the conclusion that 3,4 benzpyrene is complexed with DNA as individual molecules and preclude the possibility that the red shift is due to self-interaction of hydrocarbon molecules within colloidal particles.

4. The Physical Binding of [^{3}H]3,4-Benzpyrene to Calf Thymus Nucleohistone

The results of the binding of [^{3}H]3,4-benzpyrene to nucleohistone as measured by solubility methods are reported in Table 1. The [^{3}H]3,4-benzpyrene-nucleohistone physical complex was further examined by sucrose gradient centrifugation and electrophoresis. After centrifugation, the pattern of the distribution of [^{3}H]3,4-benzpyrene in the centrifuge tube followed closely the pattern of the nucleohistone represented by absorbance at 260 nm. This is a strong indication of the existence of a [^{3}H]3,4-benzpyrene-nucleohistone complex. Most of the [^{3}H]3,4-benzpyrene migrates with the nucleohistone on electrophoresis, again indicating the existence of a [^{3}H]3,4-benzpyrene-nucleohistone complex. The amount of [^{3}H]3,4-benzpyrene solubilized by nucleohistone is about 60% of that solubilized by DNA. However, the [^{3}H]3,4-benzpyrene/base ratio for nucleohistone was almost the same as that for DNA when the binding experiment was carried out in a solution of 1.5 M NaCl (Table 1). This high concentration of salt would cause an almost complete dissociation of the histones from DNA. Most of the histones can be solubilized and extracted from nucleohistone by cold 0.5 N H_2SO_4. When the [^{3}H]3,4-benzpyrene-nucleohistone complex was so treated with H_2SO_4, 97% of the radioactivity was found in the precipitate of the insoluble DNA after centrifugation at 3000 rpm for 10 min in a clinical centrifuge. Treatment of a [^{3}H]3,4-benzpyrene-DNA complex with 0.5 N H_2SO_4 caused precipitation of 99% of the radioactivity. Together these experiments indicate that very little of the hydrocarbon originally bound to the nucleohistone remains in association with the histones after they are extracted. This conclusion received further support from the following experiment with isolated histones. Preparations of unfractionated histones were shaken together with [^{3}H]3,4-benzpyrene for 14 days at 5°. After filtration through a sintered glass filter, hydrocarbon was found to be solubilized to the extent of 0.23 mμmole per mg of histone or about $^1/_6$ of the amount solubilized by one mg of DNA. However, only a small fraction of the [^{3}H]3,4-benzpyrene migrated with the histones toward the cathode on electrophoresis, suggesting a rather loose association between the two (LESKO et al., 1968).

The results show that the binding of [^{3}H]3,4-benzpyrene to nucleohistone is about 50 to 60% of that to DNA. In addition, [^{3}H]3,4-benzpyrene is not associated very strongly with the histones in the nucleohistone complex. The precise reason for the reduction in the binding of [^{3}H]3,4-benzpyrene to DNA in nucleohistone as compared with isolated DNA in dilute salt solution is not obvious. Two explanations are that of reduction of the number of binding sites due to stereochemical arrangement of the histones on the double helix, and that of general tightening of the double helix due to the binding of positively charged histones. The second explanation is analogous to the reduced binding of [^{3}H]3,4-benzpyrene to DNA in concentrated salt solutions. The polyamine, spermidine, also reduces the binding of 3,4-benzpyrene to DNA (LIQUORI, ASCOLI, DESANTIS and SAVINO, 1967). Therefore, in regard to this question, more information about the detailed structure of the nucleohistones is needed.

III. Chemical Linkage

Physical binding studies have shown that the affinity and the extent of interaction of carcinogenic and noncarcinogenic polycyclic hydrocarbons with nucleic acids cannot be correlated with carcinogenicity. However, chemical complexes of polycyclic hydrocarbons with DNA have been found in biological systems (HEIDELBERGER and DAVENPORT, 1961; BROOKES and LAWLEY, 1964; GOSHMAN and HEIDELBERGER, 1967; BROOKES and HEIDELBERGER, 1969) and the available data indicate that the carcinogenic potency of these compounds can be correlated to the extent of interaction (BROOKES and HEIDELBERGER, 1964). More recently, 3,4-benzpyrene has been covalently linked to DNA through the action of a microsomal enzyme system involving NADPH (GROVER and SIMS, 1968; GELBOIN, 1969). Since the amount of chemical complex formed in biological and enzymic systems is very small, it is necessary that a model *in vitro* system be found which will provide larger quantities of chemical complex. The first attempt from this laboratory to find such a model system involved the use of radiation energy and the second attempt was by chemical activation.

1. Procedure for the Establishment of a Covalent Linkage between DNA and Polycyclic Hydrocarbons

When DNA is precipitated from an aqueous solution containing a [^{3}H]3,4-benzpyrene-DNA physical complex by ethanol, and the precipitate is washed repeatedly with ethanol, about 99.7% of the physically-bound hydrocarbon can be extracted (Ts'o and LU, 1964; LESKO, Ts'o and UMANS 1969). However, when [^{3}H]3,4-benzpyrene is covalently linked to DNA, it can no longer be removed by precipitation and extraction with organic solvent (LESKO et al., 1969). The possibility of a mere exchange of tritium between the [^{3}H]hydrocarbon and DNA can be excluded when no radioactivity is found in the extracted DNA from a reaction mixture containing [^{3}H]water instead of [^{3}H]3,4-benzpyrene (RAPAPORT and Ts'o, 1966). In addition, as shown in a later section, the same conclusion was reached with the chemical linkage of [^{3}H]3,4-benzpyrene and [^{14}C]3,4-benzpyrene to poly *G*.

In order to show that the chemical complex is not contaminated by any [^{3}H]3,4-benzpyrene reaction products insoluble in organic solvents, solutions of washed chemical complex are analyzed by physicochemical techniques. It can be demonstrated that under the influence of a force field (electrical or centrifugal), or in a system of molecular-sieve chromatography, that the movement of [^{3}H]3,4-benzpyrene is coincident with the movement of DNA. These findings provide a strong argument that the [^{3}H]3,4-benzpyrene assayed after the washing procedure is actually attached to DNA.

Finally the [^{3}H]3,4-benzpyrene-DNA chemical complex can be enzymatically hydrolyzed and examined by sucrose gradient electrophoresis and other chromatographic techniques. However, we do not exclude the possibility that attachment of polycyclic hydrocarbons to DNA will inhibit complete enzymic degradation. Demonstrating that the [^{3}H]3,4-benzpyrene migrates with the hydrolytic products provides conclusive proof of a covalent attachment. Selective examples of the results of these procedures are shown throughout this communication.

2. Formation of [^{3}H]3,4-Benzpyrene Complexes with Radiation Energy

Since most of our results have been published previously (TS'O and LU, 1964; RAPAPORT and TS'O, 1966), only a brief account will be given here together with our new findings on deoxyribonucleohistone. Our efforts in this area can be classified into two related programs, induction by photoirradiation and induction by X-ray irradiation.

Photoirradiation of [^{3}H]3,4-benzpyrene-DNA physical complexes at wavelengths above 340 nm, where nucleic acids absorb very little photo energy, induces the formation of a chemical complexes with little or no damage to DNA as indicated by sedimentation analyses. The sample for irradiation was placed under a nitrogen atmosphere to reduce most of the photodynamic effect. The light source was a one kilowatt, air cooled mercury arc lamp. Under the experimental conditions, 30% of the [^{3}H]3,4-benzpyrene-DNA physical complexes are converted to chemical complexes after 1 h, and the conversion reaches a maximal value of about 50% in 3 to 4 h. The [^{3}H]3,4-benzpyrene-DNA physical complex usually contains about one hydrocarbon molecule/1000 to 1500 nucleotides and the chemical complex contains about one hydrocarbon molecule/2000 to 3000 nucleotides. However, this [^{3}H]3,4-benzpyrene-DNA chemical complex can bind more [^{3}H]3,4-benzpyrene, first physically and then chemically after a second irradiation, when a fresh amount of hydrocarbon is supplied to the solution. By this procedure, we have obtained a [^{3}H]3,4-benzpyrene-DNA chemical complex which contains a hydrocarbon/nucleotide ratio of 1/600 to 800. This photo-induced chemical linkage can be achieved with both native and denatured DNA.

The induction of chemical complexes between [^{3}H]3,4-benzpyrene and DNA by X-ray irradiation has also been studied (RAPAPORT and TS'O, 1966). The mechanism of covalent linkage by X-ray irradiation is most likely due to the formation of free radicals by direct or indirect action, while the mechanism of photoirradiation is most likely due to the formation of exited states and perhaps biradicals. The energy transfer process involved in X-ray irradiation is far less selective and consumes much more energy than photoirradiation. Consequently, more destruction of DNA is to be expected. The X-ray instrument was operated at 250 Kv with dose rates varying from 100 r/min to about 2500 r/min. The efficiency for conversion to chemical linkage is higher for denatured DNA (2 ×) in comparison to native DNA. Generally speaking, 4 to 15% and 10 to 30% of the [^{3}H]3,4-benzpyrene-DNA physical complex is converted to a chemical complex after X-ray irradiation at 20 and 40 Kr, respectively. There was also degradation of DNA as shown by loss of viscosity and by decrease in sedimentation coefficient.

Since most of the DNA in higher organisms is in the form of nucleohistone complexes, the formation of [^{3}H]3,4-benzpyrene-nucleohistone chemical complexes as induced by photoirradiation was investigated. When [^{3}H]3,4-benzpyrene-nucleohistone physical complexes were photoirradiated for 1 h, about 25 to 30% of the hydrocarbon became covalently linked to the nucleohistone. Efforts have been made to see whether the radioactive was associated with the DNA component or with the histone component of the nucleohistone. Five procedures have been employed to separate the DNA and protein moieties: (1) cold acid extraction; (2) Sevag procedure of shaking with detergent and chloroform; (3) 3 M CsCl density gradient equilibrium sedimentation; (4) sucrose gradient electrophoresis at pH 12.2; (5) molecular sieve chromatography (Bio-gel A-5 M agarose column) with 2 M NaCl-5 M urea as eluant. The results clearly indicate that about 10% of the radioactivity is chemically linked to isolated histones and that about 10% of the radioactivity is chemically linked to the pure nucleic acid components. The evidence suggests that most of the radioactivity (60—80%) is chemically bound to a small fraction of the original nucleohistone containing both DNA and protein. It is interesting to speculate that photoirradiation causes a cross-link between DNA and protein mediated in some manner through 3,4-benzpyrene. Purified histone [^{3}H]3,4-benzpyrene physical complexes can be converted to chemical complexes with an efficiency of 44% by photoirradiation, indicating that a covalent linkage can be formed between these proteins and [^{3}H]3,4-benzpyrene.

Preliminary experiments have been undertaken to investigate the specificity of the photoirradiation reaction. For the same dosage of irradiation at wavelengths above 300 nm, the percentage of conversion from physical complex to chemical complex revealed no clear-cut specificity for carcinogenic compounds over noncarcinogenic hydrocarbons.

3. Induction of Polycyclic Hydrocarbon-DNA Chemical Complexes by Chemical Activation

The success in the X-ray experiments suggested that the covalent linkage between [^{3}H]3,4-benzpyrene and DNA may be mediated by hydroxyl radicals formed during X-ray irradiation. When research in the direction of chemical activation was started in this laboratory, Rochlitz (1967) reported that 3,4-benzpyrene could be linked to pyridine via activation by iodine. He proposed that the reaction mechanism involves the cationic radical of 3,4-benzpyrene and that the product was a 5-benzopyrenyl pyridinium salt. Therefore, both iodine and hydrogen peroxide have been used as activating agents in our study. In addition, Boyland, Kimura and Sims (1964) have compared the reaction products of several polycyclic hydrocarbons obtained in a rat liver microsomal enzyme with those obtained in the model hydroxylating system of Udenfriend, Clark, Axelrod and Brodie (1954). The similarity of products obtained in the two systems prompted us to test the ability of the model hydroxylating system to link polycyclic hydrocarbons to DNA. Most of the results obtained by using these 3 systems have been published (Lesko et al., 1969; Umans, Lesko and Ts'o, 1969) and only the more salient observations will be presented here.

Table 4 shows the specificity of chemical complex formation and the influence of the conformational states of DNA on the iodine reaction. The iodine-induced reaction is very specific; [^{3}H]3,4-benzpyrene becomes covalently linked to DNA while [^{3}H]1,2-benzpyrene reacts only to a very limited extent. The values shown for [^{3}H]1,2-benzpyrene in Table 4 are about the same as background level values obtained from incubation mixtures containing no iodine. In the reaction with [^{3}H]3,4-benzpyrene, the percentage of conversion is much higher for denatured DNA (about 30%) than for native DNA (10%). This is a very effective procedure for linking 3,4-benzpyrene to DNA under very mild conditions. The reaction is highly specific and the extent of covalent linkage can be correlated with the carcinogenicity of this pair of isomers, viz., carcinogenic 3,4-benzpyrene and noncarcinogenic 1,2-benzpyrene. As shown in the following paragraph, essentially the same results were observed for the reaction with H_2O_2.

Table 4 shows the effect of DNA secondary structure and hydrocarbon specificity on the H_2O_2 reaction in the presence and in the absence of $FeCl_2$. As indicated, the presence of Fe^{++} ion greatly enhances the rate and the extent of the reaction. A concentration of 1.5×10^{-2} M H_2O_2 was used in these experiments because higher concentrations did not increase the rate appreciably and also because excessive H_2O_2 can lead to DNA degradation. The data clearly indicate that the H_2O_2-induced reaction is highly specific for [^{3}H]3,4-benzpyrene and that [^{3}H]1,2-benzpyrene does not react to any great extent. The reaction proceeds more rapidly and more extensively with

Table 4. *Percentage of physically bound [^{3}H] benzpyrene that becomes chemically linked to DNA in reactions induced by iodine (1×10^{-4} M, 2 h, room temperature) and by H_2O_2 (1.5×10^{-2} M, 37°, 24 h) in HMP*[a]

	Native DNA		
	Iodine	H_2O_2 1×10^{-2} M citrate	1×10^{-3} M $FeCl_2$
3,4-benzpyrene	10.5	4.5	15.5
1,2-benzpyrene	1.0	2.0	3.0
	Denatured DNA		
	Iodine	H_2O_2 1×10^{-2} M citrate	1×10^{-3} M $FeCl_2$
3,4-benzpyrene	31.5	15.0	40.0
1,2-benzpyrene	1.0	1.0	2.0

[a] HMP = 1×10^{-2} M phosphate buffer, pH 6.8.

denatured DNA than with native DNA. The specificity of the reaction can also be correlated with carcinogenicity.

When a physical complex of [^{3}H]3,4-benzpyrene-denatured DNA was placed in the model hydroxylating system (6.8×10^{-3} M EDTA, 1.4×10^{-3} M $FeSO_4$, 1.5×10^{-2} M ascorbic acid) for 24 h, there was a 42% conversion to a chemical complex. The reaction is specific with 32% of the physically bound [^{3}H]3,4-benzpyrene being covalently linked after 7 h while only 7% of the [^{3}H]1,2-benzpyrene is linked in the same time period. This linkage is of considerable interest because of the ability of the model hydroxylating system to catalyze reactions similar to those in enzymic systems.

The [^{3}H]3,4-benzpyrene-DNA chemical complexes induced by iodine and H_2O_2 were also characterized by sucrose gradient electrophoresis and gel filtration chromatography. Gel filtration through Sephadex G 200 showed that the radioactivity and DNA were eluted together at the void volume. The complex was then hydrolyzed to mononucleotides with DNAse I and snake venom phosphodiesterase and placed on the same column of Sephadex G 200. The radioactivity now eluted from the column with the same effuent volume as the hydrolytic products, about 25 ml after the elution of DNA as described above. These data indicate that before hydrolysis the [^{3}H]3,4-benzpyrene was attached to DNA in an iodine-induced reaction, and after hydrolysis of the DNA, the [^{3}H]hydrocarbon was attached to the resultant hydrolytic products. When the [^{3}H]3,4-benzpyrene-DNA chemical complex was subjected to sucrose

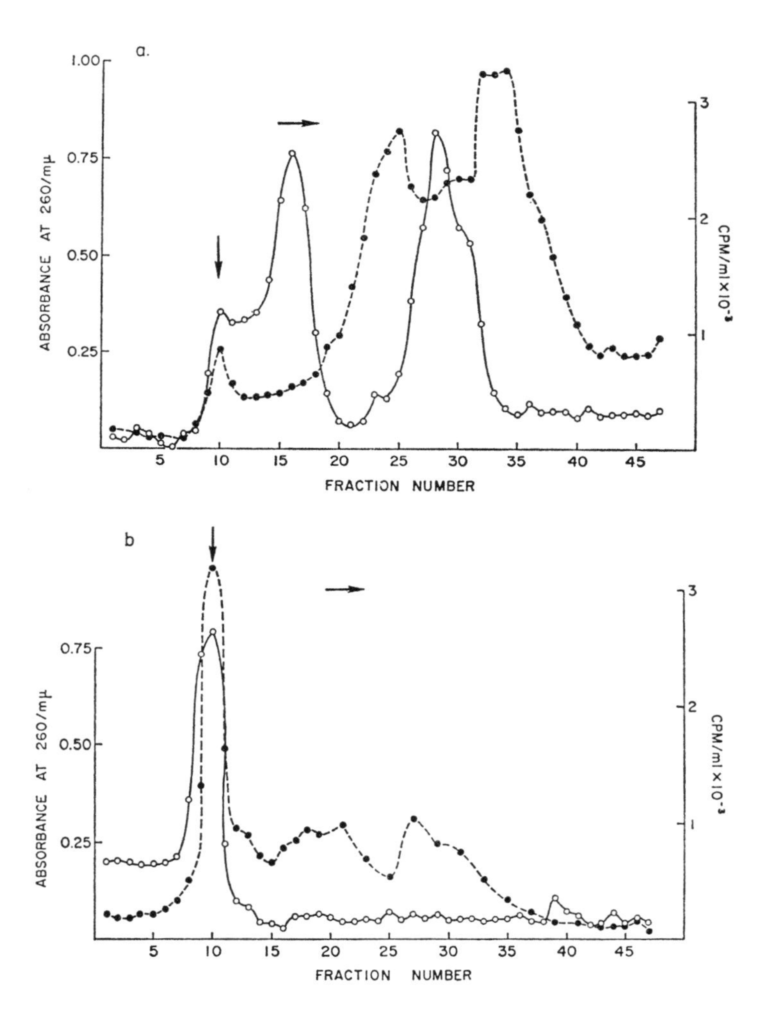

Fig. 2. Sucrose gradient electrophoresis patterns of nucleotide and nucleoside hydrolysates resulting from enzymic digestion of a denatured DNA-[^{3}H]3,4-benzpyrene chemical complex induced by H_2O_2. One ml fractions were collected and absorbance at 260 nm (O——O) and radioactivity (●- - - -●) were measured. (a) Nucleotides obtained by digestion with DNase I for 17 h at 37°, pH 6.6, and venom phosphodiesterase for 6 h at 37°, pH 9.2. Electrophoresis was for 5 h at 4.5 ma, 2250 v, using 5×10^{-3} M sodium citrate buffer (pH 3.5). (b) Nucleosides obtained by treatment of the nucleotide hydrolysate with alkaline phosphatase for 3 h at 37° at pH 9.2. Conditions for electrophoresis are the same as in (a)

gradient electrophoresis at pH 7, the radioactivity had the same electrophoretic mobility as DNA, thus indicating an attachment of [^{3}H]3,4-benzpyrene to DNA even after organic solvent extraction.

The nucleotide hydrolysate was also examined by sucrose gradient electrophoresis at pH 3.5 and the pattern is shown in Fig. 2 a. Most of the radioactivity left the origin and migrated toward the anode with a mobility similar (but not identical) to mononucleotides. The mononucleotide digest was treated with alkaline phosphatase to produce nucleosides and also examined by sucrose electrophoresis at pH 3.5. Fig. 2 b shows that a large portion of the radioactivity now remains at the origin with the neutral nucleosides, however, some radioactivity still migrates toward the anode. These observations are consistent with the idea that the [^{3}H]3,4-benzpyrene must be chemically coupled to the base or sugar moiety of nucleotides and that the migrating radioactivity in the nucleoside digest is linked to oligonucleotides.

The extent of DNA degradation produced by these chemical reactions has been examined. No diminution in the sedimentation coefficient of DNA was found after the iodine reaction. There was, however, a reduction in the sedimentation of DNA after reaction with H_2O_2-$FeCl_2$. No detectable change in optical density was observed when denatured DNA was incubated with H_2O_2-$FeCl_2$ for 24 h at 37°. These data indicate that there was degradation of DNA leading to chain scission in the H_2O_2-$FeCl_2$ reaction. There was also DNA degradation in reactions involving the use of the model hydroxylating system and this should be expected since this system also relies on hydroxy or perhydroxy radicals for its mode of action.

We have been able to improve the reaction yield of the iodine system by conducting the reaction in a solution composed of 1×10^{-2} M phosphate buffer (pH 6.8)-ethanol (2:1 *v/v*). In this solvent the concentration of iodine and hydrocarbon can be substantially increased and the number of hydrocarbon molecules linked to DNA is dependent on the concentration of the reactants. The DNA concentration has been reduced so that the ratio of hydrocarbon/DNA base in the reaction mixture becomes as high as possible. The kinetics of the reaction indicate that the [^{3}H]3,4-benzpyrene concentration becomes depleted within half an hour, therefore, the reaction yield can also be increased by adding more [^{3}H]hydrocarbon at intervals during the course of the reaction. Finally, the DNA-[^{3}H]3,4-benzpyrene adduct is isolated and then reacted again under identical conditions. With a combination of these approaches and after 3 reaction sequences, about 3.2 molecules of [^{3}H]3,4-benzpyrene have been linked per 10^3 bases of calf thymus DNA. In reactions with *B. subtilis* transforming DNA, about 10 molecules of [^{3}H]3,4-benzpyrene have been covalently linked per 10^3 bases by a similar procedure (LESKO, HOFFMANN, TS'O and MAHER, 1970). This is a 10 to 30-fold increase in reaction yield compared to the previous procedure in the aqueous system.

The specificity of the iodine-induced covalent linkage of polycyclic hydrocarbons to DNA is maintained in the 33% ethanol-phosphate buffer system. Two carcinogenic hydrocarbons, [^{3}H]3,4-benzpyrene and [^{3}H]9,10-dimethyl-1,2-benzanthracene, are manyfold (4 to 14) more reactive (hydrocarbon/base $\times 10^4$ is 8.9 and 3.0 respectively) than their noncarcinogenic analogs, [^{3}H]1,2-benzpyrene and [^{3}H]1,2-benzanthracene (hydrocarbon/base $\times 10^4$ is 0.65 and 0.88 respectively). The ratio $\times 10^4$ for carcinogenic [^{3}H]20-methylcholanthrene is 3.6 in this iodine-induced reaction.

4. Base Specificity of the Chemically-Induced Covalent Linkage

The base specificity of the chemically-induced covalent linkage of polycyclic hydrocarbons to nucleic acids has been examined using homopolynucleotides and their double-stranded complexes as model systems. Table 5 shows that the reaction with [^{3}H]3,4-benzpyrene is indeed base specific and that the degree of specificity depends on the activation system. Poly *G* is much more reactive than the other polynucleotides in the iodine system. This is true even when poly *G* is complexed with

Table 5. *Chemical yields (3,4-benzpyrene/base × 10^4) between [^{3}H]3,4-benzpyrene and ribosyl homopolynucleotides in reactions induced by iodine (1 × 10^{-4} M, 2 h) and by H_2O_2/Fe^{++} (1.5 × 10^{-2} M H_2O_2, 1 × 10^{-3} M $FeSO_4$, 24 h) in HMP buffer or in HMP/ethanol (33%)*

Polynucleotide	I_2 reaction in HMP buffer at 5 °C	H_2O_2/Fe^{++} reaction in HMP buffer at 37 °C
A	0.10	0.50
U	0.02	0.03
A + *U*	0.08	0.08
G	10.00	8.00
C	0.04	0.07
G + *C*	4.70	3.10
I	—	—
I + *C*	—	—

Polynucleotide	I_2 reaction in HMP/ethanol (33%) at 5 °C	H_2O_2/Fe^{++} reaction in HMP ethanol (33%) at 37 °C
A	0.50	2.10
U	0.07	0.08
A + *U*	0.13	0.29
G	18.00	15.00
C	0.19	0.30
G + *C*	1.10	0.90
I	0.40	0.35
I + *C*	0.21	0.70

[a] HMP = 1 × 10^{-2} M phosphate buffer (pH 6.8).

poly *C* in a double helical conformation. When [^{3}H]3,4-benzpyrene is activated in a H_2O_2/Fe^{++} system, the hydrocarbon is bound preferentially to purine polynucleotides with poly *G* still being the most reactive. The specificity is the same in both solvent systems, viz., 1 × 10^{-2} M phosphate buffer or buffer-ethanol (2:1).

The data in Table 5 show that [^{3}H]3,4-benzpyrene was covalently linked to the double-stranded poly (*G*) · poly (*C*) complex as well as to the double-stranded poly (*A*) · poly (*U*) complex. Sucrose gradient electrophoresis experiments at pH 4 showed that 95% of the [^{3}H]3,4-benzpyrene in the double-standed complex was associated with poly *A* after it was separated from poly *U*. The procedure of Haselkorn and Fox (1965) was used to separate the double-stranded complex of poly *G* and poly *C*. Essentially all the radioactivity was associated with poly *G* after the double-straned complex was dissociated at alkaline pH and the poly C was digested during neutrlization with acetate buffer containing pancıeatic ribonuclease.

In order to exclude the possibility of tritium exchange between [^{3}H]3,4-benzpyrene and poly G and to further substantiate the existence of a covalent linkage, the following experiment was performed. Poly G was reacted with 3,4-benzpyrene containing a

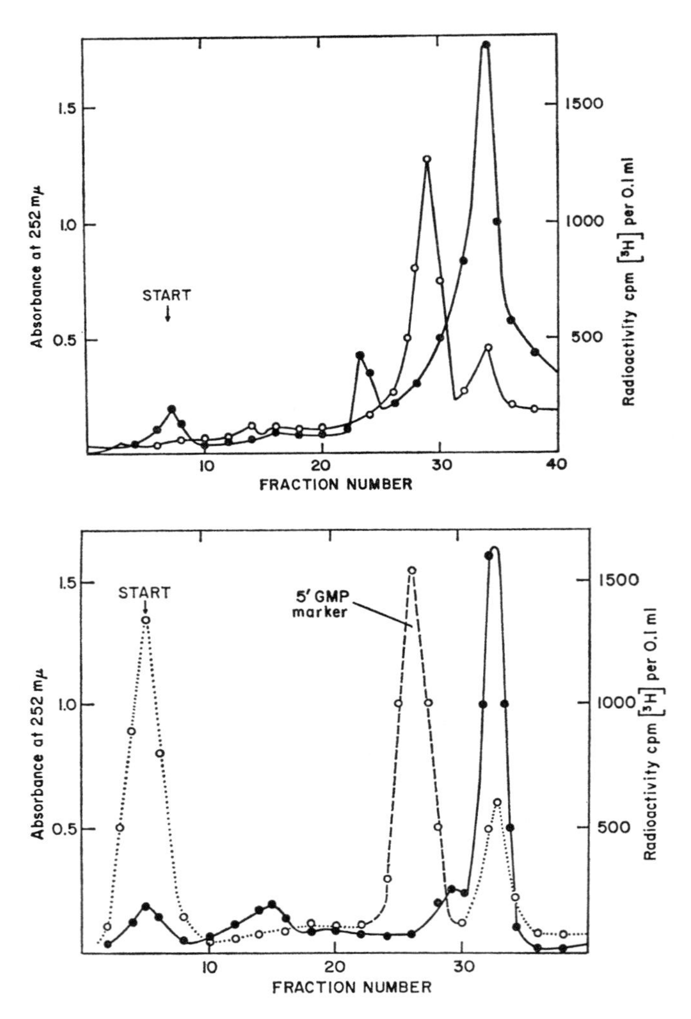

Fig. 3. Sucrose gradient electrophoresis patterns of nucleotide and nucleoside hydrolysates resulting from enzymic digestion of a poly G-[^{3}H]3,4-benzpyrene chemical complex induced by iodine. One ml fractions were collected and absorbance at 252 nm (○——○) and radioactivity (● - - - - ●) were measured. (a) Nucleotides obtained by treatment with T_1 in 2×10^{-2} M Tris buffer (pH 7.5) at 37° for 24 h. Electrophoresis was for 4.5 h at 50 v/cm using 5 ma in 1×10^{-2} M sodium acetate buffer (pH 4.7). (b) Nucleosides obtained by treatment of the nucleotide hydrolysate with alkaline phosphatase in 2×10^{-2} M Tris (pH 8.5) at 37° for 24 h. Conditions for electrophoresis are the same as in (a). After removal of alkaline phosphatase, 5′-GMP was added as a marker (○ - - - - ○)

mixture of [^{3}H] and [^{14}C] labeled hydrocarbons in a ratio of 18:1. After precipitation and washing, the isolated 3,4-benzpyrene-poly *G* complex was found to contain radioactivity with a [^{3}H]/[^{14}C] ratio of 15:1. This finding completely excludes the possibility that the radioactivity associated with nucleic acids, in experiments using only [^{3}H]3,4-benzpyrene, could have originated from tritium exchange. The reduction of [^{3}H]/[^{14}C] ratio is equivalent to a loss of about 16% of the tritium of [^{3}H]3,4-benzpyrene in reacting with poly *G*. Formation of a covalent bond between [^{3}H]3,4-benzpyrene and poly *G* should remove at least one hydrogen atom (therefore, also the corresponding amount of tritium) from the hydrocarbon. If the [^{3}H]3,4-benzpyrene is labelled randomly with the tritium, the loss of tritium should be 8.3%. Further oxidation or non-random labelling can lead to a greater percentage of loss as observed.

The [^{3}H]3,4-benzpyrene-poly *G* chemical complex was degraded by ribonuclease T_1 and then by alkaline phosphatase. Each enzymic hydrolysate was analyzed by sucrose gradient electrophoresis at pH 4.7 and the patterns are shown in Fig. 3. As shown in Fig. 3 a, most of the poly *G* was degraded to 3′-GMP by T_1 ribonuclease, however, the pattern also indicates the presence of a component which has a higher electrophoretic mobility and is believed to be an oligonucleotide fraction. Most of the radioactivity from the [^{3}H]3,4-benzpyrene is associated with this oligonucleotide component. After incubation with alkaline phosphatase, the 3′-GMP was converted to guanosine which remains at the origin after electrophoresis (Fig. 3 b). The "oligonucleotide fraction" continues to move ahead of the added 5′-GMP marker and contains most of the radioactivity (Fig. 3 b). These observations are consistent with the idea that the attachment of [^{3}H]3,4-benzpyrene to poly *G* has inhibited the action of the T_1 enzyme which has base specificity. Therefore, after the T_1 digestion, most of the radioactivity is found attached to the oligonucleotide fragments; and the phosphate groups of the oligonucleotides are resistant to alkaline phosphatase. The average size of the oligonucleotides must be larger than that of a dimer since the calculated nucleotide/hydrocarbon ratio is much greater than two. Preliminary experiments indicate that these [^{3}H]3,4-benzpyrene-oligo *G* fragments are also resistant to the action of snake venom diesterase.

5. Mechanism of Iodine and H_2O_2 Activation of Polycyclic Hydrocarbons

While the chemical mechanisms in the I_2-induced and the H_2O_2/Fe^{++}-induced reactions are not known in detail, there exists a considerable amount of pertinent information. A strong interaction between iodine and 3,4-benzpyrene is indicated by the formation of black "charge-transfer" complexes (Szent-Györgyi, Isenberg and Baird, 1960; Epstein, Bular, Kaplan, Small and Mantel, 1964) and the appearance of an EPR signal (Szent-Györgyi et al., 1960). Formation of 5,5′-dimers of 3,4-benzpyrene and various quinones of 3,4-benzpyrene, as well as the formation of N-substituted 5-benzopyrenyl pyridinium salts have been reported when 3,4-benzpyrene alone or with pyridine absorbed in a silica gel surface was exposed to iodine vapor (Wilk, Bez and Rochlitz, 1966; Rochlitz, 1967). Reaction of 3,4-benzpyrene with various bases and nucleotides in the silica gel-iodine vapor system has been shown by chromatography (Wilk and Girke, 1969). These findings led Wilk and coworkers to propose that the "radical cation" of 3,4-benzpyrene is the active intermediate in these 3,4-benzpyrene reactions induced by iodine vapor (for review, see Wilk and

Girke, 1969). In our preliminary experiments, the 5-benzopyrenyl pyridinium salt was also found as the reaction product in the [^{3}H]3,4-benzpyrene-pyridine (50%)-HMP buffer-I_2 system (Hoffmann, Lesko and Ts'o, 1969). This finding suggests that when the reaction is carried out in a solution of water and organic solvents, the radical cation of 3,4-benzpyrene also is likely to be the active intermediate. It is well known that alkylating agents preferentially react with guanine residues of DNA at the N-7 position. Therefore, our finding of the overwhelming preference of [^{3}H]3,4-benzpyrene to form a chemical complex with poly *G* also supports the notion that [^{3}H]3,4-benzpyrene reacts with nucleic acids via a radical cation intermediate in the iodine-induced reaction. The chemical mechanism in the reaction induced by H_2O_2/Fe^{++} is unlikely to be identical to that induced by I_2. As shown in Table 5, [^{3}H]3,4-benzpyrene reacts with poly *G* and poly *A* (although to a lesser extent) in the reaction induced by H_2O_2/Fe^{++}, while the [^{3}H]3,4-benzpyrene reacts mainly with poly *G* in the I_2-induced reaction. In our preliminary experiment with the [^{3}H]3,4-benzpyrene-pyridine (50%)-HMP$=H_2O_2/Fe^{++}$ system, a product other than the N-substituted 5-benzopyrenyl pyridinium salt was found (Hoffmann et al., 1969). It is tempting to speculate that the active intermediate of [^{3}H]3,4-benzpyrene in the H_2O_2/Fe^{++} reaction is a neutral radical either a neutral radical of the hydrocarbon itself generated by the abstraction of an H-atom from the hydrocarbon by an OH radical or a neutral phenoxy radical which can be formed from the hydroxylated hydrocarbon by further oxidation. The latter type of radical was proposed by Nagata, Inomata, Kodama and Tagashira (1968) to exist in the incubation mixture of 3,4-benzpyrene with skin homogenates based on EPR experiments. Hoffmann and Müller (1969) report that the product of 3,4-benzpyrene with guanine nucleotides formed by X-ray irradiation is an adduct of 3,4-benzpyrene to the C-8 position of guanine residues. The neutral aryl radicals are known to be less specific in their reaction with the bases and they react with the C-8 atom of both guanine and adenine derivatives (Hoffmann and Müller, 1966).

The electrochemical oxidation of 3,4-benzpyrene has been studied by Pysh and Yang (1963) and more recently by Jeftic and Adams (1970) in a more extensive and thorough manner. The initial oxidation of 3,4-benzpyrene is a one-electron process to the radical cation which can form a neutral radical by reacting with H_2O and by removal of a proton. This neutral radical can be further oxidized to be a hydroxyl hydrocarbon. The hydroxylated 3,4-benzpyrene upon further oxidation produces a neutral phenoxy radical which then becomes a quinoid derivative. Thus, these authors have demonstrated a relationship between the radical-cation, the neutral phenoxyl radical and other intermediates in a stepwise oxidative process starting from the neutral hydrocarbons. The cellular oxidation process of 3,4-benzpyrene may occur in a similar manner and may produce similar radicals and intermediates. As mentioned before, the active intermediates of the hydrocarbon in the model chemical reactions are most likely to be radicals of various types and at present, it is not certain which type of radical intermediate is more important for the reaction in the biological process. Therefore, the work of Jeftic and Adams (1970) is very helpful for the future correlation between the chemical studies and the biological process. A comprehensive and thoughtful review on the theoretical aspects of the formation of radical cations of polycyclic hydrocyrbons and the possible relevance of such a reaction to the mechanism of chemical carcinogenesis has been written recently by Pullman, Pull-

MAN, UMANS and MAIGRET (1969). This article is of particular interest to the present paper, since the authors have attempted to calculate the stereochemical interaction of the 3,4-benzpyrene cation radical to the *G-C* base pair.

The inability of 1,2-benzpyrene to form a covalent linkage with nucleic acids in the iodine and H_2O_2-induced reactions might be explained by the following alternative reasons: (1) the hydrocarbon will not form a stable radical intermediate; (2) the radical produced will not react with nucleic acids. Experiments are now in progress to distinguish between these two possibilities.

6. Biological Consequences of the Covalent Linkage of Polycyclic Hydrocarbons to DNA

Procedures have been described earlier in this communication for obtaining a relatively high number of polycyclic hydrocarbons covalently linked to DNA. This

Table 6. *Inactivation and mutagenesis of transforming DNA from B. subtilis by covalent linkage of carcinogenic polycyclic hydrocarbons*

Hydrocarbon	$\frac{\text{Hydrocarbon}}{\text{DNA base}} \times 10^3$	% Survival	$\frac{\text{Fluorescing colonies}}{\text{Transformant colonies}}$ =	Frequency of mutants (× 10⁴)
Control	—	100	5/50,860	1.0
Iodine	--	90	7/17,000	4.1
DMBA[a]	13.2	60	6/5,360	11.0
	12.3	40	6/5,625	10.7
	17.2	35	2/2,180	9.0
	16.5	30	6/2,800	18.0
3,4-BP[b]	11.7	37	3/2,568	11.6
	15.5	24	5/5,260	9.5
	8.9	6	3/2,516	12.0
	12.7	16	4/3,309	12.0

[a] DMBA = 9,10-dimethyl-1,2-benzanthracene. [b] 3,4-BP = 3,4-benzpyrene.

increased yield provides one with the material necessary to examine directly the biological effect of the covalent linkage of carcinogenic hydrocarbons to DNA in model systems. The mutagenic effect of covalent attachment to *B. subtilis* SB 19 transforming DNA of 3,4-benzpyrene and 9,10-dimethyl-1,2-benzanthracene in an iodine-induced reaction has been assayed with the linked genes in the tryptophan operon. MAHER, MILLER, MILLER and SZYBALSKI (1968) have used this system to measure the mutagenic activity of derivatives of 2-acetylaminofluorene.

As can be seen in Table 6, covalent attachment of 3,4-benzpyrene and 9,10-dimethyl-1,2-benzanthracene to DNA caused a measurable decrease in the ability of the DNA to transform the recipient strain to tryptophan-independence as compared to control DNA which has been incubated in the buffer-ethanol solution in the absence of iodine and hydrocarbon and has been precipitated and washed in accordance with the isolation procedure. Incubation with iodine alone caused a slight loss in transforming ability as well as a fourfold increase in mutation frequency over background. Mutagenicity in this system is assayed by determining increases in the frequency of mutations in the genes linked to tryptophan synthetase. The mutant transformant colonies are blocked in the synthesis of indole and when grown on suboptimal concentrations of indole accumulate fluorescent compounds. The spontaneous mutation frequency in this system was found to be about 1×10^{-4}. Table 6 also shows that covalent linkage of 3,4-benzpyrene to the extent of about one hydrocarbon per 100 bases in DNA and the linkage of 9,10-dimethyl-1,2-benzanthracene to the extent of about one hydrocarbon per 80 bases caused about a 10 to 20-fold increase in mutation frequency over background or a 6 to 16-fold increase when the iodine effect is subtracted. In these preliminary experiments, the number of lethal hits observed was never much over 1.5 even though the number of hydrocarbons covalently linked was as high as 17 per 10^3 bases in DNA. To assess the effect of cellular repair mechanisms, inactivation of transforming ability was also assayed in a uv-sensitive mutant. Samples which showed 37% and 24% survival when assayed against normal recipient strains exhibit only 12% and 10% survival respectively when assayed against uv-sensitive mutants. Therefore, it appears that the recipient strains of *B. subtilis* have a system for repairing some of the damage caused by the attachment of 3,4-benzpyrene to transforming DNA. However, when [^{3}H]3,4-benzpyrene-DNA or [^{3}H]3,4-benzpyrene-[^{32}P] DNA were treated with the excision enzymes isolated from *Micrococcus leuteus* by KAPLAN, KUSHNER and GROSSMAN (1969), no breakage of phosphodiester linkages or excision of [^{3}H] or [^{32}P] products was observed. These latter experiments were conducted in collaboration with Dr. L. GROSSMAN at Brandeis University.

It was shown in an earlier section that the attachment of 3,4-benzpyrene to nucleic acids may cause an inhibition of the action of a base-specific endonuclease (T_1 ribonuclease) and of a general exonuclease (venom phosphodiesterase). Therefore, it is entirely possible that attachment of carcinogenic hydrocarbons to DNA may alter or affect in some way the enzymatic processes of transcription and replication within the cell.

IV. Concluding Remarks

The above discourse indicates that the present research program has led us a long way in understanding the interaction between carcinogenic polycyclic hydrocarbons and nucleic acids, the genetic material. Through hydrophobic-stacking interactions, polycyclic hydrocarbon molecules are bound tightly to the bases of nucleic acids via a face-to-face mode. While this type of physical binding is not very specific with respect to various hydrocarbons, and certainly not related to their carcinogenic activity, it may have a certain selectivity with respect to the secondary structure and the different bases of nucleic acids. Employing homopolynucleotides as model systems, poly *G* was found to have a much higher affinity for [^{3}H]3,4-benzpyrene than all other single-stranded or double-stranded polynucleotides.

With a closely overlapping contact, the bases and the polycyclic hydrocarbons may efficiently form covalent linkages with each other if the system is properly activated. Apparently, activation can be obtained through a proper supply of radiation energy or through free radical formation. Owing to the extensive delocalized π-electron system possessed by these polycyclic hydrocarbons, they can be easily activated to form long-lived free radicals. In aqueous solution, at neutral pH and 5 to 37 °C, the formation of covalent linkages between these hydrocarbons and nucleic acids exhibits two interesting specificities. First, under identical conditions, carcinogenic hydrocarbons (3,4-benzpyrene, 9,10-dimethyl-1,2-benzanthracene, and 20-methylcholanthrene) are manyfold (4 to 14) more reactive than their respective noncarcinogenic isomers or analogues (1,2-benzpyrene and 1,2-benzanthracene). Second, reactions of 3,4-benzpyrene with ribosyl homopolynucleotides (poly *A*, poly *G*, poly *C*, poly *U*, poly *I*, poly *X*, poly *A* · poly *U*, poly *G* · poly *C*) indicate that the hydrocarbon is linked principally to poly *G* (even in the poly *G* · poly *C* complex) in the iodine-induced reaction; and the hydrocarbon is linked mainly to the purine polynucleotides in the peroxide induced reaction. At present, transforming DNA can be covalently linked with the carcinogenic hydrocarbons to an extent of about 1 hydrocarbon per 75 bases in the DNA with virtually no degradation of the nucleic acid.

The biological consequences of the attachment of these carcinogenic hydrocarbons to transforming DNA are that there is a measurable loss of transforming ability (up to 1.5 lethal hits) and a significant increase in mutation frequency (up to 16-fold) above background when assayed in the *B. subtilis* tryptophan operon system. Preliminary data also indicate the existence of repair systems in the recipient strain for this kind of insult to DNA which are absent or reduced in activity in a uv-sensitive mutant.

These results suggest a close relationship among the hydrophobic-stacking properties of the bases of nucleic acid, the ease of free radical formation of polycyclic hydrocarbons, the specificity of the chemical reaction of these free radicals in model systems, the mutagenicity of these polycyclic hydrocarbons on transforming DNA and carcinogenesis.

Note Added after Completion of Manuscript

Most recently, we have synthesized 5-hydroxy-3,4-benzpyrene starting from a mixture of generally labelled [^{3}H]3,4-benzpyrene and [^{14}C]3,4-benzpyrene by the method of Fieser and Horshberg [J. Am. Chem. Soc. *61*, 1565 (1939)]. Incubation of this labelled 5-hydroxy-3,4-benzpyrene (about 7 mg/ml) with DNA (about 1 mg/ml) and poly *G* (3.5×10^{-5} M) in 1×10^{-2} M phosphate buffer (pH 6.8)-ethanol (2:1 *v*/*v*) at 37 °C for 24 h, led to a covalent linkage of 5-hydroxy-3,4-benzpyrene with these nucleic acids in the absence of any activation system. After the standard precipitation and washing procedure, the amount of radioactivity remaining with the nucleic acids indicates that the ratios of chemically linked hydrocarbon to the bases of nucleic acid are about 1:800 for DNA and about 1:300 for poly *G*. The product of the 5-hydroxy-3,4-benzpyrene and poly *G* reaction was degraded by the T^1 ribonuclease. The hydrolytic products were analyzed by sucrose gradient electrophoresis in a manner similar to that in the aforementioned experiments. The electrophoretic pattern suggests the presence of hydrocarbon-oligoguanine nucleotides. This preliminary experiment, on one hand, offers a new and mild procedure for linking 3,4-benzpyrene covalently to nucleic acid without any additional activation, and on the other hand provides new understanding of the reaction mechanism.

Acknowledgement

Research at Johns Hopkins University was supported in part by an Atomic Energy Commission Contract No. AT (30-1) 3538.

References

Booth, J., Boyland, E., Orr, S. F. D.: A spectroscopic study of the nature of the complexes of purines with aromatic compounds. J. chem. Soc. **1954**, 598.

Boyland, E., Green, B.: The interaction of polycyclic hydrocarbons and nucleic acids. Brit. J. Cancer **16**, 507 (1962).

— Kimura, M., Sims, P.: The hydroxylation of some aromatic hydrocarbons by the ascorbic acid model hydroxylating system and by rat-liver microsomes. Biochem. J. **92**, 631 (1964).

Brookes, P., Heidelberger, C.: Isolation and degradation of DNA from cells treated with tritium labeled 7, 12-dimethylbenz(a)anthracene: Studies on the nature of the binding of this carcinogen to DNA. Cancer Res. **29**, 157 (1969).

— Lawley, P. D.: Evidence for the binding of polynuclear aromatic hydrocarbons to the nucleic acids of mouse skin: Relation between carcinogenic power of hydrocarbons and their binding to deoxyribonucleic acid. Nature (Lond.) **202**, 781 (1964).

Broom, A. D., Schweizer, M. P., Ts'o, P. O. P.: Interaction and association of bases and nucleosides in aqueous solutions. V. Studies of the association of purine nucleosides by vapor pressure osmometry and by proton magnetic resonance. J. Amer. chem. Soc. **89**, 3612 (1967).

Chan, S. I., Schweizer, M. P., Ts'o, P. O. P., Helmkamp, G. K.: Interaction and association of bases and nucleosides in aqueous solutions. III. A nuclear magnetic study of the self-association of purine and 6-methylpurine. J. Amer. chem. Soc. **86**, 4182 (1964).

Damiani, A., de Santis, P., Giglio, E., Liquori, A. M., Puliti, R., Ripamonti, A.: The crystal structure of the 1:1 molecular complex between 1, 3, 7, 9-tetramethyluric acid and pyrene. Acta Cryst. **19**, 340 (1965).

— Giglio, E., Liquori, A. M., Puliti, R., Ripamonti, A.: Molecular geometry of a 2:1 crystalline complex between 1, 3, 7, 9-tetramethyluric acid and 3,4-benzpyrene. J. molec. Biol. **20**, 211 (1966).

Epstein, S., Bular, I., Kaplan, J., Small, M., Mantel, N.: Charge-transfer complex formation, carcinogenicity and photodynamic activity in polycyclic compounds. Nature (Lond.) **204**, 750 (1964).

Freese, E., Strack, H. B.: Induction of mutations in transforming DNA by hydroxylamine. Proc. nat. Acad. Sci. (Wash.) **48**, 1796 (1962).

Gelboin, H. V.: A microsome-dependent binding of benzo(a)pyrene to DNA. Cancer Res. **29**, 1272 (1969).

Goshman, L. M., Heidelberger, C.: Binding of tritium labelled polycyclic hydrocarbons to DNA of mouse skin. Cancer Res. **27**, 1678 (1967).

Grover, P. L., Sims, P.: Enzyme-catalyzed reactions of polycyclic hydrocarbons with deoxyribonucleic acid and protein *in vitro*. Biochem. J. **110**, 159 (1968).

Haselkorn, R., Fox, C. F.: Synthesis and properties of a complex of polyriboguanylic acid and polyribocytidylic acid. J. molec. Biol. **13**, 780 (1965).

Heidelberger, C., Davenport, G. R.: Local functional components of carcinogenesis. Acta Un. int. Cancr. **17**, 55 (1961).

Hoffmann, H. D., Lesko, S. A., Ts'o, P. O. P.: Covalent linkage of 3,4-benzpyrenze to DNA in aqueous solution. Report in the Division of Medicinal Chemistry, 158th National Meeting of American Chem. Soc., New York 1969.

— Muller, W.: Schonende Arylierung von Guanosin in 8-Stellung durch Diazoniumsalze. Biochim. biophys. Acta (Amst.) **123**, 421 (1966).

— — Reactions of carcinogens with guanine nucleotides. The Jerusalem Symposia on Quantum Chemistry and Biochemistry: Physicochemical Mechanisms of Carcinogenesis **1**, 183 (1969).

Isenberg, I., Baird, S. L., Bersohn, R.: Interaction of polynucleotides with aromatic hydrocarbons. Biopolymers **5**, 477 (1967).

Jardetzky, O.: Proton magnetic resonance of purine and pyrimidine derivatatives. XI. Proton magnetic resonance studies of nucleotide interactions. Biopolymers Symp. **1**, 501 (1964).

Jeftic, L., Adams, R. N.: Electrochemical oxidation pathways of benzo(a) pyrene. (1970) J. Amer. chem. Soc. **92**, 1332 (1970).

KAFLAN, J. C., KUSHNER, S. R., GROSSMAN, L.: Enzymatic repair of DNA, 1. Purification of two enzymes involved in the excision of thymine dimers from ultraviolet-irradiated DNA. Proc. nat. Acad. Sci. (Wash.) **63**, 144 (1066).

KODAMA, M., TAGASHIRA, Y., IMAMURA, A., NAGATA, C.: Effect of secondary structure of DNA upon solubility of aromatic hydrocarbons. J. Biochem. **59**, 257 (1966).

LERMAN, L. S.: The combination of DNA with polycyclic aromatic hydrocarbons. Proceedings of the Fifth National Cancer Conference, p. 39. Philadelphia, Pa.: Lippincott 1964.

LESKO, S. A., HOFFMANN, H. D., TS'O, P. O. P., MAHER, V.: Chemical linkage of carcinogenic hydrocarbons to DNA and its mutagenic effect. Biophys. J. Abstracts **1970**, 171a.

— SMITH, A., TS'O, P. O. P.: UMANS, R. S.: Interaction of nucleic acids. IV. The physical binding of 3,4-benzpyrene to nucleosides, nucleotides, nucleic acids, and nucleoprotein. Biochemistry **7**, 434 (1968).

— TS'O, P. O. P., UMANS, R. S.: Interaction of nucleic acids. V. Chemical linkage of 3, 4-benzpyrene to deoxyribonucleic acid in aqueous solution. Biochemistry **8**, 2291 (1969).

LIQUORI, A. M., ASCOLI, F., DE SANTIS SAVINO, M.: Competitive effect of spermidine on the solubilization of 3,4-benzpyrene in DNA solutions. J. molec. Biol. **24**, 123 (1967).

— DE LERMA, B., ASCOLI, F., BOTRE, C., TRASCIATTI, M.: Interaction between DNA and polycyclic aromatic hydrocarbons. J. molec. Biol. **5**, 521 (1962).

MAHER, V. M., MILLER. E. C., MILLER, J. A., SZYBALSKI, W.: Mutations and decreases in density of transforming DNA produced by derivatives of the carcinogens 2-acetylaminofluorene and N-methyl-4- amino azobenzene. Molec. Pharmacol. **4**, 411 (1968).

NAGATA, C., INOMATA, M., KODAMA, M., TAGASHIRA, Y.: Electron spin resonance study on the interaction between the chemical carcinogens and tissue components. III. Determination of the structure of the free radical produced either by stirring 3,4-benzpyrene with albumin or incubating it with liver homogenates. Gann **58**, 289 (1968).

— KODAMA, M., TAGASHIRA, Y., IMAMURA, A.: Interaction of polynuclear aromatic hydrocarbons, 4-nitroquinoline l-oxides, and various dyes with DNA. Biopolymers **4**, 4096 (1966).

PULLMAN, A., PULLMAN, B., UMANS, R. S., MAIGRET, B.: A few afterthoughts. The Jerusalem Symposia on Quantum Chemistry and Biochemistry: Physicochemical Mechanisms of Carcinogenesis **1**, 325 (1969).

PYSH, E. S., YANG, N. C.: Polarographic oxidation potentials of aromatic compounds. J. Amer. Chem. Soc. **85**, 2124 (1963).

RAPAPORT, S. A., TS'O, P. O. P.: Interaction of nucleic acids. III. Chemical linkage of the carcinogen 3,4-benzpyrene to DNA induced by χ-ray irradiation. Proc. nat. Acad. Sci. (Wash.) **55**, 381 (1966).

RESKE, G., STAUFF, J.: Über Photoreaktionen und spektrale Veränderungen von 3,4-benzpyren in wabrigen Protein- und DNA-Lösungen. Z. Naturforsch. **20 b**, 15 (1965).

RICH, A., DAVIS, D. R., CRICK, F. H. C., WATSON, J. D.: The molecular structure of polyadenylic acid. J. molec. Biol. **3**, 716 (1961).

ROCHLITZ, J.: Neue Reaktionen der carcinogenen Kohlenwasserstoffe. II. Tetrahedron **23**, 3043 (1967).

SCHWEIZER, M. P., BROOM, A. D., TS'O, P. O. P., HOLLIS, D. P.: Studies of inter- and intramolecular interaction in mononucleotides by proton magnetic resonance. J. Amer. chem. Soc. **90**, 1042 (1968).

— CHAN, S. I., TS'O, P. O. P.: Interaction and association of bases and nucleosides in aqueous solutions. IV. Proton magnetic resonance studies of the association of pyrimidine nucleosides and their interaction with purine. J. Amer. chem. Soc. **87**, 5241 (1965).

SZENT-GYORGYI, A., ISENBERG, I., BAIRD, S. L.: On the electron donating properties of carcinogens. Proc. nat. Acad. Sci. (Wash.) **46**, 1444 (1960).

TS'O, P. O. P., CHAN, S. I.: Interaction and association of bases and nucleosides in aqueous solutions. II. Association of 6-methylpurine and 5-bromouridine and treatment of multiple equilibria. J. Amer. chem. Soc. **86**, 4176 (1964).

— HELMKAMP, G. K., SANDER, C.: (1) Interaction of nucleosides and related compounds with nucleic acids as indicated by the change of helix-coil transition temperature. Proc. nat. Acad. Sci. (Wash.) **48**, 686 (1962).

— — — (2) Secondary structures of nucleic acids in organic solvents. II. Optical properties of nucleotides and nucleic acids. Biochim. biophys. Acta (Amst.) **55**, 584 (1962).

— — — STUDIER, F. W.: (1) Secondary structure of nucleic acids in organic solvents. IV. Effects of electrolytes. Biochim. biophys. Acta (Amst.) **76**, 54 (1963).

— MELVIN, I. S., OLSON, A. C.: (2) Interaction and association of bases and nucleosides in aqueous solutions. J. Amer. chem. Soc. **85**, 1289 (1963).

— LU, P.: Interaction of nucleic acids. II. Chemical linkage of the carcinogen 3,4-benzpyrene to DNA induced by photoirradiation. Proc. nat. Acad. Sci. (Wash.) **51**, 272 (1964).

UDENFRIEND, S., CLARK, C. T., AXELROD, J., BRODIE, B.: Ascorbic acid in aromatic hydroxylation. I. A model system for aromatic hydroxylation. J. biol. Chem. **208**, 731 (1954).

UMANS, R. S., LESKO, S. A., TS'O, P. O. P.: Chemical linkage of carcinogenic 3,4-benzpyrene to DNA in aqueous solution induced by peroxide and iodine. Nature (Lond.) **221**, 763 (1969).

VAN DUUREN, B. L.: Fluorescence of 1, 3, 7, 9-tetramethyluric acid complexes of aromatic hydrocarbons. J. phys. Chem. **68**, 2544 (1964).

WILK, M., BEZ, W., ROCHLITZ, J.: Neue Reaktionen der carcinogenen Kohlenwasserstoffe 3,4-benzpyren, 9, 10-dimethyl-1, 2-benzanthracen und 20-methylcholanthren. Tetrahedron **22**, 2599 (1966).

— GIRKE, W.: Radical cations of carcinogenic alternant hydrocarbons, amines and azo dyes, and their reactions with nucleo bases. The Jerusalem Symposia on Quantum Chemistry and Biochemistry: Physicochemical Mechanisms of Carcinogenesis **1**, 91 (1969).

Interaction of N-2-Acetylaminofluorene with RNA

DEZIDER GRUNBERGER, I. BERNARD WEINSTEIN, LOUIS M. FINK, JAMES H. NELSON and CHARLES R. CANTOR

I. Introduction

Several carcinogens bind *in vivo* to nucleic acids, proteins, and carbohydrates (MILLER and MILLER, 1967; FARBER, 1968). The binding of these agents to nucleic acids is of particular interest, owing to the central role of DNA and RNA in perpetuation and expression of genetic information. Binding of carcinogens to DNA and protein has been investigated in considerable detail, but there are relatively few studies which examine the binding of carcinogens to cellular RNAs.

Previous studies from this and other laboratories suggest that tRNA may be a critical target during chemical carcinogenesis (AXEL, WEINSTEIN and FARBER, 1967; WEINSTEIN, 1968; NOVELLI, 1970). In addition, evidence is accumulating that the tRNA population of certain tumor cells may differ qualitatively from that of normal mammalian cells (BERGQUIST and MATHEWS, 1962; TSUTSUI, SRINIVASAN and BOREK, 1966; BALIGA, BOREK, WEINSTEIN and SRINIVASAN, 1969; GOLDMAN, JOHNSTON and GRIFFIN, 1969). *In vivo* modification of liver RNA has been described with the hepatic carcinogen ethionine (FARBER, MCCONOMY, FRANZEN, MARROQUIN, STEWART and MAGEE, 1967; AXEL et al., 1967; ROSEN, 1968; ORTWERTH and NOVELLI, 1969).

N-2-acetylaminofluorene (AAF) is a potent hepatic carcinogen which also binds to liver tRNA when administered *in vivo* (HENSHAW and HIATT, 1963; IRVING, VEAZEY and WILLIARD, 1967; AGARWAL and WEINSTEIN, 1970). The drug also binds *in vivo* to liver rRNA, DNA and protein (MARROQUIN and FARBER, 1965; KRIEK, 1968; MILLER and MILLER, 1967; KRIEK, MILLER, JUHL and MILLER, 1967). This carcinogen, and certain other aromatic amines, require metabolic activation as a prerequisite for combination with macromolecules (MILLER, 1968; KING and PHILLIPS, 1968). The primary step in this activation is *N* hydroxylation (MILLER, 1968). The final metabolite, or proximate carcinogen, has not been identified *in vivo*. It is thought that it may be an ester of N-OH-AAF since a synthetically prepared ester, N-acetoxy-AAF, complexes directly with RNA and DNA at neutral pH *in vitro*. Hydrolysis of these nucleic acids indicated that the major nucleoside derivative is 8-(N-2-fluorenylacetamido)-guanosine (MILLER, JUHL and MILLER, 1966; KRIEK et al., 1967).

The present studies were undertaken to explore the mechanism by which attachment of AAF residues to ribonucleic acids involved in the translation process might distort their structure and thereby their biological activity.

1. Binding of AAF to tRNA

We have examined the effects of *in vitro* modification of *E. coli* tRNA by N-acetoxy-AAF with respect to amino acid accepting capacity and ribosomal binding (FINK,

NISHIMURA and WEINSTEIN, 1970). In this model system, the carcinogen selectively modifies the function of specific tRNAs. The accepting capacity for 15 amino acids, of control tRNA and tRNA previously reacted with 1.5×10^{-3} M N-acetoxy-AAF, is listed in Table 1. The activity of arginine and lysine tRNAs were inhibited to the greatest extent; there was a lesser inhibition or no inhibition for several other tRNAs, and there was actual stimulation of valine tRNA acceptance, when compared to equivalent amount of control tRNA. In contrast to the selective effects obtained with 1.5×10^{-3} M N-acetoxy-AAF, when the tRNA was reacted with high concentrations

Table 1. *Amino acid acceptance capacity of E. coli tRNA after treatment with N-acetoxy-AAF (1.5×10^{-3} M)*

Amino acid	Amino acid acceptance (pmoles/assay system) Control tRNA	AAF tRNA	% of control[a]
Arginine	35	12	34
Lysine	17	9	52
Leucine	63	36	57
Isoleucine	15	9	60
Threonine	19	12	62
Glycine	17	11	62
Histidine	6	4	63
Phenylalanine	18	12	65
Proline	17	12	69
Aspartic	20	15	74
Tyrosine	21	17	80
Serine	9	8	88
Methionine	45	41	91
Valine	37	51	136

[a] Acceptance AAF tRNA × 100/acceptance control tRNA. All tRNAs were tested at a limiting concentration, i.e. 0.61 A_{260} units/0.1 ml assay system. For additional details see FINK et al., 1970.

(10^{-2} M) of drug, there was extensive inactivation of the accepting capacity of tRNA for all amino acids.

The functional properties of AAF-tRNA with respect to ribosomal binding and codon recognition are listed in Table 2. When AAF tRNA was aminoacylated with lysine both the poly *A*- and the poly *AG*-stimulated ribosomal binding of this tRNA were less than 40% of that obtained with control lysyl-tRNA. On the other hand, no significant difference between AAF and control tRNA was observed in the binding of phenylalanyl-tRNA stimulated by poly *U*.

The above studies indicate that low concentrations of the carcinogen N-acetoxy-AAF, at neutral pH *in vitro*, can selectively react with tRNA. The attachment of AAF to tRNA impairs the ability of certain tRNAs to accept amino acids and also produces impairment in the function of specific tRNAs during ribosomal binding and codon recognition.

2. Modification of Oligonucleotides and Polymers by AAF

In the course of protein synthesis the specificity of translation depends on base pairing between a nucleotide region of tRNA, the anticodon, and a corresponding triplet of nucleotides (the codon) in mRNA (CRICK, 1966). In contrast to certain

Table 2. *Codon response of AAF tRNA*

	Template	Aminoacyl-tRNA bound to ribosomes Control tRNA	AAF tRNA
^{14}C-Lys-tRNA		Δpmoles	
	Poly *A*	1.58	0.61
	Poly (*A*, *G*) (3:1)	2.06	0.45
		pmoles	
	None	0.28	0.21
^{14}C-Phe-tRNA		Δpmoles	
	Poly *U*	2.80	2.55
		μμmoles	
	None	0.14	0.27

E. coli control tRNA and *E. coli* tRNA previously reacted with N-acetoxy-AAF (1.5×10^{-3} M) were charged with the indicated ^{14}C-amino acids and tested at comparable concentrations in the ribosomal binding assay. Details of the assay system are described by FINK et al., 1970.

other types of chemical modifications, the presence of an AAF residue on the 8 position of guanine would not be expected to directly interfere with hydrogen bonding and base pairing (KRIEK et al., 1967). Data obtained with AAF modified tRNA led to the suggestion that the observed changes in biologic activity might be due to a conformational change in the nucleic acid. We then simplified the analysis of this problem by examining the functional properties of certain oligonucleotides and polymers, previously modified with N-acetoxy-AAF, in a ribosomal binding assay, and also obtained information of the conformation of AAF containing oligonucleotides, by examining their circular dichroism (CD) spectra.

The ability of trinucleotides, which code for amino acids, to induce tRNA binding to ribosomes (NIRENBERG and LEDER, 1964) made it possible to study the functional effects of AAF modification of G residues in trinucleotides. We modified the triplets GUU and AAG with N-acetoxy AAF under conditions similar to those previously described (KRIEK et al., 1967; FINK et al., 1970), extracted the reaction mixture with ether and separated the products in the aqueous phase by paper chromatography (GRUNBERGER, NELSON, CANTOR and WEINSTEIN, 1970). Fig. 1 compares the UV absorption spectra of the two chromatographic fractions obtained in the case of GUU. The compound with the greater mobility (AAF-GUU), showed a characteristic shift in the absorption maxima from 258 to 265 nm and a new shoulder at 300 to 310 nm. The slower moving compound had a spectrum characteristic of GUU. After T_2 ribonuclease digestion of the former compound, and thin-layer chromatography of the products, Up, U and Gp-AAF, but no Gp were detected. From the spectral results and the base composition we conclude that the compound with lower Rf was the unreacted GUU whereas the faster compound was GUU with AAF bound

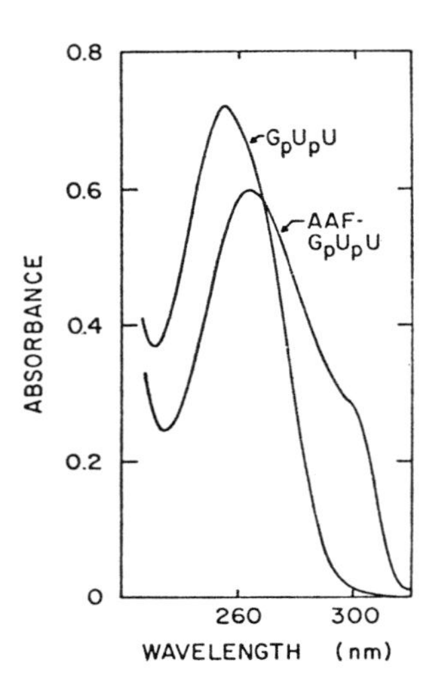

Fig. 1. Ultraviolet absorption spectra of GUU and AAF-GUU

to the G residue. Very similar results were obtained after the reaction of AAG, one of the lysine codons, with N-acetoxy-AAF.

Since the trinucleotide codons for valine contain G in the 5′ terminal position, it was of interest to study the effect of AAF modification of this G on the template activity of GUU. Table 3 indicates that, whereas ^{14}C-valyl-tRNA recognized the GUU codon, there was no response of the ^{14}C-valyl-tRNA to the AAF-containing triplet. To determine whether modification of *G* at the 3′ terminal position of a triplet also affects codon recognition, we have compared the response of lysyl-tRNA to AAG and AAG-AAF. Table 3 demonstrates that AAF modification of *G* in AAG led to complete inactivation of the ability of this triplet to stimulate ribosomal binding of ^{14}C-lysyl-tRNA. These results indicate that modification of *G* in either the 5′ or 3′ end of the triplet totally inhibits the normal template activity of that triplet. There remained the possibility that the AAF-modified *G* in codons might behave as adenosine or uridine during base pairing and thereby produce miscoding during the recognition process. To test this, we have measured the effect of modified GUU on the stimulation of ^{14}C-Ile-tRNA (normal codon, AUU), and of modified GUU on the stimulation of ^{14}C-Phe-tRNA (normal codon, UUU), binding to ribosomes. The results in Table 3 demonstrate that neither GUU nor AAF-GUU were recognized by either Ile-or-Phe-tRNA. It appears, therefore, that AAF modification of *G* leads to inactivation rather than to mistakes in base pairing.

Since poly (*U*, *G*) normally directs the binding of valyl- and phenylalanyl-tRNA to ribosomes, we have tested the ability of the AAF-modified polymer to function in this system. The randomly ordered polymer, poly (*U*, *G*), was reacted at neutral pH with N-acetoxy-AAF, under conditions similar to those described above, repurified by ether extraction and ethanol precipitation. The modified polymer displayed a shift

Table 3. *The effect of AAF modification of polynucleotides upon their stimulation of ^{14}C-aminoacyl-tRNA binding to ribosomes*

Polynucleotide (0.1 A_{260} nm)	^{14}C-aminoacyl-tRNA bound Valyl- (12.2 pmol)		Lysyl- (17 pmol)		Phenylalanyl- (6.5 pmol)		Isoleucyl- (13.4 pmol)	
	pmol	Δpmol	pmol	Δpmol	pmol	Δpmol	pmol	Δpmol
None	0.35	—	0.67	—	0.68	—	0.96	—
GUU	1.70	1.35	—	—	0.62	−0.06	0.83	−0.13
AAF-GUU	0.36	0.01	—	—	0.53	−0.15	0.75	−0.21
AAG	—	—	1.46	0.79	—	—	—	—
AAG-AAF	—	—	0.66	0.01	—	—	—	—
Poly (*U*, *G*)	5.87	5.52	—	—	4.26	3.58	0.87	−0.09
poly (*U*, *G*-AAF)	3.44	3.09	—	—	4.32	3.64	0.69	−0.30
Poly (*U*, *A*)	—	—	—	—	—	—	2.53	−1.57

The incubation mixture (0.05 ml) contained 0.10 M Tris-acetate (pH 7.2), 0.05 M KCl, 0.03 M magnesium acetate, and 2 to 2.5 A_{260} units of *E. coli* ribosomes. ^{14}C-aminoacyl-tRNA, trinucleotides and polymers were added as specified in the Table. Incubation was carried out at 24 °C for 20 min, and samples processed and counted as described by NIRENBERG and LEDER (1964).

in absorption maximum from 258 to 265 nm and a new shoulder at 300 to 310 nm, reflecting the presence of bound AAF. Nucleotide analysis after T_2 ribonuclease digestion of modified poly (*U*, *G*), and separation on two-dimensional cellulose TLC plates, revealed about 60% decrease of guanylic acid and the appearance of a new compound which exhibited a bright blue fluorescence and developed a blue-green color upon exposure to ultraviolet light. The UV spectrum of this compound at neutral pH was identical with that of 8-(N-2-fluorenyl-acetamido)-guanylic acid described by KRIEK et al. (1967). Thus, N-acetoxy-AAF reacted with *G* residues in poly (*U*, *G*) in the same manner as with free guanosine or deoxyguanosine. The extent of AAF modification of *G* residues in poly (*U*, *G*) was dependent upon the concentration of N-acetoxy-AAF in the reaction mixture. Under our reaction conditions the polymer contained about 40% of modified *G*.

As shown in Table 3 the stimulation in binding of *E. coli* ^{14}C-valyl-tRNA to ribosomes by poly (*U*, *G*)-AAF was lower than the stimulation obtained with unmodified poly (U, G). The decrease in stimulatory effect of the modified polymer corresponded approximately to the extent of G modification by AAF.

The stimulatory effect of poly (U, G) on the binding of ^{14}C- phenylalanyl-tRNA to ribosomes is due to sequences of UUU (LEDER and NIRENBERG, 1964). Therefore, the modification of G residues in this polymer should not alter its stimulation of phenylalanyl-tRNA binding to ribosomes. Results in Table 3 confirm this, thus indicating that the AAF modification is specific and does not result in generalized inactivation of the polymer.

Since binding of tRNA to codons and ribosomes represents only one step in the translation process, it was also important to assay the behavior of AAF modified

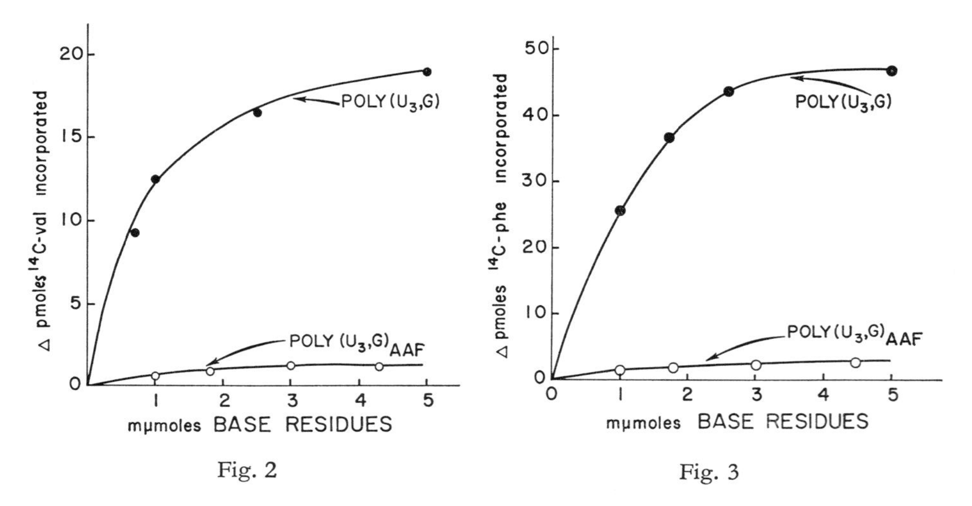

Fig. 2 Fig. 3

Fig. 2. Effect of poly (U, G) and poly (U, G)-AAF on ^{14}C-valine incorporation into protein in S-30 *E. coli* extract

Fig. 3. Effect of poly (U, G) and poly (U, G)-AAF on ^{14}C-phenylalanine incorporation into protein by the S-30 fraction of *E. coli*

polymers during polypeptide chain formation. For these reasons we tested the ability of poly (U, G), previously modified with AAF, to direct the incorporation of ^{14}C-valine and ^{14}C-phenylalanine into protein employing the S-30 extract from *E. coli*. We found that our preparation of poly (U, G)-AAF, in which 40% of the G residues were modified by AAF, was non-active in stimulating the incorporation of ^{14}C-valine into protein (Fig. 2). In contrast to the results obtained in the ribosomal binding assay (Table 3), the modified poly (U, G) was also markedly impaired in its ability to stimulate the incorporation of ^{14}C-phenylalanine into protein (Fig. 3).

There are two possible explanations for the lack of incorporation of both valine and phenylalanine into protein in the presence of modified poly (U, G): (1) the modified polymer is not bound to ribosomes and therefore cannot direct protein synthesis or (2) the modified polymer is bound to ribosomes, but since modified G residues are not recognized by tRNA, translation and polypeptide chain growth is terminated each time the reading comes to a modified *G*. The first explanation can be excluded because poly (*U*, *G*)-AAF was active in stimulating phenylalanyl-tRNA binding to ribosomes, so we know that the modified polymer is bound to ribosomes. The second explanation therefore, applies and is quite plausible since the relatively high content of modified *G* residues would permit the synthesis of only short chains of polyphenylalanine which would escape precipitation with trichloroacetic acid. Studies are in progress to test this hypothesis.

3. CD Spectra of AAF-Modified Oligomers

The results of the codon recognition experiments suggested that nucleotide sequences containing AAF-substituted guanosine residues may undergo substantial conformational changes as a result of the AAF substitution. Since circular dichroism spectra of oligonucleotides were expected to be sensitive indicators of such conformational changes, CD offered a possible means of investigating the effect of AAF-modification on the conformational properties of these nucleic acid derivatives. Accordingly, we have studied the influence of AAF-substitution on the CD spectra of several oligonucleotides.

Fig. 4 and 5 present the molar ellipticity, $[\theta]$, as a function of wavelength, for the compounds: GMP, ApG, GpA, UpG, GUU, and their corresponding AAF-substituted derivatives. One must be cautious in interpreting these spectra, because it is difficult to distinguish between: (1) optical activity induced in AAF because of its attachment to a guanosine residue, or because of interaction of AAF with other chromophores of the oligonucleotide, and (2) optical activity resulting from changes in conformation of the nucleoside residues themselves. It was, therefore, necessary to examine the CD spectrum of AAF-substituted GMP (AAF-GMP). Fig. 4b indicates that AAF substitution does alter the CD spectral characteristics of GMP. The dichroism of GMP itself is fairly weak, and there is little optical activity at wavelength greater than 270 nm. AAF-GMP, however, has a relatively strong negative dichroism in the spectral region of 240 to 310 nm. Since AAF has a high extinction coefficient in this region, it is likely that the strong dichroism of AAF-GMP at these wavelengths results from optical activity induced in AAF by covalent attachment to guanosine. The CD spectrum of AAF-GMP is also much different from that of guanosine at lower wavelengths (200 to 240 nm).

Fig. 4a shows that the spectra of unsubstituted ApG and GpA, are quite dissimilar. These differences have been observed previously and may be attributed to the difference in the interactions of the transition of the base chromophores in the respective base-stacked sequence isomers (Warshaw and Cantor, 1970). The spectral differences between ApG and GpA clearly illustrate the effect on the CD spectra of changes in the relative conformation of the two bases.

Of particular interest, are the striking changes in the spectra of ApG and GpA produced by AAF-substitution (Fig. 4a). Somewhat similar changes were also observed with UpG and GpUpU (Fig. 5). These large CD changes are not simply the result of covalent bonding of AAF to guanosine residues, because the differences between the spectra of each oligonucleotide and its derivative is much greater than the corresponding difference between guanosine and AAF-GMP. This observation, together with the fact that the dichroism of the AAF-modified oligonucleotides is very strong at wavelength greater than 290 nm, where only the AAF moeity absorbs significantly, indicates that AAF interacts with the base adjacent to the substituted guanine. The spectrum of ApG-AAF is particularly striking. The strength of the circular dichroism observed for this substituted dinucleotide is an order of magnitude greater than that observed for either unsubstituted oligonucleotides or the substituted monomer. Although the dichroism of UpG-AAF is also unusually

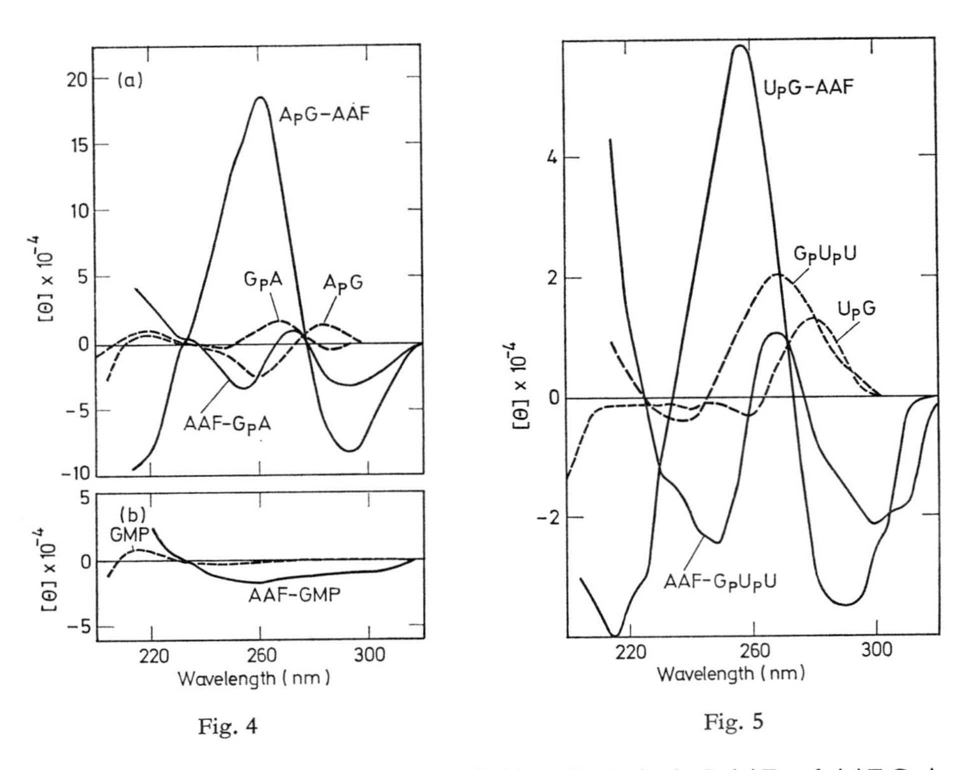

Fig. 4

Fig. 5

Fig. 4a and b. Circular dichroism curves of: (a) ApG, GpA, ApG-AAF and AAF-GpA; (b) GMP and AAF-GMP

Fig. 5. CD curves of UpG, GpUpU, UpG-AAF and AAF-GpUpU

strong it is only about one-third that of ApG-AAF in the region of the most intense CD band. The spectacularly large bands in ApG-AAF probably arise from a strong interaction of AAF with the adjacent adenine residue. Although the spectra of ApG-AAF and AAF-GpA are substantially different, those of ApG-AAF and UpG-AAF are qualitatively quite similar. The same observation is also true for the spectra of AAF-GpA and AAF-GpUpU. Apparently AAF determines the basic spectral characteristics of the substituted oligonucleotides, and the quality of the alteration is largely dependent on whether the substituted guanosine is at the 5′ or 3′ end of the oligomer.

II. Discussion

Although it is not possible at the present time to determine which conformation most closely approximates the average preferred configuration of the modified oligonucleotides, it is evident that the binding of AAF introduces drastic conformational changes in these molecules. A study of CPK molecular models of AAF-modified dinucleoside phosphates suggests that the most probable conformational change is a rotation of the modified guanosine about the glycosidic bond, followed by stacking of AAF with the base adjacent to this guanine. This seems quite plausible for two reasons: Firstly, attachment of AAF to the 8 position of guanine is sterically hindered, if the guanosine remains in the normal "anti" conformation, with torsion angle values of $\phi CN \sim -30°$ (DONOHUE and TRUEBLOOD, 1960). However, rotation about the glycosidic bond to torsional angles $\phi CN \sim +50°$ to $150°$ will relieve this hindrance (see Fig. 6). Secondly, the CD data suggest that AAF strongly interacts with the neighboring base. This is quite reasonable since a large hydrocarbon like AAF should exhibit a strong tendency to stack with non-polar bases. These conformational changes would result in decreased base-stacking between guanine and the adjacent base with a change in the orientation of the two dinucleoside bases with respect to each other. These changes have important implications because it is reasonable to assume that similar events occur at sites of AAF-substitution in coding triplets such as ApApG and GpUpU. Such changes would make it difficult for the normal hydrogen bonding sites of the respective bases along the nucleotide chain to become sufficiently aligned, with respect to orientation and distance, to effect binding to complementary sites on the anticodon of tRNA. A similar effect would also be expected for higher molecular weight polynucleotides. In fact, the substitution of AAF on guanine residues of natural nucleic acids could alter the base-stacking properties of the polymer over a considerable distance.

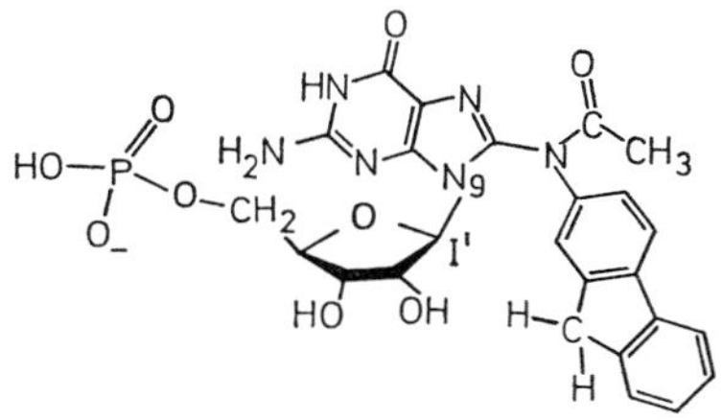

Fig. 6. Schematic representation of AAF-GMP. In contrast to the normal configuration of GMP, the guanine base has been rotated approximately 180° about the N(9)-C(1′) bond to minimize steric hindrance with the bulky AAF residue

The relationship of these findings to the carcinogenic activity of AAF is not clear at the present time. It is apparent that conformational changes in nucleic acids produced by AAF could profoundly alter the biologic properties of cellular DNA, as well as RNA. The relative susceptibility of specific guanosine residues to AAF modification, as well as the effects of base sequence on the type of conformational distortion which occurs, might play an important role in determining the biological specificity of this carcinogen. If attack on the DNA were the critical event, than a conformational change would be more likely to produce small deletions rather than single base changes during DNA replication [see MAHER, MILLER, MILLER and SZYBALSKI (1968) for a summary of conflicting data on AAF mutagenesis]. A conformational change in specific regions of the DNA might also impair the transcription of certain genetic loci. With respect to RNA, the presence of an AAF residue on tRNA would be expected to distort the conformation of that region of the molecule, thereby altering interaction with aminoacyl-tRNA synthetases, codons and/or ribosomes. It is not known whether or not AAF interacts with messenger RNAs *in vivo*, but the present results with synthetic codons suggest that if this does occur then it would inhibit their function in protein synthesis. These, as well as additional effects on cellular RNAs, might lead to rather widespread disturbances in the translation apparatus with secondary consequences with respect to cell differentiation, regulation and autonomy (WEINSTEIN, 1968).

Acknowledgement

This research was supported by Grants No. CA-02332 and GM-14825 from the U.S. Public Health Service.

References

AGARWAL, M. K., WEINSTEIN, I. B.: Modification of RNA by chemical carcinogens. II. *In vivo* reaction of N-acetylaminofluorene with rat liver RNA. Biochemistry **9**, 503 (1970).

AXEL, R., WEINSTEIN, I. B., FARBER, E.: Patterns of transfer RNA in normal rat liver and during hepatic carcinogenesis. Proc. nat. Acad. Sci. (Wash.) **58**, 1255 (1967).

BALIGA, B. S., BOREK, E., WEINSTEIN, I. B., SRINIVASAN, P. R.: Differences in the transfer RNA's of normal liver and Novikoff hepatoma. Proc. nat. Acad. Sci. (Wash.) **62**, 899 (1969).

BERGQUIST, P. L., BURNS, D. J., PLINSTON, C. A.: Participation of redundant transfer ribonucleic acids from yeast in protein synthesis. Biochemistry **7**, 1751 (1968).

CRICK, F. H. C.: Codon-anticodon pairing: The Wobble hypothesis. J. molec. Biol. **19**, 548 (1966).

DONOHUE, J., TRUEBLOOD, K. N.: Base pairing in DNA. J. molec. Biol. **2**, 363 (1960).

FARBER, E.: Biochemistry of carcinogenesis. Cancer Res. **28**, 1859 (1968).

— MCCONOMY, J., FRANZEN, B., MARROQUIN, F., STEWART, G. A., MAGEE, P. N.: Interaction between ethionine and rat liver ribonucleic acid and protein *in vivo*. Cancer Res. **27**, 1761 (1967).

FINK, L. M., NISHIMURA, S., WEINSTEIN, I. B.: Modifications of RNA by chemical carcinogens. I. *In vitro* modification of transfer RNA by N-acetoxy-2-acetylaminofluorene. Biochemistry **9**, 496 (1970).

GRUNBERGER, D., NELSON, J. H., CANTOR, C. R., WEINSTEIN, I. B.: Coding and conformational properties of oligonucleotides modified with the carcinogen N-2-Acetylaminofluorene. Proc. nat. Acad. Sci. (Wash.) **66**, 488 (1970).

HENSHAW, E. C., HIATT, H. H.: Binding of fluorenylacetamide to rat liver ribonucleic acid (RNA) *in vivo*. (Abstract) Proc. Amer. Ass. Cancer Res. **4**, 27 (1963).

IRVING, C. C., VEAZEY, R. A., WILLIARD, R. F.: On the significance and mechanism of the binding of 2-acetylaminofluorene and N-hydroxy-2-acetylaminofluorene to rat-liver ribonucleic acid *in vivo*. Cancer Res. **27**, 720 (1967).

KING, C. M., PHILLIPS, B.: Enzyme-catalyzed reactions of the carcinogen N-hydroxy-2-fluorenylacetamide with nucleic acid. Science **159**, 1351 (1968).

Kriek, E.: Difference in binding of 2-acetylaminofluorene to rat liver deoxyribonucleic acid and ribosomal ribonucleic acid *in vivo*. Biochim. biophys. Acta (Amst.) **161**, 273 (1968).
— Miller, J. A., Juhl, U., Miller, E. C.: 8-(N-2-fluorenylacetamido) guanosine, and arylamidation reaction product of guanosine and the carcinogen N-acetoxy-N-2-fluorenylacetamide in neutral solution. Biochemistry **6**, 177 (1967).
Leder, P., Nirenberg, M.: RNA codewords and protein synthesis. II. Nucleotide sequence of valine RNA codeword. Proc. nat. Acad. Sci. (Wash.) **52**, 420 (1964).
Maher, V. M., Miller, E. C., Miller, J. A., Szybalski, W.: Mutations and decrease in density of transforming DNA produced by derivatives of the carcinogen 2-acetylaminofluorene and N-methyl-4-aminoazobenzene. Molec. Pharmacol. **4**, 411 (1968).
Marroquin, F., Farber, E.: The binding of 2-acetylaminofluorene to rat liver ribonucleic acid *in vivo*. Cancer Res. **25**, 1262 (1965).
Miller, E. C., Juhl, U., Miller, J. A.: Nucleic acid guanine: reaction with the carcinogen N-acetoxy-2-acetylaminofluorene. Science **153**, 1125 (1966).
Miller, J. A.: Summary of informal discussion on the mechanisms involved in carcinogenesis. Cancer Res. **28**, 1875 (1968).
— Miller, E. C.: Activation of carcinogenic aromatic amines and amides by N-hydroxylation *in vivo*. Carcinogenesis: A broad critique (The University of Texas M.D. Anderson Hospital and Tumor Institute, 20th Annual Symposium on Fundamental Cancer Research, 1966, pp. 397—420). Baltimore, Maryland: The Williams & Wilkins Co.
Nirenberg, M., Leder, P.: RNA codewords and protein synthesis: The effect of trinucleotides upon the binding of sRNA to ribosomes. Science **145**, 1399 (1964).
Novelli, G. D.: Action of carcinogen on nucleic acids. Genetic concepts and neoplasia (The University of Texas M.D. Anderson Hospital and Tumor Institute, 23rd Annual Symposium on Fundamental Cancer Research, 1969). Baltimore, Maryland: The Williams & Wilkins Co.
Ortwerth, B. Y., Novelli, G. D.: Studies on the incorporation of L-ethionine-ethyl-1-^{14}C into the transfer RNA of rat liver. Cancer Res. **23**, 380 (1969).
Rosen, L.: Ethylation *in vivo* of purines in rat-liver tRNA by L-ethionine. Biochem. biophys. Res. Commun. **33**, 546 (1968).
Tsutsui, R. P., Srinivasan, R., Borek, E.: tRNA methylases in tumors of animal and human origin. Proc. nat. Acad. Sci. (Wash.) **56**, 1003 (1966).
Warshaw, M. M., Cantor, C. R.: Oligonucleotide interaction, 4. Conformational differences between deoxy- and ribonucleoside phosphates. Biopolymers **9**, 1079 (1970).
Weinstein, I. B.: A possible role of transfer RNA in the mechanism of carcinogenesis. Cancer Res. **28**, 1871 (1969).

Intercalability, the ψ Transition, and the State of DNA in Nature

LEONARD S. LERMAN

I. Natural Vs. Pure DNA

There are a variety of observations which suggest that the structure of part, if not all, of the DNA contained in cells and viruses differs from the conformations that have been recognized in solutions and fibers of the pure substance. Known, or partly known, conformations include the A, B, and C fiber structures, the superhelically twisted conformations of closed, circular DNA, the single-stranded state (which obtains above the melting temperature), and the denatured state (corresponding to imperfect base pairing at ordinary temperatures). The properties of DNA that are observable despite its state of association with other substances and can usefully be compared include the gross spatial disposition of the molecule, the absorption spectrum, the optical activity as measured by optical rotatory dispersion and circular dichroism, the response to environmental variables such as temperature and pH, the non-covalent interaction with small molecules including water, the chemical reactivity, the molecular effects of radiation, etc. Any particular test is useful if the contributions to the results of the other cell components can adequately be segregated from the direct contribution of DNA.

Although the lack of simplicity in the system and the lack of a highly ordered array of molecules impose severe limitations on the investigation of structure, certain features of the state, or states, of DNA in nature are clear, and are not easily reconciled with the various conformations just mentioned. For the present discussion we shall be concerned principally with the compactness of natural DNA, its optical activity, and its dye binding capacity. The compactness is exemplified by the requirement that bacteriophage DNA be accommodated within a small globular volume offering hardly more space than required by the atoms comprising the DNA itself; obviously some deviation from stiff, rod-like fiber structures must obtain. DNA in microbial nucleoids as well as in eukaryotic nuclei and in metaphase chromosomes is also clearly far more compact than in its protein-free state, as can be seen by the dramatic hydrodynamic changes when suspensions of these particles are dissociated.

The optical activity of natural DNA is known for bacteriophage from studies of optical rotatory dispersion (MAESRTE and TINOCO, 1967), circular dichroism measurements on bacteriophage (MAESTRE, 1969; JORDAN, LERMAN, and VENABLE, 1971), and to a limited extent for mammalian nuclei (WAGNER, 1970). These systems show a severe weakening or disappearance of the zone of positive circular dichroism on the long wavelength side of the DNA absorption band (or the equivalent changes in optical rotation) as compared with free DNA. Model systems in which the combination of DNA with proteins is simulated by complexes with polylysine (COHEN and

Kidson, 1968; Shapiro, Leng and Felsenfeld, 1969) show very similar properties. The incompatibility of these changes with the various familiar conformations is indicated by the failure of their spectra to match or even approach these changes. The positive circular dichroism of B DNA is enhanced in the A structures, and also in the strongly superhelically twisted conformation (Maestre and Wang, 1969), and is only relatively slightly changed by the transition to the single-stranded state. It may be noted also that the A and C structures require conditions of low hydration that seem quite incompatible with the interior of any cell.

X-ray diffraction studies of nucleohistone fibers (Pardon, 1966; Pardon, Wilkins and Richards, 1967) reveal low angle scattering corresponding to several long spacings, of which the longest is about 110 Å, that are presumably related to the fibrils of 100 to 150 Å in diameter seen in chromatin preparations by electron microscopy. A superhelical model including two plectonemically twisted double helices has been proposed to account for these differences from fibers of pure DNA, but the structural characteristics that would confer the unusual properties with respect to circular dichroism are not evident.

II. Specificity in Dye Binding

Another suggestion that the conformation of some DNA *in vivo* differs from its *in vitro* state can be inferred from the changes in response to intercalating dyes that can be seen within cells—among different chromosomes in one cell, at different sites in the same chromosome, or in the same interphase nucleus as a function of physiological activation of the cell. These changes have been observed in various organisms by means of the localization of acridine orange, ethidium, actinomycin, and alkylating derivatives of acridine related to quinacrine. In general it is found that the molecules that intercalate without covalent reaction or without a formal charge on a side chain tend to bind substantially less to condensed chromatin than would be expected from their *in vitro* binding either to DNA or to purified nucleoprotein. The alkylating compounds and quinacrine, on the other hand, tend to concentrate somewhat in condensed chromatin (Caspersson, Zech, Modest, Foley, Wagh and Simonsson, 1969). The participation of alkylation in the binding was shown by an excess of chromosome breakage in the same regions.

Inactivity of the nucleus appears to be related to condensation of chromatin even where there are no microscopically visible changes. The staining of human leucocyte nuclei by acridine orange has been examined by Rigler (1966) using emission spectroscopy to distinguish the molecular state of bound dye. The measurements readily distinguish between the yellow-green fluorescence of intercalated acridine orange in DNA and the red-orange fluorescence emitted from the aggregated dye bound to RNA. Freshly isolated leucocytes ordinarily do not engage in DNA synthesis or in cell division *in vitro*, but can be induced to do so by exposure to phytohemagglutenin. Within 10 min, binding of intercalated acridine orange in the nucleus increases rapidly, approaching within 70 min a level value containing about four times as much dye as before phytohemagglutenin treatment; DNA synthesis is not initiated until more than a day later. The maximum intercalation after activation is usually significantly below the amount that would be expected for the nucleoprotein isolated from chromatin or for purified DNA.

While these differences have been conventionally attributed to interference with dye binding by the proteins associated with DNA, that explanation still begs the question, since the basis for interference by protein is far from clear. Two arguments, (1) the complementary specificities of the alkylating and nonalkylating dyes, both cationic, and (2), the failure of the bound protein in purified calf thymus deoxyribonucleoprotein to diminish *in vitro* binding, indicate that the effect is not simply an electrostatic consequence of the proximity of positive charges from the proteins.

The irrelevance of a direct effect of the protein charge is also supported by the results on the binding of actinomycin D to the chromosomes of mealy bugs (BERLOWITZ, PALLOTTA and SIBLEY, 1969). In many cells of the male mealy bug, the paternal members of each of the five pairs of chromosomes remain condensed during interphase and are found to bind only a third as much actinomycin as their extended maternal partners. While histone extraction increases actinomycin binding in both the paternal and maternal DNA, and effects a relatively greater increase in the condensed DNA, the binding ratio remains significantly in favor of the extended DNA even when no detectable traces of histone remain. Thus it seems appropriate to ask whether interaction with the nuclear environment, including the bound proteins, stabilizes one or more conformations of DNA different from that of its free state and less amenable to the further structural alteration required for intercalation.

III. The ψ Transition in DNA Solutions

We have recently (LERMAN, 1971; JORDAN, LERMAN and VENABLE, 1971) observed that DNA undergoes a cooperative structure transition when present in aqueous solution together with relatively high concentrations of certain synthetic neutral or anionic polymers at higher than critical concentrations of simple salts. Since the new conformation defined by the transition has some properties that are strikingly similar to those of DNA in nature, it has been of interest to explore the system with respect to those variables that might permit its detection *in vivo*. For example, the interesting relations established by RIGLER and others based on the intercalation properties of acridine orange suggest that the relation of the structure transition to intercalability is pertinent.

The main features of the polymer-and-salt-induced transition (which will be designated in the following by the acronym, the ψ transition) can be summarized from our studies on the sedimentation rates, the ultraviolet absorption spectra, and the optical activity of DNA in the presence of the polymers and varying concentrations of sodium chloride. (1) Although most of the work has been carried out with poly(ethylene oxide) (abbreviated PEO) of nominal molecular weight 6000 as the polymer component, sample experiments with other molecular weights or other polymers indicate that the results are at least qualitatively independent of these parameters and of the detailed chemical nature of the polymer. That the transition is induced also by sodium polyacrylate at pH 7 in the presence of sufficient salt is of particular interest, since the polymer is at least 80% ionized, and its high negative charge would apparently preclude any intimate association with DNA. We tentatively infer that the interactions between DNA and the polymers that are responsible for the transition are principally repulsive, as is usually the case for the two polymers in a common solvent in the absence of Coulombic attraction between opposite formal charges. (2) In an arbitrarily chosen, standard concentration of PEO, 100 mg per ml, high molecular weight bacteriophage DNA travels in the ultracentrifuge at roughly its normal sedimentation rate, allowing for ordinary viscosity and density corrections, so long as the concentration of univalent salts is below about a quarter molar. When the salt concentration exceeds that critical value, the sedimentation pattern changes abruptly to a much faster, broad band with a peak at roughly one third the sedimentation coefficient of the virus from which the DNA was extracted,

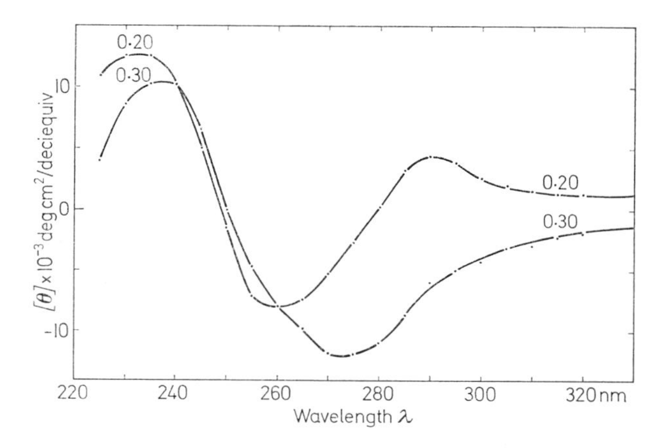

Fig. 1. Optical rotatory dispersion for T4 DNA in PEO solutions. The spectra shown here for two values of sodium chloride concentration are representative of the curves found below and above the critical salt concentration. The solution contained 100 mg/ml poly (ethylene oxide), 0.01 M sodium phosphate, and 22×10^{-6} equiv/L T4 DNA in addition to the specified concentration, either 0.20 or 0.30 M, sodium chloride. The curve at the lower salt concentration is essentially the same as that for DNA in the absence of PEO, where it is independent of salt concentration in the same range (JORDAN and LERMAN, 1969)

and extending to an upper limit corresponding to the maximum sedimentation coefficient observed for virus heads. These values are the concentration-independent limits obtained at very low DNA concentration. In less dilute solutions the sedimentation rate is significantly concentration-dependent and at high DNA concentrations, visible aggregation is observed when the salt concentration exceeds its critical value. (3) At low salt concentrations the ultraviolet spectrum of DNA and polymer is the same as that of DNA in polymer-free solutions. When the salt concentration is at any value that gives an abnormally high sedimentation in the presence of polymer, the absorption band in the vicinity of 260 nm is shifted slightly towards longer wavelengths. (4) At low salt concentrations, the optical rotation and circular dichroism spectra of DNA are the same in the presence or absence of polymer. When the salt concentration exceeds the same critical value found for sedimentation and UV absorption, certain radical changes appear in these spectra, and these remain almost unaltered by further increase in salt concentration. As shown in Fig. 1, the peak in the optical rotation spectrum at 290 nm disappears, and the wavelength minimum of the trough, originally near 260 nm in low salt, moves closer to 270 nm when the salt concentration is more than about 1/4 M. The curves for salt concentrations below 0.2 M or above 0.3 M differ only slightly from the pair of curves in the figure. The normal circular dichroism spectrum of T4 DNA, which consists of a peak near 280 nm and a trough of almost equal area with a minimum near 247 nm, changes when the sodium exceeds .25 M to a single trough with no indication of a positive lobe, and a minimum near 265 nm. Measurement of these spectra below 240 nm becomes increasingly difficult because of light absorption of the concentrated polymer solution. (5) The transition is rapidly reversible. When solutions (below the concentrations at which precipitation occurs) are diluted and remeasured promptly, the spectra are seen to be the same as those of native DNA. When sedimenting material passes into a region in the centrifuge tube of subcritical salt concentration, it immediately returns to its normal slow sedimentation rate. The bouyant density of DNA returned from the ψ state is that of helical DNA. (6) The high molecular weight of DNA is essential to the reaction; fragments produced by enzymatic hydrolysis show neither absorption shifts nor circular dichroism alterations. (7) Examination of the circular dichroism spectra in the narrow range of salt concentrations giving intermediate levels of transition fail to reveal the spectrum of any hypothetical intermediate differing from the normal or ψ forms; all of the intermediate spectra are accountable as mixtures of the two fundamental types. (8) Much lower concentrations of divalent than univalent cations are efficacious.

IV. Comparison of ψ and Natural DNA

Correspondence between ψ DNA and DNA *in vivo* is inferred from the similar compactness, absorption shift, and optical activity. The sedimentation patterns show that some of the ψ DNA molecules spontaneously reach a configuration that is about as compact as the DNA in a phage head. The frictional coefficients for simple rod-like superhelices, which would be appreciably more compact than extended DNA, are not compatible with the sedimentation rates at the peaks or above in these patterns.

The ultraviolet absorption spectrum of DNA in intact phage including light scattering corrections has been reported by TIKCHONENKO (1970). A shift in the maximum absorption from the value of 259 nm for free DNA to 262 nm is found, Also, action spectra for the photoinactivation of intact virus (RAUTH, 1965) show a red shift with respect to the absorption of free DNA.

According to MAESTRE and TINOCO (1967), the rotation spectrum of intact T4 phage shows a trough minimum at 271 nm similar to the minimum found at 270 nm for the closely related phage T2. The 290 nm peak seen for both T2 and T4 free DNA is entirely absent from the rotation spectra of both phages, which remain negative at all wavelengths above the trough. It will be seen that the features of the higher salt curve in Fig. 1 are closely similar to those of the intact phage. It may be noted that at wavelengths longer than 250 nm the rotation contribution of the phage protein is small and decreases monotonically to longer wavelengths so that it hardly affects the positions of the peak and trough. With the better discrimination obtained by measurement of circular dichroism than optical rotation, we find (JORDAN et al.) that the DNA component of the intact phage spectrum is not entirely identifiable with the ψ spectrum, but that rather, the phage spectrum is accountable as mixtures of B DNA and ψ DNA for both T4 and T7.

While there is no data in the literature on the optical activity of DNA in chromosomal complexes, it may be supposed that the properties of these complexes are simulated by the complexes of DNA with poly-L-lysine. Optical rotation spectra have been reported by COHEN and KIDSON (1968) showing again the absence of the rotational peak at 290 nm and the shift in the position of the trough to 284 nm. Similar results using a lower molecular weight polylysine have also been reported by SHAPIRO, LENG and FELSENFELD (1969). The qualitative resemblance to the ψ results is quite strong, although the red shift of the trough is slightly greater, and there are unexplained large increases in the amplitude of the curves. WAGNER (this symposium, 1970) has examined the circular dichroism of whole nuclei, which includes large contributions from protein over most of the spectrum. However, if most of the DNA were present in the B form, the intensity of its peak near 280 nm would be sufficient to impress a peak on the spectrum of the nuclei. As with ψ DNA, this peak is not found.

V. Intercalability into ψ DNA

The properties of the ψ configuration suggest its appropriateness for physiologically inert DNA in the nucleus. If one supposes that the staining changes observed by RIGLER reflect a transition of a substantial part of the leucocyte nuclear DNA from an inert to a functional state, and that the former resembles the ψ conformation, diminished intercalation into ψ DNA should be demonstrable. It has been more convenient to attempt to determine this property conversely by measuring the effect of intercalation of the stability of DNA toward the ψ transition. If there is less intercalation in the ψ state, then forcing intercalation by introducing a relatively high concentration of free dye into the environment should assist the DNA in remaining in the non-ψ conformation under conditions that would otherwise favor the transition. A set of experiments in which the effect of the presence of ethidium on the transition is detected by means of sedimentation rate measurements is shown in Fig. 2. These experiments have been carried out with the intercalating dye, ethidium, because its tendency to bind to DNA in an aggregated state, in addition to the intercalated binding, is smaller than that of acridine orange.

The curves in Fig. 2 show the distribution of DNA in gradients containing two concentrations of ethidium, or none, under standard conditions with 100 mg/ml PEO, and also with only 80 mg/ml PEO, and with 0.45 M Na^+ throughout. The curves have been drawn in the sequence in which fractions were collected; thus the meniscus is on the right, and sedimentation has occurred from right to left. The amount of DNA is sufficiently low to provide a concentration-independent result. It will be seen that at both PEO concentrations the controls lacking ethidium exhibit the usual ψ pattern (LERMAN, 1971), presenting a broad distribution with very little DNA in the region of 60 s, (the normal sedimentation coefficient of T4 DNA), but extending through larger values to an upper limit near 500 s. (Because of inaccuracies in the calibration, calculated values of s near the meniscus may be taken with a grain

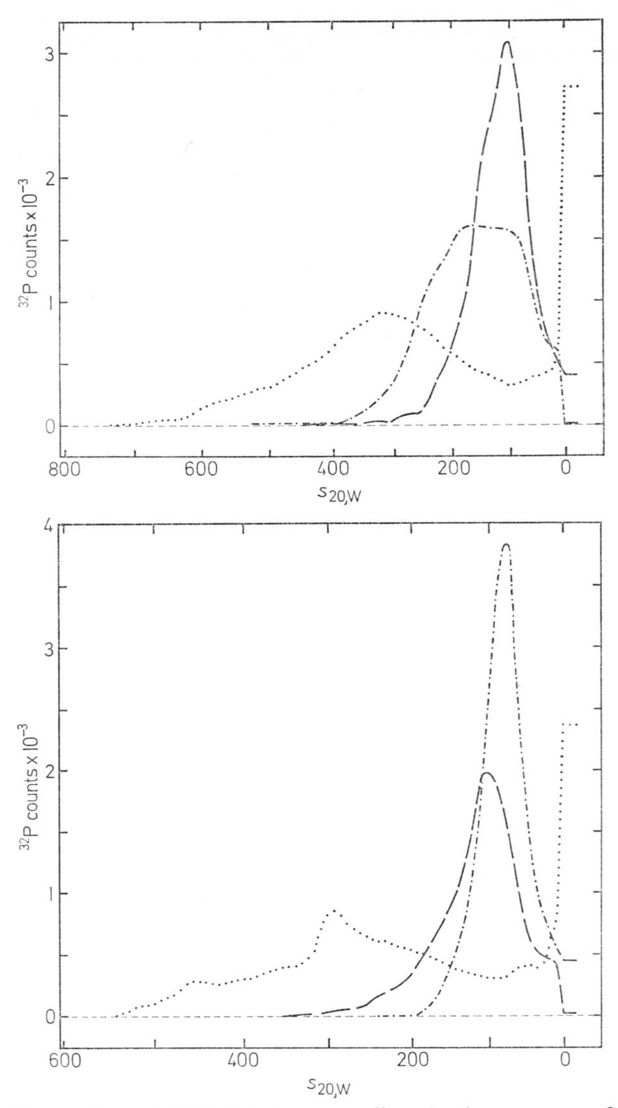

Fig. 2. Zone sedimentation of T4 DNA in a ψ medium in the presence of ethidium. Each tube held 4.5 ml of a solution containing 0.4 M NaCl, 0.01 M phosphate, and either 100 mg per ml poly(ethylene oxide)—upper panel, or 80 mg/ml—lower panel, together with 0.3 mg per ml ethidium bromide (solid line), 0.1 mg/ml (long dashes), or no ethidium (dotted line), pH 7.0. The stabilizing density gradient was established by linear variation of D_2O concentration from about 90% at the bottom to about 4% at the top. About 0.04 ml of solution containing fresh, ^{32}P-labeled T4 DNA was applied to the top of each tube—3.5×10^{-12} nucleotide equivalents—over an intermediate 0.1 ml layer of the same composition as the gradient but with no PEO or D_2O. The tubes were centrifuged for 45 min at 30,000 RPM in the SW 39 rotor at 25°. By means of bottom puncture, fractions of 0.112 ml each were deposited directly on filter paper, dried, and counted in a liquid scintillator. Weak statistical smoothing was applied to each series of data, with no corrections larger than reasonably probable according to Poisson statistics. Rather than present the resulting values as a histogram, they were transformed into a smooth curve which has the same integral over the volume equivalent to each fraction as the histogram itself (VENABLE, 1970). Position in the gradient at the end of centrifugation was converted to $s_{20,w}$ by a conventional calculation including the variation of density and viscosity through the gradient and assuming a partial specific volume of 0.55 for DNA. (The necessary density and viscosity measurements were carried out by Mr. LEIF GERJUOY.) The amounts of DNA in each tube were not precisely equal, and the curves have not been normalized

of NaCl.) In both 80 mg/ml PEO containing 100 µg/ml ethidium and 100 mg/ml PEO containing 300 µg/ml ethidium, the ψ pattern appears to be almost completely suppressed; the DNA has moved only a short distance from the meniscus, as though the salt concentration were below the critical level for the transition. Except in terms of a structural effect on DNA, the consequences of the presence of ethidium are hard to understand. Its contribution to the ion concentration is less than 10^{-3} M, and a significant effect on the polymer is implausible when there is at most one dye per 13 PEO molecules. The amount of intercalation in 0.45 M sodium may be estimated from the equilibrium studies of LE PECQ and PAOLETTI (1967), from which an equilibrium constant of about 2×10^{-5} M^{-1} in simple solutions at 25° is interpolated. The presence of 100 mg/ml PEO effectively decreases this value by a factor of about 7.4, presumably because of weak binding between ethidium and PEO, independent of salt concentration (GREEN and LERMAN, 1970). Nevertheless, the free dye concentration (which is here essentially equal to the total concentration) is so much larger than $1/K$ that the sites for intercalated binding to the normal DNA are almost saturated —87% and 95% saturated at 100 and 300 µg/ml, respectively.

It may be concluded that the intercalation of ethidium and the ψ change are competitive structural alterations. The balance between the effects is seen in Fig. 2, where an ethidium concentration of 100 µg/ml prevents the ψ transition when the PEO concentration is 80 mg/ml, but gives only about 50% inhibition when the PEO concentration is 100 mg/ml. At the higher PEO concentration, 300 µg/ml ethidium is as effective in suppressing ψ as was 100 µg/ml at the lower.

A small, seemingly paradoxical effect may be noted in the more dilute PEO, in that 300 µg/ml ethidium allows some of the DNA to sediment a little more rapidly than the peak at lower ethidium concentration. While this effect may bear further investigation, it is not difficult to reconcile with familiar dye-binding results. In general, as may be seen in a number of other contributions to this symposium, high concentrations of cationic dyes bring about external, non-intercalated binding of dye aggregates to DNA. Externally bound ethidium can be expected to be equivalent to a high local concentration of sodium ions, and in that way to promote, rather than to inhibit, the ψ transition.

A simple interpretation of the competition between intercalation and the change to the ψ structure imples that affinity for the dye is stronger in non-ψ than in ψ DNA, and indeed the difference must be considerable to provide the drastic changes seen in Fig. 2. It can be supposed that the ψ structure is characterized either by an obligate level of regularity or by a backbone configuration that cannot be altered without strain, precluding intercalation. Presumably other differences in the properties of ψ and ordinary DNA will be found to be attributable to the structure change, including perhaps differences in the reactivity of the bases toward chemical attack.

It is possible that a structural feature of ethidium, the out-of-plane benzene ring in the 9 position, may be important in the competition with the ψ state. It is implausible to suppose that the benzene ring can in any sense be accommodated in the space between nucleotide pairs into which the phenanthridine ring system fits, and therefore, it must be supposed to protrude from that site, out from the double helix. On the other hand, the ψ conformation appears to require close association between parts of the DNA molecule that are not covalently adjacent. One such structure might be a simple superhelix, although some repeated, additional folding would be necessary to achieve the observed compactness. The self association is inferred not only from the compactness, but also from the high molecular weight requirement of the transition in the sense that one major structural difference of low molecular weight DNA (which fails to undergo the ψ change) is its stiffness, which may easily be sufficient to preclude loops and knots, etc. It can be imagined that the irregularity at the surface of the helix due to the protruding benzene rings interferes with the close proximity of helices that may be necessary for stability of the ψ form. Whether it is this structural feature of ethidium or intercalation *per se* that provides the basis for the competition should be determinable by studies with other intercalating materials.

VI. Suggested Interpretation of Nuclear Staining Differences

If the effects found with ethidium can be generalized with respect to the other simple intercalating dyes, we may have a basis for understanding the staining difference discussed earlier. The ψ conformation may speculatively be identified with a functionally inert state of DNA, not only for DNA as contained in a virus head, but also DNA in the nucleus under storage conditions, when it does not participate in either replication or transcription. The roles of the polymer and sodium ion in the nucleus may be fulfilled by the histones, although more specific structural constraints due to the histones may also operate. In the leucocyte system, response to phytohemagglutenin requires retrieval of at least some of the DNA from dormancy. It is imaginable that acetylation and phosphorylation of the histones may constitute some part of the mechanism, and some DNA returns from ψ to the ordinary B-like state to become available for staining by acridine orange.

The observations of CASPERSSON et al. on the relative exclusion of ethidium from heterochromatic regions of *Trillium* chromosomes suggests that heterochromatin in its condensed state is more ψ-like than the remainder of the chromosomal DNA. However, the preferential binding of alkylating derivatives of quinacrine, and to some extent quinacrine itself, to the heterochromatic regions necessitates the corollary hypothesis that the ψ structure favorably predisposes the nucleotides toward a chemical reaction with the alkylating agents and also toward non-covalent binding of those acridines that carry a cationic nitrogen on an aliphatic side chain substituted on the 9 amino group. (The virtues of that side chain are also discussed elsewhere in the Symposium.) The correlation of the binding of the alkylating compounds with their chemical reactivity has been shown by CASPERSSON et al. in terms of the concentration of chromosome breakage due to these agents in the same region as their visible binding. The human Y chromosome has been shown to be selectively stained by the same agents (PEARSON, BOBROW and VOSA, 1970; GEORGE, 1970), although the chromatin of the inert member of an XX pair is also condensed (the Barr body), as is demonstrated by routine staining with triphenylmethane dyes. The alteration of patterns of chemical reactivity due to secondary structure is not novel in principle; an enhancement of the relative reactivity of the 3-nitrogen of adenine in double helical DNA, as compared with the free base has been shown (LAWLEY and BROOKES, 1963), and examples involving other types of reagents are known.

These studies suggest that continued exploration of the conformation of DNA in nucleus and its physiological control through observations of dye binding are likely to be of interest.

Acknowledgements

I am grateful to Mr. S. ALLEN (Lt. Col., USAF, Retired), for the experimental work described here. This work was supported by a research grant from the National Science Foundation (GB-4119) and by a grant from the National Institutes of Health (GM-13767).

References

BERLOWITZ, L., PALLOTTA, D., SIBLEY, G. H.: Chromatin and histones: Binding of tritiated actinomycin D to heterochromatin in mealy bugs. Science **164**, 1527 (1969).

CASPERSSON, T., ZECH, L., MODEST, E. J., FOLEY, G. E., WAGH, U., SIMMONSSON, E.: Chemical differentiation with fluorescent alkylating agents in vicia faba metaphase Chromosomes. Exp. Cell Res. **58**, 128, 141 (1969).

COHEN, P., KIDSON, C.: Conformational analysis of DNA-poly-L-lysine complexes by optical rotatory dispersion. J. molec. Biol. **35**, 241 (1968).

GEORGE, K. P.: Cytochemical differentiation along human chromosomes. Nature (Lond.) **226**, 80 (1970).

GREEN, R., LERMAN, L. S.: Unpublished, 1970.

JORDAN, C. F., LERMAN, L. S.: Unpublished, 1969.

— — VENABLE, J. H.: In preparation, 1971.

LAWLEY, P. D., BROOKES, P.: Further studies on the alkylation of nucleic acids and their constituent nucleotides. Biochem. J. **89**, 127 (1963).

LERMAN, L. S.: Proc. nat. Acad. Sci. (1971) (in press).

MAESTRE, M. F., TINOCO, I.: Optical rotatory dispersion of viruses. J. molec. Biol. **23**, 323 (1967).

— WANG, J. C.: The circular dichroism of supercoiled helices of DNA. 3rd Int. Biophysics Congress, p. 181, Abstracts. Cambridge 1969.

PARDON, J. F.: PhD Thesis, King's College, University of London.

— WILKINS, M. H. F., RICHARDS, B. M.: Super-helical model for nucleohistone. Nature (Lond.) **215**, 508 (1967).

PEARSON, P. L., BOBROW, M., VOSA, C. G.: Technique for identifying Y chromosomes in human interphase nuclei. Nature (Lond.) **226**, 79 (1970).

RAUTH, A. M.: The physical state of viral nucleic acid and the sensitivity of viruses to ultraviolet light Biophys. J. **5**, 201 (1969).

RIGLER, R., Jr.. Microfluorometric characterization of intracellular nucleic acids and nucleoproteins by acridine orange. Acta physiol. scand. **67**, Supp. **267**, 1—122 (1966).

SHAPIRO, J. T., LENG, M., FELSENFELD, G.: Deoxyribonucleic acid-poly-lysine complexes. Structure and nucleotide specificity. Biochemistry **8**, 3219 (1969).

TIKCHONENKO, T. I.: Conformation of viral nucleic acids in situ. Advanc. Virus Res. **15**, 201 (1969).

VENABLE, J. H.: Unpublished, 1970.

Subject Index